Alcohol Related Diseases in Gastroenterology

Edited by
H. K. Seitz and B. Kommerell

With 71 Figures and 38 Tables

Springer-Verlag
Berlin Heidelberg New York Tokyo

Priv. Doz. Dr. med. Helmut Karl Seitz
Professor Dr. med. Burkhard Kommerell

Medizinische Universitätsklinik Heidelberg
Innere Medizin IV (Gastroenterologie)
Bergheimer Strasse 58
D-6900 Heidelberg 1

Cover photograph:
Evidence of excessive alcohol consumption in early Egypt. These hieroglyphics, with English subtitles, are taken from "The Precept of Ani" – a papyrus of etiquette, which dates from about 1500 B.C.
(Cited by H. O. Conn (1982) Cirrhosis. In: Schiff L, Schiff ER (eds) Diseases of the Liver. Lippincott, Philadelphia, 847–978)

ISBN-13:978-3-642-70050-7 e-ISBN-13:978-3-642-70048-4
DOI: 10.1007/978-3-642-70048-4

Library of Congress Cataloging in Publication Data.
Main entry under title. Alcohol related diseases in gastroenterology. Includes bibliographies and index. 1. Gastrointestinal system–Diseases. 2. Alcoholism–Complications and sequelae. 3. Alcohol–Toxicology. 4. Liver–Diseases. I. Seitz, H. K. (Helmut Karl), 1950. II. Kommerell, B. [DNLM: 1. Alcohol, Ethyl–pharmacodynamics. 2. Alcoholism–complications. 3. Gastrointestinal System–drug effects. 4. Liver Diseases, Alcoholic. [QV 84 A3546] RC802.A36 1985
616.3'385-2788 ISBN-13:978-3-642-70050-7 (U.S.)

Softcover reprint of the hardcover 1st edition 1985

2123/3140–543210

To Charles S. Lieber
and
To our Wives

However, if someone asks – to what purpose should we help one another, make life easier for each other, make beautiful music or have inspired thoughts? – he would have to be told: If you don't feel it, no-one can explain it to you.

A. Einstein

Foreword

Alcohol abuse ranks among the most common and also the most severe environmental hazards to human health. Its significance is heightened by the possibility of prevention by elimination of the habit, however, rarely exerted. The incidence of deleterious effects on human health has relentlessly risen in the past years for a variety of factors. They include migration of populations and, particularly, increased urbanization. Thus, in some parts of the world, population groups previously spared have become involved, which is also reflected in the increasing number of breweries and distilleries in the developing countries. Social, religious, and gender-related barriers to alcohol consumption are loosening, and the financial improvement of some segments of populations now enable them to buy alcoholic beverages. Thus the greatest percentage rise in the United States has recently been in black women. Adolescents and young people drink more alcoholic beverages than ever, and growing alcohol abuse by pregnant women has let to an increase of the incidence of the fetal alcohol syndrome.

While the social and behavioral, including psychiatric, consequences of alcoholism are staggering, the gastrointestinal and, particularly, hepatic manifestations are the most widespread somatic effects, and chronic hepatic disease in alcoholics appears to cause the greatest cost to society. Indeed, mortality from liver cirrhosis is considered a reliable index of alcohol consumption in a country. No wonder that the gastrointestinal and, again especially, hepatic manifestations of alcoholism continue to attract great interest from both practical and academic points of view, and call for repeated review of available information.

This book represents a new attempt to marshal existing information and to look at it, at least in part, from a new perspective. It thus supplements a number of preceding reviews and monographs. Scholarly presentation is guaranted by the excellence of the contributors from various countries in Europe and North America and from Japan. All have written authoritatively on their subjects before, and some are authors of previous monographs on the medical disorders of alcoholism in general and its hepatic manifestations. Taking stock anew of gastrointestinal problems related to alcohol increases the knowledge in the field and spreads it to wider circles. This, in itself, is a first step in improved management of a health hazard of growing, indeed alarming dimensions. Thus, we find new data on the extent of the problem by epidemiologic studies and learn of its geographic variations; for instance, in Japan alcoholic liver disease seems to have increased to a conspicuous degree in the last decade. Information is also being offered on a series of unresolved questions in the management of the multifold gastrointestinal disorders related to alcohol abuse. To list some of both clinical and investigative interest, the character of the predisposition to alcoholism as well as to the progression of the diseases in alcoholics are still not established. The recognition of the diseases and their differential diagnostic separation from other diseases, which may even complicate the alcoholic disorders, remain a problem. The pathogenesis of the consequences exemplified by carcinomatous transformation, particularly of the liver, is still not clear. The recognition of continued alcohol abuse by laboratory tests has significantly advanced but is not yet established. Some major problems deserve to be stressed. One is the mechanism of the evolution of the various diseases related to alcohol. In the liver, the pathogenesis of the steatosis and associated metabolic disorders is well understood. They are due to the metabolism of ethanol, which is a biologic phenomenon not only in the mammalian liver, but indeed in yeast and other microorganisms; the latter accounts for the alcohol in beverages. Far less do we about alcohol-induced cell injury, for instance of hepatocytes (the hepatitis), where hypotheses still prevail. By contrast, knowledge of the mechanisms of the hepatic fibroplasia and cirrhosis has recently far advanced. The understanding of these underlying processes is the basis of the progress in the management of diseases when withdrawal of alcohol either does not succeed or, in the rare instances where it does, they are no longer reversible. Then, modulation of the disturbed multiple pathways endowed with many feedback loops offers a more promising approach than the application of a single drug supposed to stop one particular event, for instance the formation of collagen in fibrosis.

From the thorough review of clinical, experimental, biochemical, immunologic, and pathologic information in the various diseases, the answers to the various questions raised should emerge. Thus the hope is expressed that this comprehensive and up-to-date review of a major field in gastroenterology will inform as well as stimulate a wide range of persons who require this knowledge in their professional activities.

Hans Popper, M.D., Ph.D.
New York

Contents

List of Contributors*

* The address of the principal author is given on the first page of each contribution

[1] Page, on which contribution commences

1 Epidemiology of Alcohol Use and Its Gastrointestinal Complications

W. K. Lelbach

Medizinische Universitäts-Klinik Venusberg,
Sigmund-Freud-Strasse 25, 5300 Bonn 1, Federal Republic of Germany

Trends in Alcohol Consumption

In most European countries aggregate alcohol consumption was quite high during the latter half of the nineteenth century (Sulkunen 1976), but except in France and Italy consumption began to decrease after the turn of the century. It reached a comparatively low level in the years shortly before the Second World War, during which a further drop occurred. Since the end of World War II, however, alcohol consumption has been increasing at an alarming rate in nearly all parts of the world, irrespective of the enormous variation in acceptable drinking habits which initially prevailed both within and between countries. As a consequence, the prevalence of alcohol-related disorders and medical complications has almost reached epidemic proportions. At present, alcohol problems rank among the world's major public health concerns (World Health Organization 1980).

A comparison of the annual average alcohol consumption per capita of total population in 26 countries throughout Europe and the Anglo-American world between 1950 and 1981 (Table 1) reveals that there was an overall increase of more than 100% during this period. However, the increase was particularly high in most of those countries with a low initial level (less than 5 liters absolute alcohol per capita). A stabilization of per capita alcohol consumption (or even a decrease, as in Italy, Spain, and Austria) was observed during the first half (West Germany, Switzerland, Belgium, Netherlands, Finland, Ireland) or the second half of the 1970s (Czechoslovakia, Poland, Luxembourg, Australia), whereas in France, which had the worlds' highest per capita intake of alcohol, consumption began to decline gradually much earlier after an unprecedented peak level had been reached in 1955 (Table 2). Larger fluctuations occurred in Portugal during this 30-year period, where consumption had always ranged above 10 liters per capita. Consumption continued to increase up to 1981 in the United States, Canada, New Zealand, Japan, Hungary, Rumania, and Denmark. It seems that drinking habits and patterns are becoming more uniform throughout the developed countries and that, in general, alcohol consumption tends to converge on an annual rate of between 10 and 15 liters absolute alcohol per capita of total population. It could be speculated that this range of consumption constitutes a kind of saturation level beyond which the requirements of an increasingly complex modern industrial world could no longer be met. In low-consumption countries the proportion of abstainers tends to be higher and

Alcohol Related Diseases in Gastroenterology
Edited by H. K. Seitz and B. Kommerell

Table 1. Increase in average annual alcohol consumption (liters of absolute alcohol per head of total population) between 1950 and 1981 (initial level < 5 liters vs > 5 liters) (Brown et al. 1982)

	1950	1981	Increase
< 5 liters			
Netherlands	2.07	8.75	+ 323%
West Germany	3.29	12.52	+ 281%
Finland	2.23	6.50	+ 191%
East Germany[a]	3.9 (1955)	10.30	+ 164%
Denmark	3.78	9.92	+ 162%
Hungary	4.91	12.34	+ 151%
Rumania	3.35 (1960)	7.90	+ 136%
Austria	4.79	10.75	+ 124%
Poland	3.11	6.63	+ 113%
		(1980: 8.73)	(+ 181%)
Canada	4.41	8.80	+ 100%
Republic of Ireland	3.69	7.18	+ 95%
New Zealand	4.57 (1953)	8.86	+ 94%
Norway	2.17	4.16	+ 92%
United Kingdom	3.94	7.31	+ 86%
Japan	3.60 (1962)	5.94	+ 65%
Sweden	3.94	5.44	+ 38%
> 5 liters		(Average increase	= 138%)
Czechoslovakia	5.48	10.84	+ 98%
Luxembourg	8.71	16.77[b]	+ 93%
Belgium	5.40	10.31	+ 91%
Spain	7.74[b]	12.72	+ 64%
USA	5.52	8.55	+ 55%
Australia	6.62	10.19	+ 54%
Switzerland	8.77	11.11	+ 27%
Italy	9.49	10.42	+ 10%
Portugal	15.60 (1955)[b]	13.03	− 16%
France	18.73	14.94	− 20%
		(Average increase	= 46%)

[a] From Statistisches Jahrbuch der DDR 1981; Ziegler 1984
[b] Consumption of spirits estimated

Table 2. Peak level of alcohol consumption and subsequent decrease

	Peak level		Consumption in 1981 (liters)	% decrease
	Year	Liters abs. alc. per capita		
France	1955	20.28	14.94	26%
Portugal	1955	15.60[a]	13.03	16%
	(1960)	(10.80)[a]		
	(1971)	(14.16)		
Italy	1973	14.20	10.42	26%
Austria	1973	12.00	10.75	10%
Spain	1975	13.84	12.72	8%
West Germany	1979	12.74	12.24 (1982)	4%
Switzerland	1973	11.22	11.11	1%

[a] Consumption of spirits estimated

increases in consumption come about by more abstainers joining the ranks of the drinking population, whereas in high-consumption countries abstainers are a minority and with an increase in consumption a greater proportion of alcohol users would adopt drinking styles which would be "hazardous" (Davies and Walsh 1983). Official consumption statistics have shortcomings insofar as they do not include untaxed, illicit, or otherwise unregistered production, so that consumption may in fact be even higher in countries where home-made alcoholic beverages play a role.

A complex pattern of factors has been responsible for the almost universal increase in alcohol consumption during the last 30 years. Of particular importance were postwar economic recovery and growing affluence on the one hand and the decline in the relative price of alcoholic beverages (ratio of real price of absolute alcohol to real personal disposable income) on the other. A close inverse relationship between fluctuations of relative price over time and the level of consumption has been observed in a number of geographic areas which differ widely in ethnic composition, drinking patterns, and beverage preference (Lelbach 1985). Additional factors have been the liberalization of sales regulations and licensing laws, greater availability due to promotion of off-premise sales (by grocery shops, supermarkets, and vending machines), longer leisure hours, gradual changes in prevailing attitudes toward drinking practices, acquaintance with hitherto unfamiliar drinking styles through tourism and foreign travel, with the introduction and addition of nontraditional beverages and drinking customs (for instance, wine with meals), expansion of international trade and advertising, opening up of new markets, recruitment of new segments of the population (young people, women), and so forth. Within the framework of a public health perspective, governments alarmed by the growing number of alcohol-related problems have discussed and used a variety of alcohol measures and policies, such as state monopolies, sales restrictions, levy of special taxes and excise duties (Kendell et al. 1983), introduction and enforcement of legislation on drinking and driving, health education, etc. Such control policies and their degrees of effectiveness differ widely from country to country, depending on consumption habits, public acceptance of drinking, problem awareness and, of course, economic weight and influence of national alcohol industries (Armyr et al. 1982; Brown et al. 1982; Davies and Walsh 1983).

An alarming escalation of alcohol consumption has also been recorded for many parts of the Third World during the last two decades (Edwards 1979), due particularly to the growth of indigenous industrial production and also to import of European types of beer (Anonymous 1981b) and, to a lesser degree, of spirits. From 1960 to 1972 recorded world production of beer increased by 68%, but indigenous production of European-type beer in Africa and Asia increased by 187% and 256% respectively during this period, whereas spirits production in Africa, Asia, and South and Central America (+69%, +51%, +62% respectively) rose parallel to world production (+61%) (Smith 1982a). Since independence, European types of beer and spirits have become a desirable status symbol in a number of sub-Saharan African states, and are now competing with traditional alcoholic beverages. In connection with rapid changes in traditional societies, breakdown of old cultural ties and customs, migration to cities with overpopulation of slum areas, and easier accessibility to alcoholic beverages, the abuse of alcohol is becoming a major problem in Black Africa. This problem has been studied so far only in a few preliminary investigations, but these have yielded alarming results (Acuda 1982; Isaacson 1978, 1982). Available information and literature on alcohol consumption and alcohol-related medical problems in the less developed parts of the world were recently reviewed by Lelbach (1985).

Alcohol-Related Gastrointestinal Disease

It was largely empirical evidence that established an association between alcohol consumption and certain gastrointestinal disorders, notably alcoholic liver disease, alcoholic pancreatitis, and cancer of the upper gastrointestinal tract.

Alcoholic Liver Disease

In most countries which permit a direct comparison with the degree of alcohol consumption, epidemiological data (annual mortality statistics) are only available for cirrhosis of the liver, the most advanced stage of liver disease. In spite of large geographical variations in drinking habits, proportion of abstainers and occasional users, volume of unrecorded consumption, and the extent of exposure to hepatitis B and non-A, non-B hepatitis viruses, a close linear correlation emerges when total cirrhosis mortality (unspecified) is plotted against average annual consumption (liters of absolute alcohol) per capita of drinking age population (15 years and over). Latest available figures for both consumption and cirrhosis death rates are shown in Fig. 1 for 29 European and non-European countries, with allowance for a time lag of 1 year. (Consumption per head of drinking age population was calculated on the basis of the annual WHO statistics for age structure.) It can safely be assumed that the resulting correlation would be even closer if consumption were recorded separately for male and female alcohol users – considering the differences in male and female cirrhosis mortality (between 1.5 : 1 and 3 : 1) in most countries – but no such records exist. However, women develop liver disease at lower consumption levels than men, as was confirmed by a recent large study of 510 patients admitted for alcoholic liver disease (from fatty liver to active cirrhosis) in Scotland and North Eastern England during the years 1974–1979 (Hislop et al. 1983). In males the mean daily alcohol intake had been 58% higher (190 vs 120 g/day) than in female patients, a difference which could not be attributed solely to differences in body weight, since average body weight (with respect to height and age) in men is only 7%–12% higher than in women (Diem and Lentner 1968). Although alcohol use and alcoholism among females, especially younger women, increased in most countries during the period 1950–1980 due to increasing acceptance, the ratio of male to female abstainers is still roughly in the range of 1 : 2 to 1 : 3 (Armyr et al. 1982). Recent representative survey data on consumption patterns, however, are available only for a few countries, and it is generally accepted that such surveys only account for about 40%–60% of the actual aggregate alcohol consumption because heavy (at risk)

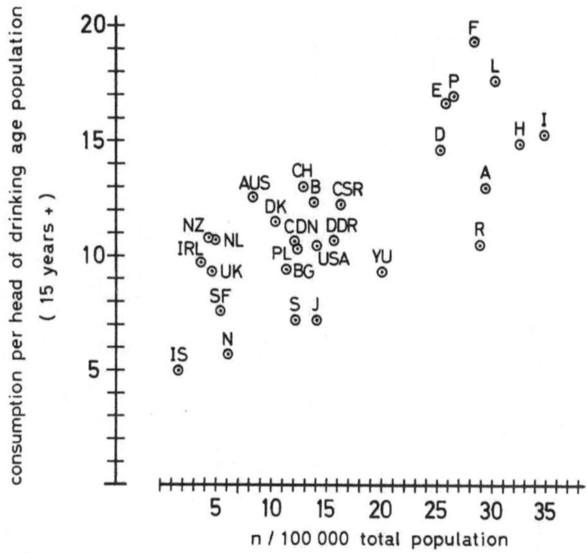

Fig. 1. Average annual alcohol (A) consumption in liters of absolute alcohol per capita of drinking age population plotted against cirrhosis death rates (C) per 100,000 total population 1 year later for 29 countries. Latest available figures. Countries are indicated by their international letter symbols. A 1974/C 1975: CSR; 1977/1978: B, CDN, I, IRL; 1978/1979: DDR, P, SF, UK; 1979/1980: AUS, E, F, N, NZ, PL, R, S, USA, YU; 1980/1981: A, BG, CH, DK, H, IS, J, L, NL; 1981/1982: D

drinkers are notoriously underrepresented in these surveys (Davies and Walsh 1983; Salonen et al. 1983; Tuyns et al. 1983).

The development of average aggregate alcohol consumption in relation to cirrhosis mortality is shown for West Germany and France in Figs. 2 and 3. In West Germany, consumption figures, which had spectacularly increased in an unprecedented manner after 1950, began to level off in 1970 and reached a plateau in the mid-1970s. The steady rise in consumption between 1950 and 1970 was conspicuously paralleled by an equally steady but even steeper increase in cirrhosis mortality. It also becomes obvious that the subsequent stabilization of the level of consumption was followed, with a time lag of about 2–3 years, by a similar levelling off of cirrhosis mortality and, finally, when consumption dropped slightly after 1979, even by a slightly downward trend.

In France, a three-phase parallel development of consumption and cirrhosis mortality can be discerned from Fig. 3 for the period 1940 through 1980. As a consequence of rigorous wine rationing during the Second World War, both consumption per head of adult French population (20 years and over) and cirrhosis mortality per 100,000 total population dropped precipitously until 1946. After 1946 alcohol consumption rose at a steady rate until it levelled off in 1955. Ten years later, from 1965 onward, alcohol consumption began to decrease slowly for a number of reasons, not least being influenced by a noticeable increase in activities designed to improve health education and other measures to combat alcohol abuse. Again, cirrhosis mortality followed these alternating trends closely, with a time lag but at a clearly disproportionately exaggerated pace. Nevertheless, cirrhosis mortality in France ranks among the highest in the world even today.

Fairly reliable data on alcohol-induced cirrhosis morbidity and/or mortality are registered separately from those on nonalcoholic cirrhosis in only a few countries, mostly those where aggregate alcohol consumption is still moderate or even

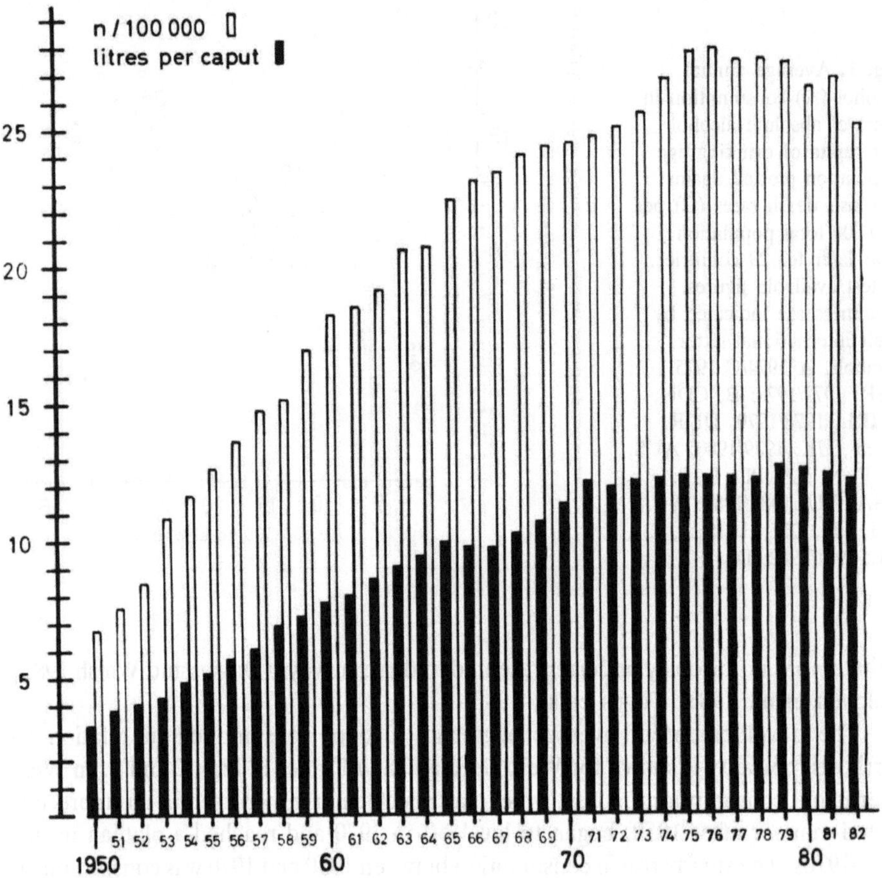

Fig. 2. Cirrhosis mortality per 100,000 total population (*white columns*) and annual consumption in liters of absolute alcohol per caput of total population (*black columns*) in West Germany for the period 1950–1982 (modified from Biel 1975; Ziegler 1984; Statistisches Bundesamt, personal communication)

comparatively low (The Netherlands, Sweden, Norway, Finland, Denmark, Switzerland, East Germany). Mortality from nonalcoholic cirrhosis remained more or less unchanged during the period under observation in The Netherlands and the four Scandinavian countries but, as Table 3 shows, mortality from alcoholic cirrhosis rose dramatically between 1960 and 1978, especially among males in The Netherlands, Finland, and Denmark. Among young and middle-aged males in Denmark a threefold increase in total mortality from cirrhosis of the liver was observed during the period 1965–1978, following a twofold rise in annual adult consumption of alcohol (from 6 to 12 liters pro capita, 1964–1976) (Prytz and Skinhøj 1981). The incidence of hospital admissions (new cases) for cirrhosis (clinical morbidity) during 1976–1977 was 27 per 100,000 total population in Denmark, of which cases with alcoholic cirrhosis comprised two-thirds (80% in males, 60% in females) (Prytz and Skinhøj 1980). In The Netherlands the clinical morbidity of

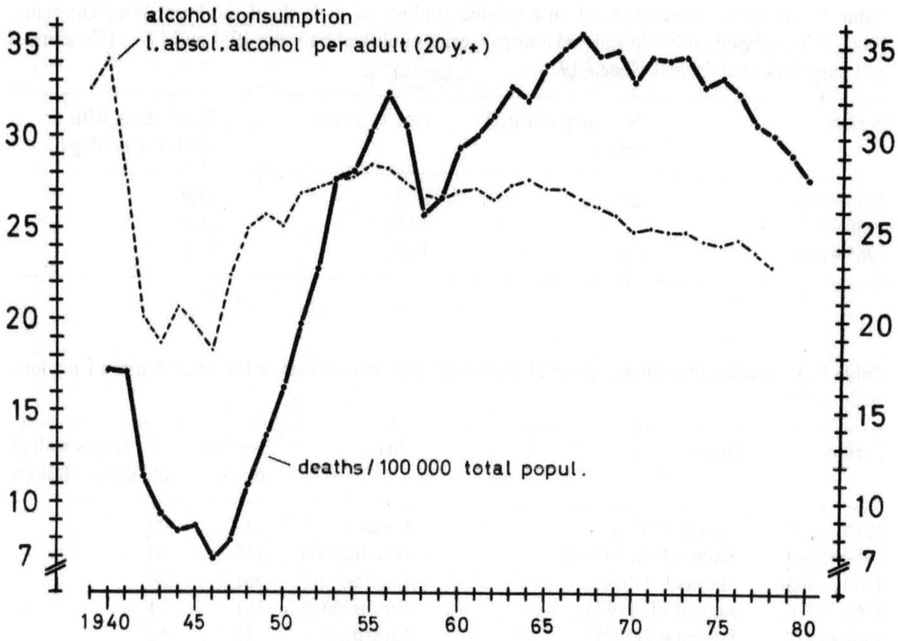

Fig. 3. Alcohol consumption (adults) and cirrhosis mortality (total population) in France 1939–1980 (Lelbach 1984)

Table 3. Development of mortality from alcohol-induced cirrhosis of the liver, 1960–1978

Country		Mortality from alcoholic cirrhosis per 100,000 inhabitants		Increase	Reference
		1960	1975		
Netherlands	M	0.4	3.2	700%	Gips (1978)
	F	0.1	0.8		
		1969	1978		
East Germany	M+F (inpatients only)	~ 2.25	~ 4.8	113%	Jorke and Reinhardt (1982)
		1963	1974		
Sweden	M	1.5	4.7	213%	Svendsen and Mosbech (1977)
	F	0.3	0.9	200%	
Norway	M	0.8	1.8	125%	
	F	0.3	0.7	133%	
Finland	M	0.4	3.9	875%	
	F	0.1	0.4	300%	
Denmark	M	0.7	4.0	471%	
	F	0.2	0.9	450%	

Table 4. Increasing importance of an alcoholic etiology in cirrhosis of the liver: West Germany, 1966–1974. (Based on 22 clinical and autopsy series published between 1955 and 1976) (Condensed and supplemented from Lelbach 1976)

Period	No. of published series	No. of cases	% of cases with an alcoholic etiology
1946–1959	12	2,517	17.9
1958–1968	5	997	38.9
1965–1974	5	1,173	52.8

Table 5. Increasing importance of an alcoholic etiology in cirrhosis of the liver: United Kingdom, 1919–1976

Period	Series	Place	No. of cases	% of cases with an alcoholic etiology
1919–1955	Parker (1957)	London	134	22
1959–1964	Stone et al. (1968)	Birmingham	155	34
1959–1965	Sherlock (1966)	London	561	24
1965–1970	Jain et al. (1973)	Birmingham	181	51
1961–1972	Forshaw (1972)	Liverpool	87	63
1968–1974	Hodgson and Thompson (1976)	London	78	65
1971–1976	Saunders et al. (1981)	Birmingham	176	66

alcoholic cirrhosis (rate of hospital admissions) increased threefold for males from (6.5 to 18.4 per 100,000 male population) and twofold for females (from 3.7 to 7.6 per 100,000) during the 7-year period from 1969 to 1976; simultaneously, unspecified cirrhosis morbidity rose from 3.9 to 5.9 per 100,000 (+51%) for males and from 4.4 to 5.4 per 100,000 for females (+23%) (Hoogendoorn 1978). Hospital admission rates for alcoholic cirrhosis in 1977–1979 were 17.5 per 100,000 for males and 18.0 per 100,000 for females (Hoogendoorn 1983). A highly significant linear correlation between total cirrhosis (unspecified) mortality and proportion of alcoholic cirrhosis was also observed for the 10-year period 1968–1977 in Japan by Takeuchi et al. (1979), by screening records from 94 Japanese hospitals. This indicated that the increase in total cirrhosis mortality was predominantly due to an increase in alcoholic liver disease.

The increasing significance of alcohol as an etiologic factor in cirrhosis of the liver is also borne out by its proportional representation in autopsy and clinical series published during the last 30 years in West Germany (see Table 4), but also in the United Kingdom (Table 5), where aggregate alcohol consumption and overall cirrhosis mortality is still comparatively low, although alcoholism has particularly increased in Scotland as compared to England and Wales (Hisop et al. 1983). In typical "wine countries" such as Portugal, on the other hand, where a more uniform pattern of daily alcohol intake prevails and there is a low proportion of total abstainers, the prevalence of an alcoholic etiology of cirrhosis of the liver is now even higher than 90% of all cases. Thus in a recent Portuguese study (Pinto et al. 1983), 93% of 498 patients with cirrhosis of the liver hospitalized for acute upper gastrointestinal bleeding during the 7-year period 1975–1981 in the Lisbon area had had a daily intake exceeding 160 g ethanol.

In Poland, a paradigm for European socialist countries, death rates from cirrhosis of the liver more than trebled between 1960 and 1978 (from 3.4 to 12.0 per 100,000 total population) concomitantly with a steep rise in alcohol consumption from 3.81 to 8.0 liters absolute alcohol (more than 60% of it being consumed as hard liquor) when economic conditions and real wages started to rise faster than the price of alcoholic beverages after the late 1960s (Smith 1982b). Statistical evidence revealed that this increase in cirrhosis mortality could certainly not be explained by an increase in posthepatitic cirrhosis (Jędrychowski and Flak 1982). An example of conspicuous regional differences in both consumption and morbidity/mortality from alcoholic liver disease is the gradient between northern industrialized regions of Italy with their comparatively high socioeconomic status and the notoriously backward south (Coltorti and Lucchelli 1977; Lelbach 1984). In 1955 cirrhosis mortality was 21 per 100,000 total population in northern Italy, 11.5 per 100,000 in central Italy, and 8.5 per 100,000 in southern Italy, with annual wine consumption per person amounting to 82.1, 47.6, and 21.4 liters respectively (Bonfiglio et al. 1979). A comparison of two cohorts of patients admitted for treatment of cirrhosis of the liver to hospitals in Brescia (north) and Naples (south) in 1973–1974 showed that the incidence of an alcoholic etiology was 68.8% and 10.5% respectively (Coltorti et al. 1974). Differing degrees of urbanization and differing socioeconomic levels are also responsible for considerable geographical differences in alcohol consumption and cirrhosis mortality between the various states of the United States (Welte and Russell 1982).

Therefore, it seems reasonable to assume that an inhomogenous population such as that living in North America would be less suited for a study of the individual risk of cirrhosis in relation to the level of alcohol consumption than, for instance, France, where overall consumption is high and drinking habits are more uniform. Such investigations of the relative risk of cirrhosis as a function of alcohol consumption were carried out by Péquignot et al. (1974, 1978) for two French *départements* (Bouche-du-Rhône, Ille-et-Vilaine) with different degrees of "alcoholization," and by Durbec et al. (1979) for a hospital population in Marseille. Both groups of investigators came to the conclusion that a linear correlation exists between the logarithm of the relative individual risk of cirrhosis and the mean daily alcohol intake. In addition, Durbec et al. found out that this correlation is valid up to an intake of 180 g ethanol per day, above which the risk levels off; further, that duration of consumption up to the first clinical symptoms and mean daily intake are negatively correlated; and that the mean duration of alcohol consumption necessary for the development of cirrhosis is longer than that for chronic pancreatitis. A similar linear correlation has been established between cancer of the esophagus and alcohol consumption in Normandy (Tuyns et al. 1979), but this problem will be dealt with in another chapter.

In clinical medicine we are increasingly confronted with patients suffering from advanced stages of alcoholic liver disease who no longer fit into the type of the "derelict alcoholic" but who have managed to sustain a continual alcohol intake of substantial magnitude for many years, even decades, without necessarily failing grossly in their normal social integration. They even may never have been overtly drunk. It is apparently not the degree of dependence that determines the development of alcoholic liver disease (Anonymous 1981a; Johnson et al. 1984; De Marchi et al. 1984; Wodak et al. 1983), although the prevalence of psychiatric disorders (predominantly affective disorders and neurosis, often covert) in these patients is high (Ewusi-Mensah et al. 1983).

Alcoholic Pancreatitis

According to statistics compiled by Dürr (1978), histopathologic evidence of chronic damage to the pancreas (five series; 1952–1972) could be found at autopsy in 17%–45% of chronic alcoholics (mean: 28%; 162 out of 584). On the other hand, clinically manifest pancreatitis (four series; 1971–1977) was observed during life (either current or in the past) in only 0.8%–9.4% (mean: 2%; 51 out of 2,424). This indicates that in the majority of cases damage to the pancreas as a consequence of heavy alcohol abuse develops insidiously. It also explains the long latency period between onset of alcohol abuse (usually more than 80–120 g/day) and first clinical

Table 6. Clinical morbidity of acute and chronic pancreatitis: changing pattern of incidence and etiology

	Hospital admissions per 100,000 males and females per year		Increase	Source
Netherlands	1969–1971	1979–1981		Hoogendoorn (1983)
Acute pancreatitis	M 7.2	M 13.2	83%	
	F 7.9	F 7.4	–6%	
Chronic pancreatitis	M 1.8	M 4.7	161%	
	F 1.3	F 2.1	62%	
Total	M 15.8	M 35.4	124%	
	F 12.5	F 17.5	40%	
Finland	1969	1975		Poikolainen (1980)
Pancreatitis	M 41.7	M 83.4	100%	
	F 29.4	F 40.8	39%	
Denmark (Copenhagen)	1970–1974 M+F	1975–1979 M+F		Andersen et al. (1982)
Alcoholic chronic pancreatitis	2.8	7.7	175%	
Nonalcoholic chronic pancreatitis	4.1	2.3		
All cases	6.9	10.0	45%	
Sweden (Göteborg)	1960 M > 15 years	1975 M > 15 years		Svensson et al. (1979)
First attack of pancreatitis	0.08/1,000	0.40/1,000	500%	
Italy (Bologna)	1967–1970 M+F	1971–1975 M+F		Gullo et al. (1977)
Chronic pancreatitis	0.6/1,000	8/1,000[a]		

[a] Although this increase was largely due to the systematic use of pancreatic function tests, it is the author's impression that the increase in incidence is a real one and that the progressive rise in alcohol consumption in Italy may explain part of it

symptoms of pancreatitis in a progressively destroyed gland (3–20 years; mean: 8–10 years). Interpretation of the numerous worldwide statistics concerning the respective prevalence of the alcohol-induced type of acute and chronic pancreatitis is rendered difficult because it is impossible to distinguish between first attacks of chronic relapsing pancreatitis, which is the characteristic type of alcoholic pancreatitis, and acute pancreatitis of nonalcoholic origin unless pancreatic calcification is present, which may be indicative of prior subclinical attacks. Sarles et al. (1979) believe that in Caucasians noncalcified chronic pancreatitis (as opposed to acute relapsing pancreatitis) is probably only an early stage of calcified chronic pancreatitis in those regions where the diet is rich in alcohol, proteins, and lipids.

In general, about 60%–90% of all cases of chronic pancreatitis are alcohol related, as larger studies from various geographic regions in Europe, North and South America, South Africa, and Australia between 1960 and 1979 have shown (Ammann and Sulser 1976; Bernades et al. 1983; Dani et al. 1974; Lehnert 1979;

Table 7. Increasing prevalence of an alcoholic etiology in acute and chronic pancreatitis

Country (city)	Periods			Source
Finland (Tampere)	1967–1968 (*n* = 88) M+F	1977–1978 (*n* = 120) M+F		Mero (1982)
Acute pancreatitis: % of cases (first attack) with alcoholic etiology	15.9%	57.5%		
Sweden (Göteborg)	1956–1960 (*n* = 454) M+F	1968–1969 (*n* = 105) M+F	1974–1975 (*n* = 204) M+F	Svensson et al. (1979)
First attack of pancreatitis: % of cases with alcoholic etiology	14%	68%	66%	
England (Manchester)	1963 (*n* = 44) M+F	1968 (*n* = 54) M+F	1968–1974 (*n* = 31) M+F	Howat (1963, 1968) Read et al. (1976)
Chronic pancreatitis: % of cases with alcoholic etiology	4.5%	20.4%	54.8%	
England (London)			1968–1973 (*n* = 107) M+F	James et al. (1974)
Chronic pancreatitis: % of cases with alcoholic etiology			42%	
Denmark (Copenhagen)	1970–1974 (*n* = 50) M+F	1975–1979 (*n* = 76) M+F		Andersen et al. (1982)
Chronic pancreatitis: % of cases with alcoholic etiology	42%	75%		

Neves et al. 1983; Sarles et al. 1979; Dani and Nogueira 1976). The relative frequency of this association depends on the prevalence of alcoholism in the populations studied. This is also valid for acute pancreatitis (or first attacks), in which an alcoholic etiology was found in 0–75% (Goebell and Hotz 1976).

When different periods after the end of World War II are compared (Tables 6 and 7), it becomes evident that in a number of countries where alcohol consumption has increased during the last 30 years the pattern of incidence and etiology of acute and chronic pancreatitis have been changing and that alcohol-related pancreatic disease is now predominant even where its prevalence was initially low as, for instance, in England.

In view of the differing pathogenetic pathways of alcohol-induced lesions of liver and pancreas [although recent observations of sequential histopathologic changes in alcoholic pancreatitis seem to suggest a common initial pathogenesis of the toxic-metabolic type parallel to that seen in alcoholic liver disease (Bordalo et al. 1982; Noronha et al. 1981a, b)], it is of interest to study the rate of coincidence of the two lesions in chronic alcoholics. In five autopsy series with a total of 618 cases of alcoholic cirrhosis of the liver studied between 1936 and 1969, coexistent chronic pancreatitis was found in slightly over one-third of the cases (mean: 36%; range: 28%–49%) (Greiner et al. 1983). A retrospective study of a well-defined but highly selected cohort of 112 alcoholics with clinical symptoms suggestive of chronic

Table 8. Coincidence of liver disease and alcohol-induced pancreatitis

Country (city)	Type of study	Rate of coincidence of liver disease		Reference
USA (Baltimore)	Retrospective study (males; alcohol-induced pancreatitis) (*n* = 50)	23 (46%) clinical liver disease	8 (16%) cirrhosis (biopsy)	Dutta et al. (1976)
	Prospective study (males; alcohol-induced chronic pancreatitis) (*n* = 46; liver biopsy in 30)	15 (33%) 3 (6.5%) 2 (4.3%) 10 (22%)	cirrhosis with and without alcohol. hepatitis alcohol. hepatitis severe fatty liver normal or minimal fat	Dutta et al. (1978)
Italy (Verona)	Prospective study (males; alcohol-induced chronic pancreatitis) (*n* = 30; liver biopsy in all)	5 (16.7%) 4 (13.3%) 9 (30.0%) 3 (10.0%)	cirrhosis cholestasis inflammatory lesions steatosis	Angelini et al. (1982)
Brazil (São Paulo)	Prospective study (alcohol-induced chronic calcifying pancreatitis) (*n* = 60; liver biopsy in all)	8 (13.3%)	cirrhosis	Dani et al. (1983)
West Germany (Wuppertal)	Retrospective study (alcohol-induced chronic pancreatitis; see text) (*n* = 65; liver biopsy in all)	20 (30.8%) 17 (26.2%) 20 (30.8%) 8 (12.3%)	cirrhosis fatty liver/hepatitis fatty liver normal liver histology	Greiner et al. (1983)

pancreatitis, for whom both endoscopic retrograde pancreatography and liver biopsy specimens were available, showed lesions of the pancreatic duct compatible with chronic pancreatitis in 65 (58%) patients. In only eight (12.3%) of these patients was liver histology normal whereas cirrhosis of the liver was diagnosed in 30.8% (20 out of 65) (Greiner et al. 1983). Similar results were obtained in other recent studies from different regions of the world (Table 8). If, on the other hand, cohorts of chronic alcoholics are screened for simultaneous presence of both hepatic and pancreatic lesions, it emerges that there is no close conformity between the two types of organ involvement (Table 9). However, it appears that the coincidence proves to be considerably more frequent than has hitherto been assumed, provided that adequate functional and morphological methods are applied. It should also be kept in mind that the mean duration of alcohol use before onset of first symptoms is shorter in chronic pancreatitis than in cirrhosis of the liver (Durbec et al. 1979).

Within the framework of a multicenter survey, Durbec and Sarles (1978) carried out two retrospective case-control studies – one in Marseille and the other in collaboration with nine out of 36 participating international centers from countries with Caucasian populations – and tried to analyze the quantitative aspects of the association of chronic pancreatitis and alcohol use in males. Employing log-linear

Table 9. Chronic alcoholics: hepatic and pancreatic lesions

Country (city)	Type of study	Liver	Pancreas	Reference
Brazil (São Paulo)	Autopsy results of 46 chronic alcoholics	16 (34.8%) cirrhosis (Laennec's) 19 (41.3%) cirrhosis (postnecrot.) 3 (6.5%) steatosis 8 (17.4%) normal liver	27 (58.7%) fibrosis of the pancreas[a]	Mincis et al. (1973)
East Germany (Dresden)	129 chronic alcoholics admitted for alcohol-induced liver disease[b]	Biopsy 33 (25.6%) cirrhosis 34 (26.3%) fatty liver and fibrosis 62 (48.6%) moderate to marked steatosis	Secretin-pancreozymin test 86 (66.7%) normal 43 (33.3%) pathologic; severe: 13 (10%) calcification (plain abdom. film) 12 (9.3%)	Poegel et al. (1981)

[a] 19 (41.3%) had both hepatopathy and pancreatic fibrosis
[b] There was no confirmity between the degree of hepatic and of pancreatic damage

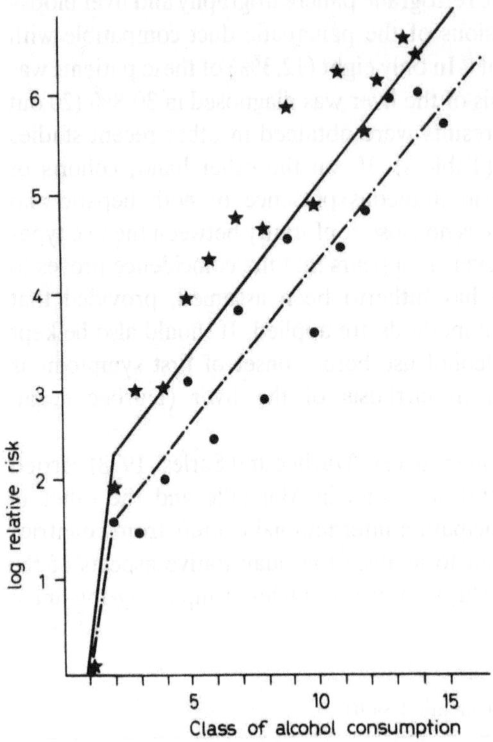

Fig. 4. Log-relative risk of chronic pancreatitis in relation to increasing mean daily alcohol intake (class 1 = 20 g/day). Marseille survey: ✱ = observed; ——— = estimated; international survey: ● = observed; –●– = estimated (Durbec and Sarles 1978)

models of contingency tables, they arrived at the conclusion that there is no apparent threshold of ethanol toxicity on the human pancreas. In approximation, a linear increase in the logarithm of the relative risk of chronic pancreatitis was observed as a function of the mean daily alcohol (Fig. 4) and protein intake. The increase varied quadratically as a function of the mean daily fat intake. The effects, which were additive, proved each to be independent of the other two intakes, and alcohol was the predominant factor. A significant association of chronic relapsing pancreatitis with the frequent use of alcohol was also found in eastern Massachusetts and Rhode Island for the period 1975–1979 for males aged 35–64 years, but not for females (Yen et al. 1982).

As a sidelight on pancreatitis as a consequence of biliary tract disease, it should be mentioned that a recent Australian case-control study suggested that within a low consumption range (0–40 g/day) an inverse relation exists between alcohol intake and the risk of developing gallstone disease (Scragg et al. 1984). No clear-cut direct association emerged when alcohol use and the relative risk of cancer of the pancreas were analyzed (Durbec et al. 1983).

Other Gastrointestinal Disorders

The epidemiology of cancer of the upper gastrointestinal tract, notably of the esophagus, will be dealt with in a separate chapter. Acute erosive/hemorrhagic gastritis as a consequence of heavy bouts of alcohol overindulgence is a common and well-known self-limited disorder, but epidemiologic data are not to my knowledge available. No convincing evidence has been forthcoming up to now to substantiate a presupposed association between chronic alcoholism and chronic (atrophic) gastritis or peptic ulcer, although superimposed acutely bleeding erosions may be of alcoholic origin (Berges and Wienbeck 1981; Cheli et al. 1981; Piper et al. 1982; Valencia-Parparcén 1981). The epidemiology of alcohol-induced impairment of the absorptive capacity of the small intestine has not been explored.

Note Added in Proof

The limited information available on alcohol consumption and its trends in the *Soviet Union* during the 20 year period 1960 to 1980 provides remarkably low figures[1], especially in regard to the consumption of spirits (predominantly vodka and some brandy), not in keeping with occasional reports on a growing concern in the USSR about alcohol problems, mainly among the Russian population.

In a recent confidential study of the Siberian section of the Soviet Academy of Sciences in Novosibirsk – as was disclosed by the French news agency *Agence France Presse* – figures are mentioned that seem more likely to reflect the actual trend situation[2]. It was stated that – irrespective of a two- and threefold increase in beer and wine consumption between 1960 and 1980 (beer, from 11.9 to 23.1 l annually per capita of total population; wine, from 4.6 to 14.4 l[1]) – the annual per capita consumption of *vodka* increased from 5 l in 1952 to 30 l in 1983. This would mean that the annual per capita consumption of pure alcohol, derived from vodka alone, rose from 2 to 12 l during this period. According to this study, 40 million alcoholics and heavy drinkers were officially registered in 1980, and it was estimated that there are now 17 million clinical cases of alcoholism and another 23 million heavy drinkers in the Soviet Union.

References

Acuda SW (1982) Drug and alcohol problems in Kenya today: a review of research. East Afr Med J 59: 642–644

Ammann R, Sulser H (1976) Die "senile" chronische Pankreatitis: eine neue nosologische Einheit? Schweiz Med Wochenschr 106: 429–437

Andersen BN, Thorsgaard Pedersen N, Scheel J, Worning H (1982) Incidence of alcoholic chronic pancreatitis in Copenhagen. Scand J Gastroenterol 17: 247–252

---

1 Anonymus (1960–1981) Hoeveel alcoholhoudende dranken worden in de wereld gedronken? Produktschap voor Gedistilleerde Dranken. 1st to 20th ed., Schiedam, Netherlands

2 Dragović M (1985) Russland: Fataler Griff zur Wodkaflasche. Die Zeit (Hamburg), No. 2, p 8. 4 Jan 1985
Anonymus (1984) Alkoholismus in der UdSSR: Tragödie des Jahrtausends. Die Presse (Wien), No. 11036, 28 Dec 1984

Angelini G, Pederzoli P, Bovo P, Antolini G, Camplani N, Sandrini T, Scuro LA (1982) Liver and pancreatic involvement in chronic pancreatitis (CP) and liver cirrhosis (LC) of alcoholic aetiology. Gastroenterology 82: 1704 (abstract)

Anonymous (1981a) Alcohol and alcoholism – the report of a special committee of the Royal College of Psychiatrists. Tavistock, London

Anonymous (1981b) The Brewers' Society: international statistical handbook, 2nd edn. Brewing, London

Armyr G, Elmer A, Herz U (1982) Alcohol in the world of the 80s. Habits, attitudes, preventive policies and voluntary efforts. Sober, Stockholm

Berges W, Wienbeck M (1981) Alkohol und Gastrointestinaltrakt. In: Teschke R, Lieber CS (eds) Alkohol und Organschäden. Epidemiologische, klinische, biochemische und therapeutische Aspekte. Witzstrock, Baden-Baden

Bernades P, Belghiti J, Athouel M, Mallardo N, Breil P, Fekete F (1983) Histoire naturelle de la pancréatite chronique: étude de 120 cas. Gastroenterol Clin Biol 7: 8–13

Biel W (1975) Jahrbuch zur Frage der Suchtgefahren. Neuland-Verlagsgesellschaft, Hamburg

Bonfiglio G, Falli S, Pacini A (1979) Alcoholism in Italy: an outline highlighting some special features. Br J Addict 72: 3–12

Bordalo O, Noronha M, Baptista A, Dreiling DA, Schaffner F (1982) Do the liver and the pancreas have a similar behaviour facing alcoholic injury? Scand J Gastroenterol 17 [Suppl 78]: 1587 (abstract)

Brown MM, Dewar MF, Wallace P (1982) International survey: alcoholic beverage taxation and control policies, 5th edn. Brewers, Ottawa

Cheli R, Giacosa A, Marenco G, Canesa M, Dante GL, Ghezzo L (1981) Chronic gastritis and alcohol. Z Gastroenterol 19: 459–463

Coltorti M, Lucchelli PE (Coordinazione) (1977) Patologia cronica del fegato in Italia. Ichiesta retrospettiva multicentrica. Recent Prog Med 63: 447–470

Coltorti M, Calonghi G, Vigasio T et al. (1974) Differenze geografiche nell'eziologia di epatopatie croniche in Italia. Confronto tra una casistica del Nord ed una del Sud. Il Fegato 20: 61–72

Dani R, Nogueira CED (1976) Chronische kalzifizierende Pankreatitis in Brasilien – eine Analyse von 92 Fällen. Leber Magen Darm 6: 272–275

Dani R, Nogueira CED, Furtado M, Leal S (1974) Epidemiology and aetiology of chronic calcifying pancreatitis in Belo Horizonte, Brazil. Rendic Gastroenterol 6: 153–156

Dani R, Nogueira CED, Ribeiro M, Nunes A (1983) Concomitância de lesões hepáticas com pancreatitite crônica calcificante de alcoólatra. Estudo de 60 casos. Arq Gastroenterol 20: 103–107

Davies P, Walsh D (1983) Alcohol problems and alcohol control in Europe. Croom Helm, London

De Marchi S, Basile A, Grimaldi F, Macor C, Vitale G, Cecchin E (1984) Fractures and hypercalcuria: two markers of severe dependence in alcoholics. Br Med J 288: 1457–1458

Diem K, Lentner C (1968) Documenta Geigy, Wissenschaftliche Tabellen 7th edn. Geigy AG, Basel, p 701

Dürr HK (1978) Alkoholschädigung des Pankreas. Internist 19: 123–130

Durbec JP, Sarles H (1978) Multicenter survey of the aetiology of pancreatic disease. Relationship between the relative risk of developing chronic pancreatitis and alcohol, protein and lipid consumption. Digestion 18: 337–350

Durbec JP, Bidart JM, Sarles H (1979) Etude des variations du risque de cirrhose du foie en fonction de la consommation d'alcool. Gastroenterol Clin Biol 3: 725–734

Durbec JP, Chevillotte G, Bidart JM, Berthezene P, Sarles H (1983) Diet, alcohol, tobacco and risk of cancer of the pancreas: a case-control study. Br J Cancer 47: 463–470

Dutta S, Mobrahan S, Iber FL (1976) Alcoholic liver disease in association with alcoholic pancreatitis. Gastroenterology 70: 980 (abstract)

Dutta S, Mobrahan S, Iber FL (1978) Associated liver disease in alcoholic pancreatitis. Am J Dig Dis 23: 618–622

Edwards G (1979) Drinking problems: putting the third world on the map. Lancet 2: 402–404

Ewusi-Mensah I, Saunders JB, Wodak AD, Murray RM, Williams R (1983) Psychiatric morbidity in patients with alcoholic liver disease. Br Med J 287: 1417–1419

Forshaw J (1972) Alcoholic cirrhosis of the liver. Br Med J 4: 608–609
Gips CH (1978) Alcohol, disease of alcoholics and alcoholic liver disease. Neth J Med 21: 83–90
Goebell H, Hotz J (1976) Die Aetiologie der akuten Pankreatitis. In: Forell MM (ed) Pankreas. (Handbuch der Inneren Medizin, vol 3, chapter 6). Springer, Berlin Heidelberg New York, pp 615–675
Greiner L, Schubert E, Franken FH (1983) Koinzidenz von chronischer Pankreatitis und Leberzirrhose bei Alkoholabusus. Eine röntgen- und histomorphologische Studie. Z Gastroenterol 21: 526–532
Gullo L, Costa PL, Labó G (1977) Chronic pancreatitis in Italy. Aetiological, clinical and histological observations based on 253 cases. Rendic Gastroenterol 9: 97–104
Hislop WS, Bouchier IAD, Allan JG (1983) Alcoholic liver disease in Scotland and northeastern England: presenting features in 510 patients. Q J Med (New Series) 52: 232–243
Hodgson HJF, Thompson RPH (1976) Cirrhosis in South London. Lancet 2: 118–121
Hoogendoorn D (1978) Het toenemende gebruik van alcohol en de stijgende frequentie van enkele (mede) door alcohol veroorzaakte ziekten. Ned Tijdschr Geneeskd 122: 1275–1280
Hoogendoorn D (1983) Opnamen in ziekenhuizen wegens overmatig gebruik van alcohol. Ned Tijdschr Geneesk 127: 1011–1017
Howat HT (1963) Chronic pancreatitis. Practitioner 191: 42
Howat HT (1968) Chronic pancreatitis: medical aspects. Postgrad Med J 44: 703–736
Isaacson C (1978) The changing pattern of liver disease in South African blacks. S Afr Med J 53: 365–368
Isaacson C (1982) Pathology of a black African population. Springer, Berlin Heidelberg New York
Jain S, Paton A, Wansbrough-Jones MH (1973) Alcoholic cirrhosis of the liver. Br Med J 1: 116
James O, Agnew JE, Bouchier IAD (1974) Chronic pancreatitis in England: a changing picture? Br Med J I: 34–38
Jędrychowski W, Flak E (1982) Trendy roswojowe umieralności z powodu marskości wątroby w polsce na tle innych krajów europejskich. Patol Pol 33: 269–277
Johnson RD, Davidson S, Saunders JB, Williams R (1984) Fractures on chest radiographs as indicators of alcoholism in patients with liver disease. Br Med J 288: 365–366
Jorke D, Reinhardt M (1982) Contributions to the epidemiology of liver cirrhosis and chronic hepatitis. Dtsch Z Verdau Stoffwechselkr 42: 129–137
Kendell RE, de Roumanie M, Ritson EB (1983) Influence of an increase in excise duty on alcohol consumption and its adverse effects. Br Med J 287: 809–811
Lehnert P (1979) Ätiologie und Pathogenese der chronischen Pankreatitis. Internist 20: 321–330
Lelbach WK (1976) Epidemiology of alcoholic liver disease. In: Popper H, Schaffner F (eds) Progress in liver diseases, vol V. Grune and Stratton, New York, chapter 30, pp 494–515
Lelbach WK (1984) Epidemiology of alcoholic liver disease in continental Europe. In: Hall P (ed) Alcoholic liver disease. Arnold, London
Lelbach WK (1985) Epidemiology of alcohol use and its hazards. In: Morgan MY (ed) Alcohol and disease. Livingstone, London (to be published)
Mero M (1982) Changing aetiology of acute pancreatitis. Ann Chir Gynaecol 71: 126–129
Mincis M, Barollo CR, Filho BH, Saad FA (1973) Alterações hepáticas e pancreáticas em alcoólatras crônicos. Rev Assoc Med Brasil 19: 365–370
Neves MM, Borges DR, Vilela MP (1983) Exocrine pancreatic hypersecretion in Brazilian alcoholics. Am J Gastroenterol 78: 513–516
Noronha M, Salgadinho A, Ferreira De Almeida MJ, Dreiling DA, Bordalo O (1981a) Alcohol and the pancreas: I. clinical associations and histopathology of minimal pancreatic inflammation. Am J Gastroenterol 76: 114–119
Noronha M, Bordalo O, Dreiling DA (1981b) Alcohol and the pancreas: II. pancreatic morphology of advanced alcoholic pancreatitis. Am J Gastroenterol 76: 120–124
Parker RGF (1957) The incidence of primary hepatic carcinoma in cirrhosis. Proc R Soc Med 50: 145–147
Péquignot G, Chabert C, Eydoux H, Courcoul MA (1974) Augmentation du risque de cirrhose en fonction de la ration d'alcool. Rev alcool 20: 191–202

Péquignot G, Tuyns AJ, Berta JL (1978) Ascitic cirrhosis in relation to alcohol consumption. Int J Epidemiol 7: 113–120
Pinto Correia J, Carvlainhos A, Silveira J, Vasconcelos Esteves A, Grima N, Rosario Batista MA (1983) Hémorragie gastro-intestinale dans la cirrhose hépatique. Acta Endoscopica 13: 11–20
Piper DW, McIntosh JH, Greig M, Shy M (1982) Environmental factors and chronic gastric ulcer. Scand J Gastroenterol 17: 721–729
Poegel K, Riedel H, Arnold K, Friedrich P (1981) Häufigkeit simultaner äthylischer Pankreasschäden bei alkoholtoxischen Lebererkrankungen. Dtsch Ges-Wesen 36: 269–271
Poikolainen K (1980) Increase in alcohol-related hospitalizations in Finland 1969–1975. Br J Addict 75: 281–291
Prytz H, Skinhøj P (1981) Alkoholisk og ikke-alkoholisk cirrose. Udviklingen i mortalitet 1965–1978 sammenholdt med alkoholforbruget. Ugeskr Laeger 143: 1175–1178
Prytz H, Skinhøj P (1980) Morbidity, mortality, and incidence of cirrhosis in Denmark 1976–1978. Scand J Gastroenterol 16: 839–844
Read G, Braganza JM, Howat HT (1976) Pancreatitis – a retrospective study. Gut 17: 945–952
Salonen JT, Puska P, Nissinen A (1983) Intake of spirits and beer and risk of myocardial infarction and death in a longitudinal study in Eastern Finland. J Chronic Dis 36: 533–543
Sarles H, Cros RC, Bidart JM, International Group for the Study of Pancreatic diseases (1979) A multicenter inquiry into the etiology of pancreatic diseases. Digestion 19: 110–125
Saunders JB, Walters JRF, Davies P, Paton A (1981) A 20-year prospective study of cirrhosis. Br Med J 282: 263–266
Scragg RKR, McMichael AJ, Baghurst PA (1984) Diet, alcohol, and relative weight in gall stone disease: a case-control study. Br Med J 288: 1113–1119
Sherlock S (1966) Waldenström's chronic active hepatitis. Acta Med Scand [Suppl 445] 179: 426–433
Smith R (1982a) Alcohol in the Third World: a chance to avoid a miserable trap. Br Med J 284: 183–185
Smith R (1982b) Polish lessons on alcohol policy. Br Med J 284: 98–101
Statistisches Jahrbuch der Deutschen Demokratischen Republik (1981) 26: 265
Stone WD, Islam NRK, Paton A (1968) The natural history of cirrhosis: experience with an unselected group of patients. Q J Med 37: 119–132
Sulkunen P (1976) Drinking patterns and the level of alcohol consumption: an international overview. In: Gibbins RJ, Israel Y, Kalant H, Popham RE, Schmidt W, Smart RG (eds) Research advances in alcohol and drug problems, vol 3. Wiley, New York, p 223
Svendsen HO, Mosbech H (1977) Alcoholic cirrhosis of the liver in the Scandinavian countries 1961–1974. Int J Epidemiol 6: 345–348
Svensson JO, Norbäck B, Bokey EL, Edlund Y (1979) Changing pattern in aetiology of pancreatitis in an urban Swedish area. Br J Surg 66: 159–161
Takeuchi J, Ojudaira M, Takada A et al. (1979) The increasing incidence of alcoholic liver diseases in Japan (1969–1977). Jpn J Gastroenterol 76: 2178–2185
Tuyns AJ, Pequignot G, Abbatucci JS (1979) Oesophageal cancer and alcohol consumption: importance of type of beverage. Int J Cancer 23: 443–447
Tuyns AJ, Hu MX, Péquignot G (1983) Alcohol consumption patterns in the department of Calvados (France). Rev Epidemiol Santé Publique 31: 179–197
Valencia-Parparcén J (1981) Alcoholic gastritis. In: Leevy CM (ed) Alcohol and the GI tract. Clin Gastroenterol 10: 389–399
Welte JW, Russell M (1982) Regional variations in the consumption of alcohol in the USA. Drug Alcohol Depend 10: 248–249
Wodak AD, Saunders JB, Ewusi-Mensah I, Davis M, Williams R (1983) Severity of alcohol dependence in patients with alcoholic liver disease. Br Med J 287: 1420–1423
World Health Organization (1980) Technical report series 650: problems related to alcohol consumption. Report of a WHO Expert Committee, Geneva
Yen S, Hsieh CC, Macmahon B (1982) Consumption of alcohol and tobacco and other risk factors for pancreatitis. Am J Epidemiol 116: 407–414
Ziegler H (ed) (1984) Jahrbuch zur Frage der Suchtgefahren. Neuland-Verlagsgesellschaft, Hamburg

2 Ethanol Metabolism and Pathophysiology of Alcoholic Liver Disease

C. S. Lieber

GI-Liver Training Program and Alcohol Research and Treatment Center,
Veterans Administration Medical Center, 130 West Kingsbridge Road, Bronx, NY 10468, USA

The hepatocyte contains three main pathways for ethanol metabolism, each located in a different subcellular compartment: the alcohol dehydrogenase pathway of the cytosol or the soluble fraction of the cell, the microsomal ethanol-oxidizing system located in the endoplasmic reticulum, and catalase located in the peroxisomes (Fig. 1).

The Alcohol Dehydrogenase Pathway

Chemical Characterization of ADH-Mediated Ethanol Oxidation

A major pathway for ethanol disposition involves alcohol dehydrogenase (ADH), an enzyme of the cell sap (cytosol) that catalyzes the conversion of ethanol to acetaldehyde. The raison d'etre of this enzyme might be to rid the body of the small amounts of alcohol produced by fermentation in the gut. Another possible explanation for the presence of ADH in the liver is the fact that this enzyme has a broad substrate specificity, which includes dehydrogenation of steroids (Okuda and Takigawa 1970) and omega oxidation of fatty acids (Björkhem 1972). These compounds may represent the "physiological" substrates for ADH, although the small amount of endogenous ethanol could play such a role.

In ADH-mediated oxidation of ethanol, hydrogen is transferred from the substrate to the cofactor nicotinamide-adenine dinucleotide (NAD), converting it to its reduced form (NADH) (see Fig. 1), and acetaldehyde is produced. As a net result, the first step in the oxidation of ethanol generates an excess of reducing equivalents in the cytosol, primarily as NADH. Thus in normal rats given ethanol there is a marked shift in the redox potential of the cytosol, as measured by changes in the lactate and pyruvate ratio (Veech et al. 1972; Domschke et al. 1974). The altered redox state, in turn, is responsible for a variety of metabolic abnormalities. Some of these, such as hyperlactacidemia, are linked to the utilization of the excess NADH in the cytosol (Fig. 1).

Alcohol Related Diseases in Gastroenterology
Edited by H. K. Seitz and B. Kommerell

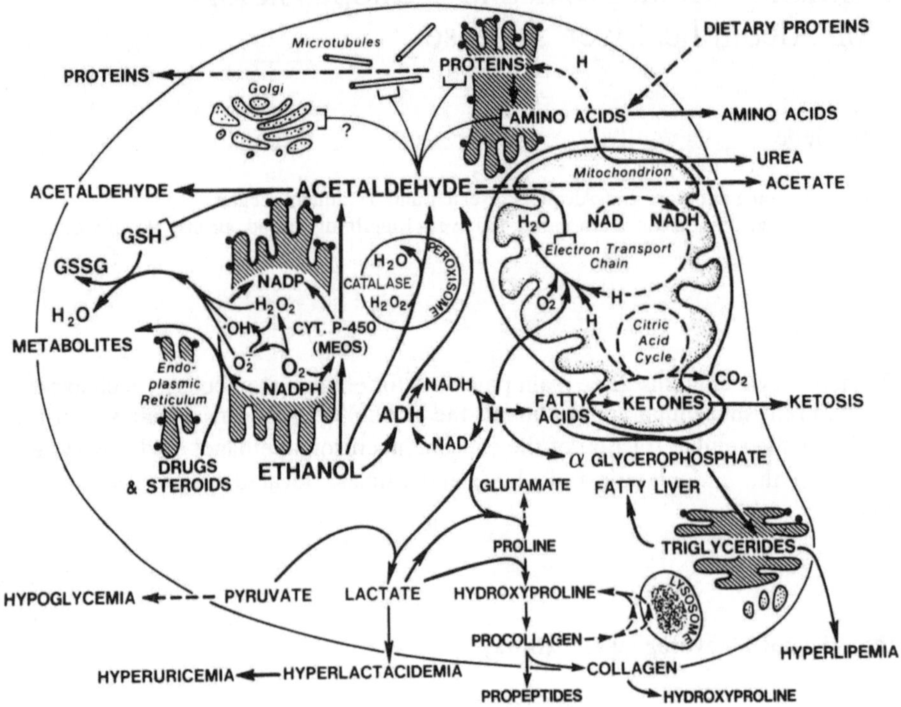

Fig. 1. Oxidation of ethanol in the hepatocyte and link of the two products (acetaldehyde and H) to disturbances in intermediary metabolism, including abnormalities of lipid, carbohydrate, and protein metabolism. *GSH,* reduced glutathione; *GSSG,* oxidized glutathione; *NAD,* nicotinamide-adenine dinucleotide; *NADH,* reduced NAD; *NADP,* nicotinamide-adenine dinucleotide phosphate; *NADPH,* reduced NADP; *MEOS,* microsomal ethanol-oxidizing system; *ADH,* alcohol dehydrogenase. The *broken lines* indicate pathways that are depressed by ethanol. The *symbol* denotes interference or binding

Hyperlactacidemia, Lactic Acidosis, Ketosis, and Hyperuricemia

The redox changes associated with the oxidation of ethanol result in a shift of pyruvate to lactate. This leads to increased lactate levels, resulting from either increased hepatic lactate production (Jorfeldt and Juhlin-Dannfelt 1978) or, depending on the metabolic state of the liver, decreased utilization by the liver of lactate derived from extrahepatic tissues (Krebs 1967; Kreisberg et al. 1971) or both. As a consequence, lactate rises in the blood. The levels usually achieved are moderate and are much lower than those described in so-called lactic acidosis. However, ethanol may strikingly exacerbate the hyperlactacidemia resulting from some other causes, for instance, essential chronic lactic acidosis (Sussman et al. 1970), and among diabetic patients (Daughaday et al. 1962), particularly those treated with phenformin, a drug known to cause lactic acidosis in its own right. In the United States, it was ordered off the general market. However, phenformin, or its close congener, metformin, are still widely used in other countries.

In addition to lactic acidosis, the hyperlactacidemia also has clinically significant consequences with regard to uric acid metabolism. Indeed, the rise in blood lactate decreases urinary uric acid output, which leads to an increase in serum uric acid concentration (Lieber et al. 1962). One can assume that the fasting state (via ketosis) and ingestion of alcohol (via hyperlactacidemia and ketosis) could be additive and produce a significant increase in serum uric acid concomitant with a decrease in urinary uric acid. Indeed, ketosis occurs in the alcoholic not only because of fasting, but also because of a selective rise in β-hydroxybutyrate (Lefevre et al. 1970).

Two additional factors are worth mentioning in the context of alcoholism and hyperuricemia. These factors are the influence of delirium tremens on urate metabolism and the probable occurrence of increased serum uric acid concentrations during and after a seizure. Furthermore, various hepatotoxic agents result in an increased breakdown of liver nucleoproteins and enhanced release of uric acid into the bloodstream. It is possible that such a mechanism may contribute to the hyperuricemia of the alcoholic. Indeed, in gouty subjects with apparent hyperproduction of uric acid, ethanol was found to exacerbate this process (Faller and Fox 1982).

Alterations in Lipid Metabolism

The interaction of ethanol and lipid metabolism are described in detail by Baraona in another chapter and therefore only highlights will be mentioned here. In the liver, the increased NADH/NAD ratio raises the concentration of α-glycerophosphate (Nikkila and Ojala 1963), which favors hepatic triglyceride accumulation by trapping fatty acids (Johnson 1974). Hydrogen equivalents are also transferred into the mitochondria by various "shuttle" mechanisms (Fig. 2). Normally, fatty acids are oxidized via β-oxidation and the citric acid cycle of the mitochondria, which serves as "hydrogen donor" for the mitochondrial electron transport chain. When ethanol is oxidized, however, the generated hydrogen equivalents, which are shuttled into the mitochondria, supplant the citric acid cycle as a source of hydrogen. Depressed fatty acid oxidation by ethanol has been demonstrated in liver slices (Lieber and Schmid 1961), perfused livers (Lieber et al. 1967), isolated hepatocytes (Ontko 1973), and human liver biopsy tissue (Fischel and Oette 1974) and in vivo (Blomstrand et al. 1973). This change results in the deposition in the liver of dietary fat, when available, or fatty acids derived from endogenous synthesis in the absence of dietary fat (Lieber and Spritz 1966; Lieber et al. 1967; Mendenhall 1972), and can be regarded as a major cause of the development of alcoholic fatty liver, the first stage of alcoholic liver injury.

Theoretically, lipids which accumulate in the liver can originate from three main sources: dietary lipids (which reach the bloodstream as chylomicrons), adipose tissue lipids (which are transported to the liver as free fatty acids, FFA), and lipids synthesized in the liver itself. These fatty acids from various sources can accumulate in the liver because of a large number of metabolic disturbances, primarily (a) decreased lipid oxidation in the liver, (b) enhanced hepatic lipogenesis, (c) decreased hepatic release of lipoproteins, (d) increased mobilization of peripheral fat, and (e) enhanced hepatic uptake of circulating lipids. Depending on the experimental

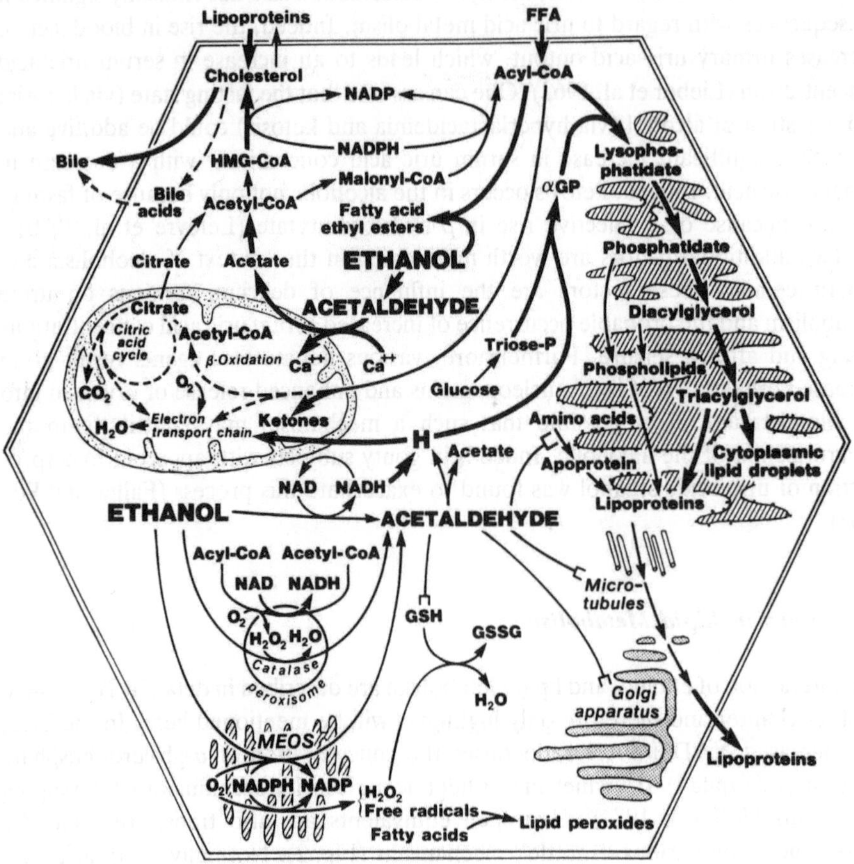

Fig. 2. Oxidation of ethanol in the hepatocyte and link to disturbances in lipid metabolism. *NAD*, nicotinamide-adenine dinucleotide; *NADH*, reduced NAD; *NADP*, nicotinamide-adenine dinucleotide phosphate; *NADPH*, reduced NADP; *FFA*, unesterified ("free") fatty acid; *αGP*, *sn*-glycerol-3-phosphate; *triose-P*, dihydroxyacetone phosphate or glyceraldehyde 3-phosphate; *GSH*, reduced glutathione; *GSSG*, oxidized glutathione; *H*, reducing equivalent; *MEOS*, microsomal ethanol oxidizing system (Lieber and Savolainen 1984)

conditions, any of the three sources and the various mechanisms can be implicated.

Alterations in the Metabolism of Protein, Including Collagen

In addition to changes in carbohydrate and lipids, the abnormal redox state may also affect protein metabolism. Inhibition of protein synthesis has been observed after addition of ethanol to various preparations in vitro (Jeejeebhoy et al. 1975; Rothschild et al. 1971). In vivo, the acute effects of ethanol on protein synthesis have been less consistent than those described in vitro. No changes in the synthesis of total liver protein were found after administration of ethanol to well-fed naive rats

(Baraona et al. 1980). It would be of great interest to identify in vivo conditions that mimic the alterations of ethanol metabolism occurring in isolated liver preparations and that induce inhibitory effects of ethanol on protein synthesis. Indeed, under such conditions, the toxicity of ethanol should be greatly enhanced. The perivenular area of the hepatic lobule, which is somewhat hypoxic already in the normal state, may represent such an area of exaggerated toxicity (Jauhonen et al. 1982). Indeed, this zone has a striking exaggeration of the ethanol-induced redox changes; the latter are sufficient to impair protein synthesis. Not all proteins, however, are necessarily affected the same way. The constituent protein of fibrous tissue, namely collagen, may in fact undergo increased synthesis.

The accumulation of hepatic collagen during the development of alcoholic liver injury could theoretically be accomplished by increased synthesis, decreased degradation, or both. The mechanisms of collagen degradation in the liver are complex. Preliminary results suggest a paradoxical increase in the activity of neutral collagenase in animals fed ethanol, at least during the early stage of alcoholic liver injury (Okazaki et al. 1977). Subsequently, collagenase activity may decrease and contribute to the collagen accumulation (Maruyama et al. 1982). On the other hand, the role of increased collagen synthesis is suggested by increased activity of hepatic peptidylproline hydroxylase in rats and primates and increased incorporation of proline C^{14} into hepatic collagen in rat liver slices (Feinman and Lieber 1972). A possible mechanism whereby alcohol consumption may be linked to collagen formation is the increase in tissue lactate secondary to alcohol metabolism, discussed before (Fig. 1). Elevated concentrations of lactate have been associated with increased peptidylproline hydroxylase activity both in vitro (Green and Goldberg 1964) and in vivo (Lindy et al. 1971). The hepatic free proline pool size, which has been incriminated in the regulation of collagen synthesis (Chvapil and Ryan 1973; Rojkind and DeLeon 1970), may be increased by ethanol (Hakkinen and Kulonen 1975) and is expanded in human portal cirrhosis (Kershenobich et al. 1970). In patients with alcoholic cirrhosis, increased serum free proline and hydroxyproline have been reported (Mata et al. 1975). More recently, it has again been postulated that lactate may play a role (Kershenobich et al. 1979), this time through inhibition of proline oxidase (Kowaloff et al. 1977). In patients with sepsis and cirrhosis, the reduced mitochondrial state has been incriminated for the associated rise in blood lactate and proline (Cerra et al. 1979). However, in other patient populations, plasma proline was found to be normal or even decreased and lactate elevations were less common (Shaw et al. 1983a). Nevertheless, even in the absence of proline and lactate increases in the peripheral blood, the postulated mechanism could be important, since it is related to the alcohol-induced redox change in the liver, which was found to be selectively exacerbated in the perivenular zone (zone III of Rappaport) (Jauhonen et al. 1982) where the alcohol-induced liver lesions occur first and foremost. In cultured myofibroblasts, both lactate and acetaldehyde (the first product of ethanol metabolism) were found to stimulate collagen synthesis (Savolainen et al. 1984). A detailed discussion of the interaction of ethanol with collagen metabolism and fibrogenesis is provided in the chapter by Hahn in this volume.

Miscellaneous Redox-Related Effects

As discussed in subsequent chapters and also elsewhere (Lieber 1982), other metabolic effects of alcohol that can also be attributed to the generation of NADH include interference with glucose, galactose, serotonin, and other amine metabolism. The increased availability of NADH also results in alterations of hepatic steroid metabolism in favor of the reduced compounds (Admirant et al. 1970; Chronholm and Sjovall 1970).

Metabolic Derangements Produced by Acetaldehyde

Metabolism and Levels of Acetaldehyde

Acetaldehyde is the first major "specific" oxidation product of ethanol, whether the latter is oxidized by the classic alcohol dehydrogenase of the cytosol or by the microsomal system (Lieber and DeCarli 1970b; Teschke et al. 1974, 1975). It is generally accepted that acetaldehyde oxidation proceeds via aldehyde dehydrogenase, much of the activity of which is located in the mitochondria, at least in experimental animals. Since metabolism of acetaldehyde via aldehyde dehydrogenase results in the generation of NADH, some of the acetaldehyde effects may be attributed to the NADH generation, as discussed in the case of ethanol.

Acetaldehyde, however, is a very reactive compound that may exert some toxic effects of its own, discussed by Salaspuro and Lindros in another chapter. In summary, acetaldehyde was found capable of "binding" to proteins, an effect enhanced by chronic ethanol consumption (Nomura and Lieber 1981). Covalent binding of "active" metabolites has been incriminated in the hepatotoxicity of many drugs; this could also pertain to acetaldehyde, especially after chronic ethanol consumption.

Little was known about blood acetaldehyde levels after alcohol consumption until Korsten et al. (1975) demonstrated a difference in blood acetaldehyde level between alcoholic and nonalcoholic subjects after comparable ethanol challenges. Recent methodological advances have revealed that the concentration of acetaldehyde is lower than the level previously reported. However, even at the lower level, blood acetaldehyde is increased after chronic alcohol consumption (Lindros et al. 1980; Pikkarainen and Lieber 1980), probably reflecting increased production coupled with decreased disposition.

Swelling of the Hepatocyte and Possible Relation to Microtubular Alterations, Portal Hypertension, and Necrosis

Two of the earliest and most conspicuous features of the hepatic damage produced by alcohol are the deposition of fat and the enlargement of the liver. This hepatomegaly was traditionally attributed to the accumulation of lipids. However, in animals fed alcohol-containing diets it was shown that lipids account for only half the increase in liver dry weight (Lieber et al. 1965), while the other half is almost totally accounted for by an increase in proteins (Baraona et al. 1975), possibly secondary to acetaldehyde-induced impairment of microtubular-mediated protein secretion. Indeed, microtubules are shortened and thickened (Matsuda et al. 1979). It is noteworthy that impairment of microtubules was found to affect hepatocellular transport of biliary lipids (Gregory et al. 1978) and be associated with engorgement of the Golgi (Matsuda et al. 1979). More recently, Berman et al. (1983) could not confirm the ethanol effect using morphometry. Inexplicably, however, the value for volume density of microtubules for control rats of Berman et al. (1983) is 21 times higher than that reported by Reaven et al. (1977), and 15 times higher than that of Matsuda et al. (1979). Such gross overestimation of the microtubular mass could obviously obscure the decrease produced by ethanol. Moreover, crucial items of experimental design to test the effects of ethanol independently of associated nutritional changes (such as the techniques of pair-feeding and equalization of the food intake prior to the measurements) are not described by Berman et al. (1983). Alterations in the nutritional status are known to alter microtubules (Pipeleers et al. 1977) and may also explain why the effect of ethanol may have been missed by Berman et al. (1983). Indeed, reevaluation of the effects of alcohol consumption on rat liver microtubules by four biochemical assays for liver tubulin confirmed that the feeding of the ethanol-containing diet decreased hepatic microtubule-derived tubulin and the ratio of polymerized to total tubulin, indicating a reproducible effect of ethanol on this cytoskeletal organelle (Baraona et al. 1984).

The increase in hepatic protein observed after ethanol was not associated with changes in concentration (Baraona et al. 1977), indicating that water was retained in proportion to the increase in protein. The mechanism of the water retention is not fully elucidated, but the rise in both protein and amino acids, plus a likely increase in associated small ions, could account osmotically for a large fraction of the water increase. The increases in lipid, protein, amino acid, water, and electrolytes result in increased size of the hepatocytes. The number of hepatocytes and the hepatic content of desoxyribonucleic acid did not change after alcohol treatment, and thus the hepatomegaly is entirely accounted for by the increased cell volume (Baraona et al. 1975, 1977). There is also an increase in the number of hepatic mesenchymal cells after ethanol feeding (Baraona et al. 1975), but this increase does not significantly contribute to the hepatomegaly. The swelling of the hepatocytes after chronic alcohol administration was found to be associated with a reduction of the intercellular space and with portal hypertension (Orrego et al. 1981). Although perivenular and perisinusoidal fibrosis may be more important for the increase in portal pressure in precirrhotic stages, swelling of the hepatocytes can play a contributing role (Miyakawa et al. 1985). One suspects that ballooning and associated gross distortion of the volume of the hepatocytes may also result in severe impairment of key cellular functions. In alcoholic liver disease, some cells have a diameter which is increased two to three times, and thereby volume is increased about four- to tenfold. One may wonder to what extent this type of cellular disorganization, with protein retention and ballooning, may promote progression of the liver injury and eventual cell necrosis in the alcoholic.

Structural and Functional Alterations of the Mitochondria

Studies with the electron microscope have revealed striking morphological alterations, including swelling and abnormal cristae in the liver mitochondria of alcoholics. Controlled studies in animals and man (Iseri et al. 1966; Lane and Lieber 1966; Lieber and Rubin 1968; Rubin and Lieber 1967; Arai et al. 1984) have shown that these changes are caused by alcohol itself rather than by other factors, such as a poor diet. These structural abnormalities are associated with functional impairments, especially decreased oxidation of fatty acids and of a variety of other substrates, including acetaldehyde (Hasumura et al. 1976). Mitochondria of alcohol-fed animals have a reduction in cytochrome *a* and *b* content (Koch et al. 1977; Rubin et al. 1970) and in succinic dehydrogenase activity (Oudea et al. 1970; Rubin et al. 1970). The respiratory capacity of the mitochondria was found to be depressed (Gordon 1973; Hasumura et al. 1975b; Kiessling and Pilstrom 1968; Rubin et al. 1972) using pyruvate, succinate, and acetaldehyde as substrates. After chronic alcohol consumption, the liver mitochondria are unusually susceptible to the toxic effects of acetaldehyde, and a variety of important mitochondrial functions, such as fatty acid oxidation, are depressed, even in the presence of relatively low acetaldehyde concentrations (Matsuzaki and Lieber 1977).

Increased Lipid Peroxidation

Aldehydes react quite readily with mercaptans, and L-cysteine could complex with acetaldehyde to form a hemiacetal. Binding of acetaldehyde with cysteine and/or glutathione (Fig. 3) may contribute to a depression of liver glutathione (Shaw et al. 1981). Glutathione offers one of the mechanisms for the scavenging of toxic free radicals; a severe reduction in glutathione favors peroxidation (Wendell and Thurman 1979), and the damage may possibly be compounded by the increased generation of active radicals by the "induced" microsomes following chronic ethanol consumption. It is well known that the microsomal pathway, which requires O_2 and reduced nicotinamide-adenine dinucleotide phosphate (NADPH), is capable of generating lipid peroxides. Enhanced lipid peroxidation, possibly mediated by acetaldehyde (DiLuzio and Stege 1977), has been proposed as a mechanism for ethanol-induced fatty liver (DiLuzio and Hartman 1967), but its role has been controversial. Theoretically, increased activity of microsomal NADPH oxidase following ethanol consumption (Lieber and DeCarli 1970c; Reitz 1975) could result in enhanced H_2O_2 production, thereby also favoring lipid peroxidation.

In any event, it was found that in in naive rats, very large amounts of ethanol (5–6 g/kg) are required to produce lipid peroxidation (Farrell et al. 1978; MacDonald 1973), whereas a smaller dose (3 g/kg) had no effect (Shaw et al. 1981). By contrast, after chronic ethanol administration to the rat, even the smaller dose of ethanol administered acutely induced liver peroxidation and this effect could be prevented, at least in part, by the administration of methionine, a precursor of glutathione (Shaw et al. 1981). The ethanol-induced lipid peroxidation was even more striking in the baboon: administration of relatively small doses of ethanol (1–2 g/kg) after 5–6 h produced lipid peroxidation and depletion of reduced glutathione (GSH). Additional evidence for increased peroxidation was provided by

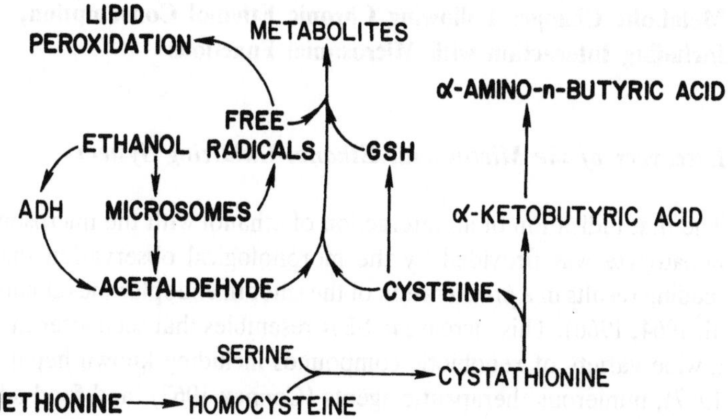

Fig. 3. Hypothetical link between accelerated acetaldehyde production, increased free radical generation by the "induced" microsomes and enhanced lipid peroxidation and possibly increased α-amino-n-butyric acid (AANB) production *GSH,* reduced glutathione (Lieber 1980)

finding increased dienes conjugates in lipids in liver biopsies of patients with alcoholic liver disease (Shaw et al. 1983b). There was an associated reduction in hepatic GSH which may contribute to peroxidation. It is tempting to speculate that the propensity of primates to develop more severe lesions than the rat after chronic ethanol consumption may in some way be related, at least in part, to the greater susceptibility to glutathione depletion and the initiation of lipid peroxidation. It is apparent, however, that glutathione depletion per se does not suffice to produce liver damage (Siegers et al. 1977). As mentioned before, concomitant enhanced production of active radicals may be required, possibly resulting from the microsomal "induction" to be discussed subsequently.

Metabolic Derangements Associated with Acetate Production

The role of acetate is less clear than that of acetaldehyde. Acetate was found (Liang and Lowenstein 1978) to increase cardiac output, myocardial contractility, and coronary blood flow. The effects of a rise of circulating acetate on intermediary metabolism in various tissues have not been well defined. In adipose tissue, acetate inhibits lipolysis (Nilson and Belfrage 1978), and it was found to be responsible, at least in part, for the decreased release of FFA and the fall of circulating FFA (Crouse et al. 1968). A fall in FFA, a major fuel for peripheral tissues, may have significant metabolic consequences. In the liver, acetate was also shown to promote steatosis (Morgan and Mendenhall 1977).

In addition, because of its conversion to acetyl CoA, acetate promotes the breakdown of ATP to AMP. The latter can be either recycled to ATP or degraded to purines and uric acid. It has been postulated that the latter process may contribute to the hyperuricemia (Faller and Fox 1982) (see above). Furthermore, it is possible that the high demand for ATP induced by ethanol may contribute to liver toxicity in a way akin to that produced by fructose (Bode et al. 1973a, b).

Metabolic Changes Following Chronic Ethanol Consumption, Including Interaction with Microsomal Functions

Discovery of the Microsomal Ethanol-Oxidizing System

The first indication of an interaction of ethanol with the microsomal fraction of the hepatocyte was provided by the morphological observation that in rats, ethanol feeding results in a proliferation of the smooth endoplasmic reticulum (SER) (Iseri et al. 1964, 1966). This increase in SER resembles that seen after the administration of a wide variety of xenobiotic compounds including known hepatotoxins (Meldolesi 1967), numerous therapeutic agents (Conney 1967), and food additives (Lane and Lieber 1967). Most of these substances that induce a proliferation of the SER are metabolized, at least in part, in the microsomal fraction of the hepatocyte that comprises the SER. The observation that ethanol produced proliferation of the SER raised the possibility that, in addition to its oxidation by ADH in the cytosol, ethanol may also be metabolized by the microsomes. A microsomal system capable of methanol oxidation had been described (Orme-Johnson and Ziegler 1965), but its capacity for ethanol oxidation was extremely low. Furthermore, this system could not oxidize long-chain aliphatic alcohols such as butanol and was exquisitely sensitive to the catalase inhibitors azide and cyanide. Therefore, Ziegler concluded that this system is clearly different from the cytochrome-*P*-450-dependent system and involves the H_2O_2-mediated ethanol peroxidation by catalase (Ziegler 1972). However, a microsomal ethanol oxidizing system (MEOS) with a rate of ethanol oxidation ten times higher than reported by Orme-Johnson and Ziegler (1965) was described (Lieber and DeCarli 1968, 1970b; Lieber et al. 1974). The system required NADPH and O_2 and was relatively insensitive to catalase inhibition. Furthermore, the MEOS was differentiated from the system reported by Orme-Johnson and Ziegler (1965) and from catalase by its ability to oxidize long-chain aliphatic alcohols (Teschke et al. 1975), which are not substrates for catalase (Chance and Oshino 1971). The striking increase in the non-ADH fraction of ethanol metabolism with increasing ethanol concentrations (Thieden 1971; Grunnet et al. 1973; Matsuzaki et al. 1981) is consistent with the known K_m for MEOS and ADH. Whereas the former has a value of 8–10 m*M* (Lieber and DeCarli 1970b), the latter has a K_m varying from 0.26 to 2.0 m*M* (Reynier 1969; Makar and Mannering 1970; Feytmans and Leighton 1973; Lindros et al. 1974). The in vitro K_m of MEOS agrees well with the corresponding value of 9 m*M* for the pyrazole-insensitive pathway in vivo (Lieber and DeCarli 1972) and with a similar value in isolated hepatocytes (Matsuzaki et al. 1981), suggesting that MEOS may play a significant role in ethanol metabolism.

Differentiation of MEOS from ADH and Catalase

Differentiation of MEOS in total microsomes from alcohol dehydrogenase was achieved by subcellular localization, pH optimum in vitro (7.4 versus 10), cofactor requirements, and effects of inhibitors such as pyrazole (Lieber and DeCarli 1973; Lieber et al. 1970). Studies with inhibitors have also indicated that a major fraction of the ethanol-oxidizing activity in microsomes is independent of catalase (Lieber and DeCarli 1970b, 1973; Lieber et al. 1970). Subsequently, MEOS was solubilized and separated from alcohol dehydrogenase and catalase activities by diethylaminoethyl cellulose column chromatography (Teschke et al. 1972, 1974; Mezey et al. 1973).

More recently, the reconstitution of the ethanol-oxidizing activity with the three microsomal components cytochrome *p*-450, NADPH-cytochrome *c* reductase, and lecithin was demonstrated (Ohnishi and Lieber 1977). Successful reconstitution of MEOS was confirmed by Miwa et al. (1978). Ethanol oxidation can be mediated either by reductase or cytochrome *P*-450 (Ohnishi and Lieber 1978; Winston and Cederbaum 1983). It was shown by Ohnishi and Lieber (1977) that reductase-mediated ethanol oxidation is less than oxidation catalyzed by ethanol-induced *P*-450 (used in saturating amounts), but is equal to oxidation catalyzed by noninduced *P*-450. Superoxide dismutase inhibits the reductase-mediated reaction (Winston and Cederbaum 1983). Indeed, there is now evidence that the reductase operates via the hydroxyl radical mechanism (Bosterling and Trudel 1981) and can therefore be expected to be dismutase sensitive, whereas the ethanol-induced *P*-450 mechanism can be expected to be dismutase insensitive. Thus, when one used noninduced cytochrome *P*-450 (Ingelman-Sundberg and Johansson 1981), reductase-mediated inhibition by dismutase may be more apparent than when the more active ethanol-induced cytochrome *P*-450 (Ohnishi and Lieber 1978) is used.

Alteration in the Metabolism of Ethanol After Chronic Ethanol Consumption

Regular drinkers tolerate large amounts of alcoholic beverages, mainly because of central nervous system adaptation. In addition, alcoholics develop increased rates of blood ethanol clearance, that is, metabolic tolerance (Kater et al. 1969; Ugarte et al. 1972). Experimental ethanol administration also results in an increased rate of ethanol metabolism (Feinman et al. 1978; Lieber and DeCarli 1970b, 1972; Misra et al. 1971; Pikkarainen and Lieber 1980; Tobon and Mezey 1971). The progressive acceleration of ethanol metabolism after chronic ethanol consumption is not to be confounded with the rise that occurs after an acute dose of ethanol (the so-called "swift increase" in alcohol metabolism), which appears to results from a stress-associated adrenaline discharge (Yuki and Thurman 1980). The mechanism of the chronic acceleration is still the subject of discussion. Part of the effect is related to ADH-mediated metabolism, whereas part is due to microsomal alterations.

ADH-Related Acceleration of Ethanol Metabolism

The long-standing debate about whether or not chronic alcohol consumption enhances the activity of liver ADH has now been finally settled. Contrary to past findings of enhanced ADH activity after chronic alcohol consumption (Hawkins et al. 1966; Hawkins and Kalant 1972), it has now been reported (Kalant et al. 1975; Videla et al. 1973) that ADH activity does not increase after chronic ethanol feeding, a finding consistent with the observations of other groups (Raskin and Sokoloff 1972;

deSaint-Blanquat et al. 1972; Singelvich and Barboriak 1971). In some studies, there was actually a decrease in ADH-activity in the liver after chronic ethanol consumption (Brighenti and Pancalid 1970; Lieber and DeCarli 1970b; Salaspuro et al. 1975), a finding in keeping with the observation that alcoholics may display decreased hepatic ADH activity even in the absence of liver damage (Ugarte et al. 1967). In view of the observation of Crow and co-workers (1977) that, under certain circumstances, the level of the activity of ADH may become a major rate-limiting factor for the metabolism of ethanol, the decrease in ADH after chronic ethanol consumption may acquire functional significance.

One of the mechanisms that could contribute to the acceleration of ADH-dependent ethanol metabolism after ethanol consumption (based on increased NADH reoxidation) involves enhanced adenosine triphosphatase (ATPase) activity (susceptible to ouabain inhibition) (Bernstein et al. 1973) and the creation of a hypermetabolic state akin to hyperthyroidism (Bernstein et al. 1975; Israel et al. 1975). Israel et al. (1975) reported that in liver slices, ouabain, an inhibitor of the Na^+-K^+-activated ATPase, can completely block the extra ethanol metabolism elicited by chronic ethanol treatment. Oxygen consumption was also found to be increased in the livers of animals chronically treated with ethanol (Bernstein et al. 1973; Israel et al. 1973; Thurman et al. 1976), mimicking the effects of thyroxine. Thus it was proposed that the hypermetabolic condition that occurs in the livers of animals chronically treated with ethanol may in some aspects be similar to that found in the livers of animals treated with thyroid hormones, in which the hypermetabolic state also appears to be associated with an increased hydrolysis of ATP by the (Na^+-K^+)-ATPase system (Ismail-Beigi and Edelman 1970, 1971) and a resulting lowering of the phosphorylation potential. However, there is controversy in the literature concerning the effects of thyroid treatment on rates of alcohol metabolism, as discussed in detail elsewhere (Lieber 1982). Furthermore, not all investigators found the increase in oxygen consumption (Cederbaum et al. 1977; Gordon 1977). Moreover, under conditions that mimic the clinical situation with development of fatty liver, chronic ethanol consumption was not found to be associated with increased ATPase activity (Gordon 1977), and the increase in the rates of ethanol metabolism after chronic ethanol consumption could not be abolished by ouabain (Cederbaum et al. 1977), which indicates that the theory of enhanced ATPase activity may not be applicable to the situation that normally prevails after chronic alcohol consumption. Indeed, it has now been acknowledged that the ouabain effect is nonspecific and does not imply involvement of ATPase (Yuki and Thurman 1980).

Among other mechanisms that enhance metabolic activity, one should mention the increase in the release of epinephrine from the adrenal glands after ethanol administration, both in animals and in man (Anton 1952; Klingman and Goodall 1957; Perman 1958, 1960, 1961). This may indirectly accelerate ethanol metabolism.

If, following chronic ethanol consumption, changes affecting the ADH pathway (such as ATPase activity) were exclusively responsible for the acceleration of ethanol metabolism, the latter should be fully abolished by pyrazole treatment. However, this was not the case (Lieber and DeCarli 1970b, 1972; Salaspuro et al. 1975). This raised the possibility of the involvement of non-ADH pathways, especially at higher ethanol concentrations, at which ADH inhibitors only partially inhibit the accelerated metabolism (Matsuzaki et al. 1981), whereas at very low (1 m*M*) ethanol concentrations an almost complete inhibition was reported (Thurman et al. 1976). Since all these experiments were carried out in animals, the higher K_m pyrazole-insensitive ADH, which has been described in humans but not in animals (Bosron and Li 1977; Salaspuro and Lieber 1978), does not have to be taken into account. Furthermore, in baboons fed alcohol chronically, there was significant acceleration of ethanol metabolism and a significant reduction in ADH activity, whether expressed per gram of liver or per total liver, despite the associated hepatomegaly (Nomura et al. 1983). By contrast, MEOS activity was significantly increased. It is noteworthy that the decrease in ADH activity was associated with a reduction in mitochondrial functions, including respiratory chain activity. It is unlikely, therefore, that the ADH pathway can account for the accelerated ethanol metabolism under these conditions. Two factors probably underlie the acceleration of ethanol metabolism. The hepatomegaly, to the extent that it reflects an increase in active tissue mass, obviously enhances the capacity of the body to rid itself of ethanol whatever the pathways involved. Furthermore, a non-ADH mechanism appears to be contributory (see below).

Non-ADH-Related Acceleration of Ethanol Metabolism

Following chronic ethanol consumption, MEOS significantly increases in activity (Lieber and DeCarli 1968, 1970b). this is associated with an increase in various constitutents of the smooth fraction of the membranes involved in drug metabolism, such as phospholipids, cytochrome *P*-450 reductase, and cytochrome *P*-450 (Ishii et al. 1973; Joly et al. 1973a; Sato et al. 1978). It is noteworthy that the MEOS activity increased not only when expressed per gram of liver, or per milligram of microsomal protein, but also when expressed per nanomole of cytochrome *P*-450, suggesting the emergence of an "ethanol-specific" form of cytochrome *P*-450. This was confirmed by the shift of the cytochrome reduced spectrum maximum to higher wavelengths. A cytochrome *P*-450 species showing high affinity for cyanide has been reported as preferentially induced by ethanol (Joly et al. 1972; Comai and Gaylor 1973; Hasumura et al. 1975a). Evidence in favor of an increase in a special species of cytochrome *P*-450 after ethanol treatment was also derived from inhibitor studies (Ullrich et al. 1975). More direct proof was obtained from studies of microsomal proteins (Ohnishi and Lieber 1977). The rise in cytochrome *P*-450 involved a hemoprotein different from those induced by phenobarbital or 3-methylcholanthrene treatment. Studies by Joly et al. (1976, 1977) also showed that chronic ethanol administration to rats is associated with the appearance of a form of cytochrome *P*-450 with spectral and catalytic properties different from those of the cytochrome *P*-450 of control, phenobarbital-treated, and methylcholan-threne-treated rats. Although an ethanol-specific form of cytochrome *P*-450 has been revealed by sodium dodecyl sulfate gel electrophoresis in the rat (Ohnishi and Lieber 1977) and the deer mouse (Shigeta et al. 1984), the only species in which this form has been purified is the rabbit (Koop et al. 1982).

That achronic ethanol feeding results in an increased activity in liver tissue of a non-ADH and noncatalase pathway was also shown in liver slices and in isolated hepatocytes. Ethanol oxidation was enhanced in isolated liver tissue by increasing the ethanol concentration employed in vitro from 10 to 30 m*M*. Of particular interest was the observation that this phenomenon was more pronounced in ethanol-fed rats than in their pair-fed controls (Teschke et al. 1977). Similarly, when a relatively constant blood ethanol level is maintained through continuous infusion in the baboon, the acceleration of ethanol metabolism with higher blood levels is more pronounced in alcohol-fed than in control animals (Pikkarainen and Lieber 1980). All these data indicate that a non-ADH pathway, most likely MEOS, represents a major mechanism for the acceleration of ethanol metabolism at high ethanol concentrations. A similar change was shown in man: in volunteers, alcohol consumption resulted in a progressive acceleration of blood ethanol clearance, particularly at high ethanol concentrations (Salaspuro and Lieber 1978).

There is also some debate about whether ethanol feeding enhances catalase activity in rats. Both an increase (Carter and Isselbacher 1971) and no change (von Wartburg et al. 1961; Hawkins et al. 1966; Lieber and DeCarli 1970b) have been reported. This question, however, may not be fully relevant to the rate of ethanol metabolism, since peroxidative metabolism of ethanol in the liver is probably limited by the rate of hydrogen peroxide formation rather than by the amount of available catalase (Boveris et al. 1972). Ethanol consumption, however, does enhance the activity of hepatic NADPH oxidase (Lieber and DeCarli 1970c; Carter and Isselbacher 1971; Thurman 1973), which can participate in H_2O_2 generation. It is conceivable that this mechanism contributes to ethanol metabolism in vivo (and to its

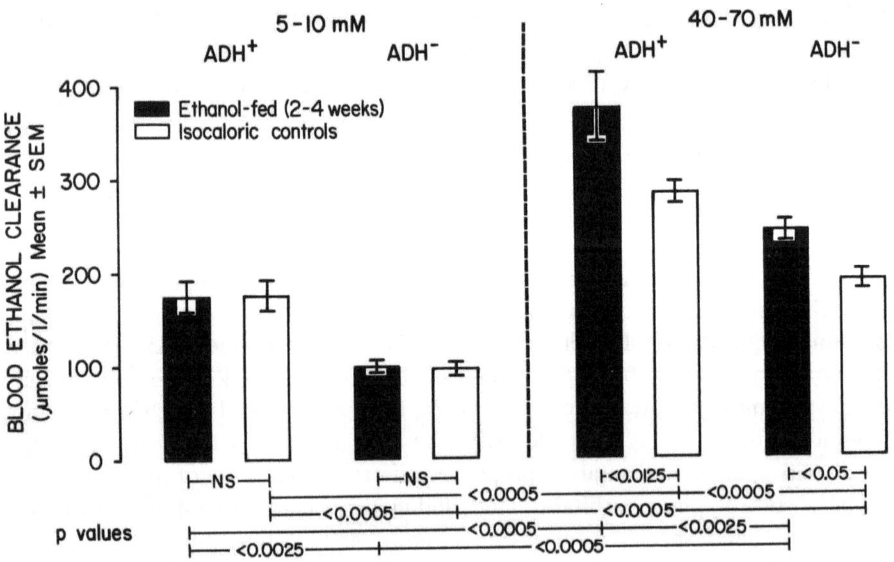

Fig. 4. Effect of chronic ethanol feeding on blood ethanol elimination rate in ADH-positive and ADH-negative deer mice given ethanol. Ethanol, 3 g/kg body wt. (10% w/v solution), was injected; each blood ethanol elimination curve was plotted and elimination rates were calculated at 5–10 m*M* and 40–70 m*M* in each ethanol-fed and pair-fed control animal of both strains. (Shigeta et al. 1984)

increase after chronic ethanol consumption) by furnishing the H_2O_2 needed for the oxidation mediated by the $\dot{O}H$ generated by the reductase (Ohnishi and Lieber 1977, 1978; Winston and Cederbaum 1983). In the presence of the ethanol-induced form of cytochrome *P*-450, however, such a mechanism may be less important than the cytochrome-*P*-450-(non-$\dot{O}H$) dependent oxidation of ethanol (Ohnishi and Lieber, 1977, 1978). In any event, such a mechanism does not involve catalase, the role of which is probably limited to ethanol metabolism by H_2O_2 generated inside the peroxisomes. Catalase, however, appears to participate in the oxidation of methanol, at least in the rat, whereas in the monkey alcohol dehydrogenase may play a greater role (Makar et al. 1968). Chronic ethanol feeding resulted in a striking increase of microsomal NADPH-dependent oxidation of methanol (Teschke et al. 1974), but there was also a response with *n*-propanol and *n*-butanol, substrates that virtually fail to react peroxidatically with catalase-H_2O_2. Recently, the capacity of MEOS to sustain increased rates of ethanol metabolism was most convincingly illustrated in deer mice lacking ADH. Rate of ethanol metabolism strikingly increased after chronic ethanol feeding (Fig. 4), with an associated rise in MEOS activity (Fig. 5) and cytochrome *P*-450 content (Fig. 6) Shigeta et al. 1984).

In summary, because of all the experimental evidence obtained (including the substrates involved, the insensitivity to the ADH and catalase inhibitors) and because the activating effect of high ethanol concentrations is consistent with the saturation of MEOS ($K_m = 8-10$ m*M* for ethanol), the adaptive increase of ethanol oxidation in the liver after chronic ethanol feeding most likely involves the activity of MEOS.

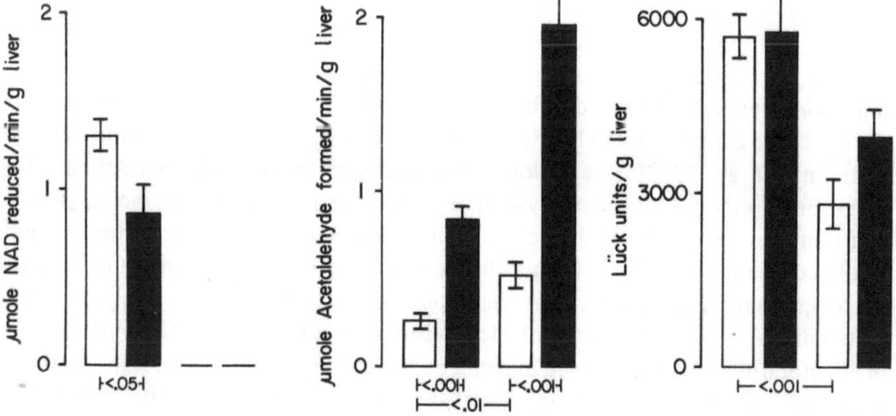

Fig. 5. Effect of chronic ethanol consumption on the activities of hepatic ADH, the microsomal ethanol-oxidizing system (MEOS), and catalase in ADH-positive and ADH-negative deer mice. Activities were measured in chronically ethanol-fed and pair-fed control deer mice of both strains. The values are expressed as mean ± SEM (per gram of liver) (Shigeta et al. 1984)

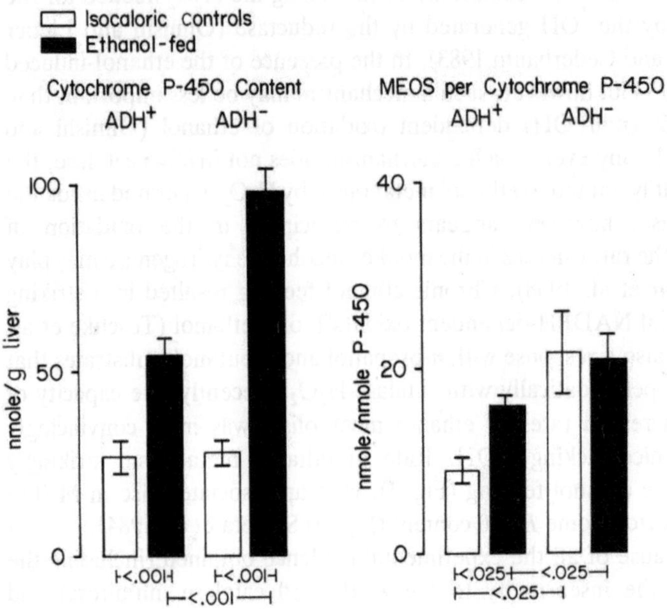

Fig. 6. Effect of chronic ethanol consumption on the microsomal cytochrome *P*-450 content and the activity of the microsomal ethanol-oxidizing system (MEOS) in ADH-positive and ADH-negative deer mice. MEOS activity is expressed per cytochrome *P*-450 content. The data represent mean ± SEM. Note the increase of an "ethanol-specific" band after ethanol consumption in both ADH^+ and ADH^- deer mice (Shigeta et al. 1984)

Increased Tolerance to Drugs and/or Potentiation of Drug Activity

Chronic ethanol consumption results in proliferation of the membranes of the smooth endoplasmic reticulum, documented by subfractionation and chemical measurements in microsomes (Ishii et al. 1973) and by electron microscopy in animals and man (Iseri et al. 1966; Lane and Lieber 1966). The proliferation is thought to be "adaptive", since it is associated with enhanced activity of the microsomal enzymes involved in lipoprotein production and thus may contribute to the increased capacity of the liver to secrete fat as lipoproteins into the bloodstream (Baraona et al. 1973) (see below). As discussed before, the activity of the microsomal ethanol oxidizing system is also increased, and this increase may contribute to the accelerated rate of alcohol metabolism that occurs after chronic alcohol consumption. The latter may be viewed as the raison d'etre of the induction. A concomitant increase in drug-metabolizing enzymes explains the enhanced metabolic tolerance of patients with alcoholism to a variety of drugs. This interaction of ethanol with drugs is discussed in detail by Sato and Hasumura in another chapter. A detailed discussion of this topic is also provided elsewhere (Lieber 1982).

Toxicity Associated with Microsomal Induction

Although many microsomal changes can be interpreted as adaptive alterations secondary to "induction" after chronic ingestion of alcohol, some injurious consequences may also ensue. Indeed, accelerated ethanol metabolism results in enhanced production of acetaldehyde and exacerbation of its various toxic manifestations discussed before. Increased microsomal activity may also enhance the oxygen requirements, thereby aggravating whatever hypoxia may be present. Furthermore, some compounds acquire hepatotoxicity only after metabolism or "activation" by the enzymes of the endoplasmic reticulum. One such compound is carbon tetrachloride, the hepatotoxicity of which is greatly increased after chronic alcohol consumption, at least partly because of enhanced activation by microsomes (Hasumura et al. 1974). It is likely that a similar mechanism increases susceptibility in these patients to the hepatotoxic actions of a variety of compounds in the environment, including many commonly used drugs such as ioniazid and acetaminophen (Sato et al. 1981), and possibly carcinogens. These aspects are discussed in detail by Sato, Hasumura, and Seitz in other chapters.

Enhanced microsomal degradation of endogenous substrates such as steroid hormones was also observed after chronic ethanol consumption; its impact on the endocrine status of the alcoholic is described by Van Thiel in this volume.

Micronutrients, such as vitamins, may also serve as substrates for the microsomes and the induction of the microsomes may therefore alter vitamin requirements and even affect the integrity of the liver. Indeed, it has been found recently that alcoholics commonly have very low vitamin A levels in their livers (Leo and Lieber 1982) (Fig. 7). In experimental animals, ethanol administration was shown to depress hepatic vitamin A levels, even when administered with an adequate diet (Sato and Lieber 1981). When dietary vitamin A was virtually eliminated, the depletion rate of hepatic vitamin A storage was two to three times faster in ethanol-fed rats than in controls, possibly because of accelerated degradation of retinoic acid by the induced microsomes (Sato and Lieber 1982). In rats, severe vitamin A depletion was associated with the appearance of multivesicular lysosomes (Leo et al. 1983a). Such lesions were commonly seen in alcoholic patients with low hepatic A levels (Leo et al. 1983a). Thus vitamin A depletion may contribute to the liver lesions of the alcoholic. Vitamin A supplementation is sometimes used to correct the problems of night blindness and sexual dysfunction of the alcoholic. Such therapy might also be useful with regard to the liver pathology. The therapeutic usage of vitamin A, however, is complicated by the fact that excessive amounts of vitamin A are known to be hepatotoxic and that alcohol consumption enhances the suspectibility to this effect (Leo et al. 1982). In control rats, amounts of vitamin A equivalent to those commonly used for the treatment of the alcoholic were found to be without significant effects on the liver, but in animals chronically fed alcohol, signs of toxicity developed, such as striking morphological and functional alterations of the mitochondria (Leo et al. 1982) and, eventually, necrosis and fibrosis (Leo and Lieber 1983). Enhanced toxicity was not associated with an increased vitamin A level in the liver. In fact, because (as mentioned before) alcohol administration tends to decrease vitamin A levels in the liver, even after vitamin A supplementation, alcohol-fed animals had vitamin A levels in the liver that were not higher than

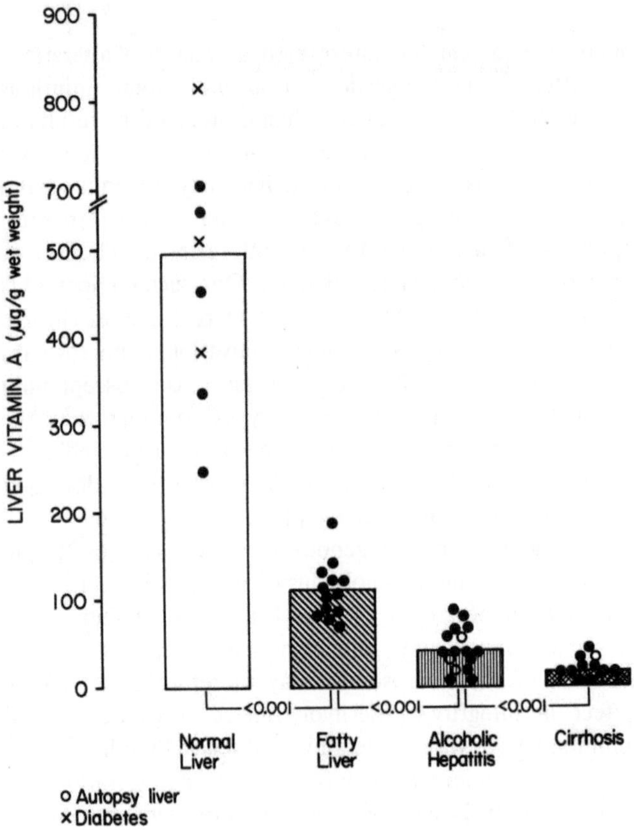

Fig. 7. Hepatic vitamin A levels in subjects with normal livers and those with various stages of alcoholic liver injury (Leo and Lieber 1982)

normal values. Nevertheless, toxicity developed. One possible explanation is that vitamin A toxicity may be mediated at least in part by the enhanced microsomal production of a toxic metabolite, as in the case of xenobiotic agents. Indeed, cytochrome *P*-450f was found recently by Leo et al. (1984) to display a selective capacity to metabolize retinoic acid to polar metabolites in a reconstituted system; in vivo, both ethanol and high A feeding appeared to increase this form of cytochrome (Leo et al. 1984), which has been newly discovered and purified (Iida et al. 1983; Ryan et al. 1984). Thus the toxicity could be related to the induction of the enzymes of the endoplasmic reticulum after chronic ethanol consumption. However, progression to severe parenchymal injury may offset the microsomal induction, as shown in baboons fed alcohol chronically (Lieber et al. 1975) and in alcoholics with advanced liver disease (Sotaniemi et al. 1977; Farrell et al. 1978).

Interaction of Ethanol with Hepatic Lipid Metabolism in Microsomes

In addition to its effects on increasing the supply of precursors, ethanol may enhance the rate of triglyceride synthesis by stimulating the activities of the enzymes catalyzing triglyceride synthesis (Fig. 2). Indeed, following an acute dose of ethanol or in animals consuming ethanol as part of an adequate diet, synthesis of triacylglycerols, a microsomal function, is increased (Joly et al. 1973b). This change in lipid metabolism is associated with an increase in the activity of the enzymes involved in this process. These as well as other interactions of ethanol with lipids at this subcellular site are discussed in detail by Baraona in another chapter.

Interaction of Ethanol with Membranes

The important rôle of membrane structures has long been recognized, but as yet it has not been clearly demonstrated that the changes observed during ethanol oxidation produce marked alterations in the lipid composition of the membranes. The presence of a physiological concentration of ethanol (20–80 m*M*) increased the fluidity in the plasma membranes from both mouse erythrocytes and synaptosomes, but had no effect on myelin (Chin and Goldstein 1977; Chin et al. 1978). The synaptosomal and erythrocyte membranes isolated from animals chronically treated with ethanol were more resistant to the in vitro fluidizing effects of ethanol than membranes prepared from their respective pair-fed controls (Chin and Goldstein 1977; Chin et al. 1978; Johnson et al. 1979). These data appear to support the theory that mammalian cells can adapt to drugs in their environment by changing the lipid composition of the membrane (Hill and Bangham 1975). In fact, membranes from the ethanol-fed animals contained a higher concentration of cholesterol than their respective controls (Chin et al. 1978; Cunningham et al. 1981). However, the small increase in cholesterol did not alter the microviscosity of the membranes unless ethanol was added in vitro. It would appear that the increase in cholesterol may have simply decreased the partition of ethanol in the membrane, as has been demonstrated with barbiturates (Miller and Yu 1977). Furthermore, in the liver, conflicting results on the interaction of ethanol with membranes have been obtained. One of the most characteristic features of alcoholic liver disease is the striking alteration in mitochondrial morphology and function. Recently, it was proposed that the functional changes produced as a consequence of prolonged ethanol consumption may be directly related to alterations in the structural properties (increased rigidity) of cellular membranes (Rottenberg et al. 1981; Waring et al. 1981). However, more recent data (Gordon et al. 1982; Yamada and Lieber 1984) indicate that no such simple correlation explains the functional changes observed in the mitochondria or plasma membranes of ethanol-fed animals. It is possible, of course, that other functional changes in the liver may be related to some alteration in membrane fluidity.

Summary

Recent advances in our knowledge of the hepatic metabolism of ethanol enable us to understand a number of metabolic and structural alterations that develop in the

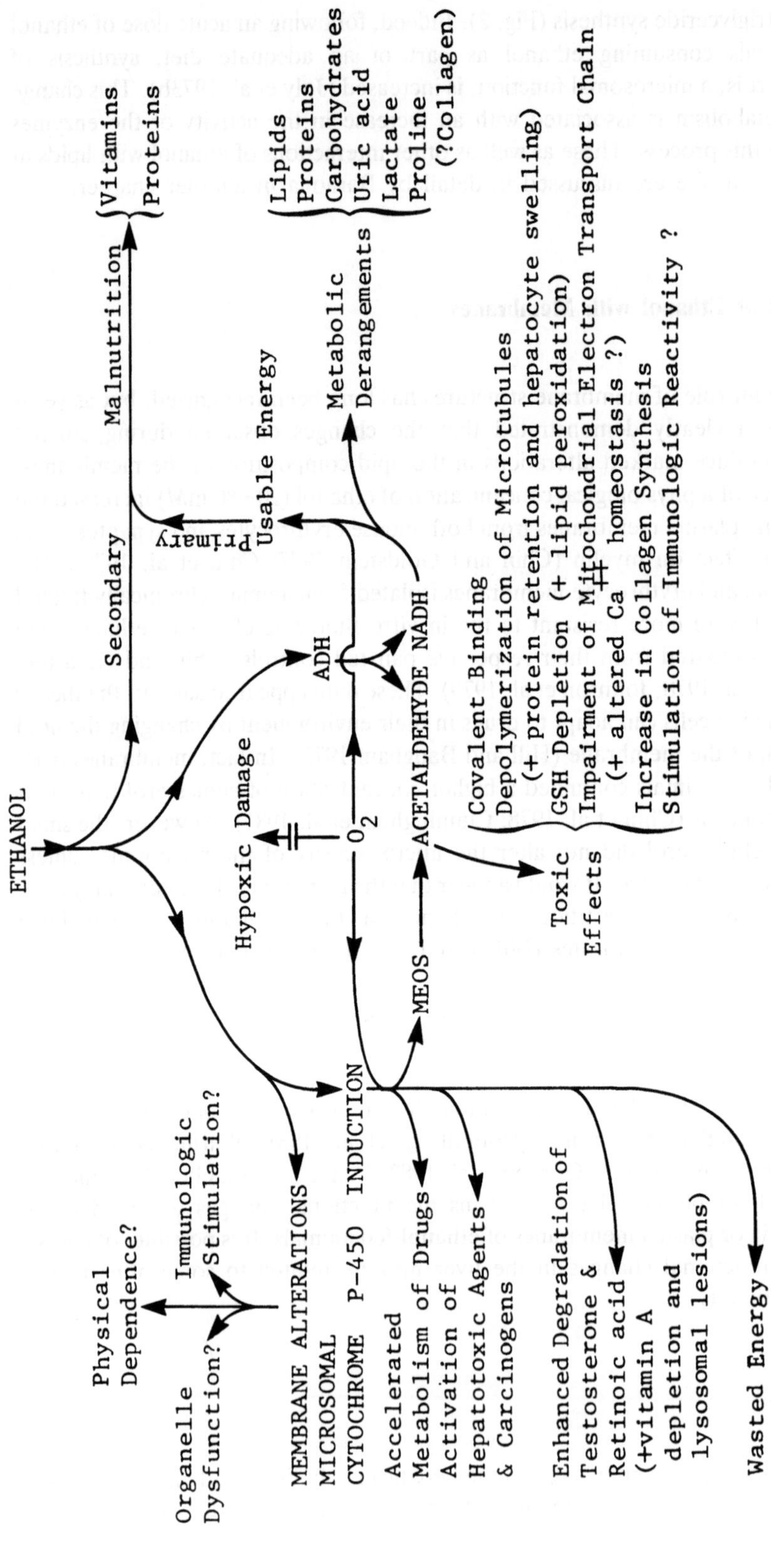

Fig. 8. Hepatic and metabolic disorders after chronic ethanol abuse. *GSH*, reduced glutathione (Lieber, 1984)

alcoholic, as illustrated in Fig. 8. Ethanol is oxidized in the liver to two products (hydrogen and acetaldehyde), to which many of the effects of ethanol can be attributed. The reduced nicotinamide-adenine dinucleotide generated from the reducing equivalents alters the redox state, and though this effect is attenuated after chronic ethanol consumption, it may still be sufficient to explain alterations in lipid and carbohydrate metabolism, possibly increased collagen deposition, and, under special circumstances, depression of protein synthesis. Acetaldehyde impairs microtubules, decreases protein secretion, and causes protein retention and ballooning of the hepatocyte. Acetaldehyde exerts toxicity also with regard to other key cellular functions, particularly in the mitochondria, and it may promote peroxidation of the cellular membranes. It is noteworthy that after chronic consumption of ethanol there is increased acetaldehyde, partly because of decreased disposition in the mitochondria and partly because of induction of an alternate pathway of ethanol metabolism, namely the microsomal ethanol-oxidizing system (MEOS). Indeed, this MEOS increases in activity after chronic ethanol consumption, with cross-induction and acceleration of the metabolism of other drugs, and increased lipoprotein production with hyperlipemia. There is also increased retinoic breakdown (possibly leading to hepatic vitamin A depletion and associated lysosomal alterations) and enhanced microsomal activation of hepatotoxic compounds, including drugs and carcinogens and, possibly, vitamin A.

References

Admirant WH, Cronholm T, Sjovall J (1970) Reduction of dehydroepiandrosterone sulfate in the liver during ethanol metabolism. Biochim Biophys Acta 202: 343–348

Anton AH (1952) Ethanol and urinary catecholamines in man. Clin Pharmacol Ther 6: 642–649

Arai M, Leo MA, Nakano M, Gordon ER, Lieber CS (1984) Biochemical and morphological alterations of baboon hepatic mitochondria after chronic ethanol consumption. Hepatology 4: 165–174

Baraona E, Pirola RC, Lieber CS (1973) The pathogenesis of postprandial hyperlipemia in rats fed ethanol-containing diets. J Clin Invest 52: 296–303

Baraona E, Leo MA, Borowsky SA, Lieber CS (1975) Alcoholic hepatomegaly: accumulation of protein in the liver. Science 190: 794–795

Baraona E, Leo MA, Borowsky SA, Lieber CS (1977) Pathogenesis of alcohol-induced accumulation of protein in the liver. J Clin Invest 50: 546–554

Baraona E, Pikkarainen P, Salaspuro M, Finkelman F, Lieber CS (1980) Acute effects of ethanol on hepatic protein synthesis and secretion in the rat. Gastroenterology 79: 104–111

Baraona E, Finkelman F, Lieber CS (1984) Reevaluation of the effects of alcohol consumption on rat liver microtubules. Res Commun Chem Pathol Pharmacol 44: 265–278

Berman WJ, Gil J, Jennett RB, Tuma D, Sorrell MF, Rubin E (1983) Ethanol, hepatocellular organelles, and microtubules. Lab Invest 48: 760–767

Bernstein J, Videla L, Israel Y (1973) Metabolic alterations produced in the liver by chronic ethanol administration. Changes related to energetic parameters of the cell. Biochem J 134: 515–522

Bernstein J, Videla L, Israel Y (1975) Hormonal influences in the development of the hypermetabolic state of the liver produced by chronic administration of ethanol. J Pharmacol Exp Ther 192: 583–591

Björkhem I (1972) On the role of alcohol dehydrogenase in ω-oxidation of fatty acids. Eur J Biochem 30: 441–445

Blomstrand R, Kager L, Lantto O (1973) Studies on the ethanol-induced decrease of fatty acid oxidation in rats and human liver. Life Sci 13: 1131–1141

Bode JC, Zelder O, Rumpelt HJ, Wittkamp U (1973a) Depletion of liver adenosine phosphates and metabolic effects of intravenous infusion of fructose or sorbitol in man and in the rat. Eur J Clin Invest 3: 436–441

Bode JC, Bode C, Rumpelt HJ, Zelder O (1973b) Loss of hepatic adenosine phosphates and metabolic consequences following fructose or sorbital administration in man and in the rat. In: Lundquist F, Tygstrup N (eds) Regulation of hepatic metabolism. Academic, New York, pp 267–284

Bosron WF, Li T-K (1977) Isolation and characterization of an anodic form of human liver alcohol dehydrogenase. Biochem Biophys Res Commun 74: 85–91

Bosterling B, Trudel JR (1981) Association of cytochrome b_5 and cytochrome P-450 reductase with cytochrome P-450 in the membrane of reconstituted vesicles. J Biol Chem 257: 4783–4787

Boveris A, Oshino N, Chance B (1972) The cellular production of hydrogen peroxide. Biochem J 128: 617–630

Brighenti L, Pancaldi G (1970) Effetto della somministrazione di alcool etilico su alcune attivita enzimatiche del fegarto di ratto. Soc Ital Di Biol Sperimen 46: 1–5

Carter EA, Isselbacher KJ (1971) The role of microsomes in the hepatic metabolism of ethanol. Ann NY Acad Sci 179: 282–294

Cederbaum AI, Dicker E, Lieber CS, Rubin E (1977) Factors contributing to the adaptive increase in ethanol metabolism due to chronic consumption of ethanol. Alcoholism 1: 27–31

Cerra FB, Caprioli J, Siegel JH, McMenamy RR, Border JR (1979) Proline metabolism in sepsis, cirrhosis and general surgery: the peripheral energy deficit. Ann Surg 190: 557–586

Chance B, Oshino N (1971) Kinetics and mechanisms of catalase in peroxisomes of the mitochondrial fraction. Biochem J 122: 225–233

Chin JH, Goldstein DB (1977) Effects of low concentrations of ethanol on the fluidity of spin-labeled erythrocyte and brain membranes. Mol Pharmacol 13: 435–441

Chin JH, Parson LM, Goldstein DB (1978) Increased cholesterol content of erythrocyte and brain membranes in ethanol-tolerant mice. Biochim Biophys Acta 513: 358–363

Chronholm T, Sjovall J (1970) Effect of ethanol metabolism on redox state of steroid sulphates in man. Eur J Biochem 13: 124–131

Chvapil M, Ryan JN (1973) The pool of free proline in acute and chronic liver injury and its effect on the synthesis of collagen-globular protein. Agents Actions 3: 38–44

Comai K, Gaylor JL (1973) Existence and separation of three forms of cytochrome P-450 from rat liver microsomes. J Biol Chem 248: 4947–4955

Conney AH (1967) Pharmacological implications of microsomal enzyme induction. Pharmacol Rev 19: 317–366

Crouse JR, Gerson CD, DeCarli LM, Lieber CS (1968) Role of acetate in the reduction of plasma free fatty acids produced by ethanol in man. J Lipid Res 9: 509–512

Crow KE, Cornell NW, Veech RL (1977) The rate of ethanol metabolism in isolated rat hepatocytes. Alcoholism 1: 43–47

Cunningham CC, Sinthusek G, Spach PT, Leathens C (1981) Effect of dietary ethanol and cholesterol on metabolic functions of hepatic mitochondria and microsomes from the monkey. Alcoholism 5: 410–416

Daughaday WH, Lipicky RJ, Rasinski DC (1962) Lactic acidosis as a cause of nonketotic acidosis in diabetic patients. N Engl J Med 267: 1010–1014

DeSaint-Blanguat G, Fritsch P, Derarche R (1972) Activité alcool deshydrogen-asigue de la mugueuse gastrigne sous l'effet de différents traitements éthanoliques chez le rat. Pathol Biol 20: 249–254

DiLuzio NR, Hartman AD (1967) Role of lipid peroxidation on the pathogenesis of the ethanol-induced fatty liver. Fed Proc 26: 1436–1442

DiLuzio NR, Stege TE (1977) The role of ethanol metabolites in hepatic lipid peroxidation. In: Fisher MM, Rankin JG (eds) Alcohol and the liver, vol 3. Plenum, New York, pp 45–62

Domschke S, Domschke W, Lieber CS (1974) Hepatic redox state: attenuation of the acute effects of ethanol induced by chronic ethanol consumption. Life Sci 15: 1327–1334

Faller J, Fox IH (1982) Ethanol-induced hyperuricemia: evidence for increased urate production by activation of adenine nucleotide turnover. N Engl J Med 307: 1598–1602

Farrell GC, Cooksley WGE, Hartand P, Powell LW (1978) Drug metabolism in liver disease. Identification of patients with impaired hepatic drug metabolism. Gastroenterology 75: 580–588

Feinman L, Lieber CS (1972) Hepatic collagen metabolism: effect of alcohol consumption in rats and baboons. Science 176: 795

Feinman L, Baraona E, Matsuzaki S, Korsten M, Lieber CS (1978) Concentration dependence of ethanol metabolism in vivo in rats and man. Alcoholism 2: 381–385

Feytmans E, Leighton F (1973) Effects of pyrazole and 3-amino-1,2,4-triazole on methanol and ethanol metabolism by the rat. Biochem Pharmacol 22: 349–360

Fischel P, Oette K (1974) Experimentelle Untersuchungen an menschlichen Leberpunktaten und Rattenleberschnitten zur Oxidation von Fettsäuren mit unterschiedlicher Kettenlänge und unterschiedlicher Zahl von Doppelbindungen. Res Exp Med (Berl) 163: 1–16

Gordon ER (1973) Mitochondrial functions in an ethanol-induced fatty liver. J Biol Chem 248: 8271–8280

Gordon ER (1977) ATP metabolism in an ethanol induced fatty liver. Alcoholism 1: 21–5

Gordon ER, Rochman J, Arai M, Lieber CS (1982) Lack of correlation between hepatic mitochondrial membrane structure and functions in ethanol-fed rats. Science 216: 1319–1321

Green H, Goldberg B (1964) Collagen and cell protein synthesis by an established mammalian fibroblast line. Nature 204: 347–349

Gregory DH, Vlahcevic ZR, Prugh MF, Swell L (1978) Mechanism of secretion of biliary lipids: role of a microtubular system in hepatocellular transport of biliary lipids in the rat. Gastroenterology 74: 93–100

Grunnet N, Quistroff B, Thieden HID (1973) Rate-limiting factors in ethanol concentration, fructose, pyruvate and pyrazole. Eur J Biochem 40: 275–282

Hakkinen H-M, Kulonen E (1975) Effect of ethanol on the metabolism of alanine, glutamic acid, and proline liver. Biochem Pharmacol 24: 199–204

Hasumura Y, Teschke R, Lieber CS (1974) Increased carbon tetrachloride hepatotoxicity, and its mechanism, after chronic ethanol consumption. Gastroenterology 66: 415–422

Hasumura Y, Teschke R, Lieber CS (1975a) Hepatic microsomal ethanol oxidizing system (MEOS): dissociation from reduced nicotinamide adenine dinucleotide phosphate-oxidase and possible role of from I of cytochrome P-450. J Pharmacol Exp Ther 194: 469–474

Hasumura Y, Teschke R, Lieber CS (1975b) Acetaldehyde oxidation by hepatic mitochondria: its decrease after chronic ethanol consumption. Science 189: 727–729

Hasumura Y, Teschke R, Lieber CS (1976) Characteristics of acetaldehyde oxidation in rat liver mitochondria. J Biol Chem 251: 4908–4913

Hawkins RD, Kalant H (1972) The metabolism of ethanol and its metabolic effects. Pharmacol Rev 24: 67–157

Hawkins RD, Kalant H, Khanna JM (1966) Effects of chronic intake of ethanol on rate of ethanol metabolism. Can J Physiol Pharmacol 44: 241–257

Hill MWK, Bangham AD (1975) General depressant drug dependency. A biophysical hypothesis. Adv Exp Med Biol 59: 19

Iida S, Levin W, Wood AW, Thomas PE, Lieber CS, Ryan DE (1983) Characterization of three highly purified cytochromes P-450 from rat liver microsomes. Fed Proc 42: 1777

Ingelman-Sundberg M, Johansson I (1981) The mechanism of cytochrome P-450-dependent oxidation of ethanol in reconstituted membrane vesicles. J Biol Chem 256: 6321

Iseri OA, Gottlieb LS, Lieber CS (1964) The ultrastructure of ethanol-induced fatty liver. Fed Proc 23: 579

Iseri OA, Lieber CS, Gottlieb LS (1966) The ultrastructure of fatty liver induced by prolonged ethanol ingestion. Am J Pathol 48: 535–555

Ishii H, Joly J-G, Lieber CS (1973) Effect of ethanol on the amount and enzyme activities of hepatic rough and smooth microsomal membranes. Biochim Biophys Acta 291: 411–420

Ismail-Beigi F, Edelman IS (1970) Mechanism of thyroid calorigenesis: role of active sodium transport. Proc Natl Acad Sci USA 67: 1071–1078

Ismail-Beigi F, Edelman IS (1971) The mechanism of the calorigenic action of thyroid hormone. J. Gen Physiol 57: 710–722

Israel Y, Videla L, MacDonald A, Bernstein J (1973) Metabolic alterations produced in the liver by chronic ethanol administration. Comparison between the effects produced by ethanol and by thyroid hormones. Biochem J 134: 523–529

Israel Y, Videla L, Fernandes-Videla V, Bernstein J (1975) Effects of chronic ethanol treatment and thyroxine administration on ethanol metabolism and liver oxidative capacity. J Pharmacol Exp Ther 192: 565–574

Jauhonen P, Baraona E, Miyakawa H, Lieber CS (1982) Mechanism for selective perivenular hepatotoxicity of ethanol. Alcoholism 6: 350–357

Jeejeebhoy KN, Bruce-Robertson A, Ho J, Sodtke U (1975) The effect of ethanol on albumin and fibrinogen synthesis in vivo and in hepatocyte suspension. In: Rothschild MA, Oratz M, Schreiber SS (eds) Alcohol and abnormal protein synthesis. Pergamon, New York, pp 373

Johnson DA, Lee NM, Cooke R, Loh HH (1979) Ethanol-induced fluidization of main lipid bilayers required presence of cholesterol in membranes for the expression of tolerance. Mol Pharmacol 15: 739–746

Johnson O (1974) Influence of the blood ethanol concentration on the acute ethanol-induced liver triglyceride accumulation in rats. Scand J Gastroenterol 2: 207–213

Joly J-G, Ishii H, Lieber CS (1972) Microsomal cyanide-binding cytochrome: its role in hepatic ethanol oxidation. Gastroenterology 62: 174

Joly J-G, Ishii H, Teschke R, Hasumura Y, Lieber CS (1973a) Effect of chronic ethanol feeding on the activities and submicrosomal distribution of reduced nicotinamide adenine dinucleotide phosphate (NADPH)-cytochrome P-450 reductase and the demethylase for aminopyrine and ethylmorphine. Biochem Pharmacol 22: 1532–1535

Joly J-G, Feinman L, Ishii H, Lieber CS (1973b) Effect of chronic ethanol feeding on hepatic microsomal glycerophosphate acyltransferase activity. J Lipid Res 14: 337–343

Joly J-G, Hetu C, Mavier P, Villeneuve JP (1976) Mechanisms of induction of hepatic drug-metabolizing enzymes by ethanol. I. Limited role of microsomal phospholipids. Biochem Pharmacol 25: 1995–2001

Joly J-G, Villeneuve J-P, Mavier P (1977) Chronic ethanol administration induces a form of cytochrome P-450 with specific spectral and catalytic properties. Alcoholism 1: 17–20

Jorfeldt L, Juhlin-Dannfelt A (1978) The influence of ethanol on splanchnic and skeletal muscle metabolism in man. Metabolism 27: 97–106

Kalant H, Khanna JM, Endrenyi L (1975) Effect of pyrazole on ethanol metabolism in ethanol-tolerant rats. Can J Pharmacol 53: 416–422

Kater RMH, Carulli N, Iber FL (1969) Differences in the rate of ethanol metabolism in recently drinking alcoholic and nondrinking subjects. Am J Clin Nutr 22: 1608–1617

Kershenobich D, Fierro FJ, Rojkind M (1970) The relationship between free pool of proline and collagen content in human liver cirrhosis. J Clin Invest 49: 2246–2249

Kershenobich D, Fierro FJ, Rojkind M (1979) The relationship between serum lactic acid and serum proline in alcoholic liver cirrhosis. Gastroenterology 77: A22

Kiessling KH, Pilstrom L (1968) Effect of ethanol on rat liver. V. Morphological and functional changes after prolonged consumption of various alcoholic beverages. Q J Stud Alcohol 29: 819–827

Klingman GI, Goodall McC (1957) Urinary epinephrine and levarterenol excretion during acute sublethal alcohol intoxication in dogs. J Pharmacol Exp Ther 121: 313–318

Koch OR, Boveris A, Sirotzky De Favelukes S, Stoppani AOM (1977) Biochemical lesions of liver mitochondria from rats after chronic alcohol consumption. Exp Mol Pathol 27: 213–220

Koop DR, Morgan ET, Tarr GE, Coon MJ (1982) Purification and characterisation of a unique isozyme of cytochrome P-450 from liver microsomes of ethanol treated rabbits. J Biol Chem 257: 8472–8480

Korsten MA, Matsuzaki S, Feinman L, Lieber CS (1975) High blood acetaldehyde levels following ethanol administration: differences between alcoholic and non-alcoholic subjects. N Engl J Med 292: 386–389

Kowaloff EM, Phang JM, Granger AS, Downing SJ (1977) Regulation of proline oxidase activity by lactate. Proc Natl Acad Sci USA 74: 5368–5371

Krebs HA (1967) Effects of ethanol on gluconeogenesis in the perfused rat liver. In: Staib W, Scholz R (eds) 3. Konferenz der Gesellschaft für biologische Chemie. Springer, Berlin Heidelberg New York, pp 216–224

Kreisberg RA, Owen WC, Siegal AM (1971) Ethanol-induced hyperlactacidemia: inhibition of lactate utilization. J Clin Invest 50: 166–174
Lane BP, Lieber CS (1966) Ultrastructural alterations in human hepatocytes following ingestion of ethanol with adequate diets. Am J Pathol 49: 593–603
Lane BP, Lieber CS (1967) Effects of butylated hydroxytoluene on the ultrastructure of rat hepatocytes. Lab Invest 16: 341–348
Lefevre AF, Adler H, Lieber CS (1970) Effect of ethanol on ketone metabolism. J Clin Invest 49: 1775–1782
Leo MA, Lieber CS (1982) Hepatic vitamin A depletion in alcoholic liver injury in men. N Engl J Med 37: 597–601
Leo MA, Lieber CS (1983) Hepatic fibrosis after long term administration of ethanol and moderate vitamin A supplementation in the rat. Hepatology 2: 1–11
Leo MA, Arai M, Sato M, Lieber CS (1982) Hepatotoxicity of moderate vitamin A supplementation in the rat. Gastroenterology 82: 194–205
Leo MA, Sato M, Lieber CS (1983a) Effect of hepatic vitamin A depletion on the liver in men and rats. Gastroenterology 84: 562–572
Leo MA, Iida S, Lieber CS (1984) Retinoic acid metabolism by a system reconstituted with cytochrome P-450. Arch Biochem Biophys 234: 305–312
Liang CS, Lowenstein JM (1978) Metabolic control of the circulation. J Clin Invest 62: 1029–1038
Lieber CS (1980) Metabolic effects of ethanol on the liver and other digestive organs. Clin Gastroenterol 10: 315–342
Lieber CS (1982) Medical Disorders of alcoholism: Pathogenesis and treatment. Saunders, Philadelphia
Lieber CS (1984) Alcohol and the liver: 1984 update. Hepatology 4: 1243–1260
Lieber CS, DeCarli LM (1968) Ethanol oxidation by hepatic microsomes: adaptive increase after ethanol feeding. Science 162: 917–918
Lieber CS, DeCarli LM (1970a) Quantitative relationship between the amount of dietary fat and the severity of the alcoholic fatty liver. Am J Clin Nutr 23: 474–478
Lieber CS, DeCarli LM (1970b) Hepatic microsomal ethanol oxidizing system: in vitro characteristics and adaptive properties in vivo. J Biol Chem 245: 2505–2512
Lieber CS, DeCarli LM (1970c) Reduced nicotinamide-adenine dinucleotide phosphate oxidase: activity enhanced by ethanol consumption. Science 170: 78–80
Lieber CS, DeCarli LM (1972) The role of the hepatic microsomal ethanol oxidizing system (MEOS) for ethanol metabolism in vivo. J Pharmacol Exp Ther 181: 279–287
Lieber CS, DeCarli LM (1973) The significance and characterization of hepatic microsomal ethanol oxidation in the liver. Drug Metab Dispos 1: 428–440
Lieber CS, Rubin E (1968) Alcoholic fatty liver in man on a high protein and low fat diet. Am J Med 44: 200–206
Lieber CS, Savolainen M (1984) Ethanol and lipids. Alcoholism: Clin Exp Res 8: 409–423
Lieber CS, Schmid R (1961) The effect of ethanol on fatty acid metabolism: stimulation of hepatic fatty acid synthesis in vitro. J Clin Invest 40: 394–399
Lieber CS, Spritz N (1966) Effects of prolonged ethanol intake in man: role of dietary, adipose, and endogenously synthesized fatty acids in the pathogenesis of the alcoholic fatty liver. J Clin Invest 45: 400–411
Lieber CS, Leevy CM, Stein SW, Georg WS, Cherrick GR, Abelmann WH, Davidson CS (1962) Effect of ethanol on plasma free fatty acids in man. J Lab Clin Med 59: 826–832
Lieber CS, Jones DP, DeCarli LM (1965) Effects of prolonged ethanol intake: production of fatty liver despite adequate diets. J Clin Invest 44: 1009–1021
Lieber CS, Lefevre A, Spritz N, Feinman L, DeCarli LM (1967) Difference in hepatic metabolism of long and medium-chain fatty acids: the role of fatty acid chain length in the production of the alcoholic fatty liver. J Clin Invest 46: 1451–1460
Lieber CS, Rubin E, DeCarli LM (1970) Hepatic microsomal ethanol oxidizing system (MEOS): differentiation from alcohol dehydrogenase and NADPH oxidase. Biochem Biophys Res Commun 40: 858–865
Lieber CS, Teschke R, Hasumura Y, DeCarli LM (1974) Interaction of ethanol with liver microsomes: In: Thurman RG, Yonetani T, Williamson JR, Chance B (eds) Alcohol and aldehyde metabolizing systems. Academic, New York, pp 243–256

Lieber CS, DeCarli LM, Rubin E (1975) Sequential production of fatty liver, hepatitis and cirrhosis in sub-human primates fed ethanol with adequate diets. Proc Natl Acad Sci USA 72: 437–441

Lieber CS, Savolainen M (1984) Ethanol and lipids. Alcoholism: Clin Exp Res 8: 409–423

Lindros KO, Oshino N, Parilla R, Williamson JR (1974) Characteristics of ethanol and acetaldehyde oxidation on flavin and pyridine nucleotide fluorescence changes in perfused rat liver. J Biol Chem 249: 7956–7963

Lindros KO, Stowell A, Pikkarainen P, Salaspuro M (1980) Elevated blood acetaldehyde in alcoholics with accelerated ethanol elimination. Pharmacol Biochem Behav 13: 119–124

Lindy S, Pedersen FB, Turto H, Uitto J (1971) Lactate, lactate dehydrogenase and protocollagen proline hydroxylase in rat skin authograft. Hoppe Seylers Physiol Chem 352: 1113–1118

MacDonald CM (1973) The effect of ethanol on hepatic lipid perxodiation and on the activities of glutathione reductase and peroxidase. FEBS Lett 35: 227–230

Makar AB, Mannering GJ (1970) Kinetics of ethanol metabolism in the intact rat and monkey. Biochem Pharmacol 19: 2017–2022

Makar AB, Tephly TR, Mannering GJ (1968) Metabolism in the monkey. Mol Pharmacol 4: 471–483

Maruyama K, Feinman L, Fainsilber Z, Nakano M, Okazaki I, Lieber CS (1982) Measurement of mammalian collagenase in alcohol-fed baboons and patients with alcoholic liver disease: decrease in activity with development of cirrhosis. In: Rautenberg J (ed) Proceedings of the conference on connective tissue of the normal and fibrotic human liver. Thieme, Stuttgart, pp 196–198

Mata JM, Kershenobich D, Villarreal E, Rojkind M (1975) Serum free proline and free hydroxyproline in patients with chronic liver disease. Gastroenterology 68: 1265–1269

Matsuda Y, Baraona E, Salaspuro M, Lieber CS (1979) Effects of ethanol on liver microtubules and Golgi apparatus. Lab Invest 41: 455–463

Matsuzaki S, Lieber CS (1977) Increased susceptibility of hepatic mitochondria to the toxicity of acetaldehyde after chronic ethanol consumption. Biochem Biophys Res Commun 75: 1059–1065

Matsuzaki S, Gordon E, Lieber CS (1981) Increased ADH independent ethanol oxidation at high ethanol concentrations in isolated rat hepatocytes: the effect of chronic ethanol feeding. J Pharmacol Exp Ther 217: 133–137

Meldolesi J (1967) On the significance of the hypertrophy of the smooth endoplasmic reticulum in liver cells after administration of drug. Biochem Pharmacol 16: 125–131

Mendenhall CL (1972) Origin of hepatic triglyceride fatty acids: quantiative estimation of the relative contribution of linoleic acid by diet and adipose tissue in normal and ethanol-fet rats. J Lipid Res 13: 177–183

Mezey E, Potter JJ, Reed WD (1973) Ethanol oxidation by a component of liver microsomes rich in cytochrome P-450. J Biol Chem 248: 1183–1187

Miller KW, Yu SCT (1977) The dependence of the lipid bilayer membrane: buffer partition coefficient of pentobarbitone on pH and lipid composition. Br J Pharmacol 61: 57–63

Misra PS, Lefevre A, Ishii H, Rubin E, Lieber CS (1971) Increase of ethanol, meprobamate and pentobarbital metabolism after chronic ethanol administration in man and in rats. Am J Med 51: 346–351

Miyakawa H, Iida S, Leo MA, Greenstein R, Zimmon DS, Lieber CS (1985) Pathogenesis of precirrhotic portal hypertension in alcohol-fed baboons. Gastroenterology 88: 143–150

Miwa GT, Levin W, Thomas PE, Lu AYH (1978) The direct oxidation of ethanol by a catalase- and alcohol dehydrogenase-free reconstituted system containing cytochrome P-450. Arch Biochem Biophys 187: 464–475

Morgan DD, Mendenhall CL (1977) The role of acetate in the pathogenesis of the acute ethanol-induced fatty liver in rats. Gastroenterology 73: 1235

Nikkila EA, Ojala K (1963) Ethanol-induced alterations in the synthesis of hepatic and plasma lipids and hepatic glycogens from glycerol-C^{14}. Life Sci 2: 717–721

Nilson NO, Belfrage P (1978) Effects of acetate acetaldehyde and ethanol on lipolysis in isolated rat adipocytes. J Lipid Res 19: 737–741

Nomura F, Lieber CS (1981) Binding of acetaldehyde to rat liver microsomes: enhancement after chronic alcohol consumption. Biochem Biophys Res Commun 100: 131–137

Nomura F, Pikkarainen PH, Jauhonen P, Arai M, Gordon ER, Baraona E, Lieber CS (1983) The effect of ethanol administration on the metabolism of ethanol in baboon. J Pharmacol Exp Ther 227: 78

Ohnishi K, Lieber CS (1977) Reconstitution of the microsomal ethanol oxidizing system (MEOS): qualitative and quantitative changes of cytochrome P-450 after chronic ethanol consumption. J Biol Chem 252: 7124–7131

Ohnishi K, Lieber CS (1978) Respective role of superoxide and hydroxyl radical in the activity of the reconstituted microsomal ethanol-oxidizing system. Arch Biochem Biophys 191: 798–803

Okazaki I, Feinman L, Lieber CS (1977) Hepatic mammalian collagenase: development of an assay and demonstration of increased activity after ethanol consumption. Gastroenterology 73: 1236

Okuda K, Takigawa N (1970) Rat liver 5-cholestane-3,7,12,26-tetrol dehydrogenase as a liver alcohol dehydrogenase. Biochim Biophys Acta 22: 141–148

Ontko JA (1973) Effects of ethanol on the metabolism of free fatty acids in isolated liver cells. J Lipid Res 14: 78–86

Orme-Johnson WH, Ziegler DM (1965) Alcohol mixed function oxidase activity of mammalian liver microsomes. Biochem Biophys Res Commun 21: 78–82

Orrego H, Blendis LM, Crossley IR, Medline A, MacDonald A, Ritchie S, Israel Y (1981) Correlation of intrahepatic pressure with collagen in the Disse space and hepatomegaly in humans and in the rat. Gastroenterology 80: 546–566

Oudea MC, Launay AN, Queneherve S, Oudea P (1970) The hepatic lesions produced in the rat by chronic alcoholic intoxication. Histological, ultrastructural and biochemical observations. Rev Eur Etud Clin Biol 15: 748–764

Perman ES (1958) The effect of ethyl alcohol on the secretion from the adrenal medulla in man. Acta Physiol Scand 44: 241–247

Perman ES (1960) The effect of ethyl alcohol on the secretion from the adrenal medulla of the cat. Acta Physiol Scand 48: 323–328

Perman ES (1961) Observations on the effect of ethanol on the urinary excretion of histamine, 5-hydroxyindole acetic acid, catecholamines and 17-hydroxycorticosteroids in man. Acta Physiol Scand 51: 62–67

Pikkarainen P, Lieber CS (1980) Concentration dependency of ethanol elimination rates in baboons: effect of chronic alcohol consumption. Alcoholism 4: 40–43

Pipeleers DG, Pipeleers-Marichal MA, Kipnis DM (1977) Physiological regulation of total tubulin and polymerized tubulin in tissues. J Cell Biol 74: 351

Raskin NH, Sokoloff L (1972) Ethanol induced adaptation of alcohol dehydrogenase activity in rat brain. Nature 236: 138–140

Reaven EP, Chen Y, Miller MD (1977) Quantitative analysis of tubulin and microtubule compartments in isolated rat hepatocytes. J Cell Biol 75: 731

Reitz RC (1975) A possible mechanism for the peroxidation of lipids due to chronic ethanol ingestion. Biochim Biophys Acta 380: 145–154

Reynier M (1969) Pyrazole inhibition and kinetic studies of ethanol and retinol oxidation catalyzed by rat liver alcohol dehydrogenase. Acta Chem Scand 23: 1119–1129

Rojkind M, DeLeon LD (1970) Collagen biosynthesis in cirrhotic rat slices. A regulatory mechanism. Biochem Biophys Acta 217: 512–522

Rothschild MA, Oratz M, Mongelli J, Schreiber SS (1971) Alcohol induced depression of albumin synthesis: reversal by tryptophan. J Clin Invest 50: 1812–1818

Rottenberg H, Waring A, Rubin E (1981) Tolerance and cross-tolerance in chronic alcoholics: reduced membrane binding of ethanol and other drugs. Science 213: 583

Rubin E, Lieber CS (1967) Early fine structural changes in the human liver induced by alcohol. Gastroenterology 52: 1–13

Rubin E, Beattie DS, Lieber CS (1970) Effects of ethanol on the biogenesis of mitochondrial membranes and associated mitochondrial functions. Lab Invest 23: 620–627

Rubin E, Beattie DS, Toth A, Lieber CS (1972) Structural and functional effects of ethanol on hepatic mitochondria. Fed Proc 31: 131

Ryan DE, Iida S, Wood AW, Thomas PE, Lieber CS, Levin W (1984) Characterization of three highly purified cytochromes P-450 from hepatic microsomes of adult male rats. J Biol Chem 259: 1239–1250

Salaspuro M, Lieber CS (1978) Non-uniformity of blood ethanol elimination: its exaggeration after chronic consumption. Ann Clin Res 10: 294–297
Salaspuro MP, Lindros KO, Pikkarainen P (1975) Ethanol and galactose metabolism as influenced by 4-methylpyrazole in alcoholics with and without nutritional deficiencies. Preliminary report of a new approach to pathogenesis and treatment in alcoholic liver disease. Ann Clin Res 7: 269–272
Sato C, Matsuda Y, Lieber CS (1981) Increased hepatotoxicity of acetaminophen after chronic ethanol consumption in the rat. Gastroenterology 80: 140–148
Sato M, Lieber CS (1981) Hepatic vitamin A depletion after chronic ethanol consumption in baboons and rats. J Nutr 111: 2015–2023
Sato M, Lieber CS (1982) Increased metabolism of retinoic acid after chronic ethanol consumption in rat liver microsomes. Arch Biochem Biophys 213: 557–564
Sato N, Kamada T, Shiohiri M, Hayashi N, Matsomura T, Abe M, Hagihara B (1978) The levels of the mitochondrial and microsomal cytochromes in drinkers' livers. Clin Chim Acta 87: 347–351
Savolainen E-R, Leo MA, Timpl R, Lieber CS (1984) Acetaldehyde and lactate stimulate collagen synthesis of cultured baboon liver myofibroblasts. Gastroenterology 87: 777–787
Shaw S, Jayatilleke E, Ross WA, Gordon ER, Lieber CS (1981) Ethanol induced lipid peroxidation: potentiation by chronic alcohol feeding and attenuation by methionine. J Lab Clin Med 98: 417–425
Shaw S, Worner TM, Lieber CS (1983a) Hyperprolinemia in alcoholic liver cirrhosis: mechanisms and pathogenetic significance. Hepatology 4: 295–299
Shaw S, Rubin KP, Lieber CS (1983b) Depressed hepatic glutathione and increased diene conjugates in alcoholic liver disease: evidence of lipid peroxidation. Dig Dis Sci 28: 585–589
Shigeta Y, Nomura F, Iida S, Leo MA, Felder MR, Lieber CS (1984) Ethanol metabolism in vivo by the microsomal ethanol oxidizing system in deermice lacking alcohol dehydrogenase (ADH). Biochem Pharmacol 33: 807–814
Siegers C-P, Schutt A, Strubelt O (1977) Influence of some hepatotoxic agents on hepatic glutathione levels in mice. Proc Eur Soc Toxicol 18: 160–162
Singlevich TE, Barboriak JJ (1971) Liver lipids after ethanol, chlorcyclizine and DDT. Fed Proc 30: 581
Sotaniemi EA, Ahlqvist J, Pelkonen RO, Pirttiaho H, Luoma PV (1977) Histological changes in the liver and indices of drug metabolism in alcoholics. Eur J Clin Pharmacol 11: 295–303
Sussman KE, Alfrey A, Kirsch WM et al. (1970) Chronic lactic acidosis in an adult. Am J Med 48: 104–112
Teschke R, Hasumura Y, Joly J-G, Ishii H, Lieber CS (1972) Microsomal ethanol-oxidizing system (MEOS): purification and properties of a rat liver system free of catalase and alcohol dehydrogenase. Biochem Biophys Res Commun 49: 1187–1193
Teschke R, Hasumura Y, Lieber CS (1974) Hepatic microsomal ethanol oxidizing system: solubilizations, isolation and characterization. Arch Biochem Biophys 163: 404–415
Teschke R, Hasumura Y, Lieber CS (1975) Hepatic microsomal alcohol oxidizing system. Affinity for methanol, ethanol and propanol. J Biol Chem 250: 7379–7404
Teschke R, Matsuzaki S, Ohnishi K, DeCarli LM, Lieber CS (1977) Microsomal ethanol oxidizing system (MEOS): current status of its characterization and its role. Alcoholism 1: 7–15
Thieden HID (1971) The effect of ethanol concentration on ethanol oxidation rate in rat liver slices. Acta Chem Scand 25: 3421–3427
Thurman RG (1973) Induction of hepatic microsomal reduced nicotinamide adenine dinucleotide phosphate-dependent production of hydrogen peroxide by chronic prior treatment with ethanol. Mol Pharmacol 9: 670–675
Thurman RG, McKenna WR, McCaffrey TB (1976) Pathways responsible for the adaptive increase in ethanol utilization following chronic treatment with ethanol: inhibitor studies with the hemoglobin-free perfused rat liver. Mol Pharmacol 12: 156–166
Tobon F, Mezey E (1971) Effect of ethanol administration on hepatic ethanol and drug-metabolizing enzymes and on rates of ethanol degradation. J Lab Clin Med 77: 110–121
Ugarte G, Pino ME, Insunza I (1967) Hepatic alcohol dehydrogenase in alcoholic addicts with and without hepatic damage. Am J Dig Dis 12: 589–592

Ugarte G, Pereda I, Pino ME, Iturriaga H (1972) Influence of alcohol intake, length of a abstinence and meprobamate on the rate of ethanol metabolism in man. Q J Stud Alcohol 33: 698–705

Ullrich V, Weber P, Wollenberg P (1975) Tetrahydrofurane – an inhibitor for ethanol-induced liver microsomal cytochrome P-450. Biochem Biophys Res Commun 64: 808–813

Veech RL, Guynn R, Veloso D (1972) The time-course of the effects of ethanol on the redox and phosphorylation states of rat liver. Biochem J 127: 387–397

Videla L, Bernstein J, Israel Y (1973) Metabolic alterations produced in the liver by chronic ethanol administration. Increased oxidative capacity. Biochem J 134: 507–514

von Wartburg JP, Rothlisberger M, Eppenberger HM (1961) Hemmung der Äthylalkoholoxydation durch Fuselöle. Helv Med Acta 28: 696–704

Waring AJ, Rottenberg H, Ohnishi T, Rubin E (1981) Membranes and phospholipids of liver mitochondria from chronic alcoholic rats are resistant to membrane disordering by alcohol. Proc Natl Acad Sci USA 78: 2582

Wendell G, Thurman RG (1979) Effect of ethanol concentration on rates of ethanol elimination in normal and alcohol-treated rats in vivo. Biochem Pharmacol 28: 273–279

Winston GW, Cederbaum AI (1983) NADPH-dependent production of oxy radicals by purified components of the rat liver mixed function oxidase system. J Biol Chem 258: 1514–1519

Yamada S, Lieber CS (1984) Decrease in microviscosity and cholesterol content of rat liver plasma membranes after chronic ethanol feeding. J Clin Invest 74: 2278–2282

Yuki T, Thurman RG (1980) The swift increase in alcohol metabolism. Time course for the increase in hepatic oxygen uptake and the involvement of glycolysis. Biochem J 186: 119–126

Ziegler DM (1972) Discussion. In: Estabrook RW, Gillette JL, Liebman KC (eds) Microsomes and drug oxidation. Williams and Wilkins, Baltimore, pp 458–460

3 Gamma-Glutamyltransferase and Other Markers for Alcoholism

R. Teschke

Medizinische Klinik D, Universität Düsseldorf,
Moorenstrasse 5, 4000 Düsseldorf 1, Federal Republic of Germany

Introduction

Alcoholism is a wide spread disease, and the growing prevalence of alcohol-related disorders creates major medical and social problems throughout the developed world. Medical complications of alcoholism may include a variety of organs such as liver (Teschke 1981), pancreas (Goebell and Singer 1981), esophagus (Wienbeck and Berges 1981), gastrointestinal tract (Berges and Wienbeck 1981), heart (Kuhn 1981), bones (Lindsell et al. 1982), endocrine glands (Kley et al. 1981), brain (Feuerlein 1981), and peripheral nerves (Schrappe 1981), and recognizing these diseases as alcohol-related ones is usually not a problem.

There is general concern that the diagnosis of alcoholism may be missed in the absence of overt clinical signs. Two main routes of approach have therefore been followed to screen for alcoholism, one involving questionnaires or interviews and the other using laboratory tests. A variety of questionnaires and interviews to be used as screening tests for alcoholism are available (Selzer 1971; Moore 1972; Mayfield et al. 1974; Feuerlein et al. 1977; Saunders and Kershaw 1980; Hurt et al. 1980; Bernadt et al. 1982). The validity of studies involving questionnaires and interviews may be criticized on the basis that full cooperation of the patient is required, and alcoholism may be undetected due to failure to complete the questionnaires.

In the course of prolonged alcohol consumption a variety of laboratory parameters may be altered in the blood, plasma, or serum. They include gamma-glutamyltransferase (Nishimura and Teschke 1983), mean corpuscular volume of the erythrocytes (Wu et al. 1974; Weill et al. 1982), aspartate aminotransferase (van Waes and Lieber 1977; Ishii et al. 1979; Nishimura et al. 1980; Morgan et al. 1981b; Teschke et al. 1983; Panteghini et al. 1983; Nalpas et al. 1984; Cushman et al. 1984), alanine aminotransferase (van Waes and Lieber 1977; Wadstein and Skude 1979; Nishimura et al. 1980; Teschke et al. 1983), glutamate dehydrogenase (van Waes and Lieber 1977; Worner and Lieber 1980; Mills et al. 1981; Teschke et al. 1983), amino acids (Shaw and Lieber 1978, 1980a), transferrin (Stibler et al. 1978, 1979, 1980), uric acid (Lieber et al. 1963; Whitfield et al. 1978; Whitehead et al. 1978; Bernadt et al. 1982; Faller and Fox 1982), urea (Weill et al. 1982), triglycerides (Lieber et al. 1963; Belfrage et al. 1977; Whitfield et al. 1978), α-lipoprotein (Belfrage et al. 1977), high density lipoprotein cholesterol (Cushman et al. 1984), ornithine carbamyl transferase (van Waes and Lieber 1977; Nishimura et

Alcohol Related Diseases in Gastroenterology
Edited by H. K. Seitz and B. Kommerell

al. 1980), hemoglobin A1 (Hoberman and Chiodo 1982), erythrocyte δ-aminolevulinic acid dehydratase (Krasner et al. 1974), lactate (Kershenobich et al. 1981), β-hexosaminidase (Hultberg et al. 1980), bile acids (Tobiasson and Boeryd 1980), and thrombocytes (Heidemann et al. 1981). Though many laboratory parameters are altered as a consequence of prolonged alcohol intake, only a few are useful as markers for alcoholism as evidenced by their sensitivity, specificity, and predictive value (Weill et al. 1982).

The ideal biological marker for detecting and monitoring alcoholism should be very sensitive and highly specific with a good predictive value for alcohol abuse. It should be reliable, quick, simple, convenient, and inexpensive to do, but at present there is no such marker available which fulfills all these criteria. This paper will therefore focus on a few markers which have nevertheless been found useful in the past or are promising for the future.

Gamma-glutamyltransferase

Gamma-glutamyltransferase (GGT; EC 2.3.2.2) is a membrane-bound enzyme with a molecular weight of 90,000 (Huseby 1977) and exhibits characteristic features of a glycoprotein (Orlowski and Meister 1965; Szewczuk and Connell 1964; Köttgen et al. 1983). Although the exact biological function is not yet known, GGT might be involved in the transport of amino acids or peptides through outer membranes into the cells by a metabolic cycle called the gamma-glutamyl cycle (Meister 1973). Enzymic activity of GGT can be found in a variety of organs including kidney, hepatobiliary tract, pancreas, spleen, and duodenum (Rutenburg et al. 1969; Kokot and Sledzinski 1974; Nishimura and Teschke 1983). A fetal, sialic acid (*N*-acetylneuraminic acid) rich and an adult, sialic acid poor form of GGT have been described in the liver (Köttgen et al. 1978) and serum (Köttgen and Gerok 1976; Teschke et al. 1980; Rauen et al. 1980; Köttgen et al. 1983).

Increased serum GGT activities are commonly observed in patients with a history of chronic alcohol consumption (Rosalki and Rau 1972; Rollason et al. 1972; Wu et al. 1976; Whitfield et al. 1978; Whitehead et al. 1978; Teschke 1980, 1981; Teschke et al. 1980, 1983; Rauen et al. 1980; Nishimura and Teschke 1983; Cushman et al. 1984), and these findings are reproducible in experimental animals (Teschke et al. 1977; Mørland et al. 1977; Ishii et al. 1978; Shaw and Lieber 1980b; Ideo et al. 1980; Nishimura et al. 1981; Nishimura and Teschke 1982; Lahrichi et al. 1982; Teschke et al. 1983). The serum changes are associated with a corresponding rise in hepatic GGT activity both in man (Schmidt 1977; Seymour and Peters 1978; Teschke 1980, 1981; Teschke et al. 1980, 1983; Ivanov et al. 1980; Satoh et al. 1980; Selinger et al. 1982; Yamauchi et al. 1984) and in experimental animals (Teschke et al. 1977; Mørland et al. 1977; Ishii et al. 1978; Shaw and Lieber 1980b; Ideo et al. 1980; Gadeholt et al. 1980; Nishimura et al. 1981; Teschke and Petrides 1982; Nishimura and Teschke 1982), although in one experimental study no change (Lahrichi et al. 1982) and in another one even a reduction of hepatic GGT activity has been described (Singer and Kaplan 1978).

The latter study has been criticized on various experimental grounds (Teschke and Petrides 1982). In the study of Lahrichi et al. (1982) alcohol was administered in drinking water rather than as part of a totally liquid diet employing the isocaloric pair-feeding technique commonly used for experimental studies regarding the effects of alcohol. Evidence has been presented more recently that the animal model with the pair-feeding technique for alcohol is an appropriate tool for studying interactions between alcohol and GGT, as alterations of the GGT enzyme pattern in the serum and liver due to alcohol intake are similar in man and rats (Teschke et al. 1983).

The observation of increased hepatic GGT activities due to prolonged alcohol intake raises the question of the cellular and subcellular site of enhanced GGT activities in the liver. Considerable variation of the cellular and subcellular localization is described in various reports for hepatic GGT, which may be explained by differences in the strain used and in the methodological approach employed. In the normal human liver GGT is found by histochemical assessment with faint staining in the sinusoidal membrane of the hepatocyte throughout the liver tissue, whereas periportal canaliculi are seldom moderately positive (Busachi et al. 1981). Moreover, the apical border of epithelial cell membranes of larger bile ducts is strongly GGT positive, and the epithelial cells of the bile ductules are weakly positive or negative. Conversely, in alcoholic liver disease a pronounced canalicular GGT activity is detected, being panlobular in most cases, and with increasing canalicular activity the enzyme staining becomes stronger in the sinusoidal membranes (Busachi et al. 1981). In other human studies staining for GGT activity appeared as a canalicular network in the cell plate of hepatocytes in normal livers, whereas an increased canalicular staining was found in liver specimens obtained from volunteers after several weeks on an alcohol diet (Belfrage et al. 1977). Moreover, in the normal human liver GGT activity was present on the inner surface and in the apical cytoplasm of small bile ducts and ductules (Uchida and Peters 1983). Proliferation of bile ductules is a common feature in advanced alcoholic liver disease, and GGT activity was positive extensively and strongly on the inner surface but only weakly in the cytoplasm of epithelial cells of proliferated ductules. The staining is stronger on the inner surface of proliferated ductules than in the normal ducts and the bile canaliculi of hepatocytes. Finally, upon analytical subcellular fractionation of normal human liver, GGT activity was recovered in the bile canalicular membrane of the hepatocytes, and there was a striking increase in enzyme activity in this particular subcellular fraction in the liver obtained from patients with alcoholic liver disease (Seymour and Peters 1978).

In normal rat liver GGT is localized in hepatocytes, biliary tract cells, and Kupffer cells (Wootton et al. 1977; Selden et al. 1978). However, by histological evaluation GGT of the normal rat liver is found only in the apical border of bile ductal and ductular structures, whereas parenchymal cells are negative (Rutenburg et al. 1969; Ronchi and Desmet 1973; Ideo et al. 1980) except for very few strictly periportal canaliculi (Ideo et al. 1980). After alcohol consumption GGT activity was demonstrated in the rats not only in ductular structures but also on the canalicular membranes of periportal parenchymal cells (Ideo et al. 1980). A weak cytosolic activity was also often present, and centrolobular cells were always negative. Analytical studies for subcellular fractionation have shown the presence of GGT activity in plasma membranes of the hepatocyte derived from rats (Smith and Peters 1978; Malvoisin et al. 1978; Ratanasevanh et al. 1979; Nishimura et al. 1981; Nishimura and Teschke 1982; Lahrichi et al. 1982), guinea pigs (Huseby 1979), or rabbits (Tazi et al. 1980; Inoue et al. 1983). Prolonged alcohol intake leads to a striking increase in GGT activity in the plasma membranes of the hepatocytes (Nishimura et al. 1981; Nishimura and Teschke 1982), findings subsequently confirmed by Lahrichi et al. (1982). GGT activity was found to be increased in liver plasma membranes free of bile canaliculi as well as in those rich in bile canaliculi with higher basal activity levels in the latter fractions than in the former ones (Nishimura et al. 1981; Nishimura and Teschke 1982). Moreover, prolonged alcohol administration causes increased GGT activities in the microsomal fraction of the hepatocytes (Teschke et al. 1977; Mørland et al. 1977; Ishii et al. 1978; Shaw and Lieber 1980b), the biochemical counterpart of the endoplasmic recitulum. There is no question that the microsomal fraction is contaminated by GGT derived from plasma membrane, which is unavoidable during subcellular fractionation (Widnell 1972), but the degree of contamination is not known. In the microsomal fraction of rat liver GGT activity is predominantly enhanced in smooth rather than rough microsomes following prolonged ethanol intake (Ishii et al. 1979a). There is also some evidence that GGT may be present in the cytosol of the hepatocytes with low (Teschke et al. 1977; Smith and Peters 1978) or extremely high specific activities (Malvoisin et al. 1978).

There has been considerable debate whether alcohol per se might increase the hepatic GGT activity following experimental administration of alcohol-containing diets. Indeed, it has been speculated that the low carbohydrate content of the alcohol diets alone rather than the alcohol itself may have caused the enhancement of hepatic GGT activity (Mørland et al. 1977), a suggestion opposed by others (Ishii et al. 1978; Shaw and Lieber 1980b; Ideo et al. 1980) and subsequently not substantiated by the group of Mørland (Gadeholt et al. 1980). In particular, a hypocaloric diet with a low carbohydrate content had no significant effect on the levels of hepatic GGT activity, and only the addition of ethanol to the latter diet to achieve a normocaloric diet resulted in a pronounced rise in hepatic GGT activity (Teschke and Petrides 1982). It was of interest to note, however, that a hypercaloric diet with a high carbohydrate content prevented the alcohol-mediated increase in hepatic GGT activity.

Chronic alcohol consumption enhances GGT activity in organs other than the liver such as brain (Reyes 1977) or duodenum (Ishii et al. 1978; Seitz et al. 1982) but not kidney or pancreas (Ishii et al. 1978). Release of intestinal GGT activity into lymph is enhanced following chronic alcohol consumption and may be a factor contributing to the increase of enzymic activity in the serum (Ebihara et al. 1982). It is generally believed, however, that the liver and/or biliary tract govern the activity levels of serum GGT, at least to a major extent (Naftalin et al. 1969; Kokot and Sledzinski 1974; Selinger et al. 1982). GGT is a typical glycoprotein (Köttgen et al. 1983), and various steps are involved in the hepatic synthesis and degradation of glycoproteins (Köttgen et al. 1979, 1983; Doyle and Le 1981; Evans 1981; Schwartz et al. 1981; Kolb-Bachofen 1981; Geuze et al. 1982; Zeitlin and Hubbard 1982; Debanne et al. 1982; Russel et al. 1983; Clarenburg 1983; Billington et al. 1983; Barnwell et al. 1983; Popper et al. 1983).

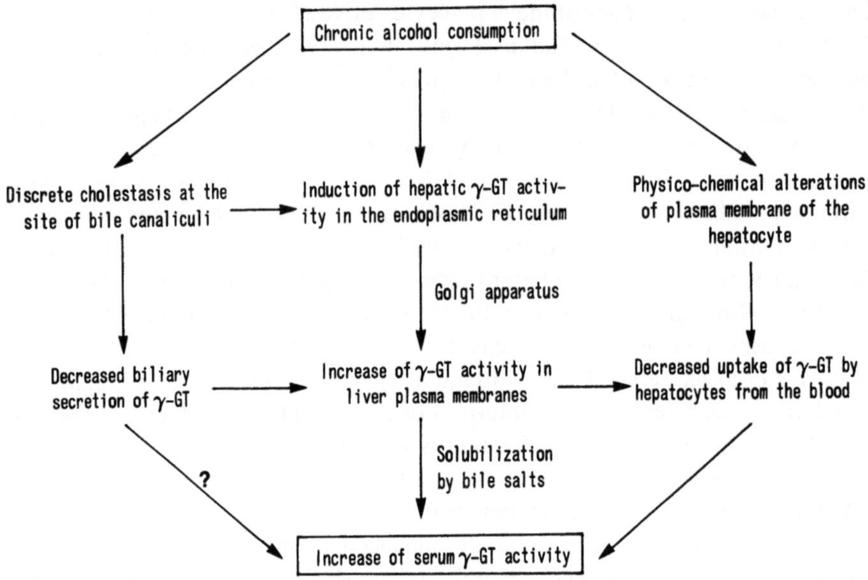

Fig. 1. Hypothetical mechanisms leading to increased serum activities of gamma-glutamyltransferase (γ-GT) following prolonged alcohol consumption

At present, proven data are not available concerning the exact mechanism involved in increased serum GGT activities associated with prolonged alcohol intake, but it appears that an increased hepatic activity alone may not be responsible, since hepatic GGT activities increased under other conditions, elicited for instance by phenobarbital administration, fail to bring about a concomitant rise of GGT activity in the serum (Bagrel et al. 1980). In any event, alcohol in vitro enhances the hepatic biosynthesis of GGT (Fig. 1), and this process occurs at the transcriptional level (Barouki et al. 1983). Hepatic enzyme induction of GGT has also been reported as a consequence of cholestasis elicited by extrahepatic biliary obstruction (Kryszewski et al. 1973). Recent studies have shown that chronic alcohol consumption leads to intrahepatic cholestasis (Fig. 1) as evidenced by impaired biliary secretion of GGT, alkaline phosphatase, and bile acids, along with enhanced serum GGT activities and augmented serum levels of bile acids, but with unaltered serum alkaline phosphatase activities, suggesting a discrete type of intrahepatic cholestasis (Teschke et al. 1984, 1985; Krukenberg et al. 1985). Enzymes present in the bile are primarily derived from the bile canalicular membrane of the hepatocyte (Reuben 1984), where GGT is located on the luminal side (Inoue et al. 1983). Since GGT activity is augmented especially in the bile canaliculi rich fraction of the hepatocytes following chronic alcohol consumption (Nishimura et al. 1981; Nishimura and Teschke 1982), this finding suggests decreased solubilization by bile acids of GGT from the luminal side of the hepatic bile canalicular plasma membrane into the bile with reduced biliary secretion of GGT. It therefore appears that impaired biliary secretion of GGT may be the most important factor in the increased serum GGT activity commonly observed following chronic alcohol consumption, but hepatic enzyme induction of GGT may also play some role (Fig. 1). In addition, it is conceivable that alcohol may impair clearance of GGT from the blood through reduced hepatocellular uptake of the enzyme via a process involving specific receptors of the plasma membranes of the hepatocytes.

The level of serum GGT activity was found to correlate with the degree of liver cell necrosis (Wu et al. 1976; van Waes and Lieber 1977; Worner and Lieber 1980). In patients with alcoholic fatty liver, GGT activity was increased in the serum (Wu et al. 1976; Schmidt 1977; Nishimura et al. 1980; Teschke et al. 1980, 1983; Rauen et al. 1980; Jenkins et al. 1982) and liver (Schmidt 1977; Teschke et al. 1980, 1983), although hepatic GGT activity was unchanged in another study (Seymour and Peters 1978). Alcoholic hepatitis was associated with increased GGT activities in the serum (Wu et al. 1976; Nishimura et al. 1980; Satoh et al. 1980; Jenkins et al. 1982) and liver (Schmidt 1977; Seymour and Peters 1978; Satoh et al. 1980). Patients with alcoholic liver fibrosis exhibited enhanced GGT activities in the serum (Wu et al. 1976; Teschke et al. 1980; Rauen et al. 1980; Yamouchi et al. 1984) and liver (Yamouchi et al. 1984). Alcoholic liver cirrhosis was found to be associated with enhanced GGT activities in the serum (Wu et al. 1976; Schmidt 1977; Teschke et al. 1980; Rauen et al. 1980; Jenkins et al. 1982) and liver (Seymour and Peters 1978; Ivanov 1980). Finally, in alcoholic patients with normal liver histology, normal activities for serum GGT (Jenkins et al. 1982) were found with increased hepatic enzymic activities (Seymour and Peters 1978). There is considerable overlap of serum GGT activities between various stages of alcoholic liver disease without the possibility of dissociation (Teschke et al. 1980; Rauen et al. 1980). Similar conclusions can be derived from studies using the absolute activity levels of either the adult or the fetal form of GGT in the serum, which are both increased in patients with alcoholic liver disease. However, on a percentage basis the fetal form of GGT in the serum is significantly higher than the adult form in patients with early stages of alcoholic liver diseases such as alcoholic fatty liver and liver fibrosis, whereas the reversed constellation was observed in patients with alcoholic liver cirrhosis. The ratio of fetal to adult form of GGT may therefore be useful to dissociate early from late stages of alcoholic liver disease in patients with a known history of chronic alcohol abuse.

A single drinking bout has been found to be insufficient to cause a noticeable alteration of serum GGT activity when tested in healthy volunteers (Luchi and Cortis 1978; Shaw and Marsh 1981; Dunbar et al. 1982; Clark et al. 1982; Gill et al. 1982), and similar data were obtained following intravenous application of ethanol in patients with alcoholic fatty liver or alcoholic hepatitis whose basal serum GGT activities were elevated (Nishimura et al. 1980). It is obvious that several days of excessive alcohol intake are commonly required to yield increased serum GGT activities (Freer and Statland 1977a, b; Belfrage et al. 1977; Shaw and Lieber 1980b). A variety of studies have shown a correlation between daily or weekly alcohol consumption and the height of the activity level of serum GGT (Rollason et al. 1972; Whitfield et al. 1978; Whitehead et al. 1978; Chick et al. 1981; Schellenberg et al. 1982), and differences between sexes with higher serum GGT activities in males than in females are apparent (Schellenberg et al. 1982). In another study a very poor correlation between alcohol consumption and serum GGT activity was observed in males but not in females (Reyes et al. 1978). It is of particular interest that serum GGT activities were highest in patients in whom alcoholism had been established for less than 5 years (Wadstein and Skude 1979). The serum concentration of ethanol, however, was lowest among these patients and gradually increased with duration of alcoholism, and no correlation was found between the serum ethanol level and serum GGT activity. The duration of the current debauch, which was shortest in cases of long-standing alcoholism, showed a positive correlation with serum GGT activity. Finally, once the serum GGT activity is elevated, several weeks of abstinence may elapse until increased values approach the normal range (Shaw and Lieber 1980; Morgan et al. 1981b; Nishimura and Teschke 1983).

Although views to the contrary have been expressed (Penn and Wortington 1983), there is general agreement that serum GGT is a useful parameter as a screening aid for alcoholism due to its high sensitivity, since up to 85% of alcoholic patients show increased values (Rosalki and Rau 1972; Morgan et al. 1981b; Weill et al. 1982). However, the specificity of the test is lower (Weill et al. 1982), since a variety of diseases unrelated to alcoholism may be associated with increased serum GGT activities. Among these are disorders of the liver and biliary tract (Eisenberg and Ortmans 1973; Kokot and Sledzinski 1974; Schmidt 1977; Selinger et al. 1982; Penn and Wortington 1983), pancreas (Kokot and Sledzinski 1974; Penn and Wortington 1983), and heart (Kokot and Sledzinski 1974). Moreover, patients treated with drugs known to induce enzyme activity may also exhibit increased serum GGT activities (Bartels et al. 1975). Most of these alcohol-unrelated diseases can easily be diagnosed upon further evaluation. Finally, the sensitivity of serum GGT for detection of alcoholism can be enhanced by combining it with other markers, such as mean corpuscular volume of red blood cells (Morgan et al. 1981b; Weill et al. 1982), aspartate aminotransferase (Morgan et al. 1981b; Weill et al. 1982; Cushman et al. 1984), glutamate dehydrogenase, and urea (Weill et al. 1982), although specificity and predictive value may slightly be reduced.

The determination of serum GGT activity can easily be performed in any routine laboratory and is reproducible and inexpensive. At present, it appears to be the best marker for alcoholism.

Mean Corpuscular Volume

Macrocytosis of red blood cells as evidenced by increased values of mean corpuscular volume (MCV) is commonly observed in alcoholic patients (Wu et al. 1974; Unger and Johnson 1974; Carney and Sheffield 1978; Whitehead et al. 1978; Whitfield et al. 1978; Shaw et al. 1979; Bagrel et al. 1979; Ryback et al. 1980, 1982; Chalmers et al. 1980, 1981; Eckardt et al. 1981; Morgan et al. 1981a, b; Heidemann et al. 1981; Chick et al. 1981; Bernadt et al. 1982; Weill et al. 1982). These changes have been attributed either to a direct toxic effect of alcohol on the bone marrow or to an underlying folate deficiency (Wu et al. 1974).

Macrocytosis in alcoholics is usually not associated with anemia (Wu et al. 1974; Unger and Johnson 1974; Carney and Sheffield 1978; Ryback et al. 1980, 1982; Chalmers et al. 1980; Eckardt et al. 1981; Heidemann et al. 1981), and megaloblastic changes are observed in a third (Wu et al. 1974; Chalmers et al. 1980) to one half (Heidemann et al. 1981) of bone marrow samples. Serum vitamin B_{12} levels have been found to be unaltered in alcoholic patients (Unger and Johnson 1974; Carney and Sheffield 1978; Chalmers et al. 1980). In less than a third of the alcoholic patients studied within the past 10 years, folate levels were decreased in the serum (Wu et al. 1974; Unger and Johnson 1974; Carney and Sheffield 1978; Heidemann et al. 1981), red blood cells (Wu et al. 1974; Chalmers et al. 1980), and liver (Wu et al. 1974). Macrocytosis disappears on alcohol withdrawal alone (Wu et al. 1974; Morgan et al. 1981b; Heidemann et al. 1981), but persists, despite folate supplement, when alcohol intake is not stopped (Wu et al. 1974). In some studies (Unger and Johnson 1974; Shaw et al. 1979) MCV values failed to decrease during the subsequent observation period, raising the question whether alcohol intake had indeed been completely stopped or folate deficiency may have been present.

In earlier studies, however, folate deficiency was a common feature in alcoholic patients due to malnutrition and low dietary folate intake (Herbert et al. 1964; Eichner and Hillman 1971). When associated with a low dietary folate intake, excess consumption of alcohol was a well-known cause of macrocytosis combined with megaloblastic anemia responsive to folic acid therapy (Klipstein and Lindenbaum 1965). These megaloblastic changes were not reversed on alcohol withdrawal, since there is a true folate deficiency (Wu et al. 1974).

Elevated values of MCV are found in up to 96% of alcoholic patients (Wu et al. 1974; Unger and Johnson 1974; Carney and Sheffield 1978; Chalmers et al. 1980; Morgan et al. 1981b; Heidemann et al. 1981), but conditions unrelated to alcoholism may also increase MCV values. The sensitivity of MCV for the detection of alcoholism can generally be enhanced by combining MCV with other parameters such as GGT (Chick et al. 1981; Morgan et al. 1981b; Chalmers et al. 1981; Weill et al. 1982), aspartate aminotransferase (Morgan et al. 1981b; Weill et al. 1982), glutamate dehydrogenase (Weill et al. 1982), urea (Weill et al. 1982), and alkaline phosphatase (Chalmers et al. 1981).

The determination of MCV is a routine method which can easily be performed in every normal clinical laboratory. The test can be done on a large-scale basis and is inexpensive. It represents a suitable marker for alcoholism, and combination with other tests may increase sensitivity.

Mitochondrial Aspartate Aminotransferase (mAST)

Serum activities of aspartate aminotransferase (AST) may be normal or slightly increased in alcoholics with or without associated liver disease (van Waes and Lieber 1977; Nishimura et al. 1980; Morgan et al. 1981b; Teschke et al. 1983; Nalpas et al. 1984), but increased activity levels do not permit precise conclusions regarding their etiology. Total AST (tAST) activity, as usually assayed in the serum or liver homogenate, consists of the activity of two isozymes (Rej 1978): a cytosolic, soluble isozyme (cAST) accounts for approximately 20% of total liver activity and a mitochondrial AST (mAST) for the remaining 80%. mAST provides a small proportion of tAST activity in the serum (Rej 1980). A variety of studies have shown that increased serum activities of mAST are commonly found in alcoholic patients

(Ishii et al. 1979b; Panteghini et al. 1983; Nalpas et al. 1984) both in the presence and absence of associated liver disease (Nalpas et al. 1984). Enhanced serum mAST activities are also described for patients with viral hepatitis (Nalpas et al. 1984). Dissociation between the alcoholic patients and those with viral hepatitis was achieved by calculating the ratio of mAST/tAST serum activities for each patient in the individual group. The mean mAST/tAST ratio in healthy controls and patients with viral hepatitis was 0.030 and 0.032, respectively, and it was four times higher in alcoholic patients with or without alcoholic liver disease. In the detection of chronic alcoholism, both mAST or the mAST/tAST ratio exhibited a high sensitivity of 93.3%–100%, but data on the specificity of these tests are not available (Nalpas et al. 1984). At present, this determination of mAST activity in serum is restricted to a few laboratories and requires an immunochemical procedure using rabbit antibodies against the soluble cAST purified from human erythrocytes.

Amino Acids

Alcoholic patients exhibit numerous plasma amino acid abnormalities, including those of aromatic and branched chain amino acids (Shaw and Lieber 1979; Lieber 1980). A striking increase in plasma α-amino-n-butyric acid (AANB) was observed following chronic alcohol consumption both in man under metabolic ward conditions (Shaw and Lieber 1978) and in experimental animals (Shaw and Lieber 1978; Dienstag et al. 1978; Shaw and Lieber 1980a). This was associated with increased levels of AANB in the liver of experimental animals, suggesting an enhanced hepatic production of this particular amino acid (Shaw and Lieber 1980a). On the other hand, protein malnutrition depresses plasma AANB, and the level of plasma AANB in alcoholic patients may therefore reflect at least two factors: chronic alcohol consumption, which tends to increase this amino acid, and dietary protein deficiency, which tends to decrease it (Shaw and Lieber 1978; Lieber 1980). Although AANB may not offer a practical tool for alcoholism screening in populations which are heterogenous with regard to nutrition, measurement of plasma AANB can be used for documentation of treatment success as well as for the detection of relapses in recovered patients having undergone rehabilitation programs; in such cases the patient can serve as his own control (Shaw et al. 1979; Lieber 1980). Under these conditions, AANB was considered to be more useful than MCV and more accurate than GGT (Shaw et al. 1979). It is noteworthy that in one study with ten chronic alcoholics drinking up to and sampled within 24 h of admission, a low plasma AANB level rather than an increased one was determined (Dienstag et al. 1978), raising the question to what extent an acute load of alcohol prior to admission may have caused the decrease. Indeed, an acute administration of alcohol leads to a striking reduction of the AANB leucine ratio in the plasma (Shaw et al. 1976), which may be caused by decreased AANB levels under these conditions.

In order to use the level of AANB in plasma as a reflection of chronic alcohol consumption, it was found necessary to control for nutritional factors (Lieber 1980). Since AANB and other branched chain amino acids such as leucine are depressed to a similar degree by dietary protein restriction, the level of AANB was expressed relative to leucine. The plasma AANB level relative to leucine was found to be increased in a large number of active alcoholics (Shaw et al. 1976; Morgan et al. 1977), findings confirmed in experimental animals after prolonged ethanol intake (Stanko et al. 1977; Shaw and Lieber 1978; Dienstag et al. 1978). This increase was reversible upon cessation of drinking, persisted for days or weeks, and did not require the presence of alcohol in the blood (Shaw et al. 1976). Since the level of the AANB/leucine ratio is not a linear relation over the range of leucine values, replacement of the simple ratio by a normal curve was subsequently recommended (Shaw et al. 1978). It was also found that determination of GGT may enhance the specificity and sensitivity of the test.

Of interest was the finding that an acute dose of alcohol decreases rather than increases the AANB/leucine ratio in the plasma (Shaw et al. 1976). This observation could explain at least in part the failure of one study showing no significant change of AANB/leucine ratio in the plasma of alcoholics drinking up to and sampled within 24 h of admission (Dienstag et al. 1978). An unaltered plasma AANB/leucine ratio has been found in alcoholics using a newly developed gas-liquid chromatographic method for assessing amino acids (Ellingboe et al. 1978), whereas all other studies were done with amino acid analyzers (Shaw et al. 1976; Morgan et al. 1977; Dienstag et al. 1978; Shaw and Lieber 1978). Obviously, there are some problems associated with accurate AANB measurements by gas-liquid chromatography, as demonstrated by the failure to resolve the AANB peak from valine and by the possible interference of contaminating substances (Ellingboe et al. 1978). Moreover, considerably lower values for AANB were found by gas-liquid chromatography than in determinations carried out with an amino acid analyzer.

Increased plasma AANB/leucine ratios were observed in experimental animals with early as well as late stages of alcoholic liver disease (Stanko et al. 1977; Shaw and Lieber 1978; Dienstag et al. 1978). In alcoholic patients the level of AANB/leucine ratio in plasma may be related to the degree of the associated liver disease. Indeed, patients with late stages of alcoholic liver disease such as alcoholic hepatitis or cirrhosis exhibit increased plasma AANB/leucine ratios, whereas unaltered ratios are found in patients with minimal alcoholic liver disease (Morgan et al. 1977; Shaw and Lieber 1978). Moreover, plasma AANB/leucine ratio was found in patients with nonalcoholic liver disease to be unaltered (Shaw et al. 1978), slightly increased (Dienstag et al. 1978), or even strikingly enhanced (Morgan et al. 1977), and these differences can be ascribed to differences of the patient groups with respect to their diagnosis.

There is no question that alcohol feeding results in increased levels of the AANB/leucine ratio in the plasma of experimental animals and man, and that this ratio is enhanced in some but not all alcoholic patients. The rise of plasma AANB/leucine ratio obviously identifies a subgroup of alcoholic patients, since factors such as the nutritional state, the presence of liver disease, the quantity of alcohol intake and recency of alcohol consumption may determine the level of the plasma AANB/leucine ratio. Moreover, the determination of AANB and leucine requires special laboratory facilities and is not inexpensive. Thus the plasma AANB/leucine ratio is at present of limited value as a screening test for alcoholism on a large-scale basis in an unselected population.

Transferrin

The serum of alcoholic patients exhibits upon analysis by isoelectric focusing an abnormally marked protein band with an isoelectric point of 5.7, and this protein appeared on immunoelectrofocusing as an increased, cathodal, microheterogeneous molecular form of transferrin (Stibler et al. 1978). Subsequently, quantitation of the abnormal transferrin was described (Stibler et al. 1980). The technique was based on analytical isoelectric focusing as the first step followed by direct immunofixation. The immunofixed transferrin is then quantified by computerized on-line densitometry, and the transferrin abnormality is calculated as a quotient, where the amount of the cathodal component is expressed as a percentage of the relative total immunofixed transferrin quantity. The sialic content of highly purified serum transferrin from alcoholics is significantly reduced compared to control transferrin, suggesting that the abnormal transferrin component with the isoelectric point of 5.7 observed in the alcoholic patients by means of isoelectric focusing represents most of the fraction of the protein with sialic acid reduction (Stibler and Borg 1981).

After a single acute alcohol load the abnormal serum transferrin was not observed (Stibler et al. 1978, 1979). Following ingestion of alcohol at a dose of 0.60 g/kg body weight per day, the transferrin abnormality was visible in some subjects after as little as 5–11 days under experimental conditions of unchanged serum activities of GGT (Stibler et al. 1979). The abnormal transferrin component was observed in 81% of patients with an admitted consumption of more than 60 g ethanol per day and normalized after at least 10 days of abstinence. It occurred in 1% of controls and in none of the cases with nonalcoholic liver diseases.

Although the determination of the abnormal microheterogeneous component of serum transferrin is a sensitive marker for alcoholism, its application is restricted to few specialized laboratories and therefore at present not useful as a general screening test for alcoholism.

References

Bagrel A, d'Houtaud A, Gueguen R, Siest G (1979) Relations between reported alcohol consumption and certain biological variables in an "unselected" population. Clin Chem 25: 1242–1246

Bagrel D, Ratanasavanh D, Siest G (1980) Influence of three inducers on rabbit gamma-glutamyltransferase. Pharmacol Res Commun 12: 557–565

Barnwell SG, Godfrey PP, Lowe PJ, Coleman R (1983) Biliary protein output by isolated perfused rat livers. Effects of bile salts. Biochem J 210: 549–557

Barouki R, Chobert MN, Finidori J, Aggerbeck M, Nalpas B, Hanoune J (1983) Ethanol effects in a rat hepatome cell line: induction of y-glutamyltransferase. Hepatology 3: 323–329

Bartels H, Evert W, Hauck W, Petersen C, Putzki H, Schulze W (1975) Significance of increased serum gamma-glutamyltransferase activity during long-term anticonvulsive treatment, clinical and experimental studies. Neuropaediatrie 6: 77–89

Belfrage P, Berg B, Hägerstrand I, Nilsson-Ehle P, Tornqvist H, Wiebe T (1977) Alterations of lipid metabolism in healthy volunteers during long-term ethanol intake. Eur J Clin Invest 7: 127–131

Berges W, Wienbeck M (1981) Alkohol und Gastrointestinaltrakt. In: Teschke R, Lieber CS (eds) Alkohol und Organschäden. Epidemiologische, klinische, biochemische und therapeutische Aspekte. Witzstrock, Baden-Baden, pp 74–88

Bernadt MW, Mumford J, Taylor C, Smith B, Murray RM (1982) Comparison of questionnaire and laboratory tests in the detection of excessive drinking and alcoholism. Lancet I: 325–328

Billington D, Juszczak R, Rahman K (1983) Chenodeoxycholate feeding increases the output of plasma-membrane marker enzymes into rat bile. Biochem Soc Trans 11: 162

Busachi C, Mebis J, Broeckaert L, Desmet V (1981) Histochemistry of γ-glutamyl transpeptidase in human liver biopsies. Pathol Res Pract 172: 99–108

Carney MWP, Sheffield B (1978) Serum folate and B12 and haematological status of in-patient alcoholics. Br J Addict 73: 3–7

Chalmers DM, Chanarin I, MacDermott S, Levi AJ (1980) Sex-related differences in the haematological effects of excessive alcohol consumption. J Clin Pathol 33: 3–7

Chalmers DM, Rinsler MG, MacDermott S, Spicer CC, Levi AJ (1981) Biochemical and haematological indicators of excessive alcohol consumption. Gut 22: 992–996

Chick J, Kreitman N, Plant M (1981) Mean cell volume and gamma-glutamyl-transpeptidase as markers of drinking in working men. Lancet I: 1249–1251

Clarenburg R (1983) Asialoglycoprotein receptor is uninvolved in clearing intact glycoproteins from rat blood. Am J Physiol 244: G247–G253

Clark PMS, Kricka LJ, Zaman S (1982) Drivers, binge drinking, and gamma-glutamyltranspeptidase. Br Med J 285: 1656–1657

Cushman P, Jacobson G, Barboriak JJ, Anderson AJ (1984) Biochemical markers for alcoholism: sensitivity problems. Alcoholism: Clin Exp Res 8: 253–257

Debanne MT, Evans WH, Flint N, Regoeczi E (1982) Receptor-rich intracellular membrane vesicles transporting asialotransferrin and insulin in liver. Nature 298: 398–400

Dienstag JL, Carter EA, Wands JR, Isselbacher KJ, Fischer JE (1978) Plasma α-amino-*n*-butyric acid to leucine ratio. Nonspecificity as a marker for alcoholism. Gastroenterology 75: 561–565

Doyle D, Le AV (1981) Regulation of the metabolism of plasma membrane proteins in hepatocytes. Hepatology 1: 184–186

Dunbar JA, Hagart J, Martin B, Ogston S, Devgun MS (1982) Drivers, binge drinking, and gamma-glutamyltranspeptidase. Br Med J 285: 1083

Ebihara Y, Ishii H, Takagi T, Miura S, Munakata Y, Takagi S, Arai M, Shigeta Y, Kono H, Tsuchiya M (1982) Role of intestinal gamma glutamyltranspeptidase (GGTP) activity in enhanced plasma GGTP level in alcoholics. Alcoholism 6: 294 A

Eckardt MJ, Ryback RS, Rawlings RR, Graubard BI (1981) Biochemical diagnosis of alcoholism. A test of the discriminating capabilities of γ-glutamyl transpeptidase and mean corpuscular volume. JAMA 246: 2707–2710

Eichner ER, Hillman RS (1971) The evolution of anemia in alcoholic patients. Am J Med 50: 218–232

Eisenberg K, Ortmans H (1973) Gamma-Glutamyl-Transpeptidase-Aktivität bei Lebererkrankungen. Med Klin 68: 798–804

Ellingboe J, Mendelson JH, Varanelli CC, Neuberger O, Borysow M (1978) Plasma alpha amino-*n*-butyric acid: leucine ratio. Normal values in alcoholics. J Stud Alcohol 39: 1467–1476

Evans WH (1981) Membrane traffic at the hepatocyte's sinusoidal and canalicular surface domains. Hepatology 1: 452–457

Faller J, Fox IH (1982) Ethanol-induced hyperuricemia. Evidence for increased urate production by activation of adenine nucleotide turnover. N Engl J Med 307: 1598–1602

Feuerlein W (1981) Psychiatrische Krankheitsbilder des Alkoholismus. In: Teschke R, Lieber CS (eds) Alkohol und Organschäden. Epidemiologische, klinische, biochemische und therapeutische Aspekte. Witzstrock, Baden-Baden, pp 225–236

Feuerlein W, Ringer C, Küfner H, Antons K (1977) Diagnose des Alkoholismus. Der Münchner Alkoholismustest (MALT). MMW 119: 1275–1282

Freer DE, Statland BE (1977a) The effects of ethanol (0.75 g/kg body weight) on the activities of selected enzymes in sera of healthy young adults: 1. Intermediate-term effects. Clin Chem 23: 830–834

Freer DE, Statland BE (1977b) Effects of ethanol (0.75 g/kg body weight) on the activities of selected enzymes in sera of healthy young adults: 2. Interindividual variations in response of γ-glutamyltransferase to repeated ethanol challenges. Clin Chem 23: 2099–2102

Gadeholt G, Aarbakke J, Dybing E, Sjöblom M, Mørland J (1980) Hepatic microsomal drug metabolism, glutamyl transferase activity and in vivo antipyrine half-life in rats chronically fed an ethanol diet, a control diet and a chow diet. J Pharmacol Exp Ther 213: 196–203

Geuze HJ, Slot JW, Strous GJAM, Lodish HF, Schwartz AL (1982) Immunocytochemical localization of the receptor for asialoglycoprotein in rat liver cells. J Cell Biol 92: 865–870

Gill GV, Baylis PH, Flear CTG, Skillen AW, Diggle PH (1982) Acute biochemical responses to moderate beer drinking. Br Med J 285: 1770–1773

Goebell H, Singer MV (1981) Alkohol und Pankreas. In: Teschke R, Lieber CS (eds) Alkohol und Organschäden. Epidemiologische, klinische, biochemische und therapeutische Aspekte. Witzstrock, Baden-Baden, pp 111–123

Heidemann E, Nerke O, Waller HD (1981) Alkoholtoxische Veränderungen der Hämatopoiese. Eine prospektive Studie bei chronischen Alkoholikern. Klin Wochenschr 59: 1303–1312

Herbert V, Zalusky R, Davidson CS (1964) Correlation of folate deficiency with alcoholism and associated macrocytosis, anemia, and liver disease. Ann Intern Med 58: 977–988

Hoberman HD, Chiodo SM (1982) Elevation of the hemoglobin A1 fraction in alcoholism. Alcoholism 6: 260–266

Hultberg B, Isaksson A, Tiderström G (1980) β-Hexosaminidase, leucine aminopeptidase, cystidyl aminopeptidase, hepatic enzymes and bilirubin in serum of chronic alcoholics with acute ethanol intoxication. Clin Chim Acta 105: 317–323

Hurt RD, Morse RM, Swenson WM (1980) Diagnosis of alcoholism with a self-administered alcoholism screening test. Results with 1,002 consecutive patients receiving general examinations. Mayo Clin Proc 55: 365–370

Huseby N-E (1977) Purification and some properties of γ-glutamyltransferase from human liver. Biochim Biophys Acta 483: 46–56

Huseby NE (1979) Subcellular localization of γ-glutamyltransferase activity in guinea pig liver. Effect of phenobarbital on the enzyme activity levels. Clin Chim Acta 94: 163–171

Idéo G, Tempini S, Bellobuono A, Bellati G, Ronchi G (1980) Biochemical and histochemical estimations of liver γ-glutamyltransferase activity in ethanol-fed rats. J Clin Chem Clin Biochem 18: 237–239

Inoue M, Kinne R, Tran T, Biempica L, Arias IM (1983) Rat liver canalicular membrane vesicles. Isolation and topological characterization. J Biol Chem 258: 5183–5188

Ishii H, Okuno F, Shigeta Y, Tsuchiya M (1979) Enhanced serum glutamic oxaloacetic transaminase activity of mitochondrial origin in chronic alcoholics. In: Galanter M (ed) Currents in alcoholism, Vol. V. Grune and Stratton, New York, pp 101–108

Ishii H, Yasuraoka S, Shigeta Y, Takagi S, Kamiya T, Okuno F, Miyamoto K, Tsuchiya M (1978) Hepatic and intestinal gamma-glutamyltranspeptidase activity: its activation by chronic ethanol administration. Life Sci 23: 1393–1398

Ishii H, Yasuraoka S, Shigeta Y, Kamiya T, Ebihara Y, Tsuchiya M (1979) Mechanism of enhanced activity of gamma glutamyl transpeptidase (GGTP) in chronic alcoholics. Alcoholism 3: 180A

Ivanov E, Adjarov D, Etarska M, Stankushev T, Brumbarov K, Kerimova M (1980) Elevated liver gamma-glutamyl transferase in chronic alcoholics. Enzyme 25: 304–308

Jenkins WJ, Rosalki SB, Foo Y, Scheuer PJ, Nemesanszky E, Sherlock S (1982) Serum glutamate dehydrogenase is not a reliable marker of liver cell necrosis in alcoholics. J Clin Pathol 35: 207–210

Kershenobich D, Garcia-Tsao G, Alvarez Saldana S, Rojkind M (1981) Relationship between blood lactic acid and serum proline in alcoholic liver cirrhosis. Gastroenterology 80: 1012–1015

Kley HK, Both H, Moreno F (1981) Alkohol und Endokrinium. In: Teschke R, Lieber CS (eds) Alkohol und Organschäden. Epidemiologische, klinische, biochemische und therapeutische Aspekte. Witzstrock, Baden-Baden, pp 124–145

Klipstein FA, Lindenbaum J (1965) Folate deficiency in chronic liver disease. Blood 25: 443–456

Köttgen E, Gerok W (1976) Isoenzymdifferenzierung der Gamma-Glutamyltransferase mit Concanavalin A and Con A-Sepharose. Klin Wochenschr 54: 439–444

Köttgen E, Reutter W, Gerok W (1978) Induction and 'superinduction' of sialylation of membrane-bound γ-glutamyltransferase during liver regeneration. Eur J Biochem 82: 279–284

Köttgen E, Bauer C, Reutter W, Gerok W (1979) Neue Ergebnisse zur biologischen und medizinischen Bedeutung von Glycoproteinen. 1. Grundlagen und interdisziplinäre Gesichtspunkte. Klin Wochenschr 57: 151–159

Köttgen E, Buchsel R, Bauer C (1983) Glycoproteins in liver disease. In: Bianchi L, Gerok W, Landmann L, Sickinger K, Stalder GA (eds) Liver in metabolic diseases. MTP Press, Boston, pp 145–153

Kokot F, Sledzinski Z (1974) Die γ-Glutamyltransferase (γ-GT). Z Klin Chem Klin Biochem 12: 374–384

Kolb-Bachofen V (1981) Hepatic receptor for asialo-glycoproteins. Ultrastructural demonstration of ligand-induced microaggregation of receptors. Biochim Biophys Acta 645: 293–299

Krasner N, Moore MR, Thompson GG, McIntosh W, Goldberg A (1974) Depression of erythrocyte δ-aminolaevulinic acid dehydratase activity in alcoholics. Clin Sci Mol Med 46: 415–418

Krukenberg S, Nishimura M, Teschke R (1985) Biliary secretion of gamma-glutamyltransferase after chronic alcohol consumption. Submitted for publication

Kryszewski AJ, Neale G, Whitfield JB, Moss DW (1973) Enzyme changes in experimental biliary obstruction. Clin Chim Acta 47: 175–182

Kuhn H (1981) Alkohol und Herz. In: Teschke R, Lieber CS (eds) Alkohol und Organschäden. Epidemiologische, klinische, biochemische und therapeutische Aspekte. Witzstrock, Baden-Baden, pp 146–158

Lahrichi M, Ratanasavanh D, Galteau M-M, Siest G (1982) Effect of chronic ethanol administration on gamma-glutamyltransferase activities in plasma and in hepatic plasma membranes of male and female rats. Enzyme 28: 251–257

Lieber CS (1980) Alcohol, protein metabolism, and liver injury. Gastroenterology 79: 373–390

Lieber CS, Jones DP, Mendelson J, DeCarli LM (1963) Fatty liver, hyperlipemia and hyperuricemia produced by prolonged alcohol consumption, despite adequate dietary intake. Trans Assoc Am Physicians 76: 289–300

Lindsell DRM, Wilson AG, Maxwell JD (1982) Fractures on the chest radiograph in detection of alcoholic liver disease. Br Med J 285: 597–599

Luchi P, Cortis G (1978) Forensic considerations on the comparison of "serum γ-glutamyltranspeptidase" ("γ-GT") activity in experimental acute alcoholic intoxication and in alcoholic car drivers who caused road accidents. Forensic Sci 11: 33–39

Malvoisin E, Mercier M, Roberfroid M (1978) Gamma glutamyl transferase. Application of a new radiochemical assay to the analysis of its subcellular distribution in the rat liver. Enzyme 23: 373–381

Mayfield D, McLeod G, Hall P (1974) The CAGE questionnaire: validation of a new alcoholism screening instrument. Am J Psychiatry 131: 1121–1123

Meister A (1973) On the enzymology of amino acid transport. Science 180: 33–39

Mills PR, Spooner RJ, Russell RI, Boyle P, MacSween RNM (1981) Serum glutamate dehydrogenase as a marker of hepatocyte necrosis in alcoholic liver disease. Br Med J 283: 754–755

Moore RA (1972) The diagnosis of alcoholism in a psychiatric hospital: a trial of the Michigan alcoholism screening test (MAST). Am J Psychiatry 128: 1565–1569

Morgan MY, Milsom JP, Sherlock S (1977) Ratio of plasma alpha amino-*n*-butyric acid to leucine as an empirical marker of alcoholism: diagnostic value. Science 197: 1183–1185

Morgan MY, Camilo ME, Luck W, Sherlock S, Hoffbrand AV (1981a) Macrocytosis in alcohol-related liver disease: its value for screening. Clin Lab Haematol 3: 35–44

Morgan MY, Colman JC, Sherlock S (1981b) The use of a combination of peripheral markers for diagnosing alcoholism and monitoring for continued abuse. Br J Alcohol Alcoholism 16: 167–177

Mørland J, Huseby N-E, Sjøblom M, Strømme JH (1977) Does chronic alcohol consumption really induce hepatic microsomal gamma-glutamyltransferase activity? Biochem Biophys Res Commun 77: 1060–1066

Naftalin L, Child VJ, Morley DA, Smith DA (1969) Observations on the site of origin of serum γ-glutamyltranspeptidase. Clin Chim Acta 26: 297–300

Nalpas B, Vassault A, Le Guillou A, Lesgourgues B, Ferry N, Lacour B, Berthelot P (1984) Serum activity of mitochondrial aspartate aminotransferase: a sensitive marker of alcoholism with or without alcoholic hepatitis. Hepatology 4: 893–896

Nishimura M, Teschke R (1982) Effect of chronic alcohol consumption on the activities of liver plasma membrane enzymes: gamma-glutamyltransferase, alkaline phosphatase and 5′-nucleotidase. Biochem Pharmacol 31: 377–381

Nishimura M, Teschke R (1983) Alcohol and gamma-glutamyltransferase. Klin Wochenschr 61: 265–275

Nishimura M, Hasumura Y, Takeuchi J (1980) Effect of an intravenous infusion of ethanol on serum enzymes and lipids in patients with alcoholic liver disease. Gastroenterology 78: 691–695

Nishimura M, Stein H, Berges W, Teschke R (1981) Gamma-glutamyltransferase activity of liver plasma membrane: induction following chronic alcohol consumption. Biochem Biophys Res Commun 99: 142–148

Orlowski M, Meister A (1965) Isolation of γ-glutamyl transpeptidase from hog kidney. J Biol Chem 240: 338–347

Panteghini M, Falsetti F, Chiari E, Malchiodi A (1983) Determination of aspartate aminotransferase isoenzymes in hepatic diseases-preliminary findings. Clin Chim Acta 128: 133–140

Penn R, Worthington DJ (1983) Is serum γ-glutamyltransferase a misleading test? Br Med J 286: 531–535

Popper H, Reutter W, Gudat F, Köttgen E (1983) Structural carbohydrates in the liver. MTP Press, Boston

Ratanasavanh D, Tazi A, Galteau MM, Siest G (1979) Localization of gamma-glutamyltransferase in subcellular fractions of rat and rabbit liver: effect of phenobarbital. Biochem Pharmacol 28: 1363–1365

Rauen J, Petrides AS, Teschke R (1980) Anstieg der adulten und fetalen Form der Gamma-Glutamyl-Transferase(GGT)-Aktivität im Serum bei Patienten mit alkoholischen Leberschäden. Verh Dtsch Ges Inn Med 86: 811–813

Rej R (1978) Aspartate aminotransferase activity and isoenzyme proportions in human liver tissues. Clin Chem 24: 1971–1979

Rej R (1980) An immunochemical procedure for determination of mitochondrial aspartate aminotransferase in human serum. Clin Chim 26: 1694–1700

Reuben A (1984) Biliary proteins. Hepatology 4: 46S–50S

Reyes E (1977) Effects of alcohol on the isoenzymes of rat brain γ-glutamyl transpeptidase. Proc West Pharmacol Soc 20: 345–349

Reyes E, Miller WR, Taylor CA, Spalding CT (1978) The activity of γ-glutamyl transpeptidase in the serum of problem drinkers. Proc West Pharmacol Soc 21: 289–297

Rollason JG, Pincherle G, Robinson D (1972) Serum gamma glutamyl transpeptidase in relation to alcohol consumption. Clin Chim Acta 39: 75–80

Ronchi G, Desmet V (1973) Histochemical study of so-called "marker enzymes of cholestasis" during extrahepatic bile duct obstruction in the rat. Beitr Pathol 149: 213–226

Rosalki SB, Rau D (1972) Serum γ-glutamyl transpeptidase activity in alcoholism. Clin Chim Acta 39: 41–47

Russel FGM, Weitering JG, Oosting R, Groothuis GMM, Hardonk MJ, Meijer DKF (1983) Influence of taurocholate on hepatic clearance and biliary excretion of asialo intestinal alkaline phosphatase in the rat in vivo and in isolated perfused rat liver. Gastroenterology 85: 225–234

Rutenburg AM, Kim H, Fischbein JW, Hanker JS, Wasserkrug HL, Seligman AM (1969) Histochemical and ultrastructural demonstration of γ-glutamyl transpeptidase activity. J Histochem Cytochem 17: 517–526

Ryback RS, Eckardt MJ, Pautler CP (1980) Biochemical and hematological correlates of alcoholism. Res Commun Chem Pathol Pharmacol 27: 533–550

Ryback RS, Eckardt MJ, Felsher B, Rawlings RR (1982) Biochemical and hematologic correlates of alcoholism and liver disease. JAMA 248: 2261–2265

Satoh T, Takenaga M, Kitagawa H, Itoh S (1980) Microassay of gamma-glutamyl transpeptidase in needle biopsies of human liver. Res Commun Chem Pathol Pharmacol 30: 151–161

Saunders WM, Kershaw PW (1980) Screening tests for alcoholism – findings from a community study. Br J Addict 75: 37–41

Schellenberg B, Arab L, Schlierf G (1982) Alkohol – "Nahrungsmittel" mit Nebenwirkungen. Inn Med 9: 122–125

Schmidt FW (1977) Enzymes in cholestasis. In: Bianchi L, Gerok W, Sickinger K (eds) Liver and bile. MTP Press, Lancaster, pp 203–214

Schrappe O (1981) Alkohol und Nervensystem. In: Teschke R, Lieber CS (eds) Alkohol und Organschäden. Epidemiologische, klinische, biochemische und therapeutische Aspekte. Witzstrock, Baden-Baden, pp 211–224

Schwartz AL, Marshak-Rothstein A, Rup D, Lodish HF (1981) Identification and quantification of the rat hepatocyte asialoglycoprotein receptor. Proc Natl Acad Sci USA 78: 3348–3352

Seitz HK, Velazquez D, Czygan P, Simon B, Waldherr R, Veith S, Götz R, Kommerell B (1982) Effect of chronic ethanol consumption on duodenal gamma-glutamyltranspeptidase activity (GGT-A) in man and in rats. Scand J Gastroenterol [Suppl 78] 17: 550A

Selden C, Wootton AM, Moss DW, Peters TJ (1978) Analytical subcellular fractionation studies on different cell types isolated from normal rat liver. Clin Sci Mol Med 55: 423–427

Selinger MJ, Matloff DS, Kaplan MM (1982) γ-Glutamyl transpeptidase activity in liver disease: serum elevation is independent of hepatic GGTP activity. Clin Chim Acta 125: 283–290

Selzer ML (1971) The Michigan alcoholism screening test: the quest for a new diagnostic instrument. Am J Psychiatry 127: 1653–1658

Seymour CA, Peters TJ (1978) Changes in hepatic enzymes and organelles in alcoholic liver disease. Clin Sci Mol Med 55: 383–389

Shaw LM, Marsh E (1981) The effect of one evening of social drinking on serum γ-glutamyltransferase activity. Clin Chem 27: 1036A

Shaw S, Lieber CS (1978) Plasma amino acid abnormalities in the alcoholic. Respective role of alcohol, nutrition, and liver injury. Gastroenterology 74: 677–682

Shaw S, Lieber CS (1979) Amino acids and ethanol. In: Majchrowicz E, Noble EP (eds) Biochemistry and pharmacology of ethanol, vol 1. Plenum, New York, pp 383–406

Shaw S, Lieber CS (1980a) Increased hepatic production of alpha-amino-*n*-butyric acid after chronic alcohol consumption in rats and baboons. Gastroenterology 78: 108–113

Shaw S, Lieber CS (1980b) Mechanism of increased gamma glutamyl transpeptidase after chronic alcohol consumption: hepatic microsomal induction rather than dietary imbalance. Subst Alcohol Actions Misuse 1: 423–428

Shaw S, Stimmel B, Lieber CS (1976) Plasma alpha amino-*n*-butyric acid to leucine ratio: an empirical biochemical marker of alcoholism. Science 194: 1057–1058

Shaw S, Lue S-L, Lieber CS (1978) Biochemical tests for the detection of alcoholism: comparison of plasma alpha-amino-*n*-butyric acid with other available tests. Alcoholism 2: 3–7

Shaw S, Worner TM, Borysow MF, Schmitz RE, Lieber CS (1979) Detection of alcoholism relapse: comparative diagnostic value of MCV, GGTP, and AANB. Alcoholism 3: 297–301

Singer JS, Kaplan MM (1978) ETOH depresses rat liver γ-glutamyl transpeptidase. Gastroenterology 74: A-263/1095

Smith GD, Peters TJ (1978) Studies on the subcellular localization of γ-glutamyl transpeptidase in rat liver. Biochem Soc Trans 6: 543–545

Stanko RT, Morse EL, Adibi SA (1977) Effect of chronic alcohol intake on amino acids. Gastroenterology 72: A-113/1136

Stibler H, Borg S (1981) Evidence of a reduced sialic acid content in serum transferrin in male alcoholics. Alcoholism 5: 545–549

Stibler H, Allgulander C, Borg S, Kjellin KG (1978) Abnormal microheterogeneity of transferrin in serum and cerebrospinal fluid in alcoholism. Acta Med Scand 204: 49–56

Stibler H, Borg S, Allgulander C (1979) Clinical significance of abnormal heterogeneity of transferrin in relation to alcohol consumption. Acta Med Scand 206: 275–281

Stibler H, Sydow O, Borg S (1980) Quantitative estimation of abnormal microheterogeneity of serum transferrin in alcoholics. Pharmacol Biochem Behav [Suppl 1] 13: 47–51

Szewczuk A, Connell GE (1964) The effect of neuraminidase on the properties of γ-glutamyl transpeptidase. Biochim Biophys Acta 83: 218–223

Tazi A, Gateau M-M, Siest G (1980) γ-Glutamyltransferase of rabbit liver: kinetic study of phenobarbital induction and in vitro solubilization by bile salts. Toxicol Appl Pharmacol 55: 1–7

Teschke R (1980) Leberschäden durch Alkohol: biochemische und klinische Studien zur Pathogenese und Früherkennung. Thieme, Stuttgart

Teschke R (1981) Alkohol und Leber. In: Teschke R, Lieber CS (eds) Alkohol und Organschäden. Epidemiologische, klinische, biochemische und therapeutische Aspekte. Witzstrock, Baden-Baden, pp 89–110

Teschke R, Petrides AS (1982) Hepatic gamma-glutamyltransferase activity: its increase following chronic alcohol consumption and the role of carbohydrates. Biochem Pharmacol 31: 3751–3756

Teschke R, Brand A, Strohmeyer G (1977) Induction of hepatic microsomal gamma-glutamyltransferase activity following chronic alcohol consumption. Biochem Biophys Res Commun 75: 718–724

Teschke R, Rauen J, Neuefeind M, Petrides AS, Strohmeyer G (1980) Alcoholic liver disease associated with increased gamma-glutamyltransferase activities in serum and liver. In: Thurman RG (ed) Alcohol and aldehyde metabolizing systems – 4. Plenum, New York, pp 647–654 (Advances in experimental medicine and biology, vol 132)

Teschke R, Neuefeind M, Nishimura M, Strohmeyer G (1983) Hepatic gamma-glutamyltransferase activity in alcoholic fatty liver: comparison with other liver enzymes in man and rats. Gut 24: 625–630

Teschke R, Krukenberg S, Nishimura M, Stremmel W, Strohmeyer G (1984) Diskrete Cholestase nach chronischem Alkoholkonsum als Ursache für die Aktivitätssteigerung der Gamma-Glutamyltransferase? Z Gastroenterologie 22: 494

Teschke R, Krukenberg S, Stremmel W, Nishimura M (1985) Alcohol, discrete cholestasis and gamma-glutamyltransferase: an experimental study in rats. Submitted for publication

Tobiasson P, Boeryd B (1980) Serum cholic and chenodeoxycholic acid conjugates and standard liver function tests in various morphological stages of alcoholic liver disease. Scand J Gastroenterol 15: 657–663

Uchida T, Peters RL (1983) The nature and origin of proliferated bile ductules in alcoholic liver disease. Am J Clin Pathol 79: 326–333

Unger KW, Johnson D Jr (1974) Red blood cell mean corpuscular volume: a potential indicator of alcohol usage in a working population. Am J Med Sci 267: 281–289

van Waes L, Lieber CS (1977) Glutamate dehydrogenase: a reliable marker of liver cell necrosis in the alcoholic. Br Med J 2: 1508–1510

Wadstein J, Skude G (1979) Serum ethanol, hepatic enzymes and length of debauch in chronic alcoholics. Acta Med Scand 205: 317–318

Weill J, Schellenberg F, LeGoff A-M, Lamy J (1982) The predictive value of gamma-glutamyl transferase and other peripheral markers in the screening of alcohol abuse. In: Siest G, Heusghem C (eds) Gammaglutamyltransferases: advances in biochemical pharmacology, 3rd series. Masson, Paris, pp 195–198

Whitehead TP, Clarke CA, Whitfield AGW (1978) Biochemical and haematological markers of alcohol intake. Lancet I: 978–981

Whitfield JB, Hensley WJ, Bryden D, Gallagher H (1978) Estimation of alcohol intake from laboratory results. Ann Clin Biochem 15: 304–306

Widnell CC (1972) Cytochemical localization of 5′-nucleotidase in subcellular fractions isolated from rat liver. I. The origin of 5′-nucleotidase activity in microsomes. J Cell Biol 52: 542–558

Wienbeck M, Berges W (1981) Oesophageal lesions in the alcoholic. Clin Gastroenterol 10: 375–388

Wootton AM, Neale G, Moss DW (1977) Enzyme activities of cells of different types isolated from livers of normal and cholestatic rats. Clin Sci Mol Med 52: 585–590

Worner TM, Lieber CS (1980) Plasma glutamate dehydrogenase: clinical application in patients with alcoholic liver disease. Alcoholism 4: 431–434

Wu A, Chanarin I, Levi AJ (1974) Macrocytosis of chronic alcoholism. Lancet I: 829–830

Wu A, Slavin G, Levi AJ (1976) Elevated serum gamma-glutamyl-transferase (transpeptidase) and histological liver damage in alcoholism. Am J Gastroenterol 65: 318–323

Yamauchi M, Kimura K, Kawase H, Watanabe Y, Kitahara T, Ogura K, Fujisawa K (1984) Hepatic gamma-glutamyl transferase and hepatic fibrosis in patients with alcoholic liver disease. Enzyme 32: 110–115

Zeitlin PL, Hubbard AL (1982) Cell surface distribution and intracellular fate of asialoglycoproteins: a morphological and biochemical study of isolated rat hepatocytes and monolayer cultures. J Cell Biol 92: 634–647

4 Ethanol and Lipid Metabolism

E. Baraona

Alcohol Research and Treatment Center, Veterans Administration Medical Center,
130 West Kingsbridge Road, Bronx, NY 10468, USA

The main alteration of lipid metabolism secondary to the consumption of alcohol is excessive accumulation of fat in both the liver and the blood, commonly referred to as fatty liver and alcoholic hyperlipemia respectively.

Alcoholic Fatty Liver

The accumulation of lipids in the hepatocyte (steatosis) is one of the earliest and most striking manifestation of alcoholic liver injury. The denomination of fatty liver is commonly applied to these early stages. Lipid accumulation, however, is also a component of more advanced stages in which the development of fibrosis, inflammation, and distortion of the hepatic acinar structure become prominent features.

The lipids that accumulate are mainly triacylglycerols, the concentration of which frequently exceeds that of phospholipids, the predominant lipids of the normal liver. Increases in phospholipids (Lieber et al. 1965; French 1967), free cholesterol (Vasdev et al. 1974), and cholesteryl esters (DeCarli and Lieber 1967) have also been found after chronic alcohol administration, but these are not constant features in all animal species. For instance, prolonged administration of alcohol failed to produce hepatic accumulation of cholesterol and phospholipids in some primates (Leathers et al. 1981; Baraona et al. 1984).

Etiologic Role of Ethanol

The concept that malnutrition is primarily responsible for the development of alcoholic fatty liver was challenged by its experimental production by administration of ethanol while maintaining adequate nutrition. The administration of ethanol to volunteers (either as a supplement of a normal or enriched diet, or in substitution for other calories) produced up to a 25-fold rise in hepatic triacylglycerol concentrations (Lieber et al. 1963, 1965; Lieber and Rubin 1968; Rubin and Lieber 1968; Wiebe et al. 1971). Also, fatty liver was produced in the rat (Lieber et al. 1965; DeCarli and Lieber 1967) and in baboons (Lieber and DeCarli 1974). The natural resistance of

Alcohol Related Diseases in Gastroenterology
Edited by H. K. Seitz and B. Kommerell

these animals to consuming amounts of alcohol comparable to those observed in alcoholics was overcome by incorporation of ethanol into nutritionally adequate liquid diets that were given as the only source of food and fluids. In rats given 36% of their energy requirement as ethanol, hepatic lipid accumulation reached a maximum in about 4 weeks and it remained stationary thereafter. Despite continuous consumption of ethanol for the life span of this species (Lieber and DeCarli 1970), there was no further progression of the liver lesions. Attempts to increase ethanol intake decreased food consumption and promoted undernutrition. In baboons, up to 50% of the energy requirement could be given as ethanol without impairing nutrition. This not only produced a more severe fatty liver but also produced most of the spectrum of alcoholic liver disease, including cirrhosis.

Role of Nutritional Factors

A more pronounced steatosis was produced in volunteers when alcohol was given with a normal than with a low-fat diet (Lieber and Spritz 1966). Also, in the rat, the steatosis was exaggerated by increasing the supply of dietary fat. But, when the dietary fat was decreased to provide for only the essential fatty acid requirement, fatty liver was not fully prevented (Lieber and DeCarli 1970). The type of fat was also important: triacylglycerols containing medium-chain fatty acids reduced liver lipid accumulation (Lieber and DeCarli 1966). This is probably due to the greater tendency of medium-chain fatty acids to undergo oxidation rather than esterification (Lieber et al. 1967). Dietary supplementation with cholesterol increased the alcohol-induced lipid changes (not only cholesterol changes) in liver and blood (Leathers et al. 1981), whereas decreased absorption of cholesterol due to supplementation with dietary fiber produced the opposite effect (Indira and Kurup 1982). The mechanism of this interaction between cholesterol and other lipids is unknown. Massive supplementation with lipotropic agents (such as choline) failed to prevent the alcohol-induced fatty liver in man (Rubin and Lieber 1968). There is no evidence that choline deficiency is deleterious to man. The fatty liver produced by choline deficiency appears to be an experimental disease of the rat with little relevance (if any) to human alcoholic liver injury. Even in rats, massive choline supplementation decreased but did not fully prevent the ethanol-induced fatty liver (Lieber and DeCarli 1966). Protein deficiency produces fatty liver in children (kwashiorkor), but its effects on adults are less well defined. Protein and choline deficiencies markedly exaggerated alcoholic fatty liver in the rat (Lieber et al. 1969). Moderate deficiency in dietary protein (7% of energy) did not affect alcoholic fatty liver in baboons (Lieber et al. 1976). Also, a high-protein diet did not prevent the fatty liver induced by ethanol in human volunteers (Rubin and Lieber 1968). Thus ethanol is the primary etiologic factor, but the magnitude of the fatty liver varies with nutritional factors, especially with the supply of dietary lipids.

Role of Associated Liver Injury

In baboons, the rate of triacylglycerol accumulation in the liver decreased with the development of fibrosis, the concentrations remaining stationary or even decreasing despite persistent ethanol intake (Savolainen et al., 1984). Thus the common impression that cirrhotic livers from alcoholics usually contain less fat than at the fatty liver stage appears to be due at least in part to an inhibitory effect of the liver injury on the rate of lipid accumulation.

Pathogenesis

The liver plays a central role in the lipid metabolism converting a discontinuous or variable supply of dietary lipids into a continuous release of lipids into the blood to satisfy cellular needs. Excessive accumulation of fat in the liver might result (a) from increased supply of lipids to the liver from either the intestine or the adipose tissue; (b) from increased synthesis in the liver itself; or (c) from decreased disposition of liver lipids by oxidation, by release into plasma lipoproteins, or by biliary excretion of cholesterol and phospholipids. These mechanisms are schematically illustrated in Fig. 1 and are discussed below.

Supply of Lipids to the Liver

Mobilization of Fatty Acids from the Adipose Tissue

The fatty acid composition of the lipids which accumulate in the liver after a single large dose of ethanol resembles that of the adipose tissue (Brodie et al. 1961; Lieber et al. 1966). This was associated with enhanced mobilization of prelabeled fatty acids

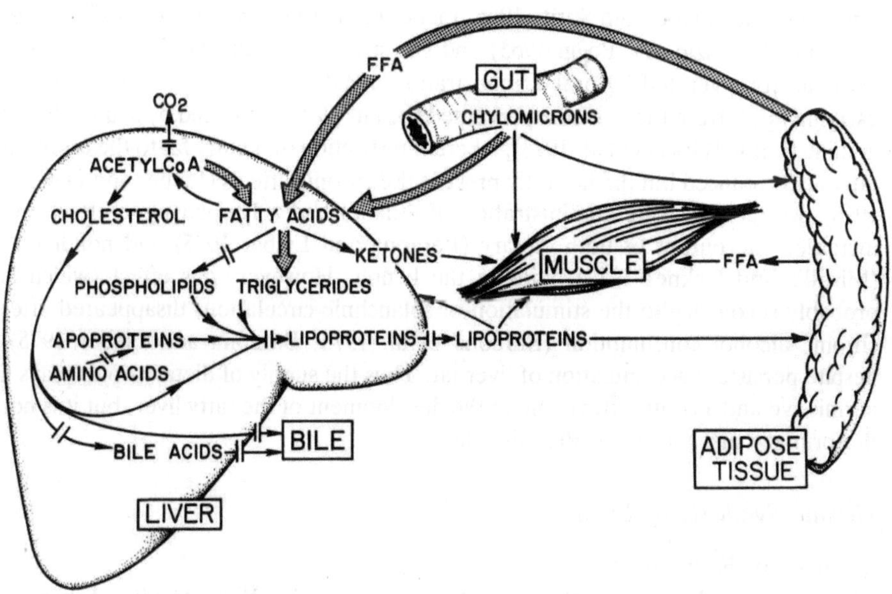

Fig. 1. Theoretical mechanism for alcoholic fatty liver. Ethanol intake could result in fatty liver either by enhancing the pathways illustrated by *wide arrows* or by blocking those illustrated by *broken lines* (Lieber 1982)

from rat epididymal fat pads and with increased serum free fatty acid (FFA) concentrations (Brodie et al. 1961; Kessler and Yalovsky-Mishkin 1966). These effects are mediated in part by the release of catecholamines during alcoholic intoxication (Mallov 1961). A number of factors which prevent the release or the action of catecholamines on FFA mobilization also prevent to a great extent the development of fatty liver after an acute ethanol dose (Rebouças and Isselbacher 1961; Kalant et al. 1975). Fasting potentiates the effect of ethanol on FFA and aggravates the acute fatty liver (Johnson 1974). However, fatty liver can be produced by acute ethanol administration under conditions in which enhanced FFA mobilization is not greater than that of fasting (Elko et al. 1961; Poggi and DiLuzio 1964), suggesting the participation of additional factors. It has been claimed, however, that although the flux of plasma FFA is not affected by a moderate dose of ethanol, the fraction of the flux taken up by the liver is increased (Abrams and Cooper 1976a, b) due to the stimulatory effect of ethanol on hepatic blood flow.

Fatty liver was also produced in man and animals by chronic administration of ethanol in the absence of significant changes of plasma FFA (Lieber et al. 1963). After chronic alcohol intake, the stimulatory effect of ethanol on splanchnic blood flow was greatly attenuated (Jauhonen et al. 1982). Thus enhanced FFA mobilization does not seem to play a major role in the development of the fatty liver produced by chronic alcohol consumption.

Contribution of Dietary and Nondietary Intestinal Lipids

After feeding ethanol in diets containing a characteristic fatty acid composition for several days, hepatic fatty acids resembled those of the diets rather than those of adipose tissue (Lieber and Spritz 1966; Lieber et al. 1966; Mendenhall 1972). Both the acute (DiLuzio and Poggi 1963) and chronic (Lieber and DeCarli 1970) fatty livers are exaggerated by oral administration of fat. Also diversion of the lymph carrying lipids from the intestine prevented the fatty liver produced by acute ethanol administration (Ockner et al. 1973), whereas restriction of dietary fat to the essential fatty acids reduced but did not fully prevent the chronic effects (Lieber and DeCarli 1970). Moreover, acute administration of ethanol had a lymphagogue effect and increased the output of both dietary (Baraona and Lieber 1975) and nondietary (Mistillis and Ockner 1972) lipids in the lymph. However, this effect (which is probably secondary to the stimulation of splanchnic circulation) disappeared after chronic alcohol consumption (Baraona et al. 1973; Baraona and Lieber 1975), despite persistent accumulation of liver fat. Thus the supply of dietary lipids plays a permissive and a contributory role to the development of the fatty liver, but it is not the primary mechanism of this disorder.

Hepatic Synthesis of Lipids

Synthesis of Fatty Acids

Chronic ethanol administration together with low-fat diets produced hepatic accumulation of fatty acids of endogenous origin (Lieber and Spritz 1966; Lieber et al. 1966, 1969). In vitro, a large fraction of the acetate derived from the oxidation of ethanol was incorporated into fatty acids (Brunengraber et al. 1974). But in vivo, most of the acetate produced from ethanol is released into the hepatic vein (Lundquist et al. 1962). Thus the conversion of ethanol into fatty acids appears to be small. The incorporation of acetate from sources other than ethanol (Lieber and Schmid 1961; Gordon 1972) was enhanced by ethanol. However, total fatty acid synthesis assessed by the incorporation of tritiated water was unaffected or even decreased (Guynn et al. 1973; Brunengraber et al. 1974; Scholz et al. 1974; Ontko et al. 1977; Savolainen et al. 1977). Moreover, the malonyl-coenzyme A (CoA) substrate and the enzyme activities involved in the generation of the reduced nicotinamide-adenine dinucleotide phosphate (NADPH) required for the cytosolic fatty acid synthetase complex were also unaffected or decreased (Guynn et al. 1973; Savolainen et al. 1977). Thus the increased incorporation of acetate into fatty acids is most likely due to stimulation of the mitochondrial system for elongation of fatty acids in response to the high reduced nicotinamide-adenine dinucleotide/nicotinamide-adenine dinucleotide (NADH/NAD) ratio generated by ethanol and acetaldehyde oxidation. This may provide a mechanism for attenuation of the redox shift, but the major cytosolic pathway for the synthesis of fatty acids does not appear to be increased by ethanol administration and it may even be slightly decreased.

Synthesis of Glycerolipids

The increased NADH/NAD ratio during ethanol oxidation favors the reduction of dihydroxyacetone phosphate to glycerol-3-phosphate (Nikkilä and Ojala 1963), one of the precursors for the synthesis of glycerolipids (triacylglycerols and phospholipids). Ethanol-fed rats incorporated significantly more labeled glycerol in liver triacylglycerols and phospholipids than pair-fed controls did (Mendenhall et al. 1969), despite the fact that ethanol decreases the hepatic uptake of glycerol (Lundsquist et al. 1965). This was also associated with increased incorporation of choline (Mendenhall and Wilson 1973), presumably to phosphatidylcholine, a change that, at least in the rat, could result in increased requirement for choline. Ethanol administration promoted a rapid increase in enzyme activities which are involved in the synthesis of glycerolipids. The esterification of glycerol-3-phosphate with fatty acyl-CoA involves a series of enzymatic steps which are common to the synthesis of triacylglycerols and phospholipids and result in the release of CoA. The release of CoA from fatty acyl-CoA was found to be increased in microsomes of ethanol-fed rats (Joly et al. 1973). Acute (Savolainen 1977; Pritchard et al. 1977; Lamb et al. 1979) and chronic (Lamb et al. 1979) ethanol administration to rodents increased the activity of phosphatidate phosphohydrolase (PPH), both in microsomes and cytosol. PPH catalyzes the hydrolysis of the phosphatidic acid intermediary, leading to the production of the diacylglycerol moiety of either triacylglycerols or phospholipids. Microsomal activities specifically involved in the synthesis of phospholipids also increased after ethanol administration to rats (Uthus et al. 1976). However, the accumulation of triacylglycerols is much greater than that of phospholipids. This is particularly so in species such as the baboon, in which ethanol feeding did not increase liver phospholipids (Savolainen et al. 1984). The branching of the glycerolipid synthetic pathway toward the formation of triacylglycerols is determined by the activity of another microsomal enzyme, the diacylglycerol acyltransferase (DGAT). Recently, this activity has been found to be increased both after acute and after chronic ethanol administration (Väänänen et al. 1981; Savolainen et al., 1984).

It is not known to what extent the increase in these enzyme activities is secondary to greater availability of substrates or whether this change plays a primary role in promoting fatty acid esterification. It is clear, however, that ethanol results not only in increased availability of fatty acids and glycerol-3-phosphate but also in a rapid increase in the capacity for esterification of these substrates.

Recently, the decrease in the rate of triacylglycerol accumulation during progression from fatty liver to fibrosis in alcohol-fed baboons was found to be associated with abolition of the increases in cytosolic PPH and microsomal DGAT activities. This may play a key role in determining the stabilization and sometimes the decrease of the steatosis with advancing alcoholic liver injury (Savolainen et al., 1984) at the expense of potentially toxic accumulation of fatty acids.

Synthesis of Cholesterol

After chronic ethanol administration, rat liver slices showed enhanced incorporation of labeled acetate into cholesterol (Lefevre et al. 1972). Since this occurred in the withdrawal state, it was considered to reflect proliferation of the endoplasmic reticulum induced by chronic alcohol consumption. This could lead to an increase in hydroxymethylglutaryl coenzyme A (HMG-CoA) reductase activity, the rate-limiting enzyme for cholesterol synthesis.

Another interesting possibility has been raised by the observation that chylomicron remnants prepared from alcohol-fed rat chylomicrons did not inhibit HMG-CoA reductase activity of isolated hepatocytes, as is normally done by those prepared from control rats (Lakshmanan and Ezekiel 1982). This suggests incomplete feedback inhibition of HMG-CoA reductase activity by chylomicron remnants, but the nature of this abnormality has not been clarified.

Enhanced cholesterol synthesis could contribute to the accumulation of cholesteryl esters in the rat liver. However, both alterations have been dissociated in rats fed alcohol in drinking water: both the synthesis of cholesterol (assessed by the incorporation of tritiated water) and the activity of HMG-CoA reductase were unchanged despite accumulation of cholesteryl esters (Lakshmanan and Veech 1977; Lakshmanan et al. 1978), suggesting a defect in disposition of these esters.

Disposition of Lipids from the Liver

Liver lipids are normally disposed of by oxidation, biliary excretion, or release into plasma lipoproteins. The participation of the various pathways depends on the type of lipid.

Fatty Acid Oxidation

One of the most consistent effects of ethanol on hepatic lipid metabolism is the inhibition of fatty acid oxidation. This has been documented in liver slices (Lieber and Schmid 1961; Rebouças and Isselbacher 1961; Blomstrand et al. 1973), perfused livers (Lieber et al. 1967), isolated hepatocytes (Ontko 1973), and in vivo (Blomstrand and Kager 1973). Fatty acids are oxidized predominantly in the mitochondria through a process of β-oxidation which leads to the formation of acetyl-CoA. This in turn is further oxidized to CO_2 in the citric acid cycle or, alternatively, utilized for the synthesis of ketone bodies and other products. Initially, the impairment of fatty acid oxidation is linked to the inhibition of NAD-dependent steps of the citric acid cycle, due to the excessive generation of NADH from cytosolic ethanol oxidation and mitochondrial acetaldehyde oxidation (Forsander et al. 1965a, b; Lieber et al. 1967; Williamson et al. 1969). In essence, the reducing equivalents generated by the oxidation of ethanol compete with those supplied by normal fuels such as fatty acids. In addition, the redox shift induced by ethanol oxidation inhibited β-oxidation, promoting accumulation of long chain fatty acyl-CoA that can be used for the synthesis of glycerolipids or cholesteryl esters.

After chronic alcohol consumption, however, the shift in redox state produced by ethanol oxidation was attenuated (Domschke et al. 1974; Salaspuro et al. 1981) and the acute effect of ethanol on fatty acid oxidation decreased (Salaspuro et al. 1981). An exception may take place at perivenular zones of the liver, where a greater redox shift is favored by the low oxygen tensions prevailing in that zone (Jauhonen et al. 1982; Baraona et al. 1983). These oxygen tensions become even lower with progression of the liver injury due to impairment of the sinusoidal circulation. In addition, chronic alcohol consumption produces structural (Svoboda and Manning 1964; Iseri et al. 1966) and functional alterations of the mitochondria. The mitochondria showed decreased oxidation of two-carbon fragments of fatty acids (Rubin et al. 1972; Gordon 1973) but no changes in the β-oxidation (Cederbaum et al. 1975). Thus the development of the mitochondrial damage converts an acute inhibitory effect on fatty acid oxidation (which depends on the concomitant oxidation of ethanol) into a chronic one that persists even in the absence of ethanol. In the withdrawal state, the lack of inhibitory effects on β-oxidation of fatty acids determines that an important fraction of the accumulated acetyl-CoA is diverted to the formation of ketone bodies. This leads to the ketonemia and ketonuria observed in alcoholics, particularly in the withdrawal state (Lefevre et al. 1970).

Lipolysis

The fatty acids derive not only from endogenous synthesis and uptake of plasma FFA, but also from hydrolysis of triacylglycerols, cholesteryl esters, and phospholipids either taken up from the plasma [as remnants of chylomicrons and very low density lipoproteins (VLDL)] or synthesized in the liver. The utilization of exogenous and endogenous triacylglycerols is carried out by a process of autophagocytosis and hydrolysis by an acid lipase of the lysosomes. The product of this reaction is probably monoglycerides, which are further hydrolyzed by a microsomal hydrolase. Although direct information on the effects of ethanol on these processes is lacking, indirect evidence suggests that inhibited lipolysis does not seem to play a primary role in the pathogenesis of the fatty liver. Indeed, alcoholic fatty liver is associated with prominence of autophagic vacuoles and lipolysosomes. Moreover, lysosomal enzyme activities increased after prolonged alcohol consumption (Mezey et al. 1976). The enhancement of lipolysis, especially during fasting, may aggravate alcoholic ketosis.

The accumulation of cholesteryl esters in the rat liver after alcohol feeding most likely reflects a defect in disposition. In ethanol-fed rats, the microsomal esterifying activity [presumably due to acyl cholesterol acyltransferase (ACAT)] was found to be unchanged, whereas the hydrolysis of cholesteryl esters by cytosolic fractions was decreased (Takeuchi et al. 1974). This suggests an imbalance between synthesis and hydrolysis of cholesteryl esters.

Biliary Excretion

The main lipids excreted in the bile are phospholipids and cholesterol. Part of the cholesterol is excreted unchanged, but the major fraction is converted into bile salts prior to excretion. Acute ethanol administration has consistent inhibitory effects on bile flow and output of biliary salts and cholesterol. This has been shown in perfused livers (Schapiro et al. 1964), as well as in vivo, when ethanol was given to either naive or ethanol-fed rats (Maddrey and Boyer 1973), bile fistula dogs (Marin et al. 1975), and normal or alcoholic volunteers (Marin et al. 1973). The mechanism of this effect is unknown. Partial prevention of this inhibition by bile salt infusion (Marin et al. 1975) suggests a primary alteration in bile salt synthesis. By contrast, when rats chronically fed ethanol were studied during the withdrawal state, bile flow and excretion of biliary lipids increased (Boyer 1972; Maddrey and Boyer 1973). This could be a rebound response following the suppression of the inhibitory effect of ethanol. Thus, during chronic alcohol consumption, the balance between these two opposite actions will determine the net effect.

In rats consuming alcohol-containing diets, there was a significant decrease in the turnover rate of both cholic and chenodeoxycholic acids, and in the daily excretion of cholic acid (Lefevre et al. 1972). Moreover, ethanol feeding suppressed the increase in the synthesis of both bile acids that normally follows the administration of cholesterol. This cholesterol supplementation markedly exacerbated the accumulation of cholesteryl esters in the liver in spite of the fact that cholesterol feeding inhibits hepatic synthesis of cholesterol. Thus the inhibition of bile acid production (rather than changes in cholesterol synthesis) plays a predominant role in determining the hepatic accumulation of cholesteryl esters. The mechanism of this inhibition is not fully clarified. Chronic administration of alcohol in drinking water decreased the activity of the microsomal 7α-hydroxylase (Lakshmanan and Veech 1977), which is considered to be the rate-limiting step in bile acid synthesis. However, this activity was not affected by acute ethanol administration, suggesting that the acute inhibitory effects may take place at another step of bile acid synthesis. In addition, preliminary results reported by Lefevre et al. (1972) indicate that ethanol feeding did not prevent the increase in bile acid synthesis that follows bile diversion, a condition in which 7α-hydroxylase markedly increases. Some steps in the synthesis of cholic acid appear to be catalyzed by alcohol dehydrogenase (Okuda and Okuda 1983) and it is possible that ethanol may compete with the bile salt intermediary for a common enzyme. This might explain the enlarged bile acid pool observed after chronic alcohol consumption (Lefevre et al. 1972). An alternate possibility could be increased reabsorption of bile salts from the intestine with secondary inhibition of bile acid synthesis.

There is considerable evidence that biliary cholesterol and bile acids arise from separate sterol pools. Using cholesterol as a tracer, Cohen and Raicht (1981) found that chronic ethanol feeding in liquid diets significantly increased the fecal excretion of neutral steroids (which includes cholesterol and its degradative colonic products), whereas the excretion of acidic steroids (which include bile salts and their colonic products) was unchanged. Since the balance between input and output of cholesterol was negative, this was attributed to increased cholesterol synthesis with a relative inhibition of bile acid synthesis.

There are marked variations among species in the cholesterol precursor for bile acid synthesis: while in rats newly synthesized cholesterol contributes significantly to bile acid synthesis, in man and non human primates the main precursor of both bile acids and biliary cholesterol is high-density lipoprotein (HDL) cholesterol, particularly the unesterified fraction (Schwartz et al. 1978). This may account for the lack of significant cholesteryl ester accumulation in the liver of some primates. Cholesterol balance studies indicate that alcohol consumption increases the excretion of fecal steroids, but the contribution of the neutral or acidic components varies with the species. In ethanol-fed baboons, there was a significant increase in fecal excretion of neutral steroids, with a marked tendency for an increase also in acidic steroids (Baraona et al. 1984). Administration of smaller doses of ethanol to pigs (Topping et al. 1982) and to patients with alcoholic hyperlipemia (Nestel et al. 1976) increased the fecal excretion of acidic steroids with variable changes in total steroid excretion. The latter effects were not reproduced by alcohol administration to normolipemic subjects (Nestel et al. 1976). These changes in the cholesterol/bile salt excretion ratio may be responsible for the decreased lithogenic index in the gallbladder bile of volunteers given small doses of alcohol (Thornton et al. 1983).

Secretion of Plasma Lipoproteins

Most agents that produce fatty liver (choline and protein deficiencies, protein synthesis inhibitors, carbon tetrachloride, ethionine, orotic acid, etc.) interfere with the formation and release of plasma lipoproteins and produce hypolipemia. Ethanol is an exception: it increases the output of triacylglycerol-carrying lipoproteins from the liver both in man (Wolfe et al. 1976) and experimental animals (Baraona and Lieber 1970; Baraona et al. 1973, and 1983). In fact, this is one of the major mechanisms for the alcoholic hyperlipemia (see below), and this may play some compensatory role in diminishing fat accumulation. However, this compensation is relatively inefficient, probably because ethanol also impairs the secretion of plasma proteins (Baraona et al. 1977). This impairment has been linked in part to disruption of liver microtubules (Baraona et al. 1977; Matsuda et al. 1979; Baraona et al. 1981), cytoskeletal organelles the integrity of which is required for normal secretion. This secretion defect aggravates but probably does not initiate the fatty liver.

Possible Consequences of Alcoholic Fatty Liver

Thus far, there is no convincing evidence that the accumulation of fat by itself has a deleterious effect on liver functions. When the accumulation is excessive, fat droplets apparently fuse, leading to formation of fatty cysts surrounded by several cell bodies. It has been claimed that they may become extracellular and produce fat embolism in lung and brain (Lynch et al. 1959). In addition to fat accumulation, the early manifestations of alcoholic liver injury include other alterations affecting protein metabolism, subcellular organelles, and mesenchyma, which are probably the major determinants of the progression of the lesions. Steatosis enlarges the hepatocytes and contributes to the hepatomegaly. In ethanol-fed rats, at least half of the increase in liver dry weight was due to fat, the other half being accounted for by increases in both structural and retained secretory proteins (Baraona et al. 1975, 1977). In the baboons fed alcohol for a much longer time, fat accumulation largely exceeds the increase in protein. The enlargement of the hepatocyte may increase the resistance to the sinusoidal flow and may contribute to the development of portal hypertension (Israel et al. 1979; Orrego et al. 1981).

Alcoholic Hyperlipemia

Alcohol consumption is associated with changes in all the lipid components of the plasma. Hypercholesterolemia was first reported by Ducceschi in 1915 and has recently promoted great interest because of the possible link between moderate alcohol consumption and prevention of coronary heart disease. Serum lactescence and hypertriglyceridemia after bouts of excessive drinking were first reported by Feigl in 1918. These episodes were sometimes associated with pancreatitis (Albrink and Klatskin, 1942) or with hemolytic anemia (Zieve 1958). Milder degrees of hyperlipemia were found to be rather common in alcoholics with fatty liver, but less so in patients with cirrhosis (Cachera et al. 1950).

Plasma Free Fatty Acids

The concentration of plasma FFA is very variable in alcoholics, depending mainly on blood ethanol concentrations and degree of alcoholic intoxication. Administration of moderate doses of ethanol produces a rapid fall in FFA concentration (Lieber et al. 1962), associated with decreases in both FFA turnover (Jones et al. 1965) and plasma glycerol concentration (Feinman and Lieber 1967). These effects are mediated by the increase in plasma acetate, which inhibits lipolysis of triacylglycerols in the adipose tissue (Crouse et al. 1968; Abramson and Arky 1968; Nilsson and Belfrage 1978). By contrast, the production of blood ethanol levels over 200 mg/dl (which are associated with overt intoxication) increased plasma FFA (Lieber et al. 1963; Schapiro et al. 1965). An initial fall followed by a secondary rise in FFA was also encountered (Kaffarnik and Schneider 1970). The stimulatory response was inhibited either by β-adrenergic blocking agents (Bouchier and Dawson 1964; Ammon and Estler 1966) or by nicotinic acid (Kaffarnik et al. 1978), and appears to be mediated by catecholamine-induced stimulation of lipolysis. Therefore, the net effect of alcohol will depend on the predominance of either the inhibitory or the stimulatory effects on lipolysis.

Plasma Triacylglycerols

The plasma triacylglycerols are the major determinants of the serum lactescence or turbidity found after bouts of excessive drinking. They increase in all lipoprotein fractions, but proportionally more in those lipoproteins which are normally rich in triacylglycerols: VLDL, which migrate as pre-β-lipoproteins on electrophoresis, and chylomicrons or chylomicron-like particles, which stay at the origin of the electrophoretic strip. Thus alcoholic hyperlipemia is usually classified as type IV or type V. However, the phenotypic pattern changes rapidly after alcohol withdrawal, from type V to type IV and to type II, because of the rapid clearance of chylomicrons, followed by VLDL and the slower clearance of cholesterol and phospholipids which predominate in low-density (LDL) or β-lipoproteins.

The incidence of hypertriacylglyceridemia varies with the population studied. Among hyperlipemic patients, alcohol constitutes the second major cause of secondary hyperlipemia, following diabetes (Chait et al. 1972). In epidemiologic studies, plasma triacylglycerol levels correlate with alcohol consumption (Ostrander et al. 1971; Castelli et al. 1977). But significant increases in plasma triacylglycerols are rather rarely encountered in patients hospitalized for alcoholism or its complications. Only 27%–28% of hospitalized alcoholics had fasting triacylglycerol levels over normal limits (2 m*M*) and 17% over 3 m*M*, with phenotypes IV, II, or V in order of frequency (Sirtori et al. 1973; Böttiger et al. 1976). There is an even rarer occurrence among cirrhotics (Cachera et al. 1950; Marzo et al. 1970; Patek and Earampamoorthy 1976).

Plasma Cholesterol

In patients with marked hypertriacylglyceridemia, plasma cholesterol is also increased. Cholesterol is a component of all lipoprotein fractions, but most of it is transported in LDL or β-lipoproteins and in HDL or α-lipoproteins. In hospitalized alcoholics, total plasma cholesterol was not significantly increased (Böttiger et al. 1976). However, in about 30% of alcoholics seeking medical attention (Johansson and Laurell 1969) and 86% of patients after a recent drinking bout (Johansson and Medhus 1974), a prominent α-lipoprotein band was found in electrophoresis. This alteration normalized after approximately 2 weeks of abstinence. By contrast, β-lipoproteins or LDL tend to be reduced in alcoholics (Taskinen et al. 1982). These changes in cholesterol-carrying lipoproteins are also shown in epidemiologic studies: there is a strong positive correlation between alcohol consumption and plasma HDL cholesterol, and a weaker but significant negative correlation with LDL cholesterol (Castelli et al. 1977; Hulley and Gordon 1981).

Since only few other factors (chronic administration of some drugs and vigorous physical exercise) are capable of increasing HDL cholesterol, this has been proposed as one of the useful biological markers of alcoholism (Danielsson et al. 1978; Barboriak et al. 1980; Sanchez-Craig and Annis 1981). This increase occurs in women (who normally have higher HDL cholesterol) as well as in men, indicating that this is not merely a consequence of the feminization of the alcoholics. The elevation of HDL cholesterol correlates not only with alcohol consumption, but also with mild increases in serum glutamic oxaloacetic transaminase and gamma-glutamyltranspeptidase activities, reflecting moderate liver injury (Kuller et al. 1983). By contrast, alcoholics with advanced liver disease failed to show the increase in HDL cholesterol shortly after their last drink, nor did they show a significant change after abstinence (Devenyi et al. 1981).

HDL is a heterogeneous family of lipoproteins of different sizes, composition, and metabolic characteristics. In recent alcoholics, the increase occurs predominantly in the HDL2 subfraction (Ekman et al. 1981; Taskinen et al. 1982), which can be separated at buoyant densities between 1.063 and 1.121, rather than in HDL3 (density between 1.121–1.210). However, moderate alcohol consumption (approximately 50 g/day) increased HDL3 without affecting HDL2 (Haskell et al. 1984). A minor subfraction that floats as HDL, although it contains apoprotein B (with a unique antigenic determinant), is named Lp(a) and has been found to be markedly decreased in heavy alcohol consumers (Marth et al. 1982).

Etiologic Role of Ethanol

The administration of an ethanol dose which results in blood concentrations over 100 mg/dl and mild intoxication increased triacylglycerols in the plasma of normal volunteers (Jones et al. 1963; Verdy and Gattereau 1967; Kaffarnik and Schneider 1970; Belfrage et al. 1973; Avogaro and Gazzolato 1975; Taskinen and Nikkilä 1977), whereas the administration of smaller doses did not (Friedman et al. 1965). In about 25% of the subjects drinking a "cocktail dose" of alcohol (1 g/kg) in the evening, the fasting serum samples obtained next morning can be diagnosed as type IV hyperlipemia (Taskinen and Nikkilä 1977). Thus the antecedent of alcohol ingestion has to be investigated in patients with hyperlipemia. Acute ethanol administration does not, however, produce significant changes in plasma cholesterol.

The effects on plasma triacylglycerols are enhanced and those on plasma cholesterol can be produced by chronic ethanol administration. Either as a supplement to a normal diet (Lieber et al. 1963; Schapiro et al. 1965; Belfrage et al. 1977) or in substitution for other calories (Losowsky et al. 1963), ethanol produced a fourfold increase in plasma triacylglycerols and lesser increases in cholesterol and phospholipids. This required several days of administration and blood ethanol concentrations greater than 100 mg/dl. Administration of smaller doses to normal subjects did not produce significant hyperlipemia (Glueck et al. 1980), but elevated HDL cholesterol (Belfrage et al. 1973, 1977; Fraser et al. 1983; Hártung et al. 1983). It must be pointed out that the ethanol intake experimentally effective is generally higher than that admitted by drinkers on questioning.

The lipemic response to chronic ethanol consumption in human volunteers is usually transient: after reaching maximal levels in several days or weeks, serum lipids decrease and even normalize despite continuation of the ethanol intake (Lieber et al. 1963; Schapiro et al. 1965; Belfrage et al. 1977; Schneider et al. 1983).

Most of these effects of ethanol can also be produced in experimental animals, although there seem to be differences among species. Whereas a consistent lipemic response has been produced by acute administration of ethanol to rabbits (Bezman-Tarcher et al. 1966; Warembourg et al. 1970), the acute effects in rats and mice have been variable. Only after chronic administration of ethanol-containing diets (with or without an acute ethanol dose) was significant postprandial hyperlipemia observed in rats (Baraona and Lieber 1970; Baraona et al. 1973). Lipids increased in all lipoprotein fractions, including HDL. The latter was associated with a significant increase in HDL cholesterol, which persisted during fasting (Hirayama et al. 1979). In nonhuman primates (which have lipoprotein composition and metabolism more akin to those of man), chronic ethanol administration produces a sustained increase in fasting triacylglycerol, cholesterol, and phospholipid levels in the plasma, at least during early stages of alcoholic liver injury (Vasdev et al. 1974; Leathers et al. 1981; Baraona et al. 1983).

Role of Nutritional Factors

The lipemic response to acute ethanol administration to volunteers is greatly enhanced by dietary fat (Talbott and Keating 1962; Brewster et al. 1966; Barboriak and Meade 1968). The increase includes both VLDL and chylomicrons (Wilson et al. 1970). Thus the usual fasting conditions employed for sampling tend to decrease the triacylglycerol levels in alcoholics. Cholesterol levels are much less affected by fasting. An enhanced triacylglycerol response can be provoked in alcoholics at early stages of liver injury by administration of dietary fat even when fasting levels of triacylglycerols have normalized after several days of alcohol withdrawal (Borowsky et al. 1980; Avgerinos et al. 1983). Thus the incidence and magnitude of the hyperlipemia can be greatly influenced by the dietary habits of the alcoholic.

Role of Underlying Alterations of Lipid Metabolism

The existence of individuals with enhanced susceptibility to the lipemic effects of ethanol (simulating other forms of chronic hyperlipemia) has long been recognized. This condition is generally suspected by the lack of response to the usual treatment for hyperlipemia and the improvement after alcohol withdrawal. The role of alcohol is indicated by a prompt reduction of at least 25% in triacylglycerol concentrations within 10 days of alcohol withdrawal (Lewis et al. 1973; Janus and Lewis 1978). Alcohol consumption may even unmask a subclinical defect in lipid metabolism. The defects most commonly aggravated by alcohol consumption are reduced ability to remove serum lipids and overproduction of VLDL. The first type of alteration is found in subjects with primary (familial) or secondary (usually due to diabetes) decreased activity of lipoprotein lipase (Losowsky et al. 1963; Chait et al. 1972). Occasionally, this may result in massive hypertriacylglyceridemia and manifestations of "chylomicronemia syndrome", such as abdominal pain (Bloomfield and Shenson 1947) with or without pancreatitis (Albrink and Klatskin 1957; Greenberger et al. 1966; Nestel 1967), lipemia retinalis (Grossberg 1966), and eruptive xanthoma (Nestel 1967; Chait et al. 1972).

In another group of individuals, alcohol aggravates either a primary or secondary (obesity) overproduction of VLDL (type IV) or both VLDL and LDL (type IIb) (Mendelsson and Mello 1973, 1974; Ginsberg et al. 1974; Taskinen and Nikkilä 1977). A large percentage of patients with marked and sustained alcoholic hyperlipemia have relatives with hyperlipemia (DeGennes et al. 1972). The participation of an underlying defect in lipid metabolism should be suspected in any alcoholic with intense hyperlipemia.

Role of Associated Liver Injury

The inverse relationship between the magnitude of the alcoholic hyperlipemia and the severity of the liver injury was early noticed (Cachera et al. 1950; Marzo et al. 1970). In baboons fed an alcohol-containing diet for several years, the hypertriacylglyceridemia and the hypercholesterolemia decreased or disappeared during the progression from alcoholic fatty liver to the fibrotic stages (Savolainen et al. 1984). Available chemical tests are insensitive indicators of this progression, raising the possibility of a potential usefulness of the serum lipid changes. In patients with cirrhosis, there was decreased incorporation of labeled fatty acids into serum triacylglycerols (Alexander et al. 1963). This was considered useful for the differentiation between hepatocellular and obstructive jaundice (Santos et al. 1974), a differential diagnosis that can generally be made with more reliable and less aggressive tests.

In a small series of alcoholics, it was noted that the densitometric quantitation of pre-β-lipoproteins in agarose gel electrophoresis allowed differentiation between recent alcoholics with precirrhotic stages and those with only fatty liver (Borowsky et al. 1980), but this simple test requires evaluation in much larger series. At more advanced stages, striking alterations of serum lipoproteins occur, regardless of the etiology of the liver injury. They also occur during acute impairment of the liver function, such as viral or alcoholic hepatitis (Sabesin et al. 1977). Electrophoretically, these alterations are characterized by marked decrease or disappearance of both pre-β- and α-lipoproteins with appearance of a broad β-band (Papadopoulus and Charles 1970) and, chemically, by a decreased cholesteryl ester/free cholesterol ratio. These changes are associated with decreased lecithin: cholesterol acyltransferase (L-CAT) activity (Day et al. 1979), a serum enzyme produced by the liver which catalyzes the esterification of free cholesterol in the HDL fraction. HDL becomes discoidal due to paucity of cholesteryl ester core. The mechanism of the changes is, however, multifactorial. The major apoprotein of HDL, which is an activator of L-CAT, namely the apo A-I, is markedly decreased. This is probably due to rapid degradation (Nestel et al. 1980). HDL is also enriched in apo E, which forms abnormal complexes (Tada et al. 1979); this suggests altered transfer of this apoprotein to remnants of chylomicrons and VLDL, a process which is required for the apo E receptor-mediated uptake of these remnants by the liver. As a consequence, marked hypertriacylglyceridemia (of a type III phenotype) may appear in patients with severe liver disease.

Pathogenesis

Hyperlipemia results when the rate of entry of lipoprotein lipids into the plasma exceeds the rate of exit from the plasma. Plasma triacylglycerols and cholesterol are secreted by the liver and the intestine, whereas fatty acids and cholesterol are also produced by extra splanchnic tissues. Thus we will consider the participation of these three sources in the serum lipid changes produced by alcohol.

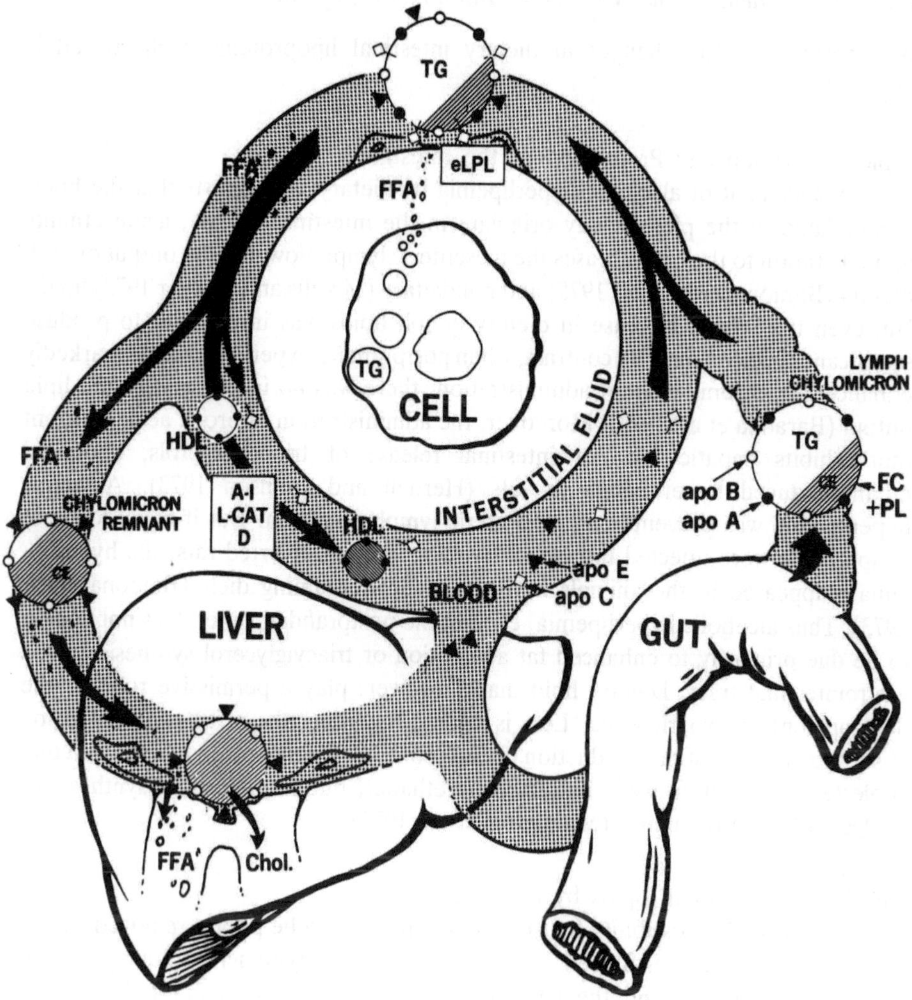

Fig. 2. Scheme of chylomicron metabolism. During fat absorption, nascent chylomicrons migrate through the lymph into the blood. They carry triacylglycerols (*TG*) and cholesteryl esters (*CE*) (of dietary origin) in a lipid core surrounded by a shell of apo A, apo B, free cholesterol (*FC*), and phospholipids (*PL*). Two apoproteins which are components of the plasma HDL fraction (a family of discrete lipoproteins) are bound to chylomicrons during circulation: the apo C, which contains the activator (apo C-II) of the extrahepatic lipoprotein lipase (*eLPL*), and the apo E. In the capillary endothelia, the activated lipase promotes hydrolysis of the triacylglycerols to free fatty acids (*FFA*) and leaves a remnant particle enriched in cholesteryl esters and apo E. This is recognized by apo E-receptors of the hepatocyte for endocytic uptake. As the lipid core of the chylomicrons shrinks, apo C and apo A with free cholesterol and phospholipids are released into the HDL fraction, where the free cholesterol is esterified with phospholipid fatty acids by the lecithin:cholesterol acyltransferase (*L-CAT*), generating a cholesteryl ester-enriched HDL. L-CAT, an enzyme secreted by the liver, circulates in the plasma as a complex with apo A-I and apo D. Thus HDL acts as an scavenger of shell components and as a source of recognition factors for hydrolysis and uptake of chylomicrons

Changes in Dietary and Nondietary Intestinal Lipoproteins

An overview of the changes in dietary intestinal lipoproteins is illustrated in Fig. 2.

Lipid Absorption and Production by the Intestine

The enhancement of alcoholic hyperlipemia by dietary fat suggests that the lipids accumulated in the plasma may originate in the intestine. In fact, acute ethanol administration to the rat increases the mesenteric lymph flow and the output of both dietary (Baraona and Lieber 1975) and nondietary (Mistilis and Ockner 1972) lipids. But even the largest increase in dietary lymph lipids was insufficient to produce significant hyperlipemia. By contrast, when postprandial hyperlipemia was markedly enhanced by chronic ethanol administration, there was no increase in lymph lipid output (Baraona et al. 1973). Moreover, the administration of orotic acid, an agent that inhibits hepatic but not intestinal release of triacylglycerols, abolished ethanol-induced hyperlipemia in rats (Hernell and Johnson 1973). Alcoholic hyperlipemia was prevented by mesenteric lymph diversion. But if equal loads of lymph lipids were injected intravenously to the lymph-diverted rats, the hyperlipemia reappeared in the animals fed the ethanol-containing diets (Baraona et al. 1973). Thus alcoholic hyperlipemia, even in the postprandial state, does not appear to be due primarily to enhanced fat absorption or triacylglycerol synthesis by the gastrointestinal tract. Dietary lipid may, however, play a permissive role for the development of hyperlipemia. Less is known regarding the effects of alcohol on absorption and intestinal production of cholesterol. Experiments in rats showed that cholesterol absorption was unaffected by ethanol, but the intestinal synthesis of cholesterol was stimulated (Middleton et al. 1971).

Removal of Intestinal Lipids by the Liver

More recently, the possibility that intestinal lipids could be poorly removed by the liver has been proposed. Indeed, after the injection of chylomicrons doubly labeled in the triacylglycerol and the cholesteryl ester moities to alcohol-fed rats, the clearance of chylomicron-cholesteryl esters was impaired to a greater extent than the clearance of chylomicron-triacylglycerols (Redgrave and Martin 1977; Lakshmanan and Ezekiel 1982). These observations are consistent with the possibility that after hydrolysis of the chylomicron-triacylglycerols by extrahepatic lipoprotein lipases, the cholesteryl ester-enriched remnants are poorly removed by the liver. These remnants are taken up by the liver by a process of endocytosis following the binding to a surface receptor specific for apoprotein E. Thus this alteration could be due to a change in the receptor or in the apo E. Alternatively, it could be due to saturation of the system by catabolic products of endogenous VLDL which share similar uptake mechanisms. To investigate this problem more directly, we recently measured the hepatic extraction of labeled chylomicron and VLDL-triacylglycerols during passage through the splanchnic vascular bed of baboons. In alcohol-fed baboons, we found that the hepatic removal of these particles was increased rather than decreased (Baraona et al. 1983). This favors the possibility that an excessive supply of products of both endogenous and dietary origin saturate the hepatic uptake, leading to their accumulation in the plasma during the postprandial state.

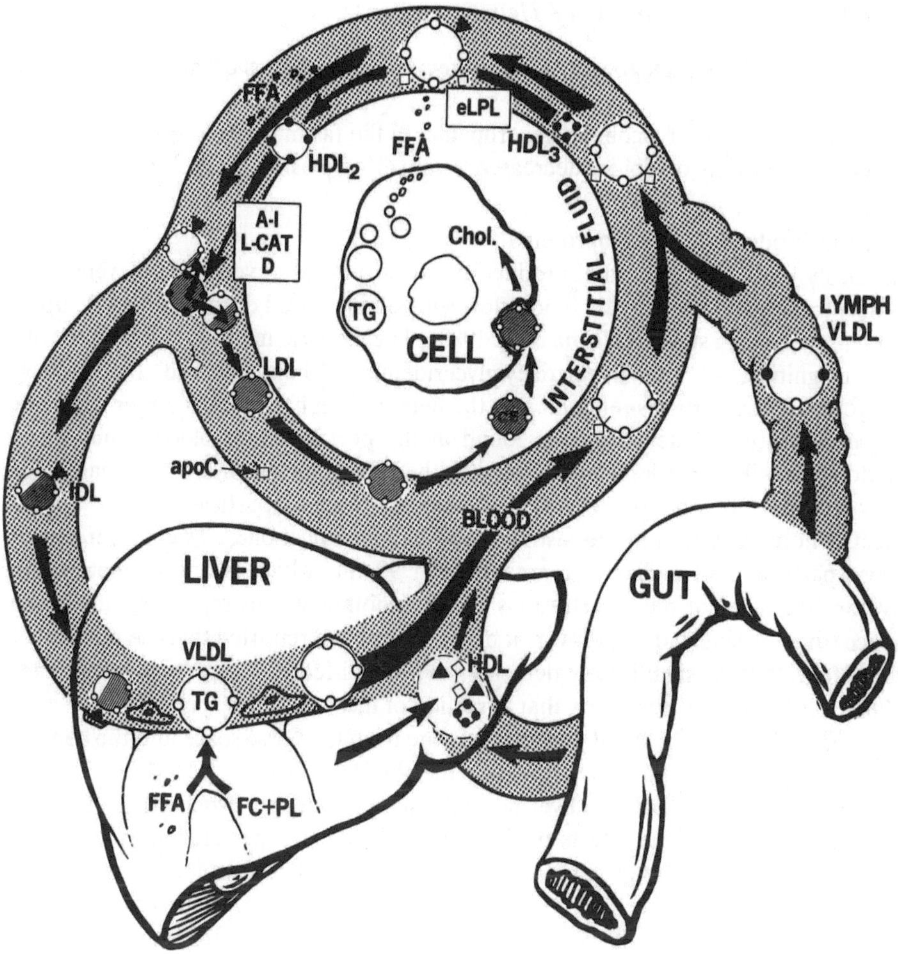

Fig. 3. Scheme of metabolism of lipoprotein of endogenous origin. The liver provides a constant supply of fatty acids, cholesterol, and phospholipids by secreting very low density lipoproteins (*VLDL*). They contain a triacylglycerol (*TG*) core surrounded mainly by apo B, free cholesterol (*FC*), and phospholipids (*PL*). Smaller amounts of VLDL (containing apo A and B) are secreted by the intestine into the lymph even in the fasting state. Both organs also produce HDL, rich either in apo E (the liver) or apo A (the intestine). The catabolism of VLDL is similar to that of chylomicrons. Particles of intermediate density between VLDL and LDL (*IDL*) are removed by the liver, but most of them complete their conversion to LDL. In man, most of the LDL-cholesteryl esters (*CE*) (major source of exogenous cholesterol for the cells) are transferred from the HDL fraction (where free cholesterol is esterified). Free cholesterol esterification and cholesteryl ester transfer are probably associated with a common L-CAT: lipoprotein complex. In rats, which lack the transfer mechanism, these esters are initially secreted in the VLDL core. LDL goes into the interstitial spaces and is recognized by apo-B receptors for endocytic uptake by most of the cells. The receptor-mediated uptake downregulates endogenous cholesterol synthesis by inhibiting HMG-Co reductase

Metabolism of Lipoproteins of Hepatic Origin

An overview of the metabolism of lipoproteins of endogenous origin is illustrated in Fig. 3.

The persistence of alcoholic hyperlipemia in the fasting state suggests increased hepatic production of lipids, decreased removal of plasma lipids, or both.

Hepatic Production of Lipoproteins

In alcohol-fed baboons, the product of arterio-hepatic venous differences in concentration and the hepatic blood flow showed increased output of VLDL lipids even in the fasting state (Baraona et al. 1983). The increase in output correlated with the magnitude of the hypertriacylglyceridemia. When assessed by electron microscopy, the particles released into the hepatic vein had a much larger size than those of controls. They were not found in the portal venous blood. Thus these chylomicron-like particles were secreted by the liver of the alcohol-fed animals. This may account for the observation of chylomicron-like particles and a type V electrophoretic pattern in the fasting blood of some alcoholics. Twenty percent of these particles had a diameter greater than 90 nm, which is approximately the average diameter of the fenestrations of the endothelial cells separating the Disse space from the sinusoid. Moreover, it has recently been reported that the number of these fenestrations significantly decreases in ethanol-fed baboons (Mak and Lieber 1984). This raises the possibility that a fraction of these particles may not be released into the hepatic venous blood, but may migrate from the Disse space into the hepatic lymphatics, reaching the blood through the thoracic duct.

In addition to VLDL, the liver normally secretes an apo-E-rich HDL. There is little information on the effects of alcohol on the hepatic production of HDL. In alcoholics, as well as after chronic administration of phenobarbital, a correlation between the plasma levels of HDL and hepatic microsomal activities has been pointed out (Cushman et al. 1982; Luoma et al. 1983). Nascent HDL may contribute to the increase in HDL3. However, a cause-effect relationship for this association has not yet been established.

The increased output of VLDL-cholesterol from the liver of alcohol-fed baboons was markedly exceeded by enhanced hepatic removal of cholesterol from other lipoprotein fractions (Karsenty et al., 1985). This suggests that alcoholic hypercholesterolemia must have predominantly an extrahepatic origin.

Removal of Plasma Lipoproteins

The mechanisms of removal are similar for hepatic and intestinal lipoproteins (Figs. 2 and 3). Most of the early studies focused on the total lipoprotein lipase activity released into the plasma after injection of heparin. In general, no changes in this activity were found after acute (Barboriak 1966; Verdy and Gattereau 1967; Wilson et al. 1970) or chronic (Kudzma and Schonfeld 1971) ethanol administration. However, heparin releases two types of enzymes (the extrahepatic and the hepatic lipoprotein lipases), which could be differently affected by ethanol. Therefore, recent studies have focused either on tissue measurements of lipoprotein lipase activity or on postheparin plasma measurements after separation of the hepatic and extrahepatic components. In man (Jaillard et al. 1974; Nikkilä et al. 1978;

Nilsson-Ehle et al. 1978; Schneider et al. 1983) and in animals (Olivecrona et al. 1972; Jaillard et al. 1973; Giudicelli et al. 1975; Johnson and Hernell 1975), a large dose of ethanol can produce a slight decrease in extrahepatic lipoprotein lipase activity and some delay in the clearance of serum triacylglycerols. Most of the delayed clearance, however, appears to be secondary to the excessive supply of endogenous and exogenous lipids competing for the lipases (Schneider et al. 1983). In keeping with this interpretation, in the rat, alcoholic hyperlipemia persisted even when the removal of lipids from the blood was completely blocked by the administration of Triton WR-1339 (p-isooctylpolyoxyethylenephenol polymer) (Baraona and Lieber 1970). Also, even in those alcoholics with intense type V hyperlipemia (in whom lipoprotein lipase could be impaired), the turnover rates of VLDL-apo B and VLDL-triacylglycerols are increased rather than decreased (Sigurdsson et al. 1976). This reflects the predominant role of the increased lipoprotein production over decreased removal.

By contrast, chronic alcohol consumption has been found to increase the activity of extrahepatic lipoprotein lipase in the majority of subjects (Belfrage et al. 1977; Ekman et al. 1981; Taskinen et al. 1982). This change may represent a compensatory adaptation to the increased supply of plasma lipids, which would increase their removal. This may account for the transient nature of the hypertriacylglyceridemia during prolonged alcohol administration (Lieber et al. 1963; Belfrage et al. 1977). Mordasini et al. (1982) recently documented that the rise in plasma triacylglycerols (mainly VLDL) is associated with an initial decrease in lipoprotein lipase activity, whereas the return of triacylglycerols to normal levels (despite continuous ethanol intake) is associated with increased lipase activity. This adaptation may also account for the modest degree of hypertriacylglyceridemia found in the majority of alcoholics, and the paradoxic increase in VLDL and LDL during the initial period of abstinence (Ekman et al. 1981). In some cases, the transiency of the hypertriacylglyceridemia may be also due to progressive impairment of the liver function.

The enhanced production and metabolism of VLDL after chronic alcohol consumption may have important consequences for plasma cholesterol, since VLDL (as well as chylomicrons) supplies free cholesterol (which is released from the shell together with some apolipoproteins) to the HDL fraction of the plasma during hydrolysis of the triacylglycerol core. This process is associated with the conversion of HDL3 into HDL2. Thus the increased lipolysis and supply of lipoprotein-free cholesterol seems to be the major determinant of the increased HDL2 in alcoholics. HDL3 is believed to originate from both nascent HDL and from HDL2. The hepatic lipoprotein lipase, an enzyme located in the surface of hepatic endothelial cells, is thought to regenerate HDL3 from HDL2 (Nikkilä et al. 1982). This hepatic activity was found to be unchanged (Ekman et al. 1981) or increased (Taskinen et al. 1982) in alcoholics with mild liver injury, but it may decrease in more severe forms of liver impairment, such as in alcoholic hepatitis (Freeman et al. 1977).

Reverse Transport of Cholesterol

An overview of the physiology of the reverse transport of cholesterol is given in Fig. 4.

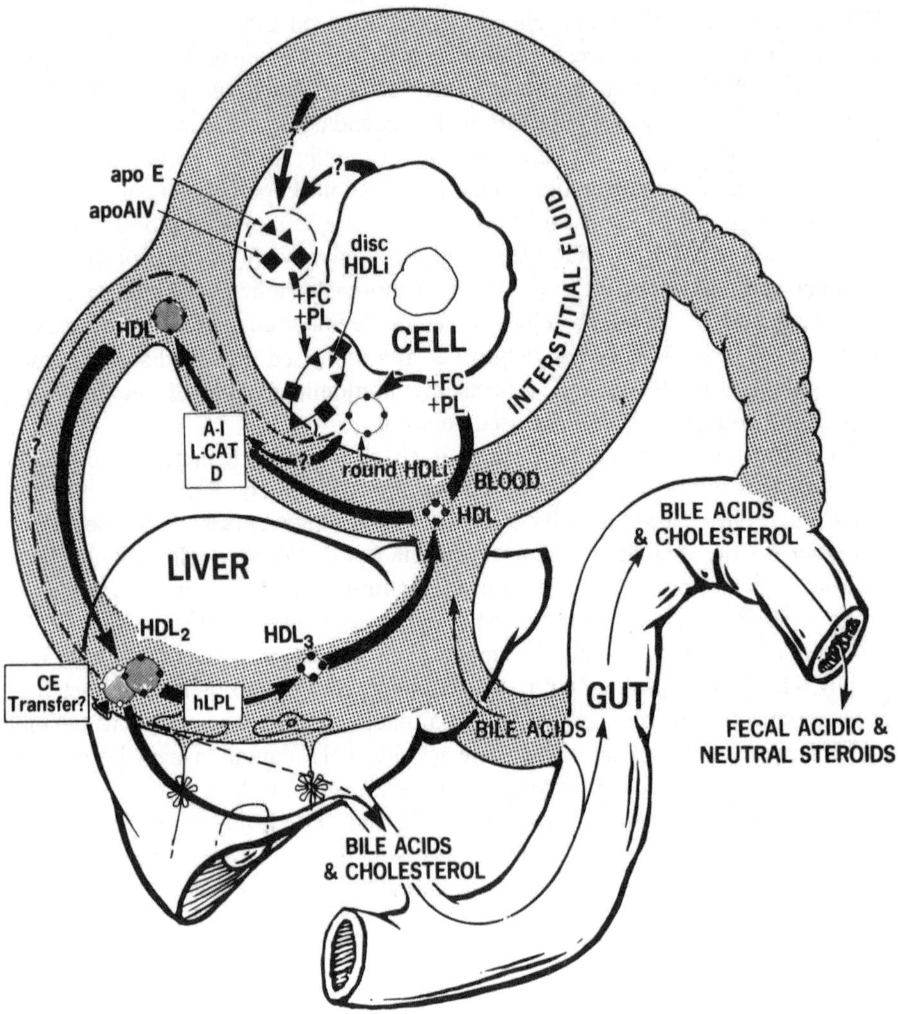

Fig. 4. Scheme for the reverse cholesterol transport. The mechanisms by which membrane cholesterol is transported to the liver for excretion as biliary cholesterol or bile acids are still highly speculative and subject to change. HDL is probably the physiological acceptor of membrane free cholesterol (*FC*). Two HDL enriched in free cholesterol (*HDLi*) have been found in peripheral lymph (which may reflect the interstitial fluid): a discoidal particle enriched in apo E and apo A-IV (which may derive from plasma or be produced by the cells) and a round form rich in apo A-I (which probably derives from the plasma HDL_3 subfraction). Some authors believe that tissue-free cholesterol is esterified by the plasma L-CAT mechanism, sharing a pathway similar to that of lipoprotein-free cholesterol, and reaching the liver as HDL-cholesteryl esters. A hepatic lipoprotein lipase (*hLPL*) located in sinusoidal endothelial cells transforms HDL_2 in HDL_3. The cholesteryl esters could be transferred to lipoproteins, which are taken up by hepatocytes through receptor-mediated endocytosis. Kinetic studies, however, support the view that HDL-free cholesterol is the direct precursor of biliary cholesterol and bile acids. *PL*, phospholipids

We recently studied in baboons the effects of chronic alcohol consumption on the reverse transport of cholesterol by measuring the turnover of autologous HDL-cholesterol labeled in vitro (Karsenty et al., 1985). Alcohol feeding significantly decreased the rate of removal of cholesteryl esters from HDL and this was associated with decreased rate of transfer of these esters to LDL. These findings indicate that one of the mechanisms by which alcohol increases HDL-cholesterol (the ester fraction being predominant) is the combination of increased input of lipoprotein-free cholesterol into HDL, with defective transfer of the esterified cholesterol back to lipoproteins of the VLDL-LDL pathway, either for supply to the tissues or removal by the liver. It is thought that tissue cholesterol may share similar pathways to lipoprotein-free cholesterol. If so, it would be unlikely that alcohol enhances tissue cholesterol mobilization through this pathway, since the flux of cholesterol was not increased.

However, when we studied the turnover of HDL-free cholesterol (which is a minor fraction of the HDL cholesterol), we found that its rate of removal as well as the flux were significantly increased in alcohol-fed baboons. This was associated with increased uptake of free cholesterol from HDL in the splanchnic vascular bed and increased losses of labeled sterols in the feces (Karsenty et al., 1985). This increased loss of fecal steroids may counteract a slow accumulation of tissue cholesterol and may provide a mechanism for the postulated prevention of atherosclerosis by alcohol.

Relationships Between Alcohol Consumption and Atherosclerosis

Since the observation by Cabot in 1904 that atherosclerosis in autopsies on patients with a history of alcoholism was remarkably low, the relationships between alcohol and atherosclerosis have been a continuous subject of controversy. Only when recognized risk factors for coronary heart disease, such as smoking and hypertension (which are positively associated with alcoholism (Klatsky et al. 1974)), were exluded, did a negative correlation between moderate alcohol consumption and coronary heart diseases emerge in most of the recent epidemiologic studies (Klatsky et al. 1974; Stason et al. 1976; Barboriak et al. 1977; Yano et al. 1977). In one of these studies (Yano et al. 1977), this prevention appears to be abolished by heavy alcohol consumption.

This epidemiological association received further support when the classical experiments by Eberhard in 1936, showing prevention of cholesterol-induced atherosclerosis by alcohol administration to rabbits, were reproduced (Goto et al. 1974) and confirmed in primates (Rudel et al. 1981). Negative or opposite results have been obtained in other species (Nichols et al. 1956; Nikkilä and Ollila 1959; Gottlieb et al. 1959). Although the number of animals was relatively small, probably the most relevant observation is the decreased extent of coronary and aortic atherosclerosis in *Macaca nemestrina* fed 36% of energy as ethanol in diets with a cholesterol content similar to those of Western human populations (Rudel et al. 1981). Moreover, ethanol prevented to a great extent the atherosclerotic lesions induced in these monkeys by a high-cholesterol diet (Rudel et al. 1981). The extent of coronary atherosclerosis correlated positively with an increase in LDL particles of high molecular weight and negatively with the increase in HDL. Human atherosclerosis also correlates positively with the plasma levels of LDL and negatively with those of HDL. Miller and Miller (1975) pointed out that the decreased levels of HDL are several times more predictive of coronary heart disease than the increased levels of LDL. However, the level of LDL may be abnormally high in persons from industrialized countries: LDL concentrations are indeed about three-fold greater than those found in newborns and in most animal species (Goldstein and Brown 1982). In studies in other populations in which LDL levels and the prevalence of coronary heart disease are lower, the HDL levels are also low (Knuiman et al. 1980). Thus the relation between LDL and HDL appears to be a better determinant. The relationship between coronary heart disease and HDL has been confirmed in epidemiologic studies (Castelli et al. 1977; Gordon et al. 1977) and extended to other vascular territories where the negative correlation is, however, weaker (Rössner et al. 1978; Gordon et al. 1981). The level of the HDL2 subfraction (which increased predominantly in alcoholics) has been estimated to be 1.5 times more predictive of coronary heart disease than total HDL cholesterol (Anderson et al. 1979). It must be pointed out, however, that moderate alcohol consumption increases HDL3 (Haskell et al. 1984).

Summary

The reducing equivalents generated by the oxidation of ethanol via alcohol dehydrogenase provide for the hepatic requirement of energy, inhibiting the utilization of normal fuels, such as fatty acids, which accumulate. Under certain conditions, this may be aggravated by stimulatory effects of ethanol on lipid absorption and/or mobilization of free fatty acids from adipose tissue. The defective oxidation of fatty acids becomes persistent when prolonged exposure to ethanol or its metabolites damages the mitochondria. Fatty acids, however, do not accumulate to a great extent because rapid compensatory changes ensue: they are released as ketone bodies or esterified to glycerolipids or cholesteryl esters, which either accumulate in the liver (fatty liver) or are released as plasma lipoproteins (alcoholic hyperlipemia). These lipoproteins compete with chylomicrons for common removal mechanisms, leading to enhanced postprandial hyperlipemia. After some time, compensatory mechanisms also develop extrahepatically: lipoprotein lipase activity increases, which tends to ameliorate hyperlipemia. Some individuals have either enhanced capacity to produce VLDL or decreased capacity to remove lipids from the blood, and develop intense and chronic hyperlipemia, which is rapidly improved by the withdrawal of alcohol. The enhanced production of VLDL increases the supply of protein and lipid components to the HDL fraction, which provides means for reutilization of cholesterol. Recent studies indicate that the recirculation of cholesterol is impaired by defective transfer of cholesteryl esters to lipoproteins that are taken up by the liver or other tissues, leading to an increase in HDL-cholesterol. Some components of the HDL fraction are also the physiological acceptor of tissue cholesterol and their increase may facilitate mobilization of tissue cholesterol.

Recent evidence in alcohol-fed baboons suggests that this may be transported as HDL-free cholesterol for enhanced excretion, both as biliary cholesterol and bile salts, the products of which are increased in the feces. This may provide a mechanism for the postulated prevention of atherosclerotic lesions, especially coronary heart disease, in moderate consumers of alcohol. But this epidemiological association still requires experimental demonstration. Moreover, as alcoholic liver damage progresses, alterations in lipoprotein metabolism and biliary secretion completely cancel possible beneficial effects.

References

Abrams MA, Cooper C (1976a) Quantitative analysis of hepatic triglycerides in ethanol-treated rats. Biochem J 156: 33–46

Abrams MA, Cooper C (1976b) Mechanism of increased hepatic uptake of unesterified fatty acid from serum of ethanol-treated rats. Biochem J 156: 47–54

Abramson EA, Arky RA (1968) Acute antilipolytic effects of ethyl alcohol and acetate in man. J Lab Clin Med 72: 105–117

Albrink MJ, Klatskin G (1957) Lactescence of serum following episodes of acute alcoholism and its probable relationship to acute pancreatitis. Am J Med 23: 26–33

Alexander RH, Friedberg SJ, Bogdonoff MD, Estes EH (1963) Hepatic cirrhosis: correlation of clinical severity and abnormality in triglyceride metabolism. Metabolism 12: 197–206

Ammon HPT, Estler C-J (1966) Tierexperimentelle Untersuchungen zur Pathogenese der Fettleber nach chronischer Alkoholverabreichung. Klin Wochenschr 11: 650–652

Anderson DW, Nichols AV, Brewer HB (1975) Ultracentrifugal characterization of the human plasma lipoprotein distribution. In: Lippel K (ed) Report of the high density lipoprotein methodology workshop. NIH, Bethesda, pp 290–300 (NIH publication no. 79-1661)

Avgerinos A, Chu P, Greenfield C, Harry DS, McIntyre N (1983) Plasma lipid and lipoprotein response to fat feeding in alcoholic liver disease. Hepatology 3: 349–355

Avogaro P, Gazzolato G (1975) Changes in the composition and physico-chemical characteristics of serum lipoproteins during ethanol-induced lipaemia in alcoholic subjects. Metabolism 24: 1231–1242

Baraona E, Lieber CS (1970) Effects of chronic ethanol feeding on serum lipoprotein metabolism in the rat. J Clin Invest 49: 769–778

Baraona E, Lieber CS (1975) Intestinal lymph formation and fat absorption: stimulation by acute ethanol administration and inhibition by chronic ethanol feeding. Gastroenterology 68: 495–502

Baraona E, Pirola RC, Lieber CS (1973) Pathogenesis of postprandial hyperlipemia in rats fed ethanol-containing diets. J Clin Invest 52: 296–303

Baraona E, Leo MA, Borowsky SA, Lieber CS (1975) Alcoholic hepatomegaly: accumulation of protein in the liver. Science 190: 794–795

Baraona E, Leo MA, Borowsky SA, Lieber CS (1977) Pathogenesis of alcohol-induced accumulation of protein in the liver. J Clin Invest 60: 546–554

Baraona E, Matsuda Y, Pikkarainen P, Lieber CS (1981) Effects of ethanol on hepatic protein secretion and microtubules. Possible mediation by acetaldehyde. In: Galanter M (ed) Currents in alcoholism, vol VIII. Grune and Stratton, New York, pp 421–434

Baraona E, Jauhonen P, Miyakawa H, Lieber CS (1983) Zonal redox changes as a cause of selective perivenular hepatotoxicity of alcohol. Pharmacol Biochem Behav [Suppl 1] 18: 449–454

Baraona E, Savolainen M, Karsenty C, Leo MA, Lieber CS (1983) Pathogenesis of alcoholic hypertriglyceridemia and hypercholesterolemia. Trans Assoc Am Physicians 96: 306–315

Barboriak JJ (1966) Effect of ethanol on lipoprotein lipase activity. Life Sci 5: 237–241

Barboriak JJ, Meade RC (1968) Enhancement of alimentary lipemia by preprandial alcohol. Am J Med Sci 255: 245–251

Barboriak JJ, Rimm AA, Anderson AJ, Schmidhoffer M, Tristani FE (1977) Coronary artery occlusion and alcohol intake. Br. Heart J 39: 289–293

Barboriak JJ, Jacobson GR, Cushman P, Herrington RE, Lipo RF, Daley ME, Anderson AJ (1980) Chronic alcohol abuse and high density lipoprotein cholesterol. Alcoholism: Clin Exp Res 4: 346–349

Belfrage P, Berg B, Cronholm T, Elmqvist D, Hägerstrand I, Johansson B, Nilsson-Ehle P, Nordén G, Sjövall J, Wiebe T (1973) Prolonged administration of ethanol to young, healthy volunteers: effects on biochemical, morphological and neurophysiological parameters. Acta Med Scand [Suppl] 552: 5–44

Belfrage P, Berg B, Hägerstrand I, Nilsson-Ehle P, Tornquist H, Wiebe T (1977) Alterations of lipid metabolism in healthy volunteers during long-term ethanol intake. Eur J Clin Invest 7: 127–131

Bezman-Tarcher A, Nestel PJ, Felts JM, Havel RJ (1966) Metabolism of hepatic and plasma triagylcerides in rabbits given ethanol or ethionine. J Lipid Res 2: 248–257

Blomstrand R, Kager L (1973) The combustion of triolein-1-^{14}C and its inhibition by alcohol in man. Life Sci 13: 113–123

Blomstrand R, Kager L, Lantto O (1973) Studies on the ethanol-induced decrease of fatty acid oxidation in rat and human liver slices. Life Sci 13: 1131–1141

Bloomfield AL, Shenson B (1947) The syndrome of idiopathic hyperlipemia with crises of violent abdominal pain. Stanford Med Bull 5: 185–191

Böttiger LE, Carlson LA, Hultman EM, Romanus V (1976) Serum lipids in alcoholics. Acta Med Scand 199: 357–361

Borowsky SA, Perlow W, Baraona E, Lieber CS (1980) Relationship of alcoholic hypertriglyceridemia to stage of liver disease and dietary lipid. Dig Dis Sci 25: 22–27

Bouchier IA, Dawson AM (1964) The effect of infusions of ethanol on the plasma free fatty acids in man. Clin Sci 26: 47–54

Boyer JL (1972) Effect of chronic ethanol feeding on bile formation and secretion of lipids in the rat. Gastroenterology 62: 294–301

Brewster AC, Lankford HG, Schwartz MG, Sullivan JF (1966) Ethanol and alimentary lipemia. Am J Clin Nutr 19: 255–259

Brodie BB, Butler WM, Horning MG, Maickel RP, Maling HM (1961) Alcohol-induced triglyceride deposition in liver through derangement of fat transport. Am J Clin Nutr 9: 432–435

Brunengraber H, Boutry M, Lowenstein L, Lowenstein JM (1974) The effect of ethanol on lipogenesis by the perfused liver. In: Thurman RC, Yonetani T, Williamson JR, Chance B (eds) Alcohol and aldehyde metabolizing systems. Academic Press, New York, pp 329–337

Cabot RC (1904) The relation of alcohol to arteriosclerosis. JAMA 43: 774–775

Cachera R, Lamotte M, Lamotte-Barrillon S (1950) Étude clinique, biologique et histologique des stéatoses de foie chez les alcooliques. Sem Hop Paris 26: 3497–3514

Castelli WP, Doyle JT, Gordon T, Hames CG, Hjortland MC, Hulley SB, Kagan A, Zukel WJ (1977) Alcohol and blood lipids. The cooperative lipoprotein phenotyping study. Lancet 2: 153–155

Cederbaum AI, Lieber CS, Beattie DS, Rubin E (1975) Effect of chronic ethanol ingestion on fatty acid oxidation by hepatic mitochondria. J Biol Chem 250: 5122–5129

Chait A, Mancini M, February AW, Lewis B (1972) Clinical and metabolic study of alcoholic hyperlipidemia. Lancet 2: 62–64

Cohen BI, Raicht RF (1981) Sterol metabolism in the rat: effects of alcohol on sterol metabolism in two strains of rats. Alcoholism: Clin Exp Res 5: 225–229

Crouse JR, Gerson CS, DeCarli LM, Lieber CS (1968) Role of acetate in the reduction of plasma free fatty acids produced by ethanol in man. J Lipid Res 2: 509–512

Cushman P, Barboriak JJ, Liao A, Hoffman NE (1982) Association between high density lipoprotein cholesterol and antipyrine metabolism in alcoholics. Life Sci 30: 1721–1724

Danielsson B, Ekman R, Fex G, Johansson BG, Kristensson H, Nilsson-Ehle P, Wadstein J (1978) Changes in plasma high density lipoproteins in chronic male alcoholics during and after abuse. Scand J Clin Lab Invest 38: 113–119

Day RC, Harry DS, Owen JS, Foo AY, McIntire N (1979) Lecithin-cholesterol acyltransferase and the lipoprotein abnormalities of parenchymal liver disease. Clin Sci 56: 575–583

DeCarli LM, Lieber CS (1967) Fatty liver in the rat after prolonged intake of ethanol with a nutritionally adequate new liquid diet. J Nutr 91: 331–336

DeGennes JL, Thomopoulus P, Truffert J, Labrousse de Tregomain B (1972) Hyperlipémies dépendantes de l'alcool. Nutr Metabol 14: 141–158

Devenyi P, Robinson GM, Kapur BM, Roncari DAK (1981) High-density lipoprotein cholesterol in male alcoholics with and without severe liver disease. Am J Med 71: 589–594

Di Luzio NR, Poggi M (1963) Abnormal lipid tolerance and hyperlipemia and acute ethanol-treated rats. Life Sci 10: 751–758

Domschke S, Domschke W, Lieber CS (1974) Hepatic redox state: attenuation of the acute effects of ethanol induced by chronic ethanol consumption. Life Sci 15: 1327–1334

Ducceschi V (1915) La colesterina del sangue nella intossicazione per alcool. Arch Fisiol 13: 147–153

Eberhard TP (1936) Effect of alcohol on cholesterol-induced atherosclerosis in rabbits. Arch Pathol 21: 616–622

Ekman R, Fex G, Johansson BG, Nilsson-Ehle P, Wadstein J (1981) Changes in plasma high density lipoproteins and lipolytic enzymes after long-term, heavy ethanol consumption. Scand J Clin Lab Invest 41: 709–715

Elko EE, Wooles WR, DiLuzio NR (1961) Alterations and mobilization of lipids in acute ethanol-treated rats. Am J Physiol 201: 923–926

Feigl J (1918) Neue Untersuchungen zur Chemie des Blutes bei akuter Alkoholintoxikation und bei chronischem Alkoholismus mit besonderer Berücksichtigung der Fette und Lipoide. Biochem Z 92: 282–317

Feinman L, Lieber CS (1967) Effect of ethanol on plasma glycerol in man. Am J Clin Nutr 20: 400–403

Forsander OA, Maenppa PH, Salaspuro MP (1965a) Influence of ethanol on the lactate/pyruvate and β-hydroxybutyrate/acetate ratios in rat liver experiments. Acta Chem Scand 19: 1770–1771

Forsander OA, Räihä N, Salaspuro M, Mäenpää P (1965b) Influence of ethanol on the liver metabolism of fed and starved rats. Biochem J 94: 259–265

Fraser GE, Anderson JT, Foster N, Goldberg R, Jacobs D, Blackburn H (1983) The effect of alcohol on serum high density lipoprotein (HDL). A controlled experiment. Atherosclerosis 46: 275–286

Freeman M, Kuiken L, Ragland JB, Sabesin SM (1977) Hepatic triglyceride lipase deficiency in liver disease. Lipids 12: 443–445

French SW (1967) Effects of chronic ethanol feeding on rat liver phospholipids. J Nutr 91: 292–298

Friedman M, Rosenman RH, Byers SO (1965) Effect of moderate ingestion of alcohol upon serum triglyceride responses of normal and hyperlipemic subjects. Proc Soc Exp Biol Med 120: 696–698

Ginsberg H, Olefsky J, Farquhar JW, Reaven GM (1974) Moderate ethanol ingestion and plasma triglyceride levels. A study in normal and hypertriglyceridemic persons. Ann Intern Med 80: 143–149

Giudicelli Y, Nordmann R, Nordmann J (1975) Modifications of plasma post-heparin lipolytic activity and tissue lipoprotein lipase activity induced in the rat by acute administration of ethanol and propan-2-ol. Clin Sci Mol Med 48: 153–156

Glueck CJ, Hogg E, Allen C, Gartside PS (1980) Effects of alcohol ingestion on lipids and lipoproteins in normal men: isocaloric metabolic studies. Am J Clin Nutr 33: 2287–2293

Goldstein JL, Brown MS (1982) Lipoprotein receptors: genetic defense against atherosclerosis. Clin Res 30: 417–426

Gordon ER (1972) Effect of an intoxicating dose of ethanol on lipid metabolism in an isolated perfused rat liver. Biochem Pharmacol 21: 2991–3004

Gordon ER (1973) Mitochondrial functions in ethanol-induced fatty liver. J Biol Chem 248: 8271–8280

Gordon T, Castelli WP, Hjortland MC (1977) High density lipoproteins as a protective factor against coronary heart disease. Am J Med 62: 707–714

Gordon T, Kannel WB, Castelli WP (1981) Lipoproteins, cardiovascular disease and death. Arch Intern Med 141: 1128–1131

Goto Y, Kikuchi H, Abe K, Nagahashi Y, Ohira S, Kudo H (1974) The effect of ethanol on the onset of experimental atherosclerosis. Tohoku J Exp Med 114: 35–43

Gottlieb LS, Broitman SA, Vitale JJ, Zamcheck NK (1959) The influence of alcohol and dietary magnesium upon hypercholesterolemia and atherogenesis in the rat. J Lab Clin Med 53: 433–441

Greenberger NJ, Hatch FT, Drummey GD, Isselbacher KJ (1966) Pancreatitis and hyperlipemia: a study of serum lipids alterations in 25 patients with acute pancreatitis. Medicine 45: 161–174
Grossberg SJ (1966) Lipemia retinalis, hyperlipoproteinemia and hepatosplenomegaly in chronic alcoholism. NY State J Med 66: 2951–2953
Guynn RW, Veloso D, Harris RL, Lawson JWR, Veech RL (1973) Ethanol administration and the relationship of malonyl-coenzyme A concentrations to the rate of fatty acid synthesis in rat liver. Biochem J 136: 639–647
Hartung GH, Foreyt JP, Mitchell RE, Mitchell JGM, Reeves RS, Gotto AM (1983) Effect of alcohol intake on high-density lipoprotein cholesterol levels in runners and inactive men. JAMA 249: 747–750
Haskell WL, Camargo C, Williams PT, Vranizan KM, Krauss RM, Lindgren FT, Wood PD (1984) The effect of cessation and resumption of moderate alcohol intake on serum high-density lipoprotein subfractions: a controlled study. N Engl J Med 310: 805–810
Hernell O, Johnson O (1973) Effect of ethanol on plasma triglycerides in male and female rats. Lipids 8: 503–508
Hirayama C, Nosaka Y, Yamada S, Yamanishi Y (1979) Effect of chronic ethanol administration on serum high density lipoprotein cholesterol in rat. Res Commun Chem Pathol Pharmacol 26: 563–569
Hulley SB, Gordon S (1981) Alcohol and high-density lipoprotein cholesterol. Causal inference from diverse study designs. Circulation [Suppl 3] 64: 57–63
Indira M, Kurup PA (1982) Effect of blackgram fiber on ethanol-induced hyperlipidemia in rats. Atherosclerosis 41: 241–246
Iseri OA, Lieber CS, Gottlieb LS (1966) The ultrastructure of fatty liver induced by prolonged ethanol ingestion. Am J Pathol 48: 535–555
Israel Y, Khanna JM, Orrego H, Rachamin G, Wahid S, Britton R, Macdonald A, Kalant H (1979) Studies on metabolic tolerance to alcohol, hepatomegaly and alcoholic liver disease. Drug Alcohol Depend 4: 109–118
Jaillard J, Sezille G, Fruchart JF, Dewailly P, Scherpereel C, Rault P (1973) Modification sous l'effet de l'alcool, de l'épuration lipidique plasmatique après hyperlipémie induite par le repas gras ou les perfusions d'émulsions lipidiques, chez l'homme. Nutr Metabol 15: 232–245
Jaillard J, Dewailly P, Sezille G, Fruchart JC, Rault C, Scherpe-Reel P (1974) Intérêt et signification du test d'hyperlipémie par voie veineuse dans l'exploration des hyperlipémies. Étude de ses variations sous l'effet de l'alcool. Atherosclerosis 19: 493–500
Janus ED, Lewis B (1978) Alcohol and abnormalities of lipid metabolism. Clin Endocrinol Metab 7: 321–332
Jauhonen P, Baraona E, Miyakawa H, Lieber CS (1982) Mechanism for selective centrolobular hepatotoxicity of ethanol. Alcoholism: Clin Exp Res 6: 350–357
Johansson BG, Laurell CB (1969) Disorders of serum α-lipoproteins after alcoholic intoxication. Scand J Clin Lab Invest 23: 231–233
Johansson BG, Medhus A (1974) Increase in plasma α-lipoproteins in chronic alcoholics after acute abuse. Acta Med Scand 195: 273–277
Johnson O (1974) Effects of ethanol on liver triglyceride concentration in fed and fasted rats. Lipids 9: 57–60
Johnson O, Hernell O (1975) Effect of ethanol on the activity of lipoprotein lipase in adipose tissue of male and female rats. Nutr Metabol 19: 41–44
Joly J-G, Feinman L, Ishii H, Lieber CS (1973) Effect of chronic ethanol feeding on hepatic microsomal glycerophosphate acyltransferase activity. J Lipid Res 14: 337–343
Jones DP, Losowsky MS, Davidson CS, Lieber CS (1963) Effects of ethanol on plasma lipids in man. J Lab Clin Med 62: 675–682
Jones DP, Perman ES, Lieber CS (1965) Free fatty acid turnover and triglyceride metabolism after ethanol ingestion in man. J Lab Clin Med 66: 804–813
Kaffarnik H, Schneider J (1970) Zur kurzfristigen Wirkung des Äthylalkohols auf die Serumlipide gesunder fastender Versuchspersonen. Z Gesamte Exp Med 152: 187–200
Kaffarnik H, Schneider J, Schubotz R, Hausmann L, Mühlfellner G, Mühlfellner O, Zöfel P (1978) Plasma lipids, triglyceride fatty acid pattern, and plasma insulin in fasted healthy volunteers during continuous ingestion of ethanol. Atherosclerosis 29: 1–7

Kalant H, Khanna JM, Seymour F, Loth J (1975) Acute alcoholic fatty liver. Metabolism or stress. Biochem Pharmacol 24: 431–434

Karsenty C, Baraona E, Savolainen M, Lieber CS (1985) Effects of chronic ethanol intake on mobilization and excretion of cholesterol in baboons. J Clin Invest 75: 976–986

Kessler JI, Yalovsky-Mishkin S (1966) Effect of ingestion of saline, glucose, and ethanol on mobilization and hepatic incorporation of epididymal pad palmitate-1-^{14}C in rats. J Lipid Res 7: 772–778

Klatsky AL, Friedman GD, Siegelaud AB (1974) Alcohol consumption before myocardial infarction: results from the Kaiser-Permanente epidemiologic study of myocardial infarction. Ann Intern Med 81: 294–301

Knuiman JT, Hermus RJJ, Hautvast JGA (1980) Serum total and high density lipoprotein (HDL) cholesterol concentrations in rural and urban boys from sixteen countries. Atherosclerosis 36: 529

Kudzma DJ, Schonfeld G (1971) Alcoholic hyperlipemia: induction by alcohol but not by carbohydrate. J Lab Clin Med 77: 384–395

Kuller LH, Hulley SB, LaPorte RE, Neaton J, Dai WS (1983) Environmental determinants, liver function, and high density lipoprotein cholesterol levels. Am J Epidemiol 117: 406–418

Lamb RG, Wood CK, Fallon HJ (1979) The effect of acute and chronic ethanol intake on hepatic glycerolipid biosynthesis in the hamster. J Clin Invest 63: 14–20

Lakshmanan MR, Ezekiel M (1982) Relationship of alcoholic hyperlipemia to the feed-back regulation of hepatic cholesterol synthesis by chylomicron remnants. Alcoholism: Clin Exp Res 6: 482–486

Lakshmanan MR, Veech RL (1977) Short and long-term effects of ethanol administration in vivo on rat liver HMG-CoA reductase and cholesterol-7 hydrolase activities. J Lipid Res 18: 325–330

Lakshman MR, Gupta AD, Veech RL (1978) Effect of chronic ethanol administration on cholesterol and bile acid synthesis in vivo. Lipids 13: 134–136

Leathers CW, Bond MG, Bullock BC, Rudel LL (1981) Dietary ethanol and cholesterol in *Macaca nemestrina* serum lipid and hepatic changes. Exp Mol Pathol 35: 285–299

Lefevre A, Adler H, Lieber CS (1970) Effects of ethanol on ketone metabolism. J Clin Invest 49: 1775–1782

Lefevre A, DeCarli LM, Lieber CS (1972) Effect of ethanol on cholesterol and bile acid metabolism. J Lipid Res 13: 48–55

Lewis B, Chait A, Sissons P (1973) Lipid abnormalities in alcoholism and chronic renal failure. Adv Exp Med Biol 38: 155–159

Lieber CS (1982) Medical disorders of alcoholism. Pathogenesis and treatment. Saunders, Philadelphia

Lieber CS, DeCarli LM (1966) Study of agents for the prevention of the fatty liver produced by prolonged alcohol intake. Gastroenterology 50: 316–322

Lieber CS, DeCarli LM (1970) Quantitative relationship between amount of dietary fat and severity of alcoholic fatty liver. Am J Clin Nutr 23: 474–478

Lieber CS, DeCarli LM (1974) An experimental model of alcohol feeding and liver injury in the baboon. J Med Primatol 3: 153–163

Lieber CS, Rubin E (1968) Alcoholic fatty liver in man on a high protein and low fat diet. Am J Med 44: 200–207

Lieber CS, Schmid R (1961) The effect of ethanol on fatty acid metabolism: stimulation of hepatic fatty acid synthesis in vitro. J Clin Invest 40: 394–399

Lieber CS, Spritz N (1966) Effects of prolonged ethanol intake in man: role of dietary, adipose, and endogenously synthesized fatty acids in the pathogenesis of alcoholic fatty liver. J Clin Invest 45: 1400–1411

Lieber CS, Leevy CM, Stein SW, George WS, Cherrick GR, Abelmann WH, Davidson CS (1962) Effects of ethanol on plasma free fatty acids in man. J Lab Clin Med 59: 826–832

Lieber CS, Jones DP, Mendelson J, DeCarli LM (1963) Fatty liver, hyperlipemia and hyperuricemia produced by prolonged alcohol consumption, despite adequate dietary intake. Trans Assoc Am Physicians 76: 289–300

Lieber CS, Jones DP, DeCarli LM (1965) Effects of prolonged ethanol intake: production of fatty liver despite adequate diets. J Clin Invest 44: 1009–1021

Lieber CS, Spritz N, DeCarli LM (1966) Role of dietary, adipose, and endogenously synthesized fatty acids in the pathogenesis of the alcoholic fatty liver. J Clin Invest 45: 51–62
Lieber CS, Lefevre A, Spritz N, Feinman L, DeCarli LM (1967) Difference in hepatic metabolism of long and medium-chain fatty acids: the role of fatty acid chain length in the production of alcoholic fatty liver. J Clin Invest 46: 1451–1460
Lieber CS, Spritz N, DeCarli LM (1969) Fatty liver produced by dietary deficiencies: its pathogenesis and potentiation by ethanol. J Lipid Res 10: 283–287
Lieber CS, Rubin E, DeCarli LM, Feinman L (1976) Respective roles of ethanol and protein deficiency in the pathogenesis of alcoholic hepatitis and cirrhosis. In: Leevy CM (ed) Diseases of the liver and biliary tract. Karger, Basel, pp 97–100
Losowsky MS, Jones DP, Davidson CS, Lieber CS (1963) Studies of alcoholic hyperlipemia and its mechanism. Am J Med 35: 794–803
Lundquist E, Tygstrup N, Winkler K, Mellengaard K, Munch-Petersen S (1962) Ethanol metabolism and production of free acetate in the human liver. J Clin Invest 41: 955–961
Lundquist F, Tygstrup N, Winkler K, Jensen B (1965) Glycerol metabolism in the human liver. Science 150: 616–617
Luoma PV, Savolainen MJ, Sotaniemi EA, Pelkonen RO, Arranto AJ, Ehnholm C (1983) Plasma high-density lipoproteins and liver lipids and proteins in man. Relation to hepatic histology and microsomal enzyme induction. Acta Med Scand 214: 103–109
Lynch MJG, Raphael SS, Dixon TP (1959) Fat embolism in chronic alcoholism. Control study on incidence of fat embolism. Arch Pathol 67: 68–80
Maddrey WC, Boyer JL (1973) The acute and chronic effects of ethanol administration on bile secretion in the rat. J Lab Clin Med 82: 215–225
Mak K, Lieber CS (1984) Alterations in endothelial fenestrations in liver sinusoids of baboons fed alcohol: a scanning electron microscopic study. Hepatology 4: 386–391
Mallov S (1961) Effect of ethanol intoxication on plasma free fatty acids in the rat. J Stud Alcohol 22: 250–253
Marin GA, Ward NL, Fischer R (1973) Effect of ethanol on pancreatic and biliary secretions in humans. Am J Dig Dis 18: 825–833
Marin GA, Karjoo M, Ward N, Rosato E (1975) Effects of alcohol on biliary lipids in the presence of a chronic biliary fistula. Surg Gynecol Obstet 141: 352–356
Marth E, Cazzolato G, Bon B, Avogaro P, Kostner GM (1982) Serum concentrations of Lp(a) and other lipoprotein parameters in heavy alcohol consumption. Ann Nutr Metab 26: 56–62
Marzo A, Ghirardi P, Sardini D, Prandini BD, Albertini A (1970) Serum lipids and total fatty acids in chronic alcoholic liver disease at different stages of cell damage. Klin Wochenschr 15: 949–950
Matsuda Y, Baraona E, Salaspuro M, Lieber CS (1979) Effects of ethanol on liver microtubules and Golgi apparatus. Lab Invest 41: 455–463
Mendelson JH, Mello NK (1973) Alcohol-induced hyperlipemia and beta lipoproteins. Science 180: 1372–1374
Mendelson JH, Mello NK (1974) Significance of alcohol-induced hypertriglyceridemia in patients with type IV hyperlipoproteinemia. Ann Intern Med 80: 270–271
Mendenhall CL (1972) Origin of hepatic triglyceride fatty acids: quantitative estimations of the relative contributions of linoleic acid by diet and adipose tissue in normal and ethanol-fed rats. J Lipid Res 13: 177–183
Mendenhall CL, Wilson NL (1973) Observations of the relationship of hepatic choline uptake to ethanolic fatty liver in the rat. Can J Biochem 51: 1010–1013
Mendenhall CL, Bradford RH, Furman RH (1969) Effects of ethanol on glycerolipid metabolism in rat liver. Biochim Biophys Acta 187: 501–509
Mezey E, Potter JJ, Ammon RA (1976) Effect of ethanol administration on the activity of hepatic lysosomal enzymes. Biochem Pharmacol 25: 2663–2667
Middleton WRJ, Carter EA, Drummey GD, Isselbacher KJ (1971) Effect of oral ethanol administration on intestinal cholestrogenesis in the rat. Gastroenterology 60: 880–887
Miller GJ, Miller NE (1975) Plasma-high-density lipoprotein concentration and development of ischemic heart-disease. Lancet 1: 16–19
Mistilis SP, Ockner RK (1972) Effects of ethanol on endogenous lipid and lipoprotein metabolism in small intestine. J Lab Clin Med 80: 34–46

Mordasini R, Kaffarnik H, Schneider J, Riesen W (1982) Alkohol und Plasmalipoproteine im Akut- und Langzeitversuch. Schweiz Med Wochenschr 112: 1928–1931

Nestel PJ (1967) Some clinical and chemical manifestations of alcohol-induced hyperlipemia. Aust Ann Med 16: 139–143

Nestel PJ, Simons LA, Homma Y (1976) Effects of ethanol on bile acid and cholesterol metabolism. Am J Clin Nutr 29: 1007–1015

Nestel PJ, Tada N, Ridge NH (1980) Increased catabolism of high density lipoprotein in alcoholic hepatitis. Metabolism 29: 101–104

Nichols CW, Siperstein MD, Gaffey W, Lindsay S, Chaikoff IL (1956) Does the ingestion of alcohol influence the development of arteriosclerosis in fowls? J Exp Med 103: 465–475

Nikkilä EA, Ojala K (1963) Role of hepatic L-glycerophosphate and triglyceride synthesis in the production of fatty liver by ethanol. Proc Soc Exp Biol Med 113: 814–817

Nikkilä EA, Ollila O (1959) Effect of alcohol ingestion on experimental chicken atherosclerosis. Circ Res 7: 588–594

Nikkilä EA, Taskinen M-R, Huttunen JK (1978) Effect of acute ethanol load on postheparin plasma lipoprotein lipase and hepatic lipase activities and intravenous fat tolerance. Horm Metab Res 10: 220–223

Nikkilä EA, Kuusi T, Taskinen MR (1982) Role of lipoprotein lipase and hepatic endothelial lipase in the metabolism of high density lipoproteins: a novel concept on cholesterol transport in HDL cycle. In: Carlson LA, Pernow B (eds) Metabolic risk factors in iscemic cardiovascular disease. Raven, New York, pp 205–215

Nilsson NO, Belfrage P (1978) Effects of acetate, acetaldehyde and ethanol on lipolysis in isolated rat adipocytes. J Lipid Res 19: 737–741

Nilsson-Ehle P, Carlstrom S, Belfrage P (1978) Effects of ethanol intake on lipoprotein lipase activity in adipose tissue of fasting subjects. Lipids 13: 433–437

Ockner RK, Mistilis SP, Poppenhausen RB, Stiehl AF (1973) Ethanol-induced fatty liver: effect of intestinal lymph fistula. Gastroenterology 64: 603–609

Okuda A, Okuda K (1983) Physiological function and kinetic mechanism of human liver alcohol dehydrogenase as 5-cholestane-3,7,12,26-tetrol dehydrogenase. J Biol Chem 258: 2899–2905

Olivecrona T, Hernell O, Johnson O, Fex G, Wallinder L, Sandgren O (1972) Effect of ethanol on some enzymes inducible by fat free refeeding. Q J Stud Alcohol 33: 1–13

Ontko JA (1973) Effects of ethanol on the metabolism of free fatty acids in isolated liver cells. J Lipid Res 14: 78–85

Ontko JA, Cook GA, Sullivan AC (1977) Actions of ethanol metabolism on fatty acid synthesis in the liver. In: Thurman RC, Williams JR, Drott H, Chance B (eds) Alcohol and aldehyde metabolizing systems. Academic Press New York, pp 137–146

Orrego H, Blendis LM, Crossley IR, Medline A, Macdonald A, Ritchie S, Israel Y (1981) Correlation of intrahepatic pressure with collagen in the Disse space and hepatomegaly in humans and in rat. Gastroenterology 80: 546–556

Ostrander LD Jr, Johnson BC, Block WD (1971) Relationship of alcohol intake to lipid levels in men from a general population. Circulation 44: II–59 (abstract)

Papadopoulus NM, Charles MA (1970) Serum lipoprotein patterns in liver disease. Proc Soc Exp Biol Med 134: 797–799

Patek AJ, Earampamoorthy S (1976) Serum lipids in alcoholic patients with and without cirrhosis of the liver, with particular reference to endogenous familial hypertriglyceridemia. Am J Clin Nutr 29: 1122–1126

Poggi M, DiLuzio NR (1964) The role of liver and adipose tissue in the pathogenesis of the ethanol-induced fatty liver. J Lipid Res 5: 437–441

Pritchard PH, Bowley M, Burditt SL, Cooling J, Glenny HP, Lawson N, Sturton RG, Brindley DN (1977) The effects of acute ethanol feeding and of chronic Benfluorex administration on the activities of some enzymes of glycerolipid synthesis in rat liver and adipose tissue. Biochem J 166: 639–642

Rebouças G, Isselbacher KJ (1961) Studies on the pathogenesis of the ethanol-induced fatty liver. I. Synthesis of fatty acids by the liver. J Clin Invest 40: 1355

Redgrave TG, Martin G (1977) Effects of chronic ethanol consumption on the catabolism of chylomicron triacylglycerol and cholesteryl ester in the rat. Atherosclerosis 28: 69–80

Rössner S, Kjelen KG, Mettinger KL (1978) Normal serum cholesterol but low HDL cholesterol concentrations in young patients with ischemic cerebrovascular disease. Lancet 1: 577–579
Rubin E, Lieber CS (1968) Alcohol induced hepatic injury in non-alcoholic volunteers. N Engl J Med 278: 869–876
Rubin E, Beattie DS, Toth A, Lieber CS (1972) Structural and functional effects of ethanol on hepatic mitochondria. Fed Proc 31: 131–140
Rudel LL, Leathers CW, Bond MG, Bullock BC (1981) Dietary-ethanol induced modifications in hyperlipoproteinemia and atherosclerosis in non-human primates *(Macaca nemestrina)*. Atherosclerosis 1: 144–155
Sabesin SM, Hawkins HL, Kuiken L, Ragland JB (1977) Abnormal plasma lipoproteins and lecithin-cholesterol acyltransferase deficiency in alcoholic liver disease. Gastroenterology 72: 510–518
Salaspuro MP, Shaw S, Jayatilleke E, Ross WA, Lieber CS (1981) Attenuation of the ethanol-induced hepatic redox change after chronic alcohol consumption in baboons: metabolic consequences in vivo and in vitro. Hepatology 1: 33–38
Sanchez-Craig M, Annis HM (1981) Gamma-glutamyl transpeptidase and high-density lipoproteins cholesterol in male problem drinkers: advantages of a composite index for predicting alcohol consumption. Alcoholism: Clin Exp Res 5: 540–544
Santos M, Friedberg SJ, Kudzma CJ, Tjoa LT (1974) Conversion of free fatty acids to triglycerides. Determination on obstructive vs hepatocellular jaundice and cirrhosis. Arch Intern Med 134: 457–460
Savolainen MJ (1977) Stimulation of hepatic phosphatidate phosphohydrolase activity by a single dose of ethanol. Biochem Biophys Res Commun 75: 511–518
Savolainen MJ, Hiltunen JK, Hassinen IE (1977) Effects of prolonged ethanol ingestion on hepatic lipogenesis and related enzyme activities. Biochem J 164: 169–177
Savolainen MJ, Baraona E, Pikkarainen P, Lieber CS (1984) Decreased hepatic triacylglycerol synthesizing capacity during progression of alcoholic liver damage in the baboon. J Lipid Res 25: 813–820
Schapiro RH, Drummey GD, Shimizu Y, Isselbacher KJ (1964) Studies on the pathogenesis of ethanol-induced fatty liver. II. Effect of ethanol on palmitate-1-C14 metabolism by isolated perfused rat liver. J Clin Invest 43: 1338–1347
Schapiro RH, Scheig RL, Drummey GD, Mendelson JH, Isselbacher KJ (1965) Effect of prolonged ethanol ingestion on the transport and metabolism of lipids in man. N Engl J Med 272: 610–615
Schneider J, Panne E, Braun H, Mordasini R, Kaffarnik H (1983) Ethanol-induced hyperlipoproteinemia. Crucial role of preceding ethanol intake in the removal of chylomicrons. J Lab Clin Med 101: 114–122
Scholz R, Kaltstein A, Schwabe U, Thurman RG (1974) Inhibition of fatty acid synthesis from carbohydrates in perfused rat liver by ethanol. In: Thurman RG, Yonetani T, Williamson JR, Chance B (eds) Alcohol and aldehyde metabolizing systems. Academic Press, New York, pp 315–328
Schwartz CS, Halloran LG, Vlahcevic ZR, Gregory DH, Swell L (1978) Preferential utilization of free cholesterol from high-density lipoproteins for biliary cholesterol secretion in man. Science. 200: 62–64
Sigurdsson G, Nicoll A, Lewis B (1976) Metabolism of very low density lipoproteins in hyperlipidaemia: studies of apolipoprotein B kinetics in man. Eur J Clin Invest 6: 167–177
Sirtori CR, Agradi E, Mariani C (1973) Hyperlipoproteinemia in alcoholic subjects. Pharmacol Res Commun 5: 81–85
Stason WB, Neff RK, Miettinen OS, Jick H (1976) Alcohol consumption and non-fatal myocardial infarction. Am J Epidemiol 104: 603–608
Svoboda DJ, Manning RT (1964) Chronic alcoholism and fatty metamorphosis of the liver. Mitochondrial alterations in hepatic cells. Am J Pathol 44: 645–662
Tada N, Fidge N, Nestel P (1979) Identification and characterization of mixed disulphide complexes of E apoprotein in high density lipoprotein of subjects with acute alcoholic hepatitis. Biochem Biophys Res Commun 90: 297–304
Talbott GD, Keating BM (1962) Effects of preprandial whiskey on postalimentary lipemia. Geriatrics 17: 802–808

Takeuchi N, Ito M, Yamamura Y (1974) Esterification of cholesterol and hydrolysis of cholesteryl ester in alcohol induced fatty liver of rats. Lipids 9: 353–357

Taskinen M-R, Nikkilä EA (1977) Nocturnal hypertriglyceridemia and hyperinsulinemia following moderate evening intake of alcohol. Acta Med Scand 202: 173–177

Taskinen M-R, Välimäki M, Nikkilä EA, Kuusi T, Ehnholm C, Ylikahri R (1982) High density lipoprotein subfractions and postheparin plasma lipases in alcoholic men before and after ethanol withdrawal. Metabolism 31: 1168–1174

Thornton J, Symes C, Heaton K (1983) Moderate alcohol intake reduces bile cholesterol saturation and raises HDL cholesterol. Lancet 2: 819–822

Topping DL, Weller RA, Nader CJ, Calvert JD, Illman RJ (1982) Adaptive effects of dietary ethanol in the pig: changes in plasma high-density lipoproteins and fecal steroid excretion and mutagenicity. Am J Clin Nutr 36: 245–250

Uthus EO, Skurdal DN, Cornatzer WE (1976) Effect of ethanol on choline phosphotransferase and phosphatidyl ethanolamine methyltransferase activities in liver microsomes. Lipids 11: 641–644

Väänänen R, Pikkukangas A, Savolainen M, Hassinen I (1981) Cytosolic factor multiplies the cortisol and ethanol effects on microsomal diacylglycerol acyltransferase. Acta Pharmacol Toxicol [Suppl 4] 49: A-42

Vasdev SC, Subrahmanyam D, Chakravarti RN, Wahi PL (1974) Effect of chronic ethanol-feeding on the major lipids of red blood cells, liver and heart of rhesus monkey. Biochim Biophys Acta 369: 323–330

Verdy M, Gattereau A (1967) Ethanol, lipase activity and serum-lipid levels. Am J Clin Nutr 20: 997–1003

Warembourg H, Biserte G, Jaillard J, Sezille G, Scherpereel P (1970) Modifications des lipides sériques et hépatiques induites par l'alcool au cours de l'intoxication éthylique chronique du lapin. Rev Europ Études Clin Biol 15: 649–653

Wiebe A, Lundquist A, Belfrage P (1971) Time course of liver fat accumulation in man after a single load of ethanol. Scand J Clin Lab Invest 27: 33–36

Williamson JR, Scholz R, Browning ET, Thurman RG, Fukami MK (1969) Metabolic effects of ethanol in perfused rat liver. J Biol Chem 244: 5044–5054

Wilson DE, Schreibman PH, Brewster AC, Arky RA (1970) The enhancement of alimentary lipemia by ethanol in man. J Lab Clin Med 75: 264–274

Wolfe BM, Havel JR, Marliss EB, Kane JP, Seymour J, Ahuja SP (1976) Effects of 3-day fast and ethanol on splanchnic metabolism of FFA, amino acids, and carbohydrates in healthy young men. J Clin Invest 57: 329–340

Yano K, Rhoads GG, Kagan A (1977) Coffee, alcohol and risk of coronary heart disease among Japanese men living in Hawaii. N Engl J Med 297: 405–409

Zieve L (1958) Jaundice, hyperlipemia and hemolytic anemia: a heretofore unrecognized syndrome associated with alcoholic fatty liver and cirrhosis. Ann Intern Med 48: 471–496

5 Alcohol Effects on Albumin Synthesis*

M. A. Rothschild, M. Oratz, and S. S. Schreiber

Nuclear Medicine Service, Veterans Administration Medical Center, First Avenue at East 24th Street, New York, NY 10010, USA

Alcoholism and alcohol-induced organ damage are two separate entities – the former, representing an abnormal syndrome, may show organ involvement early or late, while the latter may occur without true addiction. Either the effects of alcohol or its metabolite acetaldehyde or the resultant effects of complete alcohol metabolism on protein synthesis are probably organ specific, modified by genetic predisposition, altered by nutritional status, hormone inbalance, absorptive capabilities, and the presence or absence of prior alcohol-induced or associated organ damage. To try to isolate the acute and the chronic effects of ethanol exposure on a specific protein synthetic pathway in vivo is not a simple task. The picture is further complicated by the complex nature of the normal protein synthetic and cellular transport systems. It is only necessary to consider how many potential sites in the assembly line process of protein synthesis may be altered by alcohol to see how complex a scheme we are facing (Mueckler and Pitot 1981, 1982; Bantle et al. 1980a; Krieg et al. 1980; Prehn et al. 1981; Thomas et al. 1981; Emr et al. 1980; Walter and Blobel 1981; Rothman 1981; Dorling et al. 1975; Algranati and Sabatini 1979; Schreir et al. 1977; Blobel and Dobberstein 1975; Weigand et al. 1982; Peters et al. 1971; Redman and Cherian 1972; Glauman and Ericsson 1970). Following the final posttranscriptional modification, the appropriate messenger RNA (mRNA) becomes available for the translation process. In its most simplified form this process requires the presence of initiation factors which promote the formation of the ribosomal complex, a complex of the two ribosome subunits and the particular mRNA. Transfer RNAs (tRNA) insert the specific amino acids as prescribed by the mRNA into the growing peptide. For proteins destined to be secreted from the cell, the major portion of the synthetic process takes place on polysomes bound to the endoplasmic reticulum (ER). These proteins are synthesized with a *N*-terminal extension designated signal peptide, and this "signal" is synthesized on free unbound polysomes. This signal sequence is recognized by a protein in the ER (signal recognition protein), which facilitates the active threading of the nascent polypeptide chain through a channel in the ER as it is being synthesized. This portion of the growing protein and its precursor portion, not usually secreted into the plasma, are transported through the lumen of the endoplasmic reticulum. During the process,

* This work was supported by the Medical Research Service of the Veterans Administration, by grant number AA000959 awarded by the Alcohol Drug Abuse Mental Health Administration, and by the Louise and Bernard Palitz Fund

Alcohol Related Diseases in Gastroenterology
Edited by H. K. Seitz and B. Kommerell

the precursor segments are hydrolyzed from the peptide. At the conclusion of translation, the protein in its final form, for example albumin, or requiring glycosylation, for example the glycoproteins, is delivered to the Golgi apparatus, which sorts and processes peptides prior to internal cellular routing or for secretion. These numerous steps, which have only been summarized, are certainly susceptible to influence by alcohol-induced stress direct or induced by allied changes.

Alcohol consumption, acute and chronic, results in a plethora of in vivo events many of which could influence hepatic protein synthesis (Lieber 1982). Specifically, the altered redox state with the generation of excess reduced nicotinamide-adenine dinucleotide (NADH) from nicotinamide-adenine dinucleotide (NAD), from the preferred oxidation of ethanol, influences the available energy transfer potentially needed for protein synthesis and urea production. There is an accumulation of acetaldehyde, a change in mitochondrial metabolism, an accumulation of fat, and in specific instances steady and progressive liver cell damage. Hormone metabolism is altered since many hormones influence and stimulate albumin metabolism (Gordon and Southern 1982; Cain et al. 1970; Rothschild et al. 1957, 1958; Bjorneboe 1946; Bjorneboe and Schwartz 1959; Westergaard et al. 1972; Feldhoff et al. 1977). In addition, albumin synthesis is sensitive to colloidal content at or near the site of synthesis and changes in albumin distribution may occur during different stresses (Oratz et al. 1977). No organ system is immune from the effects of ethanol and the specifics for hepatocyte altered metabolism cannot be dissected in vivo.

Not only are we faced with a complexity of response to ethanol, but the methodology available for the assay of protein synthesis must be critically evaluated. Measurements of protein synthesis may be based on a specific antigen-antibody reaction or on the incorporation of a labeled precursor (amino acid or hexose) into the final product. The former will not differentiate between protein previously made and not degraded from de novo synthesis during the period of study. The isotope techniques require the knowledge of the specific activity of the immediate precursor tRNA at the site of synthesis within the cell. Many techniques are present to approximate this value and hopefully to minimize errors. However, as the ethanol and acetaldehyde levels rise and fall in vivo, fluctuations of and entrance of tracer radioactivity into the cellular amino acid pool and changes in tRNA charging may be occurring momentarily and thus influence the results. Thus the data available do not confirm a single action of ethanol or indicate specific mechanims of action.

The protein which has received the most study is serum albumin, though we shall make reference to other proteins when a specific point is to be made. Clinically, the serum albumin has long been used as an index of hepatic function, particularly in cirrhosis of the liver. Alcohol ingestion is certainly a major cause of this syndrome. It is obvious that the chronic effects of alcohol on albumin metabolism cannot be separated from those resulting from cirrhosis per se, and the data on albumin metabolism in cirrhosis pose interesting questions. A level of serum albumin below 2.5 g/dl signified an unfavorable prognosis in cirrhosis, but with the development of ascites this prognostic sign was lost (Post and Patek 1942a, b). In fact, in at least a third of the cases studied, there was an expanded total exchangeable albumin pool in patients with cirrhosis and ascites (Rothschild et al. 1970). Certainly, this must mean that the synthetic process for albumin must be capable of responding effectively even in the face of severe nutritional deficits, hepatic architectural derangement, or altered hepatic blood delivery, to say nothing of the alcohol per se. Later studies confirmed this observation (Rothschild et al. 1969a). Albumin synthesis is not a specific property of only a few hepatocytes, even though normally the total ultimate capacity to produce albumin is not tested. During stress, however, albumin can be identified diffusely in all liver cells. The maintenance of normal albumin production even with disease or decreased tissue function may be explained not only by increased synthesis by a few cells but by recruitment of the dormant capacity of the remaining cells (Feldmann 1981; Feldmann et al. 1972; Guillouzo et al. 1982). Albumin degradation, both the absolute rate (milligrams) and the fractional rate (percentage of the albumin pool degraded per day) are depressed when the serum albumin level is lowered; however, this change does not account for the increase in albumin pool or in the synthetic data, which showed a normal or increased rate of synthesis in 12 of 19 subjects with cirrhosis and ascites (Rothschild et al. 1969a). These data, which were obtained following hospitalization when the patients were not consuming ethanol and were on an adequate diet, could not aid in determining what was taking place prior to hospitalization, and thus a model system was needed. The perfused rat and rabbit liver preparations have been most useful in providing a few answers and many more questions.

Studies in vivo and in vitro revealed that albumin synthesis was depressed by as much as 50% in the presence of a short-term fast or malnutrition (Rothschild et al. 1968, 1969b; Picou and Taylor 1969; Freeman and Gordon 1964; Hoffenberg et al. 1966; Jeejebjoy et al. 1969; Flaim et al. 1982a, b; Munro 1970). This effect is associated with a decrease in ribosomal RNA and disaggregation of the bound polysomes (Enwonwu and Munro 1970; Munro et al. 1965; Rothschild et al. 1974). Further, it has been shown that the mRNA for albumin is retained in a protein complex within the cytosol, perhaps becoming readily available for reinstitution of albumin production when the nutritional stress is terminated upon refeeding (Shafritz et al. 1979). Additional studies revealed that tryptophan was essential for the maintenance of polysome reaggregation, and possibly since there is only one residue for tryptophan in albumin, this amino acid might be rate limiting. However, this was shown not to be the whole answer, for when livers from donors which had fasted were employed, some but not all amino acids at 10-m*M* levels added to the perfusate resulted in a restimulation of albumin synthesis, polysome reaggregation and, interestingly, a stimulation of urea synthesis (Oratz et al. 1976). Those amino acids (valine, methionine, leucine, and histidine) which did not stimulate albumin synthesis failed to stimulate urea production. This observation, coupled with the fact that ornithine, an amino acid not present in albumin, was one of the stimulating amino acids, suggested that the urea cycle might play a role in the albumin synthesis response to the amino acids (Oratz et al. 1980, 1983; Rothschild et al. 1977; Sidransky et al. 1982). The urea cycle generates ornithine, which is the precursor of the polyamines (putrescine, spermidine, and spermine) via the enzyme ornithine decarboxylase, and these polyamines play varied but vital roles in cellular growth and regeneration and could mediate these amino acid effects (Pegg and McCann 1982). Blocking the production of putrescine from ornithine, using a specific enzyme inhibitor, completely eliminated the stimulatory effect of ornithine on albumin synthesis and on polysome reaggregation (Oratz et al. 1983). Thus altered nutrition or fasting upsets the normal albumin synthetic mechanism and the acute effects of alcohol must be compared to these nutritionally induced changes.

When the isolated perfused rabbit liver from a fed donor was employed, the infusion of 200 mg% ethanol resulted in depressed albumin and urea synthesis and bound polysomes were disaggregated. Total RNA was not altered. As with the liver from a donor who had fasted, not stressed with alcohol, the same amino acids which reversed the fasting effects reversed these alcohol effects in livers from fed donors. Further, as with the fasting and other stresses, the polyamine spermine was synergistic in promoting bound polysome reaggregation (Oratz et al. 1976, 1980, 1983).

When the stresses of fasting and ethanol infusion were combined, albumin and urea synthesis were depressed (Rothschild et al. 1974; Oratz et al. 1976; Piccirillo and Chambers 1975). Both the bound and free polysomes were disaggregated and the amino acids which were effective in reversing either stress applied singly (alcohol or fasting) were now no longer effective when alcohol was perfused into a liver from a donor which had fasted. Similar data have been obtained in vivo in rats, where the absolute rate of albumin synthesis was decreased following a single dose of alcohol (4 ml/kg) and changes in the endoplasmic reticulum were produced by 8 ml/kg. These changes could be prevented by the simultaneous administration of an amino acid solution. Not all proteins responded in this fashion and fibrinogen synthesis was not affected by ethanol (Jeejeebhoy et al. 1972; Peters and Steele 1982; Princen et al. 1981; Baraona et al. 1980; Sidransky et al. 1980; Wunderlich et al. 1979; Burke and Rubin 1979; Morland et al. 1979a; Dich and Tonnesen 1980; Tuma and Sorrell 1980; Poso et al. 1980). Other in vitro studies using a variety of different models including perfused livers, liver slices, hepatocytes, and subcellular organelles have shown a decrease in protein synthesis. Intracellular protein degradation may also be affected (Poso 1980). The endoplasmic membrane polysomes are mainly affected, leaving the free polysomes, primarily responsible for the production of protein not destined for export from the cell, unaffected (Rothschild et al. 1971, 1974; Peters and Steele 1982; Princen et al. 1981; Khawaja and Lindholm 1978). The mechanisms producing these effects are unclear. The data on amino acid entry into hepatic cells do not permit a concise postulate (Rosa and Rubin 1980; Piccirillo and Chambers 1976; Petit and Barral-Alix 1979; Hakkinen and Kulonene 1980; Krebs et al. 1973; Morland et al. 1979b). A lack of mRNA for albumin has not been demonstrated (Bantle et al. 1980b; Zern et al. 1983). In other systems, a translational inhibitory mechanism has been postulated (Wu 1981).

Is the effect of alcohol due to alcohol per se, alcohol metabolism, or acetaldehyde? Studies to date using hepatocytes (Baraona et al. 1980) and perfused livers have shown that prevention of the oxidation of ethanol with 4-methylpyrazole (4-MP) prevented the inhibition of protein synthesis by ethanol (Baraona et al. 1980). In perfused rabbit livers, however, 4-MP was truly only effective in preventing most of the polysome disaggregation seen in the bound polysomes. Albumin synthesis may have improved slightly when livers from donors which had fasted were used. Urea synthesis remained depressed (Oratz et al. 1978). Since the urea cycle is required to generate ornithine and in turn putrescine, this finding supports the concept of the important role of the de novo synthesis of polyamines in restarting albumin production. In hepatocytes from rats which had fasted, 4-MP was effective but only at relatively low levels of ethanol (Morland et al. 1980). Thus the majority of the data indicate that alcohol metabolism is important for some of the acute protein synthesis inhibitory effects. This conclusion is also suggested by the data which fail to show an inhibitory effect on hepatic nonparenchymal cells – cells which do not metabolize ethanol (Morland et al. 1979c).

Turning now to the effects of acetaldehyde, it was observed that acetaldehyde at levels of 0.45 m*M* or 2 mg/100 ml resulted in a depression of albumin synthesis and urea synthesis in livers from fed donors, but did not result in polysome disaggregation (Oratz et al. 1978; Morland et al. 1979c, 1980). Interestingly, when the donor had fasted, acetaldehyde did not lower albumin synthesis below the low rate for donors which had fasted, namely half that of the liver from a fed donor. In another model employing hepatocytes from fed rats, acetaldehyde at levels around 0.1 m*M* did not result in a decrease in total hepatocyte protein synthesis even in the presence of a marked increase in the lactate/pyruvate ratio (Baraona et al. 1980). To complicate the picture even further, when acetaldehyde metabolism was blocked, both to alcohol and to acetate, with 4-MP and with disulfiram in vitro, employing the isolated perfused rabbit liver from fed donors, the inhibitory effects of acetaldehyde on albumin synthesis and on urea synthesis were blocked (Rothschild et al. 1980). Thus acetaldehyde does not seem to be the sole offending agent during ethanol metabolism, and further the absolute nutritional state of the donor is also important, for unlike ethanol, acetaldehyde does not inhibit albumin synthesis in vitro when a donor which has fasted is used.

To contrast these effects of acetaldehyde on albumin synthesis, it is interesting to mention the studies concerned with cardiac muscle protein synthesis (Schreiber et al. 1972; Baraona et al. 1977). Although ethanol itself may be implicated in such depression, the major causative factor may be acetaldehyde (Schreiber et al. 1972). This may be inferred from in vitro studies with hearts taken from animals unexposed to ethanol and perfused in vitro with ethanol or acetaldehyde. In such normal hearts, ethanol perfused in Krebs-Henseleit solution at levels sufficient to impair hepatic albumin synthesis did not depress cardiac protein synthesis or sufficiently alter cardiac contractile parameters. In contrast, there was a marked decrease in cardiac protein synthesis when normal hearts were perfused in vitro with low levels of acetaldehyde (Schreiber et al. 1972), in spite of significant increases in contractile action with positive inotropic and chronotropic action.

So far the term "albumin synthesis" has been used to represent de novo production of albumin, but there is another component to the appearance of albumin outside of the hepatocyte (into the medium, perfusate, or plasma), namely transport through the cell and secretion from the cell. Inhibition of proteolysis of the pre- and pro-portions of the albumin molecule has been shown to inhibit synthesis in in vitro studies, suggesting that intracellular processing and synthesis may be interrelated (Algranati and Sabatini 1979). The chronic administration of ethanol has been shown to result in an increase in hepatic intracellular nitrogen (Baraona et al. 1977; Lieber 1980). A delayed appearance of newly labeled albumin and transferrin in the plasma has also been noted, and an increase in the intracellular content of these proteins was observed. While the retention of proteins destined for secretion could account for only a small fraction of the increased cellular nitrogen, these observations certainly supported the suggestion that ethanol was interfering with secretion. Other glycoproteins have also been reported to be retained and secreted at a lower rate in rat liver slices (Sorrell and Tuma 1978, 1979; Sorrell et al. 1977). On the other hand, when hepatocytes from rats not previously exposed to alcohol were employed, an acute ethanol effect on albumin and total protein secretion was not observed (Morland et al. 1981). While these conflicting data remain to be resolved, other data show that both acute and chronic ethanol exposure result in altered microtubular content and appearance, data certainly in favor of a transport defect produced in some fashion by ethanol (Matsuda et al. 1979).

Although specific changes have been reported, these acute studies in vitro have to be contrasted with chronic studies using both in vivo and in vitro assay systems. Chronic alcohol administration does not seem to produce a definite alteration in hepatic amino acid levels, but again a causal relationship to the changes in protein synthesis has not been demonstrated (Baraona et al. 1982). Likewise, the acute administration of ethanol to animals chronically treated with ethanol has not yielded uniform results (Baraona et al. 1982; Morland 1974). Increased protein synthesis by free ribosomes has been reported, and decreased as well as increased protein synthesis by bound ribosomes freed from their endoplasmic reticulum (Zern et al. 1983). Chronic administration of alcohol in the absence of an acute alcohol load has shown both increased and decreased protein synthesis (Baraona et al. 1982). It is only necessary to return to the chronic human in vivo studies in cirrhotic ascitic subjects, where seven of 19 patients had depressed albumin synthesis and 12 showed normal or elevated production (Rothschild et al. 1969a).

Another important point to be considered, regardless of the mechanisms, is whether, if there is an acute or chronic effect on hepatic protein metabolism, this has any causal relationship to hepatic damage. The hepatic acinus may be divided into three anatomic and metabolic areas (Rappaport 1976). The third zone of Rappaport nearest the hepatic vein shows the most sensitivity to stress and reveals the lowest oxygen tensions. It is possible that ethanol toxicity may be expressed in this area initially and hepatic damage follows hepatic dysfunction (Jauhonen et al. 1982). What role decreased or increased protein synthesis by these cells plays in the development of liver disease is still only speculative. Alcohol thus appears to result in altered protein synthesis or transport and chronic effects may be attenuated. Organs respond specifically and the offending mechanism is unknown.

References

Algranati ID, Sabatini DD (1979) Effect of protease inhibitors on albumin secretion in hepatoma cells. Biochem Biophys Res Commun 90: 220

Bantle JA, Courchesne CL, Couch M (1980a) Complexity and complexity overlap in mouse liver polyadenylated and nonadenylated messenger RNA fractions. Biochem Biophys Res Commun 95: 1710–1721

Bantle JA, Edmisten GE, Doherty MD (1980b) Effect of chronic ethanol and sucrose ingestion on liver polysoma poly(A) mRNA content and incorporation of 5-^{3}H uridine into mRNA. Biochem Pharmacol 29: 2993–3001

Baraona E, Leo MA, Borowsky SA, Lieber CS (1977) Pathogenesis of alcohol-induced accumulation of protein in the liver. J Clin Invest 60: 546–554

Baraona E, Pikkarainen P, Salaspuro M, Finkelman F, Lieber CS (1980) Acute effects of ethanol on hepatic protein synthesis and secretion in the rat. Gastroenterology 79: 104–111

Baraona E, Shaw S, Lieber CS (1982) Alcohol induced changes in amino acid and protein metabolism in medical disorders of alcoholism. In: Lieber CS (ed) Medical disorder of alcoholism. Saunders, Philadelphia, pp 178–236

Bjorneboe M (1946) Studies on the serum proteins in hepatitis. The relations between serum albumin and serum globulin. Acta Med Scand 123: 393–401

Bjorneboe M, Schwartz M (1959) Investigations concerning the changes in serum proteins during immunization. The cause of hypoalbuminemia with high gamma globulin values. J Exp Med 110: 259–270

Blobel G, Dobberstein B (1975) Transfer of proteins across membranes I. Presence of proteolytically processes and unprocessed nascent immunoglobulin light chains on membrane-bound ribosomes of murine myeloma. J Cell Biol 67: 835–851

Burke JP, Rubin E (1979) The effects of ethanol and acetaldehyde on the products of protein synthesis by liver mitochondria. Lab Invest 41: 393

Cain GD, Mayer G, Jones EA (1970) Augmentation of albumin but not fibrinogen synthesis by corticosteroids in patients with hepatocellular disease. J Clin Invest 49: 2198–2204

Dich J, Tonnesen IC (1980) Effects of ethanol, nutritional status, and composition of the incubation medium on protein synthesis in isolated rat liver parenchymal cells. Arch Biochem Biophys 204: 640–647

Dorling PR, Quinn PS, Judah JD (1975) Evidence for the coupling of biosynthesis and secretion of serum albumin in rat. Biochem J 152: 341–348

Emr SD, Hall MN, Silhavy TJ (1980) A mechanism of protein localization: the signal hypothesis and bacteria. J Cell Biol 86: 701–711

Enwonwu CO, Munro HN (1970) Rate of RNA turnover in rat liver in relation to intake of protein. Arch Biochem Biophys 138: 532–539

Feldhoff RC, Taylor JM, Jefferson LS (1977) Synthesis and secretion of rat albumin in vivo, in perfused liver, and in isolated hepatocytes. J Biol Chem 252: 3611–3616

Feldmann G (1981) Morphologic aspects of hepatic synthesis and secretion of plasma proteins. In: Hand P, Schaffner F (eds) Progress in liver disease. Grune and Stratton, New York, pp 23–41

Feldmann G, Penaud-Laurencin J, Crassous J, Benhamou JP (1972) Albumin synthesis by human liver cells: its morphological demonstration. Gastroenterology 63: 1036–1048

Flaim KE, Peavy DE, Everson WV, Jefferson LS (1982a) The role of amino acids in the regulation of protein synthesis in perfused rat liver I. Reduction in rates of synthesis resulting from amino acid deprivation and recovery during flow through perfusion. J Biol Chem 257: 2932–2938

Flaim KE, Liao WSL, Peavy DE, Taylor JM, Jefferson LS (1982b) The role of amino acids in the regulation of protein synthesis in perfused rat liver II. Effects of amino acid deficiency on peptide chain initiation, polysomal aggregation, and distribution of albumin mRNA. J Biol Chem 257: 2939–2946

Freeman T, Gordon AH (1964) Metabolism of albumin and gamma globulin in protein deficient rats. Clin Sci 26: 17–26

Glaumann H, Ericsson JLE (1970) Evidence for the participation of the Golgi apparatus in the intracellular transport of nascent albumin in the liver cell. J Cell Biol 47: 555–567

Gordon GG, Southern AL (1982) The effects of alcohol and alcoholic liver disease on the endocrine system and intermediary metabolism. In: Lieber CS (ed) Medical disorder of alcoholism. Saunders, Philadelphia, pp 65–140

Guillouzo A, Beaumont C, Le Rumeur E, Rissel M, Latinier MF, Guguen-Guillouzo C, Bourel M (1982) New findings on immunolocalization of albumin in rat hepatocytes. Biol Cell 43: 163–172

Hakkinen HM, Kulonene E (1980) Effect of ethanol administration on free proline and glutamate in the intact rat liver. Biochem Pharmacol 29: 1435–1439

Hoffenberg R, Black E, Brock JF (1966) Albumin and gamma globulin tracer studies in protein depletion states. J Clin Invest 45: 143–152

Jauhonen P, Baraona E, Miyakawa H, Lieber C (1982) Mechanism for selective perivenular hepatotoxicity of ethanol alcoholism. Clin Exp Res 6: 351–357

Jeejeebjoy KN, Samuel AM, Singh B, Nadkarni GD, Desai HG, Borkar AV, Mani LS (1969) Metabolism of albumin and fibrinogen in patients with tropical sprue. Gastroenterology 56: 252–267

Jeejeebhoy KN, Phillips MJ, Bruce-Robertson A, Ho J, Sodtke U (1972) The acute effect of ethanol on albumin, fibrinogen and transferrin synthesis in the rat. Biochemistry 126: 1111–1126
Khawaja JA, Lindholm DB (1978) Differential effect of ethanol ingestion on the protein synthetic activities of free and membrane-bound ribosomes from liver of the weanling rat. Res Commun Chem Pathol Pharmacol 19: 129–139
Krebs HA, Hems R, Lund P (1973) Accumulation of amino acids by the perfused rat liver in the presence of ethanol. J Biochem 134: 697–705
Krieg L, Alonso A, Winter H, Volm M (1980) Albumin messenger RNA after partial hepatectomy and sham operation. Biochim Biophys Acta 610: 311–317
Lieber CS (1980) Alcohol protein metabolism and liver injury. Gastroenterology 79: 373–390
Lieber CS (1982) Metabolism of ethanol. In: Lieber CS (ed) Medical disorder of alcoholism. Saunders, Philadelphia, pp 1–42
Matsuda Y, Baraona E, Salaspuro M, Lieber CS (1979) Effects of ethanol on liver microtubules and Golgi apparatus. Possible role in altered hepatic secretion of plasma proteins. Lab Invest 41: 455
Morland J (1974) Effects of chronic ethanol treatment on tryptophan oxygenase, tyrosine amino transferase and general protein metabolism in the intact and perfused rat liver. Biochem Pharmacol 73: 71–75
Morland J, Bessesen A, Svendsen L (1979a) Incorporation of labelled amino acids into proteins of isolated parenchymal rat liver cells in the absence and presence of ethanol. Biochim Biophys Acta 561: 464–474
Morland J, Flengsrud R, Prydz H, Svendsen L (1979b) Hepatic amino acid levels in rats after long-term ethanol feeding. Biochem Pharmacol 28: 423–427
Morland J, Bessensen A, Svendsen L (1979c) Incorporation of labelled amino acids into proteins of isolated parenchymal and nonparenchymal rat liver cells in the absence and presence of ethanol. Biochim Biophys Acta 561: 464–474
Morland J, Bessensen A, Svendsen L (1980) The role of alcohol metabolism in the effect of ethanol on protein synthesis in isolated rat hepatocytes. J Alcoholism Clin Exp Res 4: 313–321
Morland J, Rothschild MA, Oratz M, Mongelli J, Donor D, Schreiber SS (1981) Protein secretion in suspensions of isolated rat hepatocytes: no influence of acute ethanol administration. Gastroenterology 80: 159–165
Mueckler MM, Pitot HC (1981) Structure and function of rat liver polysome populations I. Complexity, frequency distribution and degree of uniqueness of free and membrane-bound polysomal polyadenylate-containing RNA populations. J Cell Biol 90: 495–506
Mueckler MM, Pitot HC (1982) Structure and function of rat liver polysome populations II. Characterization of polyadenylate-containing mRNA associated with subpopulations of membrane-bound particles. J Cell Biol 94: 297–307
Munro HN (1970) Free amino acid pools and their role in regulation. In: Munro HN (ed) Mammalian protein metabolism. Academic, New York, pp 299–386
Munro HN, Wassington NS, Begg DL (1965) Effect of protein intake on ribonucleic acid metabolism in liver cell nuclei of the rat. J Nutr 85: 319–328
Oratz M, Rothschild MA, Schreiber SS (1976) Alcohol, amino acids and albumin synthesis. II. Alcohol inhibition of albumin synthesis reversed by arginine and spermine. Gastroenterology 71: 123–127
Oratz M, Rothschild MA, Schreiber SS (1977) Albumin, structure, function and uses. In: Rosenoer V, Oratz M, Rothschild MA (eds) Albumin – osmotic function. Pergamon, Oxford, pp 283–304
Oratz M, Rothschild MA, Schreiber SS (1978) Alcohol, amino acids, and albumin synthesis III. Effects of ethanol, acetaldehyde, and 4-methylpyrazole. Gastroenterology 74: 672–676
Oratz M, Rothschild MA, Schreiber SS et al. (1980) Spermine stimulation of CC14 cepressed protein synthesis in rabbits. Gastroenterology 79: 1165–1173
Oratz M, Rothschild MA, Schreiber SS, Burks A, Mongelli J, Matarese B (1983) The role of the urea cycle and polyamines in albumin synthesis. Hepatology 3: 567–571
Pegg AE, McCann PP (1982) Polyamine metabolism and function: a review. Am J Cell Physiol 243: C212–C221
Peters JE, Steele WJ (1982) Differential effect of chronic ethanol administration on rates of protein synthesis on free and membrane-bound polysomes in vivo in rat liver during dependence development. Biochem Pharmacol 31: 2059–2063

Peters T Jr, Fleischer B, Fleischer S (1971) The biosynthesis of rat serum albumin IV. Apparent passage of albumin through the Golgi apparatus during secretion. J Biol Chem 246:240–244

Petit MA, Barral-Alix I (1979) Effect of ethanol dose on amino acid and urea concentrations in the fed rat liver in vivo. J Biochem Physiol 28:2591–2596

Piccirillo VJ, Chambers JW (1975) The effects of ethanol and ethanol metabolites on urea production by the isolated perfused rat liver. IRCS Med Sci Alimentary System Biochem Pharmacol 3:186

Piccirillo VJ, Chambers JW (1976) Inhibition of hepatic uptake of alpha aminoisobutyric acid by ethanol: effects of pyrazole and metabolites of ethanol. Res Commun Chem Pathol Pharmacol 13:297–308

Picou D, Taylor RT (1969) The measurements of total protein synthesis and catabolism and nitrogen turnover in infants in different nutritional states and receiving different amounts of dietary protein. Clin Sci 36:283–296

Poso H (1980) Inhibition of protein degradation in regenerating rat liver by ethanol treatment. Pharmacol Biochem Behav 13:83–88

Poso H, Vaananen H, Salaspuro MP, Poso AR (1980) Effects of ethanol on liver regeneration after partial hepatectomy in rats. Med Biol 58:329–336

Post J, Patek AJ Jr (1942a) Serum proteins in cirrhosis of the liver. I Relation to prognosis and to formation of ascites. Arch Intern Med 69:67–82

Post J, Patek AJ Jr (1942b) Serum proteins in cirrhosis of the liver. II Nitrogen balance studies on five patients. Arch Intern Med 69:83–89

Prehn S, Nurnberg P, Rapoport (1981) A receptor for signal segments of secretory proteins in rough endoplasmic reticulum membranes. FEBS Lett 123:79–84

Princen JMG, Mol-Backx GPBM, Yap SH (1981) Acute effects of ethanol intake on albumin and total protein synthesis in free and membrane-bound polyribosomes of rat liver. Biochim Biophys Acta 655:119–127

Rappaport AM (1976) The microcirculatory acinar concept of normal and pathological hepatic structure. Beitr Pathol 157:215–243

Redman CM, Cherian GM (1972) The secretion pathways of rat serum glycoproteins and albumin. Localization of newly formed proteins within the endoplasmic reticulum. J Cell Biol 52:231–245

Rosa J, Rubin E (1980) Effects of ethanol on amino acid uptake by rat liver cells. Lab Invest 43:366–372

Rothman E (1981) The Golgi apparatus: two organelles in tandem. Science 213:1212–1219

Rothschild MA, Bauman A, Yallow RS, Berson SA (1957) The effect of large doses of desiccated thyroid on the distribution and metabolism of albumin I^{131} in euthyroid subjects. J Clin Invest 36:422

Rothschild MA, Schreiber SS, Oratz M, McGee HL (1958) The effects of adrenocortical hormones on albumin metabolism studied with albumin I^{131}. J Clin Invest 37:1229

Rothschild MA, Oratz M, Mongelli J, Schreiber SS (1968) The effects of a short term fast on albumin synthesis studied in vivo, in the perfused liver, and on amino acid incorporation by hepatic microsomes. J Clin Invest 47:2591–2599

Rothschild MA, Oratz M, Zimmon D, Schreiber SS, Weiner I, VanCaneghem A (1969a) Albumin synthesis in cirrhotic subjects with ascites studied in ^{14}C carbonate. J Clin Invest 48:344–350

Rothschild MA, Oratz M, Mongelli J, Fishman L, Schreiber SS (1969b) Amino acid regulation of albumin synthesis. Am J Nutr 98:395–403

Rothschild MA, Oratz M, Schreiber SS (1970) Changing concepts of albumin metabolism and distribution in cirrhosis of the liver. Scand J Gastrolenterol 7:17–23

Rothschild MA, Oratz M, Mongelli J, Schreiber SS (1971) Alcohol induced depression of albumin synthesis: reversal by tryptophan. J Clin Invest 50:1812–1818

Rothschild MA, Oratz M, Schreiber SS (1974) Alcohol amino acids and albumin synthesis. Gastroenterology 67:1200–1213

Rothschild MA, Oratz M, Schreiber SS (1977) Albumin synthesis. In: Rosenoer VM, Oratz M, Rothschild (eds) Albumin structure, function and uses. Pergamon, New York, pp 227–253

Rothschild MA, Oratz M, Schreiber SS, Mongelli J (1980) The effects of acetaldehyde and disulfiram on albumin synthesis in the isolated perfused rabbit liver. J Alcoholism Clin Exp Res 4:30–33

Schreiber SS, Briden K, Oratz M, Rothschild MA (1972) Ethanol acetaldehyde and myocardial synthesis. J Clin Invest 51:2820–2826

Schreir MH, Erni B, Staehelin T (1977) Initiation of mammalian protein synthesis. I, II. J Mol Biol 116:727–755

Shafritz DA, Yap SH, Strair RK (1979) Regulation of albumin synthesis in rat liver. Mol Biol Rep 5:71–78

Sidransky H, Verney E, Murty CN (1980) Effect of ethanol on polyribosomes and protein synthesis of transplantable hepatomas and host livers of rats. Proc Soc Exp Biol Med 163:335–339

Sidransky H, Verney E, Murty CN (1982) Effect of spermine on hepatic polyribosomes and protein synthesis in rats. Biochem Med 27:68–81

Sorrell MF, Tuma DJ (1978) Selective impairment of glycoprotein metabolism by ethanol and acetaldehyde in rat liver slices. Gastroenterology 75:200–205

Sorrell MF, Tuma DJ (1979) Effects of alcohol on hepatic metabolism: selected aspects. Clin Sci 57:481–489

Sorrell MF, Tuma DJ, Shcafer EC et al. (1977) Role of acetaldehyde in the ethanol-induced impairment of glycoprotein metabolism in rat liver slices. Gastroenterology 73:137–144

Thomas AAM, Benne R, Voorma HO (1981) Initiation of eukaryotic protein synthesis. FEBS Lett 128:177–185

Tuma DJ, Sorrell MF (1980) Effects of ethanol on the secretion of glycoproteins by rat liver slices. Gastroenterology 80:273–278

Walter P, Blobel G (1981) Translocation of proteins across the endoplasmic reticulum III. Signal recognition protein (SRP) causes signal sequence-dependent and site-specific arrest of chain elongation that is released by microsomal membranes. J Cell Biol 91:557–561

Weigand K, Schmid M, Villringer A, Birr Ch, Heinrich PC (1982) Hexa and pentapeptide extension of proalbumin: feedback inhibition of albumin synthesis by its propeptide in isolated hepatocytes and in the cell-free system. Biochemistry 21:6053–6059

Westergaard H, Jarnum S, Ramsoe K, Ranek L (1972) Albumin and immunoglobulin turnover in patients with chronic active liver disease treated with prednisone. Scand J Gastrolenterol 7:623

Wu JM (1981) Control of protein synthesis in rabbit reticulocytes. J Biol Chem 256:4164–4167

Wunderlich SM, Surendra Baliga B, Munro HN (1979) Rat placental protein synthesis and peptide hormone secretion in relation to malnutrition from protein deficiency or alcohol administration. J Nutr 109:1534

Zern MA, Chakraborty PR, Ruiz-Opazo N, Yap SH, Shafritz DA (1983) Development and use of a rat albumin cDNA clone to evaluate the effect of chronic ethanol administration on hepatic protein synthesis. Hepatology 3:317–322

6 Metabolism and Toxicity of Acetaldehyde

M. Salaspuro and K. Lindros

Research Unit of Alcohol Diseases, University Central Hospital of Helsinki and Research Laboratories of the State Alcohol Company, Alko Ltd Haartmaninkatu 4, 00290 Helsinki, Finland

Metabolism of Acetaldehyde

Acetaldehyde is the first oxidation product of ethanol, and under normal conditions it is oxidized further so rapidly that significant acetaldehyde concentrations can only be found in the liver. Aldehyde oxidase, xanthine oxidases, and aldehyde dehydrogenases are all capable of catalyzing aldehyde oxidation. The first two enzymes, however, have a broad substrate specificity and a low affinity for acetaldehyde ($K_m > 1$ mM), and consequently their involvement in the metabolism of acetaldehyde is insignificant (Lundquist 1970; Lindros 1978). The main enzyme oxidizing acetaldehyde is aldehyde dehydrogenase (ALDH), which catalyzes the oxidation of acetaldehyde in the presence of nicotinamide-adenine dinucleotide (NAD) as follows:

$$CH_3CHO + NAD^+ \xrightarrow[H_2O]{ALDH} CH_3COO^- + NADH + H^+ .$$

Most of the acetaldehyde formed from ethanol is subsequently oxidized to acetate in the liver.

Aldehyde Dehydrogenases

In 1949 Rucker described a bovine liver NAD-dependent ALDH with a broad substrate specificity for aldehydes. Human ALDH was isolated from the liver about 20 years later (Kraemer and Deitrich 1968; Blair and Bodley 1969). The enzyme has very low K_m values and is sufficiently active (Kraemer and Deitrich 1968) to explain why under normal circumstances only low concentrations of acetaldehyde are found outside the liver (Jacobsen 1952; Kiessling 1962). ALDH activity has been detected in various subcellular fractions, such as mitochondria (Walkenstein and Weinhouse 1953; Glenn and Vanko 1959), cytoplasm (Lundquist et al. 1962; Büttner 1965; Deitrich 1966), and microsomes (Tietz et al. 1964; Tottmar et al. 1973; Korsten et al. 1975). ALDH from the mitochondrial matrix has a low K_m for acetaldehyde and is therefore responsible for the oxidation of most of the acetaldehyde present in the liver during ethanol oxidation (Grunnet 1973; Marjanen 1973; Lindros et al. 1974; Parrilla et al. 1974). At higher acetaldehyde concentrations (> 0.4 mM) the increase

Alcohol Related Diseases in Gastroenterology
Edited by H. K. Seitz and B. Kommerell

in acetaldehyde oxidation is due to the activity of extramitochondrial ALDHs (Lindros et al. 1974; Parrilla et al. 1974).

Several molecular forms of ALDH exist in both the cytoplasm and the mitochondria (Marjanen 1973). More than ten ALDH isoenzymes have been demonstrated in rat liver by isoelectric focusing (Weiner et al. 1974). Multiple isoenzymes have also been found in horse and human liver (Koivula 1975; Eckfeldt et al. 1976), but the exact number of human liver isoenzymes has not been fully established (Kraemer and Deitrich 1968; Blair and Bodley 1969; Greenfield and Pietruszko 1977; Harada et al. 1980; Agarwal et al. 1981). ALDH isoenzymes can be specifically induced with various agents. Treatment with phenobarbital results in a twofold increase in the ALDH activity of mouse liver homogenate (Deitrich 1971) and in a tenfold increase in the cytosolic ALDH of rats of a certain genotype (Deitrich 1971; Deitrich et al. 1972). The induced isoenzyme has been partially purified and its properties clarified (Koivula and Kouvusalo 1975). Another cytosolic ALDH isoenzyme can be induced in rats by several drugs, regardless of the animal's genotype (Roper et al. 1976).

Deficient ALDH and Oriental Flushing

These multiple molecular forms of ALDH have certain physiological implications. About 50% of Japanese exhibit elevated blood acetaldehyde concentrations following alcohol ingestion (Mizoi et al. 1979). As a consequence of acetaldehyde-induced catecholamine release these individuals develop facial flushing and tachycardia (Ijiri 1974; Mizoi et al. 1979; Inoue et al. 1980) following alcohol ingestion. The acetaldehyde-mediated flushing occurs in individuals in whom one of the ALDH isoenzymes is physiologically inactive because of a genetically determined defect in the synthesis of the enzyme molecule (Agarwal et al. 1981; Ikawa et al. 1983). A direct relationship also exists among Japanese between the intensity of the alcohol-related facial flushing and tachycardia and the capacity of their red cells to oxidize acetaldehyde (Inoue et al. 1980). The high incidence of ALDH isoenzyme deficiency in Oriental communities and the effects of alcohol drinking in subjects who lack this isoenzyme have undoubtedly affected the epidemiology of alcohol abuse in these communities. Indeed, very few Japanese alcoholics have the inactive ALDH enzyme (Harada et al. 1982).

ALDH Inhibitors and the Disulfiram-Alcohol Reaction

A number of chemical compounds, including several naturally occurring substances such as those found in the fungus *Coprinus atramentarius,* are known to cause sensitizing reactions in the presence of ethanol (Fisher 1945; Barkman and Perman 1963; Genest et al. 1968). Calcinated bone meal fed to laboratory rats has been found to contain cyanamide – a powerful ALDH inhibitor (Lindros et al. 1975; Marchner and Tottmar 1976a, b). In 1937, a sensitization reaction to alcohol caused by tetramethylthiuram mono- and disulfide was described and the suggestion made that these compounds might be of use in the treatment of alcoholism (Williams 1937).

The first animal and clinical studies using the drug – disulfiram – appeared more than 10 years later (Asmussen et al. 1948; Hald and Jacobsen 1948; Blair and Bodley 1969). Subsequently, the trade name Antabuse was given to the compound, and the characteristics of the sensitization reaction to alcohol were extensively studied (Himwich 1956; Casier and Polet 1958; Perman 1962; Weissman and Koe 1969). In 1956 calcium cyanamide and its citrated derivative, calcium carbimide (CCC, Dipsan, Temposil) were proposed for clinical use in the treatment of alcoholism (Armstrong and Kerr 1956; Bell 1956; Ferguson 1956).

Both disulfiram and calcium cyanamide act primarily by inhibiting ALDH (Kitson 1977), thereby causing accumulation of acetaldehyde if alcohol is consumed. The ensuing response, commonly referred to as the disulfiram-alcohol reaction (DER) (Truitt and Walsh 1971; Kitson 1977; Elenbaas 1977), usually manifests itself with flushing of the face and neck, throbbing headache, palpitations, dyspnea, hyperventilation, tachycardia, hypotension, vertigo, weakness, nausea, vomiting, and changes on the electrocardiogram including T-wave flattening and ST-segment depression. Both these ALDH inhibitors are widely used in the treatment of alcoholism and the number of patients on daily Antabuse treatment in the United States alone is estimated to be several hundreds of thousands (Lundwall and Backeland 1971). Severe reactions may, however, sometimes occur when these drugs and alcohol are taken together. Respiratory depression, cardiovascular collapse, arrhythmias, myocardial infarction, acute congestive heart failure, convulsions, and coma have been described (Hotson and Langston 1976; Fried 1977; Haley 1979). At least 20 deaths have been reported in patients taking ALDH inhibitors, but in 13 of these an overdose of the drug was a contributory factor (Haley 1979). Recently, a new antialcoholic drug, nitrefazole (Altimol, 2-methyl-4-nitro-1-(4-nitrophenyl)-imidazole, also acting by inhibiting ALDH, has been introduced and its potent capacity described (Suokas et al. 1983).

No specific measures are available to treat established DER, though the patient must be generally and sometimes vigorously supported. Vitamin C, beta-adrenergic blocking agents, antihistamines, oxygen, and intravenous infusions of fluid and electrolytes are commonly used (Elenbaas 1977; Haley 1979). Recently, it has been shown that 4-methylpyrazole – an alcohol dehydrogenase inhibitor – rapidly abolishes the accumulation of acetaldehyde following alcohol ingestion in volunteers pretreated with calcium carbimide and in Antabuse-treated alcoholics (Lindros et al. 1981). 4-Methylpyrazole also attenuates other typical symptoms of the reaction, including facial flushing, tachycardia, and changes on the electrocardiogram (Fig. 1) (Lindros et al. 1981). However, this compound is not yet commercially available. Penicillamine and sodium metabisulfite may also be useful in the treatment of DER, since they have been reported to bind circulating blood acetaldehyde (Erzielev 1973; Nagasawa et al. 1977). The clinical value of these agents in the treatment of DER in man, however, has not been established.

The reaction to alcohol which occurs following use of oral antidiabetic agents of the sulfonyl urea group, such as tolbutamide, carbutamide, and chlorpropamide, closely resembles the DER and has also been attributed to the presence of elevated blood acetaldehyde concentration (Czyzyk and Mohnike 1957; Büttner and Portwich 1960; Larsen and Madsen 1962). In 1968 Podgainy and Bressler reported that chlorpropamide and tolbutamide directly inhibit ALDH, and furthermore that

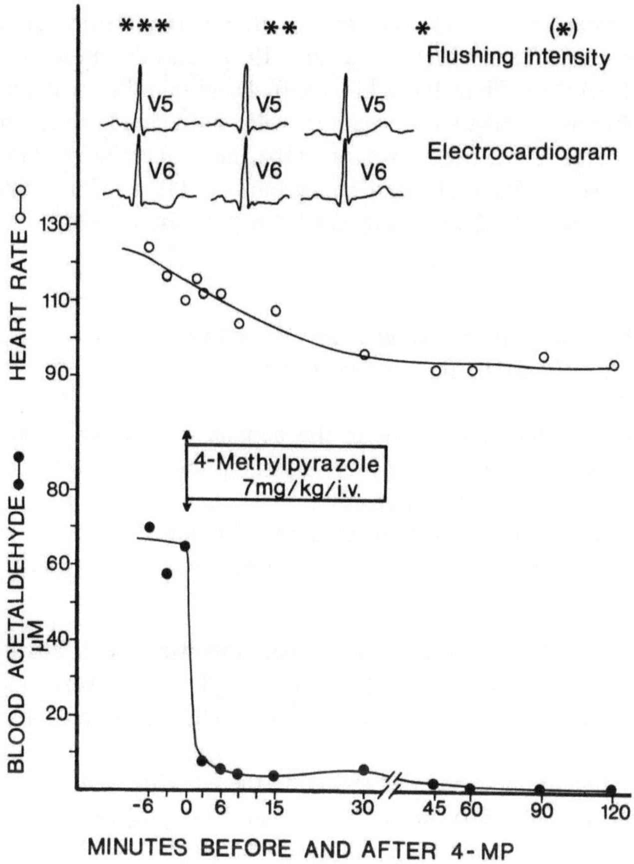

Fig. 1. Alleviation of the disulfiram-ethanol reaction by injection of 4-methylpyrazole *(4-MP)*. Flushing intensity was registered by two persons independently and was graded from marked (***) to hardly noticeable [(*)]. The electrocardiograms were taken 10 min before and 4 and 10 min after 4-methylpyrazole administration (Lindros et al. 1981)

chlorpropamide itself increases the formation of reduced metabolites of 5-hydroxytryptamine (Podgainy and Bressler 1968). Pyke and co-workers doubled interest in chlorpropamide-alcohol flushing (CPAF) in 1978 by suggesting that this phenomenon is a dominantly inherited trait found in 51% of non-insulin-dependent diabetes but in only 10% of insulin-dependent diabetes. Furthermore, non-insulin-dependent diabetics who are CPAF positive would be less likely to develop retinopathy and macroangiopathy (Leslie and Pyke 1978; Pyke and Leslie 1978; Leslie et al. 1979; Barnett and Pyke 1980). In initial studies plasma acetaldehyde concentrations in the presence of ethanol were shown to be higher in CPAF-positive than in CPAF-negative subjects (Jerntorp et al. 1980). In more recent studies, however, the results and conclusions of the earlier studies have been refuted (Köbberling and Weber 1980; de Silva and Tunbridge 1980; Köbberling et al. 1980; de Silva et al. 1981).

Other compounds known to induce alcohol-sensitizing reactions include the antioxidant and occupational agent, *n*-butyraldoxime (Lewis and Schwartz 1956; Koe and Tenen 1969; Forsander 1970), and the antibiotic metronidazole (Flagyl) (Kissin and Platz 1968; Lal 1969). Disulfiram-like activity has also been reported after use of ethanol concomitant with: sulfonamide antibacterials, chloramphenicol, furazolidone, griseofulvin, procarbazine, quinacrine, and tolazoline (Hansten 1975). However, the biochemical backgrounds of the reactions caused by these agents in the presence of ethanol have not been fully established.

Alteration in the Metabolism of Acetaldehyde After Chronic Ethanol Consumption

Three major alterations in the hepatic metabolism of ethanol and acetaldehyde secondary to chronic alcohol consumption may cause increased blood and tissue acetaldehyde concentrations:

1. The increased rate of ethanol oxidation;
2. A reduction in the capacity of mitochondria to oxidize acetaldehyde;
3. A decrease in the activity of ALDH.

In 1971 Truitt showed that following oral ethanol blood acetaldehyde concentrations in alcoholics and nonalcoholics were not significantly different, though there was a tendency toward higher concentrations in the alcoholics. In two earlier studies elevated blood acetaldehyde concentrations were found in alcoholics, but were interpreted as reflecting the acetaldehyde contained in the administered alcohol (Majchrowics and Mendelson 1970; Magrinat et al. 1973). In 1975 Korsten et al. found a difference in blood acetaldehyde concentrations between alcoholic and nonalcoholic subjects. Recent methodological advances have shown that the concentration of acetaldehyde in peripheral venous blood is much lower than reported by Korsten et al. (Pikkarainen et al. 1979; Lindros et al. 1980). However, even taking this into account, blood acetaldehyde values increase after chronic alcohol consumption both in baboons (Pikkarainen et al. 1981b) and in humans (Jenkins and Peters 1978; Lindros et al. 1980). In these reports blood acetaldehyde levels correlated positively with the rate of ethanol elimination.

Chronic alcohol consumption can also result in a significant reduction in the capacity of rat mitochondria to oxidize acetaldehyde, irrespective of the presence of substrates for NAD-linked dehydrogenase (Hasamura et al. 1975). This effect has been attributed, at least in part, to a reduction in the ability of mitochondria from ethanol-fed animals to reoxidize NADH (Cederbaum et al. 1974b). The effects of chronic alcohol consumption on ALDH activity are controversial (Eriksson and Deitrich 1983). In experimental animals decreases (Koivula and Lindros 1975; Greenfield et al. 1976; Lebsack et al. 1981), no change (Redmond and Cohen 1971; Tottmar et al. 1974; Lebsack et al. 1981), and even increases (Hasumura et al. 1975; Horton and Barrett 1975) in activity have been reported. The discrepancies in ALDH activities can be attributed to the severity of chronic alcohol treatment and to the variety of acetaldehyde concentrations used to determine ALDH activity. Acetaldehyde concentrations of less than 50 μM should be used to determine the

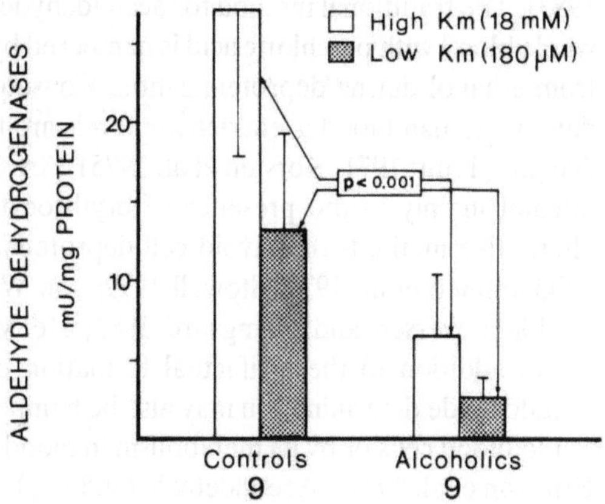

Fig. 2. The maximal activities of hepatic total and low K_m aldehyde dehydrogenase in nine alcoholics and nine controls. The activities are expressed as milliunits (nanomoles of NADH produced per minute per milligram of protein (Nuutinen et al. 1983)

activity of the low K_m mitochondrial ALDH. As stated before, this enzyme plays a major role in the metabolism of acetaldehyde, at least in rats (Eriksson et al. 1975; Petersen et al. 1977), and its activity may increase following chronic alcohol consumption. The decrease in acetaldehyde metabolism in ethanol-fed rats may sometimes reflect impaired mitochondrial oxidation of acetaldehyde rather than decreased ALDH activity (Hasumura et al. 1975). The situation may be different in other animals chronically fed alcohol and in human alcoholics. Naturally, the situation may also be influenced by nutritional factors or by associated alcoholic liver injury.

In baboons the plasma concentration of free acetaldehyde correlates positively with the production rate of acetaldehyde and negatively with liver mitochondrial ALDH activity (Pikkarainen et al. 1981b). It has also been suggested that cytosolic ALDH may play a more important role in acetaldehyde metabolism in man (Jenkins and Peters 1978). Furthermore, the results of studies in patients with alcoholic liver disease using subcellular fractionation techniques suggest that there could be a selective reduction of cytosolic high K_m ALDH (Jenkins and Peters 1980). When two acetaldehyde concentrations were used to measure ALDH, however, both total and mitochondrial enzyme activities were decreased in chronic alcoholics as compared to nonalcoholic controls (Fig. 2) (Nuutinen et al. 1983). Not only was the hepatic activity of the enzyme decreased but a significant negative correlation ($r = -0.830$, $p < 0.02$) existed between peak blood acetaldehyde values and the activity of mitochondrial low K_m ALDH (Nuutinen et al. 1983). In human alcoholics acetaldehyde concentrations in hepatic venous blood vary between 2 and 165 μM after a moderate (0.8 g/kg body weight) dose of alcohol (Nuutinen et al. 1984). If these values reflect liver acetaldehyde concentrations, then this further supports the dominant role of low K_m mitochondrial ALDH in acetaldehyde metabolism in man.

Acetaldehyde in Blood and Expired Air

Earlier attempts to examine acetaldehyde metabolism and its role in possible organ toxicity have been hampered by methodological difficulties (Eriksson 1980; Lindros

1983). The traditional method for acetaldehyde determination by deproteinization of whole blood with perchloric acid is hampered by acetaldehyde formed spontaneously from ethanol during deproteinization. Consequently, most of the earlier published data on human blood acetaldehyde levels are too high and range from near zero to 200 μ*M* (Truitt 1971; Korsten et al. 1975). Artifactual acetaldehyde is produced from ethanol mainly in the presence of erythrocytes (Stowell et al. 1977). Therefore, alternative methods that avoid cell deproteinization have recently been developed (Pikkarainen et al. 1979; Stowell 1979; von Wartburg and Ris 1979; Christensen et al. 1981; Iversen and Damgaard 1983; De Master et al. 1983).

In addition to the artifactual formation of acetaldehyde from ethanol, blood acetaldehyde determination may also be hampered by either binding of acetaldehyde to red blood cells or by its metabolism in blood cells or in plasma (Gaines et al. 1977; Eriksson et al. 1977). Acetaldehyde binding, however, has not been demonstrated in human blood (Eriksson et al. 1977). Although there is ALDH activity in human blood (Pietruszko and Vallari 1978; Inoue et al. 1978) which is responsible for some disappearance of acetaldehyde in vivo (Nuutinen et al. 1984), the reaction is too slow to influence the blood acetaldehyde determination significantly, provided the sample is processed rapidly.

The methodological difficulties involved in blood analysis can at least partly be circumvented by analysis of breath samples (Freund and O'Hollaren 1965; Fukui 1969; Zeiner et al. 1979; Stowell et al. 1980). Since, however, small amounts of acetaldehyde are formed from microbial oxidation of ethanol in the oropharynx (Pikkarainen et al. 1981a), and since lung microsomes may produce acetaldehyde from ethanol (Pikkarainen et al. 1981a), the method is of limited use at low acetaldehyde levels.

Recent studies using improved methodology for blood acetaldehyde assay show that blood acetaldehyde concentrations in peripheral venous blood of normal Caucasian subjects are extremely low (0–2 μ*M*), and are therefore often at or below the limit of detection (Pikkarainen et al. 1979; Lindros et al. 1980; Eriksson and Peachey 1980). No significant sex differences exist in blood acetaldehyde concentrations after oral administration of ethanol (Marshall et al. 1984). These findings indicate that in normal healthy Caucasian individuals almost all the acetaldehyde formed during ethanol oxidation is effectively oxidized by the liver. Recently, however, significant although modest concentrations of acetaldehyde (2–20 μ*M*) have been measured in the hepatic venous blood of moderately intoxicated nonalcoholic male Caucasians at a time when peripheral venous concentrations were less than 2 μ*M* (De Master et al. 1983; Nuutinen et al. 1984). These findings demonstrate that peripheral tissues, including blood and its constituents, have a considerable capacity to eliminate low concentrations of acetaldehyde. Significant concentrations of acetaldehyde, however, can be detected in the peripheral blood of some chronic alcoholics, in Orientals who suffer ethanol facial flushing, and in the presence of ALDH inhibitors.

As stated before, chronic alcoholics may have an enhanced rate of ethanol oxidation (metabolic tolerance) and a decreased capacity to eliminate acetaldehyde. In our studies the peripheral venous blood acetaldehyde concentration was elevated in six out of nine chronic alcoholics (Lindros et al. 1980; Nuutinen et al. 1983). In these alcoholics the main reason for blood acetaldehyde elevation was their impaired

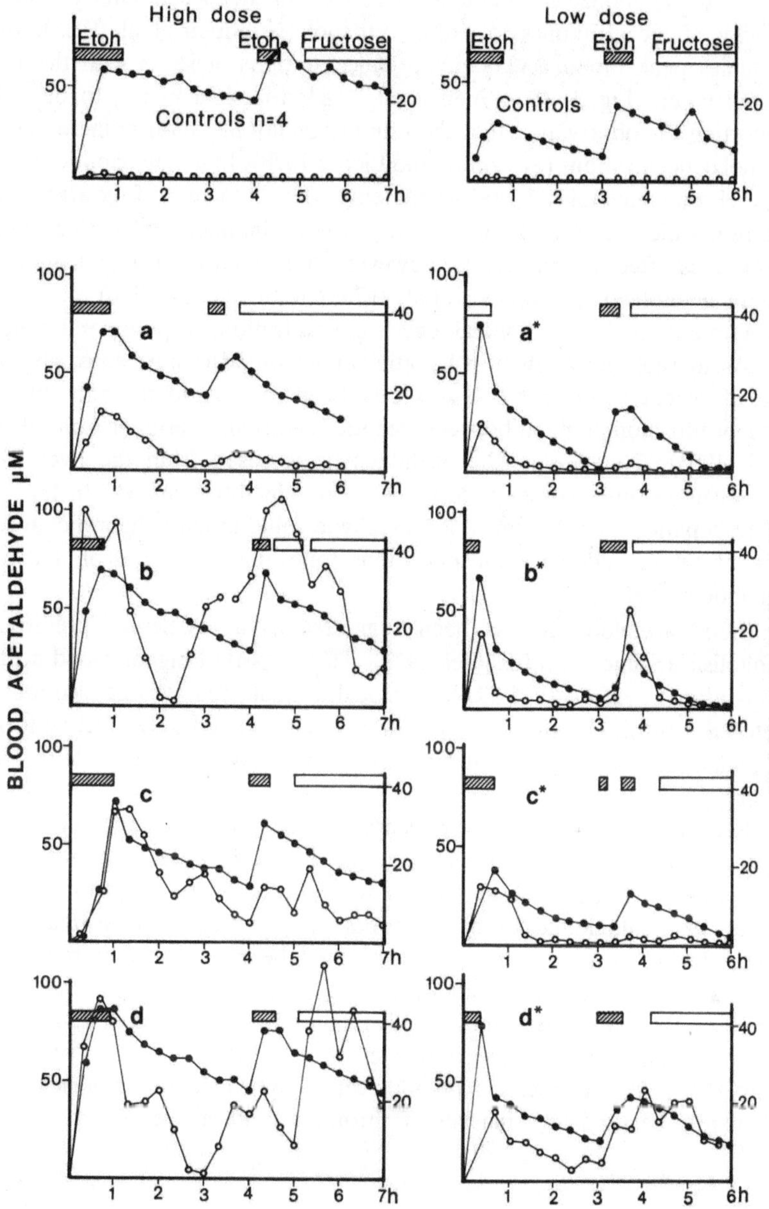

Fig. 3. The alcohol (●——●) and acetaldehyde ○——○) concentration curves for control subjects and those alcoholics who have high blood acetaldehyde levels. Note the association of peak blood acetaldehyde with peak blood ethanol and the effect of fructose infusion. High dose of alcohol was 1.2 g/kg body weight and low dose 0.6 g/kg body weight. *a–d* represent individual alcoholics (Nuutinen et al. 1983). *Etoh,* Ethanol infusion

capacity to metabolize acetaldehyde, which was further accentuated by the increased capacity of the alcoholics to oxidize ethanol (Nuutinen et al. 1983). In chronic alcoholics peak blood acetaldehyde concentrations increase with the amount of alcohol taken (Fig. 3) (Nuutinen et al. 1983; Marshall et al., to be published), supporting the observation that the rate of ethanol metabolism increases at higher ethanol concentrations (Salaspuro and Lieber 1978). In nonalcoholic controls and in some chronic alcoholics blood acetaldehyde levels increase following intravenous fructose, which accelerates the rate of ethanol elimination (Nuutinen et al. 1983, 1984). This effect of fructose is, however, more marked in nonalcoholic controls than in alcoholic patients (Bode et al. 1979; Nuutinen et al. 1984).

As mentioned above, acetaldehyde concentrations in peripheral venous blood are considerably lower than in hepatic venous blood (Nuutinen et al. 1984). Uptake of acetaldehyde in blood was calculated to account for only 30%–40% of the concentration gradient between hepatic venous and peripheral blood (Nuutinen et al. 1984). The rest of the acetaldehyde released from the liver is oxidized after diffusion into extrahepatic tissues. Consequently, changes in the peripheral – extrahepatic – metabolism of acetaldehyde could at least theoretically also influence blood acetaldehyde concentrations following chronic alcohol consumption (Lamboeuf et al. 1981).

Blood acetaldehyde has been suggested as a marker of predisposition to alcoholism (Schuckit and Rayses 1979). They reported higher blood acetaldehyde levels after a test dose of alcohol in healthy male relatives of alcoholics than in matched controls. This study was, however, confounded by artifactual formation of acetaldehyde from ethanol. Furthermore, others have not been able to confirm the finding.

Recently, it has been demonstrated that acetaldehyde may form adduct hemoglobin A, which results in an elevated concentration of minor hemoglobins (Stevens et al. 1981). The amount of these adducts was found to be significantly higher in hemoglobin from alcoholics than in that of normal volunteers. Theoretically, at least, these adducts could be used as laboratory markers of alcoholism. At the moment, the prevalence of high acetaldehyde in alcoholics is not established. On the other hand, blood and breath acetaldehyde levels are influenced by many factors, for instance blood ethanol level, smoking, nutrition, and liver injury. Therefore, an increased blood acetaldehyde level in the presence of alcohol can hardly be used as a marker of chronic alcohol abuse.

Toxicity of Acetaldehyde

Acetaldehyde is both pharmacologically and chemically a very potent and reactive compound, which for a long time has been suggested to contribute to the complications of alcoholism (Truitt and Duritz 1966; Walsh 1971). Pharmacological effects of acetaldehyde are mainly mediated through the release of vasoactive compounds such as catecholamines. Organ toxicity of acetaldehyde on the other hand can most probably be related to its capability to adduct with various tissue compounds.

Cardiovascular Effects of Acetaldehyde

Acetaldehyde stimulates the release of catecholamines (Eade 1959), which is responsible for the flushing and tachycardia associated with the DER. As mentioned earlier, a similar flushing reaction is also seen in some Orientals (Wolff 1973), who carry a genetically determined deficiency of acetaldehyde metabolism (Agarwal et al. 1981; Ikawa et al. 1983). Orientals with flushing have higher blood acetaldehyde levels than nonflushers and nonalcoholic Caucasians (Mizoi et al. 1979; Kupari et al. 1983a). Breath (Zeiner et al. 1979) but not blood acetaldehyde (Reed et al. 1976) has been reported to be higher in flushing Chinese.

During calcium cyanamide therapy, ethanol intake engenders an increase in heart rate and a decrease in diastolic blood preasure which coincide with elevated blood acetaldehyde concentrations (Peachey et al. 1981). These changes are associated with increased cardiac output and decreased peripheral vascular resistance and they can be prevented if the formation of acetaldehyde is inhibited by an alcohol dehydrogenase inhibitor (Kupari et al. 1983b). The enhancement of left ventricular function caused by acetaldehyde-induced catecholamine release and peripheral vasodilatation can also be demonstrated in flushing Orientals (Kupari et al. 1983b).

Organ Toxicity of Acetaldehyde

Chronic inhalation of acetaldehyde in mice has been related to the ethanol withdrawal reaction (Ortiz et al. 1974). The symptoms caused by acetaldehyde, however, were not completely comparable to those of ethanol. On the other hand, *t*-butanol, which is not metabolized to acetaldehyde, causes withdrawal signs similar to those of ethanol (Wallgren 1973; McComb and Goldstein 1979). The biochemical background of the acetaldehyde-induced withdrawal reaction has not been clarified. Firstly, the symptoms are associated with a transient increase in brain catecholamines (Ortiz et al. 1974), which could lead to cerebral stimulation. Secondly, acetaldehyde may condense with catecholamines and produce tetrahydroisoquinoline (TIQ) alkaloids, which could serve as false neurotransmitters (Cohen 1973). One of these alkaloids – salsolinol – is derived from dopamine and acetaldehyde (Cohen and Collins 1970). Some other more complex TIQ-alkaloids have been related to psychoactive plant products and could induce withdrawal because of their opiate-like effects (Davis and Walsh 1970; Turner et al. 1974; Santi et al. 1967). This fascinating hypothesis has, however, been criticized, since alcohol administration does not give detectable TIQ levels in vivo (Rahwan 1975; Noble and Tewari 1977).

Acetaldehyde has been shown to cross-link erythrocyte membrane proteins (Gaines et al. 1977), to brown with albumin (Mohammed et al. 1949), and to react with hemoglobin and phospholipids by adduct formation (Stevens et al. 1981;

Kenney 1982). Since high concentrations of acetaldehyde are needed for these effects, it is doubtful whether they have any relation to the hematological abnormalities seen in chronic alcoholics.

Acetaldehyde impairs myocardial protein synthesis (Schreiber et al. 1972, 1974). It also interferes with the metabolism of pyridoxal phosphate by displacing pyridoxal phosphate from its protein binding, thereby promoting the degradation of the vitamin (Veitch et al. 1975). Both the decreased plasma level of pyridoxal phosphate and the inhibition of myocardial protein synthesis may contribute to the development of alcoholic cardiomyopathy.

As mentioned earlier, only very low concentrations of acetaldehyde can be detected outside the liver and even then only in chronic alcoholics. On this basis the liver should be most vulnerable to the possible toxic effects of acetaldehyde. Experimental feeding of alcohol to rats results in significant decrease in the content of liver polymerized tubulin, the major chemical component of microtubules (Baraona et al. 1977). The decrease in and malformation of microtubules caused by alcohol has also been documented morphologically (Matsuda et al. 1979). Acetaldehyde has been demonstrated to be the major responsible agent for these effects of ethanol (Baraona et al. 1979). Acetaldehyde-mediated alteration of microtubules is responsible, at least in part, for the impaired protein secretion after ethanol (Lieber 1982; Baraona et al. 1980), which has been related to the pathogenesis of alcoholic liver injury.

High concentrations of acetaldehyde may impair mitochondrial oxidative phosphorylation (Cederbaum et al. 1974a) and fatty acid oxidation (Matsuzaki and Lieber 1977). Acetaldehyde may bind with cysteine or glutathione or both and consequently may contribute to a depression of liver glutathione (Shaw et al. 1981). Especially in rats fed chronically with alcohol, administration of alcohol enhances lipid peroxidation, which can partly be prevented with methionine, a precursor of glutathione (Shaw et al. 1981). A hypothetical link between accelerated acetaldehyde production, increased free radical generation by the "induced" liver microsomes, and enhanced lipid peroxidation has been suggested by Lieber (1980).

Acetaldehyde has been shown to have a direct toxic effect on the gonads and to decrease synthesis of testosterone (Cicero et al. 1979, 1980; Ellingboe and Varinelli 1979; Cobb et al. 1980). Acetaldehyde could be either generated locally by intratesticular metabolism of alcohol or it could reach the gonads via the bloodstream. The biochemical background of the phenomenon has not so far been clarified (Cicero and Bell 1980).

Conclusions

Reliable methods for the determination of blood acetaldehyde levels have been developed during the last few years. The results obtained indicate that small amounts of acetaldehyde can be detected in the hepatic venous blood of nonalcoholic Caucasians. In peripheral venous blood significant concentrations of acetaldehyde can be found in some chronic alcoholics, in flushing Orientals with a genetically determined deficiency in the metabolism of acetaldehyde, and after the use of aldehyde dehydrogenase inhibitors. In chronic alcoholics an increase in blood acetaldehyde is mainly caused by the decreased activity of mitochondrial aldehyde dehydrogenase, which is contributed to by the enhanced rate of ethanol oxidation. Acetaldehyde has been linked to several organ-toxic manifestations of alcohol, but its definitive role in organ toxicity has not so far been fully proved.

References

Agarwal DP, Harada S, Goedde HW (1981) Racial differences in biological sensitivity to ethanol: the role of alcohol dehydrogenase and aldehyde dehydrogenase isozymes. Alcoholism 5: 12–16

Armstrong JD, Kerr HT (1956) A new protective drug in the treatment of alcoholism. (Preliminary trial of citrated calcium carbimide). Can Med Assoc J 74: 795–797

Asmussen E, Hald J, Jacobsen E, Jørgensen G (1948) Studies on the effect of tetraethylthiuramdisulphide (Antabuse) and alcohol on respiration and circulation in normal human subjects. Acta Pharmacol Toxicol 4: 297–304

Baraona E, Leo MA, Borowsky SA, Lieber CS (1977) Pathogenesis of alcohol-induced accumulation of protein in the liver. J Clin Invest 60: 546–554

Baraona E, Matsuda Y, Pikkarainen P, Finkelman F, Lieber CS (1979) Exaggeration of the ethanol-induced decrease in liver microtubules after chronic alcohol consumption. Gastroenterology 76: 1274

Baraona E, Pikkarainen P, Salaspuro M, Finkelman F, Lieber CS (1980) Acute effects of ethanol on hepatic protein synthesis and secretion in the rat. Gastroenterology 79: 104–111

Barkman R, Perman ES (1963) Supersensitivity to ethanol in rabbits treated with *Coprinus atramentarius*. Acta Pharmacol Toxicol 20: 43–46

Barnett AH, Pyke DA (1980) Chlorpropamide-alcohol flushing and large-vessel disease in non-insulin dependent diabetes. Br Med J 2: 261–262

Bell RG (1956) Clinical trial of citrated calcium carbimide. Can Med Assoc J 74: 797–798

Blair AH, Bodley FH (1969) Human liver aldehyde dehydrogenase: partial purification and properties. Can J Biochem 47: 265–272

Bode JCh, Bode Ch, Thiele D (1979) Alcohol metabolism in man: effect of intravenous fructose infusion on blood ethanol elimination rate following stimulation by phenobarbital treatment or chronic alcohol consumption. Klin Wochenschr 57: 125–130

Büttner H (1965) Aldehyd- und Alkoholdehydrogenase. Aktivität in Leber und Niere der Ratte. Biochem Zeitsch 341: 300–314

Büttner H, Portwich F (1960) Wirkungen des *N*-(4-Methylbenzosulfonyl)-*N*-butyl-harnstoffs (D869) auf den Stoffwechsel des Äthanols. Arch Exp Pathol Pharmakol 238: 45–46

Casier H, Polet H (1958) Influence du disulfiram (Antabus) sur le metabolisme de l'alcool éthylique marqué chez la souris. Arch Int Pharmacodyn Ther 113: 439–496

Cederbaum AI, Lieber CS, Rubin E (1974a) The effect of acetaldehyde on mitochondrial function. Arch Biochem Biophys 161: 26–39

Cederbaum AI, Lieber CS, Rubin E (1974b) Effects of chronic ethanol treatment on mitochondrial functions. Arch Biochem Biophys 165: 560–569

Christensen JM, Angelo H, Knop J (1981) Determination of acetaldehyde in human blood by gas chromatographic method with negligible artefactual acetaldehyde formation. Clin Chim Acta 116: 389–395

Cicero TJ, Bell RD (1980) Effects of ethanol and acetaldehyde on the biosynthesis of testosterone in the rodent testes. Biochem Biophys Res Commun 94: 1814–1819

Cicero TJ, Meyer ER, Bell RD (1979) Effects of ethanol on the hypothalamic-pituitary-luteinizing hormone axis and testicular steroidogenesis. J Pharmacol Exp Ther 208: 210–215

Cicero TJ, Bell RD, Meyer ER, Badger TM (1980) Ethanol and acetaldehyde directly inhibit testicular steroidogenesis. J Pharmacol Exp Ther 213: 228–233

Cobb CF, Ennis MF, van Thiel DH, Gavaler JS, Lester R (1980) Isolated testes perfusion: a method using a cell and protein-free perfusate useful for the evaluation of potential drug and/or metabolic injury. Metabolism 29: 71–79

Cohen G (1973) Tetrahydroisoquinoline alkaloids: utpake, storage and secretion by the adrenal medulla and by adrenergic nerves. Ann NY Acad Sci 215: 116–119

Cohen G, Collins M (1970) Alkaloids from catecholamines in adrenal tissue: possible role in alcoholism. Science 167: 1749–1751

Czyzyk A, Mohnike G (1957) Über die Beeinflussung der Alkoholtoleranz durch blutzuckersenkende Harnstoffderivate. Dtsch Med Wochenschr 82: 1585–1586

Davis VE, Walsh MJ (1970) Alcohol, amines and alkaloids: possible biochemical basis for alcohol addiction. Science 167: 1005–1007

Deitrich RA (1966) Tissue and subcellular distribution of mammalian aldehydeoxidizing capacity. Biochem Pharmacol 15: 1911–1922

Deitrich RA (1971) Genetic aspects of increase in rat liver aldehyde dehydrogenase induced by phenobarbital. Science 173: 334–336

Deitrich RA, Collins AC, Erwin VG (1972) Genetic influence upon phenobarbital-induced increase in rat liver supernatant aldehyde dehydrogenase activity. J Biol Chem 247: 7232–7236

De Master EG, Redfern B, Weir EK, Pierpont GL, Crouse LJ (1983) Elimination of artifactual acetaldehyde in the measurement of human blood acetaldehyde by the use of polyethylene glycol and sodium azide: normal blood acetaldehyde levels in the dog and human after ethanol. Alcoholism 7: 436–442

deSilva NE, Tunbridge WMB (1980) Chlorpropamide alcohol flushing in noninsulin dependent diabetics. Diabetologia 19: 269

deSilva NE, Tunbridge WMG, Alberti KGMM (1981) Low incidence of chlorpropamide-alcohol flushing in diet-treated, non-insulin-dependent diabetes. Lancet 1: 128–131

Eade NR (1959) Mechanisms of sympathomimetic action of aldehydes. J Pharmacol Exp Ther 127: 29–34

Eckfeldt J, Mope L, Tako K, Yonetani T (1976) Horse liver aldehyde dehydrogenase. Purification and characterization of two isozymes. J Biol Chem 251: 236–240

Elenbaas RM (1977) Management of the disulfiram-alcohol reaction. N Engl Med Cent Hosp Forum 6: 8

Ellingboe J, Varinelli CC (1979) Ethanol inhibits testosterone biosynthesis by direct action on Leydig cells. Res Commun Chem Pathol Pharmacol 24: 87–102

Eriksson CJP (1980) Problems and pitfalls in acetaldehyde determinations. Alcoholism 4: 22–29

Eriksson CJP, Deitrich RA (1983) Metabolic mechanisms in tolerance and physical dependence on alcohol. In: Kissin B, Begleiter H (eds) The pathogenesis of alcoholism, vol 7. Plenum, London, pp 253–283

Eriksson CJP, Peachey JE (1980) Lack of difference in blood acetaldehyde of alcoholics and controls after ethanol ingestion. Pharmacol Biochem Behav [Suppl 1] 13: 101–105

Eriksson CJP, Marselos M, Koivula T (1975) Role of cytosolic rat liver aldehyde dehydrogenase in the oxidation of acetaldehyde during ethanol metabolism in vivo. Biochem J 152: 709–712

Eriksson CJP, Sippel HW, Forsander OA (1977) The occurrence of acetaldehyde binding in rat but not in human blood. FEBS Lett 75: 205–208

Erzielev GI (1973) Acetaldehyde and alcoholism. Pharmacogenesis of a disulfiram-alcohol reaction and its management by binding acetaldehyde with sodium metabisulfite. Sov Neurol Psych 6: 42–51

Ferguson JKW (1956) A new drug for the treatment of alcoholism. Can Med Assoc J 74: 793–795

Fischer I (1945) Säregen svampförgiftning. Svensk Läkartid 42: 2513–2515

Forsander OA (1970) Influence of ethanol and butyraldoxime on liver metabolism. Biochem Pharmacol 19: 2131–2136

Freund G, O'Hollaren P (1965) Acetaldehyde concentrations in alveolar air following a standard dose of ethanol in man. Lipid Res 6: 471–477

Fried R (1977) Comments on Antabuse therapy in alcoholism. Alcoholism 1: 275–276

Fukui Y (1969) Gas chromatograph determination of acetaldehyde in the expired air after ingestion of alcohol. Jp J Leg Med 23: 24

Gaines KC, Salhany JM, Tuma DJ, Sorrell MF (1977) Reaction of acetaldehyde with human erythrocyte membrane proteins. FEBS Lett 74: 115–119
Genest K, Coldwell BB, Hughes DW (1968) Potentiation of ethanol by *Coprinus atramentarius* in mice. J Pharm Pharmacol 20: 102–106
Glenn JL, Vanko M (1959) Choline and aldehyde oxidation by rat liver. Arch Biochem Biophys 82: 145–152
Greenfield NJ, Pietruszko R (1977) Two aldehyde dehydrogenases from human liver. Isolation via affinity chromatography and characterization of the isozymes. Biochim Biophys Acta 483: 35–45
Greenfield NJ, Pietruszko R, Lin G, Lester D (1976) The effect of ethanol ingestion on the aldehyde dehydrogenases of rat liver. Biochim Biophys Acta 428: 627–632
Grunnet N (1973) Oxidation of acetaldehyde by rat-liver mitochondria in relation to ethanol oxidation and the transport of reducing equivalents across the mitochondrial membrane. Eur J Biochem 35: 236–243
Hald J, Jacobsen E (1948) The formation of acetaldehyde in the organism after ingestion of Antabuse (tetraethylthiuramdisulphide) and alcohol. Acta Pharmacol Toxicol 4: 305–310
Haley T (1979) Disulfiram (tetraethylthiopeoxydicarbonic diamide): a reappraisal of its toxicity and therapeutic application. Drug Metab Rev 9: 319–335
Hansten PD (1975) Drug interactions, 3rd edn. Lea and Febinger, Philadelphia, pp 156–161
Harada S, Misawa S, Agarwall DP, Goedde HW (1980) Liver alcohol dehydrogenase and aldehyde dehydrogenase in the Japanese: isozyme variation and its possible role in alcohol intoxication. Am J Hum Genet 32: 8–15
Harada S, Takagi S, Agarwall DP, Goedde HW (1982) Ethanol and aldehyde metabolism in alcoholics from Japan. Abstract 1st ISBRA Congress, Munich 1982. Alcoholism 6: 298
Hasumura Y, Teschke R, Lieber CS (1975) Acetaldehyde oxidation by hepatic mitochondria. Decrease after chronic ethanol consumption. Science 189: 727–729
Himwich HE (1956) Alcohol and brain physiology. In: Thompson GN (ed) Alcoholism. Thomas, Springfield, pp 291–408
Horton AA, Barrett MC (1975) Subcellular localization of aldehyde dehydrogenase in rat liver. Arch Biochem Biophys 167: 426–436
Hotson J, Langston W (1976) Disulfiram-induced encephalopathy. Arch Neurol 33: 141–142
Ijiri I (1974) Studies on the relationship between the concentrations of blood acetaldehyde and urinary cytecholamine and the symptoms after drinking alcohol. Jpn J Stud Alcohol 9: 35–39
Ikawa M, Impraim CC, Wang G, Yoshida A (1983) Isolation and characterization of aldehyde dehydrogenase isozymes from usual and atypical human livers. J Biol Chem 258: 6282–6287
Inoue K, Ohdora Y, Yamasawa K (1978) Metabolism of acetaldehyde by human erythrocytes. Life Sci 23: 179–184
Inoue K, Rukunaga M, Yamasawa K (1980) Correlation between human erythrocyte aldehyde dehydrogenase activity and sensitivity to alcohol. Pharmacol Biochem Behav 13: 295–297
Iversen HL, Damgaard SE (1983) Determination of acetaldehyde in human blood using thiourea to inhibit ethanol interference. Clin Chim Acta 135: 151–158
Jacobsen E (1952) The metabolism of ethyl alcohol. Pharmacol Rev 4: 107–135
Jenkins WJ, Peters TJ (1978) Subcellular distribution and properties of aldehyde dehydrogenase in human liver. Clin Sci Mol Med 55: 11P–12P
Jenkins WJ, Peters TJ (1980) Selectively reduced hepatic acetaldehyde dehydrogenase in alcoholics. Lancet 1: 628–629
Jerntorp P, Öhlin H, Bergström B, Almer LO (1980) Elevation of plasma acetaldehyde: the first step in CPAF? Diabetologia 19: 286
Kenney WC (1982) Acetaldehyde adducts of phospholipids. Alcoholism 6: 412–416
Kiessling K-H (1962) The occurrence of acetaldehyde in various parts of the rat brain after alcohol injection, and its effect on the pyruvate oxidation. Exp Cell Res 27: 367–368
Kissin B, Platz A (1968) The use of drugs in the long term rehabilitation of chronic alcoholics. In: Efron DE (ed) Psychopharmacology: a review of progress 1957–1967. Public Health Service no 1836, Washington, pp 835–851
Kitson TM (1977) The disulfiram-ethanol reaction. J Stud Alcohol 38: 96–113

Koe BK, Tenen SS (1969) Blockade of ethanol metabolism and reduced alcohol selection in C57BL mice by butyraldoxime. Fed Proc 28: 546
Köbberling J, Weber M (1980) Facial flush after chlorpropamidealcohol and enkephalin. Lancet 1: 538–539
Köbberling J, Bengsch N, Brüggeboes B, Schwarck H, Tillil H, Weber M (1980) The chlorpropamide alcohol flush: lack of specificity for familial noninsulin dependent diabetes. Diabetologia 19: 359–363
Koivula T (1975) Subcellular distribution and characterization of human liver aldehyde dehydrogenase fractions. Life Sci 16: 1563–1570
Koivula T, Koivusalo M (1975) Partial purification and properties of a phenobarbital-induced aldehyde dehydrogenase of rat liver. Biochim Biophys Acta 410: 1–11
Koivula T, Lindros KO (1975) Effects of long term ethanol treatment on aldehyde and alcohol dehydrogenase activities in rat liver. Biochem Pharmacol 24: 1937–1942
Korsten MA, Matsuzaki S, Feinman L, Lieber CS (1975) High blood acetaldehyde levels after ethanol administration. N Engl J Med 292: 386–389
Kraemer RJ, Deitrich RA (1968) Isolation and characterization of human liver aldehyde dehydrogenase. J Biol Chem 243: 6402–6408
Kupari M, Eriksson CJP, Ylikahri R (1983a) Alcohol and the heart. Intense hemodynamic changes associated with alcohol flush in Orientals. Acta Med Scand 213: 91–98
Kupari M, Lindros K, Hillbom M, Heikkilä J, Ylikahri R (1983b) Cardiovascular effects of acetaldehyde accumulation after ethanol ingestion: their modification by β-adrenergic blockade and alcohol dehydrogenase inhibition. Alcoholism 7: 283–288
Lal S (1969) Metronidazole in the treatment of alcoholism. A clinical trial and review of the literature. J Stud Alcohol 30: 140–151
Lamboeuf G, de Saint Blanquat G, Betache R (1981) Mucosal alcohol dehydrogenase – and aldehyde dehydrogenase – mediated ethanol oxidation in the digestive tract of the rat. Biochem Pharmacol 30: 542–545
Larsen JA, Madsen J (1962) Inhibition of ethanol metabolism by oral antidiabetics. Proc Soc Exp Biol Med 109: 120–122
Lebsack ME, Gordon ER, Lieber CS (1981) The effect of chronic ethanol consumption on aldehyde dehydrogenase activity in the baboon. Biochem Pharmacol 30: 2273–2277
Leslie RDG, Pyke DA (1978) Chlorpropamide-alcohol flushing: a dominantly inherited trait associated with diabetes. Br Med J 2: 1519–1521
Leslie RDG, Barnett AH, Pyke DA (1979) Chlorpropamide alcohol flushing and diabetic retinopathy. Lancet 1: 997–999
Lewis W, Schwartz L (1956) An occupational agent (*n*-butyraldoxime) causing reaction to alcohol. Med Ann District Columbia 25: 485–490
Lieber CS (1980) Alcohol, liver injury and protein metabolism. Pharmacol Biochem Behav 13: 17–30
Lieber CS (1982) Alcohol and the liver. In: Lieber CS (ed) Medical disorders of alcoholism. Pathogenesis and treatment. Saunders, Philadelphia, pp 259–312 (Major problems in internal medicine, vol XII)
Lindros KO (1978) Acetaldehyde: its metabolism and role in the actions of alcohol. In: Israel Y, Glaser FB, Kalant H, Popham RE, Schmidt W, Smart RG (eds) Research advances in alcohol and drug problems, vol 4. Plenum, New York, pp 111–176
Lindros KO (1983) Human blood acetaldehyde levels: with improved methods, a clearer picture emerges. Alcoholism 7: 70–75
Lindros KO, Oshino N, Parilla R, Williamson JR (1974) Characteristics of ethanol and acetaldehyde oxidation on flavin and pyridine nucleotide fluorescence changes in perfused rat liver. J Biol Chem 249: 7956–7963
Lindros KO, Koivula T, Eriksson CJP (1975) Acetaldehyde levels during ethanol oxidation: a diet-induced change and its relation to liver aldehyde dehydrxogenases and redox states. Life Sci 17: 1589–1598
Lindros KO, Stowell A, Pikkarainen P, Salaspuro M (1980) Elevated blood acetaldehyde in alcoholics with accelerated ethanol elimination. Pharmacol Biochem Behav [Suppl 1] 13: 119–124
Lindros KO, Stowell A, Salaspuro M (1981) The disulfiram (Antabuse)-alcohol reaction in male alcoholics: its efficient management by 4-methylpyrazole. Alcoholism 5: 528–530

Lundquist F (1970) Enzymatic pathways of ethanol metabolism. In: Tremolieres J (ed) International encyclopedia of alcohol and alcoholism, section 20, vol 1. Pergamon, Oxford, pp 95–116

Lundquist F, Fugmann U, Rasmussen H, Svendsen I (1962) The metabolism of acetaldehyde in mammalian tissues. Reactions in rat liver suspensions under aerobic conditions. Biochem J 84: 281–286

Lundwall L, Backeland F (1971) Disulfiram treatment of alcoholism. J Nerv Ment Dis 153: 381–394

Magrinat G, Dolan JP, Biddy RL, Miller LD, Korol B (1973) Ethanol and methanol metabolites in alcohol withdrawal. Nature 244: 234–235

Majchrowics E, Mendelson JH (1970) Blood concentrations of acetaldehyde and ethanol in chronic alcoholics. Science 168: 633–634

Marchner H, Tottmar O (1976a) Inhibition of the acetaldehyde dehydrogenases in rat liver by a cyanamide derivative present in a commercial standard diet for small animals. Acta Pharmacol Toxicol 39: 331–343

Marchner H, Tottmar O (1976b) Influence of the diet on the metabolism of acetaldehyde in rats. Acta Pharmacol Toxicol 38: 59–71

Marjanen L (1973) Comparison of aldehyde dehydrogenases from cytosol and mitochondria of rat liver. Biochim Biophys Acta 327: 238–246

Marshall AW, Cakebread KL, Jenkins WJ, Morgan MY (to be published) Blood acetaldehyde concentrations after oral ethanol in normal males and females. Alcohol and Alcoholism

Matsuda Y, Baraona E, Salaspuro M, Lieber CS (1979) Effects of ethanol on liver microtubules and Golgi apparatus. Possible role in altered hepatic secretion of plasma proteins. Lab Invest 41: 455–463

Matsuzaki S, Lieber CS (1977) Increased susceptibility of hepatic mitochondria to the toxicity of acetaldehyde after chronic ethanol consumption. Biochem Biophys Res Commun 75: 1059–1065

McComb JA, Goldstein DB (1979) Quantitative comparison of physical dependence on tertiary butanol and ethanol in mice: correlation with lipid solubility. J Pharmacol Exp Ther 208: 113–117

Mizoi Y, Ijiri I, Tatsuno Y, Kijima T, Fujiwara S, Adachi J (1979) Relationship between facial flushing and blood acetaldehyde levels after alcohol intake. Pharmacol Biochem Behav 10: 303–311

Mohammed A, Olcott HS, Fraenkel-Conrat H (1949) The reaction of proteins with acetaldehyde. Arch Biochem 24: 270–280

Nagasawa HT, Goon DJW, Demaster EG, Alexander CS (1977) Lowering of ethanol-derived circulating blood acetaldehyde in rats by D-penicillamine. Life Sci 20: 187–194

Noble EP, Tewari S (1977) Metabolic aspects of alcoholism in the brain. In: Lieber CS (ed) Metabolic aspects of alcoholism. University Park Press, Baltimore, pp 149–187

Nuutinen H, Lindros KO, Salaspuro M (1983) Determinants of blood acetaldehyde level during ethanol oxidation in chronic alcoholics. Alcoholism 7: 163–168

Nuutinen HU, Salaspuro MP, Valle M, Lindros KO (1984) Blood acetaldehyde concentration gradient between hepatic and antecubital venous blood in ethanol-intoxicated alcoholics and controls. Eur J Clin Invest 14: 306–311

Ortiz A, Griffiths PJ, Littleton JA (1974) A comparison of the effects of chronic administration of ethanol and acetaldehyde to mice: evidence for a role of acetaldehyde in ethanol dependence. J Pharm Pharmacol 26: 249–260

Parrilla R, Ohkawa K, Lindros KO, Zimmerman U-JP, Kobayashi K, Williamson JR (1974) Functional compartmentation of acetaldehyde oxidation in rat liver. J Biol Chem 249: 4926–4933

Peachey JE, Maglana S, Robinson GM, Hemy M, Brien JF (1981) Cardiovascular changes during the calcium carbimide-ethanol interaction. Clin Pharmacol Ther 29: 40–46

Perman ES (1962) Studies on the Antabuse-alcohol reaction in rabbits. Acta Physiol Scand [Suppl 190] 55: 46

Petersen D, Collins AC, Deitrich RA (1977) Role of liver cytosolic aldehyde dehydrogenase isozymes in control of blood acetaldehyde concentrations. J Pharmacol Exp Ther 201: 471–481

Pietruszko R, Vallari RC (1978) Aldehyde dehydrogenase in human blood. FEBS Lett 92: 82–91

Pikkarainen PH, Salaspuro MP, Lieber CS (1979) A method for the determination of "free'-acetaldehyde in the plasma. Alcoholism 3: 259–261
Pikkarainen PH, Baraona E, Jauhonen P, Seitz HK, Lieber CS (1981a) Contribution of oropharynx microflora and of lung microsomes to acetaldehyde in expired air after alcohol ingestion. J Lab Clin Med 97: 631–638
Pikkarainen PH, Gordon ER, Lebsack ME, Lieber CS (1981b) Determinants of plasma free acetaldehyde level during the steady state oxidation of ethanol: effects of chronic ethanol feeding. Biochem Pharmacol 30: 799–802
Podgainy H, Bressler R (1968) Biochemical basis of the sulfonyl-urea-induced Antabuse syndrome. Diabetes 17: 679–683
Pyke DA, Leslie RDG (1978) Chlorpropamide-alcohol flushing: a definition of its relation to non-insulin-dependent diabetes. Br Med J 2: 1521–1522
Rahwan RG (1975) Toxic effects of ethanol: possible role of acetaldehyde, tetrahydroisoquinolines and tetrahydro-β-carbolines. Toxicol Appl Pharmacol 34: 3–27
Redmond G, Cohen G (1971) Induction of liver acetaldehyde dehydrogenase: possible role in ethanol tolerance after exposure to barbiturates. Science 171: 387–389
Reed TE, Kalant H, Gibbins RJ, Kapur BM, Rankin JG (1976) Alcohol and acetaldehyde metabolism in Caucasians, Chinese and Amerinds. Can Med Assoc J 115: 851–855
Roper M, Stock T, Dietrich RA (1976) Phenobarbital and tetrachlorodibenzo-*p*-dioxin induce different isoenzymes of aldehyde dehydrogenase. Fed Proc 35: 282
Rucker E (1949) Aldehyde dehydrogenase, a diphosphopyridine nucleotide-linked enzyme. J Biol Chem 177: 883–892
Salaspuro MP, Lieber CS (1978) Non-uniformity of blood ethanol elimination: its exaggeration after chronic consumption. Ann Clin Res 10: 294–297
Santi R, Ferrari M, Toth CE, Contessa AR, Fassina G, Bruni A, Luciani S (1967) Pharmacological properties of tetrahydropapaveroline. J Pharm Pharmacol 19: 45–51
Shaw S, Jayatilleke E, Ross WA, Gordon ER, Lieber CS (1981) Ethanol induced lipid peroxidation: potentiation by chronic alcohol feeding and attenuation by methionine. J Lab Clin Med 98: 417–425
Schreiber SS, Briden K, Oratz M, Rothschild MA (1972) Ethanol, acetaldehyde and myocardial protein synthesis. J Clin Invest 51: 2808–2819
Schreiber SS, Oratz M, Rothschild MA, Reff F, Evans C (1974) Alcoholic cardiomyopathy. II. The inhibition of cardiac microsomal protein synthesis by acetaldehyde. J Mol Cell Cardiol 6: 207–213
Schuckit MA, Rayses V (1979) Ethanol ingestion: differences in blood acetaldehyde concentrations in relatives of alcoholics and controls. Science 203: 54–55
Stevens VJ, Fantl WJ, Newman CB, Sims RV, Cerami A, Peterson CM (1981) Acetaldehyde adducts with hemoglobin. J Clin Invest 67: 361–369
Stowell AR (1979) An improved method for the determination of acetaldehyde in human blood with minimal ethanol interference. Clin Chim Acta 98: 201–205
Stowell AR, Greenway RM, Batt RD (1977) Acetaldehyde formation during deproteinization of human blood samples containing ethanol. Biochem Med 8: 392–401
Stowell AR, Lindros KO, Salaspuro MP (1980) Breath and blood acetaldehyde concentrations and their correlation during normal and calcium carbimide-modified ethanol oxidation in man. Biochem Pharmacol 29: 783–787
Suokas A, Kupari M, Pettersson J, Lindros KO (1983) Acetaldehyde, catecholamine and cardiovascular responses after the alcohol-aversive drug nitrefazole and ethanol. Acta Pharmacol Toxicol 53 [Suppl 2]: 7 (abstract)
Tietz A, Lindber M, Kennedy EP (1964) A new pteridine-requiring enzyme system for the oxidation of glyceryl ethers. J Biol Chem 239: 4081–4090
Tottmar SOC, Pettersson H, Kiessling K-H (1973) The subcellular distribution and properties of aldehyde dehydrogenases in rat liver. J Biochem 135: 577–586
Tottmar SOC, Kiessling KH, Forsling M (1974) Effects of phenobarbital and ethanol on rat liver aldehyde dehydrogenases. Acta Pharmacol Toxicol 35: 270–276
Truitt EB (1971) Blood acetaldehyde levels after alcohol consumption by alcoholics and nonalcoholic subjects. In: Roach MK, McIsaac WM, Creaven PJ (eds) Biological aspects of alcohol. University of Texas Press, Austin, pp 212–232

Truitt EB, Duritz G (1966) The role of acetaldehyde in the actions of ethanol. In: Mackel PP (ed) Biochemical factors in alcoholism. Pergamon, New York, pp 61–69

Truitt EB Jr, Walsh MJ (1971) The role of acetaldehyde in the actions of ethanol. In: Kissin H, Begleiter H (eds) The biology of alcoholism, vol I. Plenum, New York, pp 161–195

Turner AJ, Baker KM, Alyeri S, Figerio A, Garratini S (1974) Tetrahydropapaveroline: formation in vivo and in vitro in rat brain. Life Sci 14: 2247–2257

Veitch RL, Lumeng L, Li T-K (1975) Vitamin B_6 metabolism in chronic alcohol abuse. The effect of ethanol oxidation on hepatic pyridoxal 5-phosphate metabolism. J Clin Invest 55: 1026–1032

Von Wartburg JP, Ris MM (1979) Determination of acetaldehyde in human blood. Experientia 35: 1682–1683

Walkenstein SS, Weinhouse S (1953) Oxidation of aldehydes by mitochondria of rat tissues. J Biol Chem 200: 515–523

Wallgren H (1973) Neurochemical aspects of tolerance to and dependence on ethanol. Adv Exp Med Biol 35: 15–31

Walsh MJ (1971) Role of acetaldehyde in the interactions of ethanol with neuroamines. In: Roach MK, McIsaac WM, Creaven PJ (eds) Biological aspects of alcohol. University of Texas Press, Austin, p 233

Weiner H, King P, Hu JHJ, Bensch WR (1974) Mechanistic and enzymatic properties of liver aldehyde dehydrogenases. In: Thurman RG, Yonetani T, Williamson LR, Chance B (eds) Alcohol and aldehyde metabolizing systems. Academic, New York, pp 101–125

Weissman A, Koe BK (1969) Drugs and deterrence of alcohol consumption. In: Chain CK (ed) Annual reports in medicinal chemistry. Academic, New York, pp 246–258

Williams EE (1937) Effects of alcohol on workers with carbon disulfide. JAMA 109: 1472–1473

Wolff PH (1973) Vasomotor sensitivity to alcohol in diverse mongoloid populations. Am J Hum Genet 25: 193–199

Zeiner AR, Paredes A, Christensen HD (1979) The role of acetaldehyde in mediating reactivity to an acute dose of ethanol among different racial groups. Alcoholism 3: 11–18

7 Ethanol and Fibrogenesis in the Liver

E. G. Hahn and D. Schuppan

Medizinische Klinik und Poliklinik, Abteilung für Innere Medizin mit Schwerpunkt Gastroenterologie, Freie Universität Berlin, Hindenburgdamm 30, 1000 Berlin 45, Federal Republik of Germany

Introduction

Rapid advances in connective tissue research offer a new perspective on molecular and cellular events in alcoholic liver fibrosis, and also provide new quantitative approaches to monitoring its clinical course. Recently detected biological activities of distinct proteins of the interstitial matrix may be of fundamental importance for differentiated growth and regeneration in the liver (Rojkind and Kershenobich 1983), and the function of micro- and macrocirculation, of hepatocytes, and of Kupffer cells may be severely disturbed by the development of excess connective tissue. The tridimensional pattern of scarring in liver acini plays a pivotal role in the pathophysiology of collateral circulation and portal hypertension (Rappaport et al. 1983), and many problems of alcoholic liver disease are closely linked to hepatic fibrogenesis. This chapter will briefly review the morphologic pattern and functional implications of alcoholic liver fibrosis, describe the composition of the fibrotic matrix and its pathogenesis, and then discuss the clinical assessment of fibrogenesis in alcoholic liver disease.

Pathology

Chapter 16 of this book extensively describes the pathology of alcoholic liver disease, and only some salient points, relevant to fibrogenesis, will be mentioned here, with a special emphasis on prognosis and functional implications.

Morphologic Pattern of Fibrosis

Definition of Fibrosis and Cirrhosis

Fibrosis is defined here in morphological terms as the presence of excess connective tissue, due to new formation (Anthony et al. 1978). The definition of cirrhosis is more difficult, but a diffuse process characterized by fibrosis and the conversion of normal liver architecture into structurally abnormal nodules is a definition accepted

Alcohol Related Diseases in Gastroenterology
Edited by H. K. Seitz and B. Kommerell

by pathologists (Anthony et al. 1978; International Group 1981; Thaler 1982). Apparently "regeneration" does not play an important part in this definition, and nodules may arise from regrowth, following necrosis, dissection of nodules by fibrosis, and remodelling associated with altered vascular relationships (Anthony et al. 1978; Thaler 1982; Rappaport et al. 1983). Other sources include regeneration as part of the definition (Leevy et al. 1977; Conn 1982), but all emphasize altered vascular relationships and separation of parenchyma from vascular structures by connective tissue. Fibrosis thus interferes with the function of vascular channels as a scaffold of acinar parenchymal regeneration, and the pattern of fibrosis may ultimately mark the point of no return. A unified morphologic concept of fibrosis and cirrhosis will undoubtedly evolve once the microcirculatory hepatic unit of Rappaport is rigorously applied to chronic fibrotic liver disease of various etiologies. This is particularly well illustrated in alcoholic liver disease (Rappaport et al. 1983), which cannot be understood on the molecular and cellular level without Rappaport's admirable system.

Alcoholic Fibrosis and Cirrhosis

Histopathology and electron microscopy have revealed the fairly characteristic pattern of alcoholic liver fibrosis (Fig. 1) (Christoffersen and Poulsen 1979; Orrego 1979; International Group 1981; Thaler 1982; Nakano et al. 1982; Minato et al. 1983). It is a pericellular type of fibrosis which, as shown by electron microscopy, develops from fibrosis of Disse's space (collagenization of Disse's space). This is accompanied by conspicuous changes in number and appearance of fat-storing cells. Transitional forms between fat-storing cells and fibroblasts have been observed in alcoholic cirrhosis (Bhathal 1972; Mak et al. 1984), and up to 60% of cells in septa of alcoholic cirrhosis exhibited features of contractile fibroblasts (Rudolph et al. 1979). These cells have also been called myofibroblasts and were recently demonstrated in early alcoholic fibrosis of the perivenular type and even in normal liver (Nakano and Lieber 1982; Nakano et al. 1982). It was concluded from these observations that fat-storing cells play a major role in alcoholic fibrogenesis. Direct proof for the origin of Disse's spaces connective tissue from fat-storing cells is lacking, and endothelial cells as well as hepatocytes or Kupffer cells have been implicated.

When using classical staining methods for connective tissue ("collagen"), including silver impregnation, pericellular fibrosis in alcoholic liver disease is detected in three locations (Fig. 1b, c, f): (1) near or around the terminal hepatic veins; (2) within the acinus; and (3) periportal. There is a preponderance of necrosis and fibrosis in zone 3 of Rappaport (International Group 1981; Rappaport et al. 1983) which is particularly conspicuous in severe alcoholic hepatitis, where it may rapidly lead to venular-portal and venular-venular septa (Fig. 1c, d). Within the spectrum of fibrotic alcoholic liver disease, in all stages and all acinar locations mentioned above, there is a focal accentuation of fibrosis (Fig. 1c). The earliest fibrotic changes appear to be characteristically situated around hepatic venules (perivenular fibrosis). The heated debate on whether cirrhosis may develop with or without alcoholic hepatitis probably touches both ends of a continuum and has been extensively discussed in recent reviews (Orrego et al. 1981b; Thaler 1982; Lieber and Leo 1982). The definition of alcoholic hepatitis used in this chapter follows the one given by an international group of pathologists (International Group 1981).

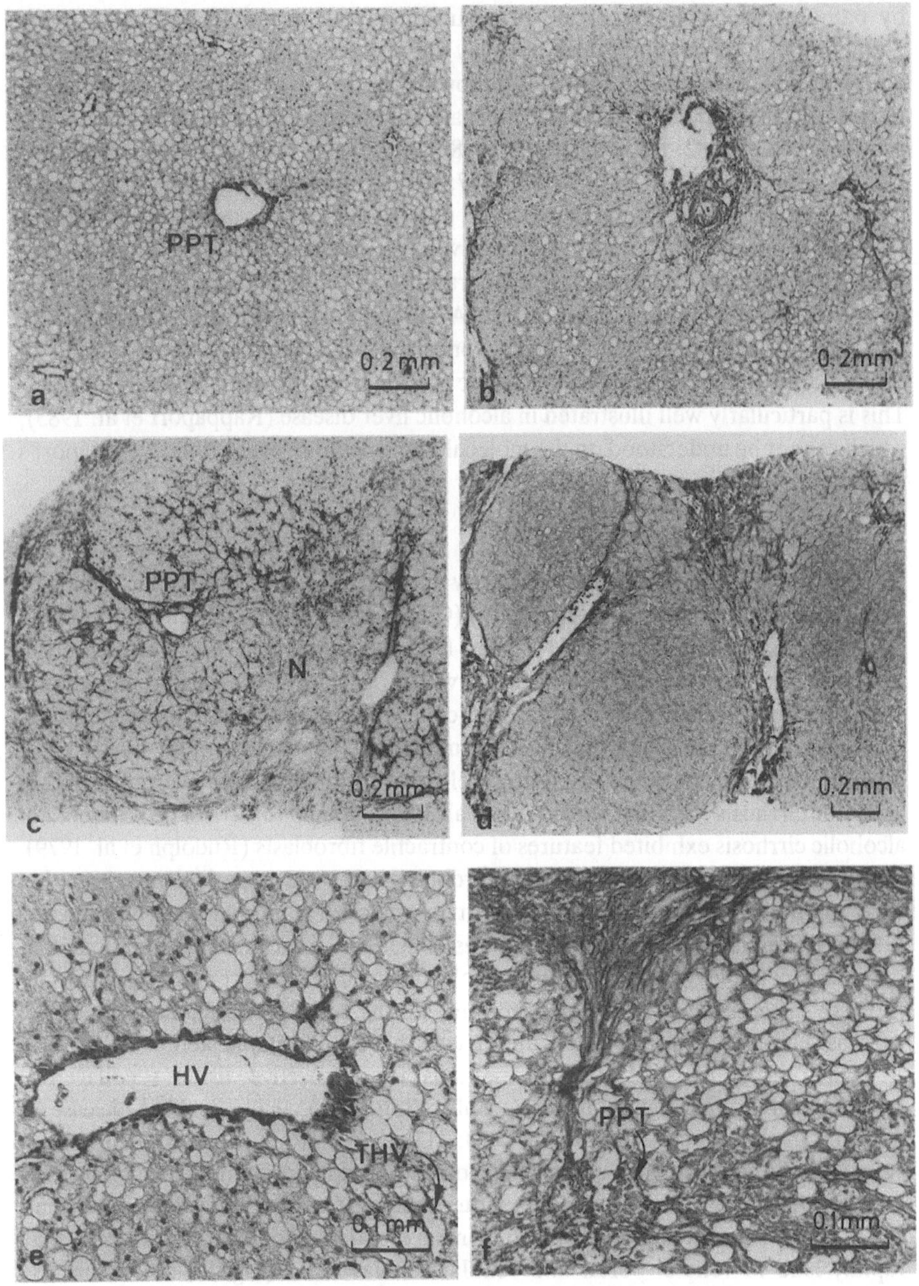

Fig. 1a–f. Stages and pattern of alcoholic liver fibrosis. **a** Alcoholic fatty liver. 37-year-old man, beer-drinking history. Domagk stain, × 60. **b** Alcoholic fatty liver plus moderate perivenular, pericellular and periportal fibrosis. 38-year-old beer and rye whiskey drinker. Domagk stain, × 60. **c** Alcoholic hepatitis with severe zone 3 necrosis and fibrosis and marked pericellular fibrosis. 38-year-old man, beer and rye whiskey drinker, biopsy taken 5 weeks after hospitalization. Gomori stain, × 60. **d** Alcoholic cirrhosis. Same patient as in **c**; biopsy taken 1 year later. Stopped drinking completely. Elastica-van Gieson stain, × 60. **e** Alcoholic fatty liver. Same biopsy as in **a**. Domagk stain, × 150. Note normal hepatic vein (*HV*) and terminal hepatic vein (*THV*). **f** Alcoholic hepatitis, fatty liver, marked zone 3 (perivenular) and moderate pericellular fibrosis. 43-year-old woman, vodka drinker. Elastica-van Gieson stain, × 150. *PPT*, preterminal portal tract; *N*, necrosis; *HV*, hepatic vein; *THV*, terminal hepatic vein. *Bars* indicate magnification

A different pattern of fibrosis has repeatedly been described in patients with alcohol abuse as the only identifiable determinant of liver disease: chronic active hepatitis (Galambos 1972; Brunt et al. 1974; Goldberg et al. 1977; Balázs et al. 1977; Crapper et al. 1983). Lymphocytic infiltration in alcoholic liver disease has been related to "healing" of alcoholic hepatitis (Galambos 1972) as well as to continued alcohol abuse (Crapper et al. 1983). It has also been observed in 8 of 48 patients with marked proliferation and enlargement of Kupffer cells and fat-storing cells (Balázs et al. 1977). Non-A-non-B hepatitis cannot be excluded in any of these patients, and there is some indication that alcohol abuse may influence the course of chronic viral liver disease, at least in hepatitis B (Ohnishi et al. 1982).

Portal fibrosis in alcoholics should direct attention to possible nonalcoholic liver disease (Morgan et al. 1978; Levin et al. 1979), particularly of the biliary tract.

Functional Implications of Fibrogenesis

Macro- and Microcirculation

It is not so much the quantity of connective tissue laid down in fibrosis or cirrhosis that determines the effect on liver function, but rather the location of the tissue. Deposits at strategic positions may cause the early development of portal hypertension as is known for perivenular fibrosis in alcoholics or for portal fibrosis in schistosomiasis. This is often aggravated by inflammatory infiltration and local edema, which is not only seen in schistosomiasis, but also in sarcoidosis, lymphoma, bile duct disease (such as nonsuppurative sclerosing cholangitis and extrahepatic bile duct obstruction), and toxic lesions caused by vinyl chloride, copper, or arsenic. When fibrosis supervenes, portal hypertension may become chronic. In these conditions, in addition to reduced portal blood flow, the regulation of microcirculation by in- and outlet sphincters (McCuskey 1966; Rappaport 1980) may be disturbed by portal fibrosis and/or by pericellular fibrosis. Liver function may be preserved over a wide range, since a 50% reduction of splanchnic blood flow in healthy individuals has no influence on galactose elimination capacity (Jakobsen et al. 1969).

The importance of gastrointestinal hormones and "hepatotrophic" factors for the regulation of cells and metabolism in the liver has repeatedly been emphasized (Starzl et al. 1980), and the influx of such substances will be diminished in chronic reduction of portal blood flow. The effect of this is difficult to define at the present time, but should be kept in mind as an early effect of hepatic fibrosis.

Quantitation of macro- and microcirculation in the human liver has been rarely conducted in relation to fibrosis because of technical limitations (Goresky 1980). One example is the increased wedged hepatic vein pressure in alcoholic fibrosis (Reynolds et al. 1969), another the elevated sinusoidal pressure in pericellular fibrosis of alcoholic liver disease (Orrego et al. 1979, 1981a). In the latter study a correlation of intrahepatic pressure with collagen in the Disse space (and hepatomegaly) was documented. Goodman and Ishak (1982) recently presented data showing a significant association of portal hypertension with the degree of venoocclusive changes and perivenular sclerosis in precirrhotic alcoholic hepatitis.

In summary, precirrhotic fibrosis in alcoholic liver disease leads, in addition to increased hepatocyte volume (see Chap. 2), to profound changes in macro- and microcirculation down to the simple acinus.

Hepatocellular Function

Sinusoidal endothelial cells are fenestrated and have no basement membrane. This allows direct communication between sinusoidal blood and Disse's space; solutes and small particles (less than 0.1 μm) have direct access to the microvilli of hepatocytes (Wisse and Knook 1979). The large surface of these microvilli is covered with receptors to selectively bind blood substances and internalize them by pinocytosis or other transport mechanisms into hepatocytes (Neufeld and Ashwell 1980). This is the basis for the regulation of hepatocytes (and other liver cells) by hormones and for the normal physicochemical activities of hepatocytes, including the uptake of dyes, bile acids, galactose, aminopyrine, antipyrine, and derivatives of lidocaine (HIDA) used as probes of liver function or for scintiscan. Increase of connective tissue in Disse's space (persinusoidal fibrosis, pericellular fibrosis, "collagenisation" of Disse's space) interferes with this free exchange between sinusoidal blood and hepatocytes. Not only does the distance between endothelial cells and hepatocytes increase, but there is also a reduction in hepatocytic microvilli (Schaffner 1970; Orrego et al. 1981a), reducing the available surface. Basement membranes are built up on the hepatocyte surface and beneath sinusoidal endothelial cells (Schaffner 1970; Hahn et al. 1980b; Orrego et al. 1981a). Single hepatocytes or groups of them may be completely isolated by pericellular fibrosis.

The changes described above most probably alter the access of blood substances to the hepatocytes quite early during the course of alcoholic liver disease, particularly when the transport is mediated by a carrier protein. Formation of basement membranes has been observed at the stage of alcoholic fatty liver (Hahn et al. 1980b), and it may be assumed that perisinusoidal fibrosis and capillarization of sinusoids impair hepatocyte function long before cirrhosis develops.

This may also influence liver function tests when the functional integrity and mass of hepatocytes are still normal. Quantitative data about the impact of fibrosis on current tests of liver function are not available. A recent issue of *Seminars in Liver Diseases* (1983) does not even mention this notion though it testifies to the limitations of conventional and more advanced assessment of liver function (McIntyre 1983). The interpretation of measurements of hepatocellular "function" or liver blood flow with the help of extraction methods (Preisig et al. 1972; Carlisle et al. 1979) is difficult without incorporation of liver fibrosis; the bad correlation with parameters of liver cell damage (transaminases) or hepatocellular synthesis (albumin, prothrombin time) may then be understood and be of little importance. Extraction methods, if influenced by fibrosis, could be a kind of "global" or "total" liver function test and carry more prognostic information than a probe of hepatocyte function or hepatocyte mass. In alcoholic liver disease, this should be particularly true, and at least the aminopyrine breath test has a prognostic value (Schneider et al. 1980).

The uncoupling of hepatocytes from sinusoidal blood by connective tissue may have other effects on the pathogenesis of alcoholic liver disease, particularly on regeneration.

Function of Kupffer Cells and Endothelial Cells

The phagocytic cell system lining the sinusoidal blood stream above all needs a normal perfusion of sinusoids and plasma fibronectin as an opsonizing protein (Bloch and McCuskey 1977; DiLucio 1979; Hahn et al. 1981). The capture of particles or molecules is exercised by all cells along the sinusoid (Bloch and McCuskey 1977), and both specific receptor systems (Neufeld and Ashwell 1980) and opsonization play a role. The ingestion of effete and sequestered material (erythrocytes, fibrin, immune complexes, endotoxin, bacteria, tumor cells) can be carried out by Kupffer and endothelial cells, but there is experimental evidence that Kupffer cells are much more active in phagocytosis than endothelial cells (Bloch and McCuskey 1977; DiLucio 1979). Hepatocytes also ingest material, and some substances appear in the lymphatics (Bloch and McCuskey 1977).

Fibrosis can disturb this function of the reticuloendothelial system in the liver by reduction of portal blood flow and interference with microcirculation in the acinus. The pulsatile blood flow in the axial portovenular and perpendicular oblique sinusoids is regulated by intermittent opening and closing of terminal arterioles which enter the sinusoids immediately after their origin from terminal portal tracts. Sinusoids also open and close with the help of sphincter cells (probably endothelial cells) observed at the origin of intersinusoidal sinusoids and before axial sinusoids empty into the terminal hepatic veins (McCuskey 1966; Rappaport 1980). Sinusoids also have an exquisite potential to contract and relax to vary their normal width of 7–15 μm up to 180 μm (Rappaport 1980). This delicate network of microcirculation may be disturbed in many ways by portal, pericellular, or perivenular fibrosis. When there are no extrahepatic or intrahepatic shunts the major effect of fibrosis may be seen in reduced adaptation of the microcirculation to peak demands on the reticuloendothelial cells (like after meals, bacteremia, endotoxinemia). In later stages of fibrosis altered vascular channels will lead to intrahepatic shunting of blood (Rappaport et al. 1983).

Composition of the Fibrotic Tissue

The characterization of the biochemical composition of the fibrotic matrix has substantially contributed to our understanding of the cellular basis of liver fibrosis and to the development of new quantitative approaches to monitor the clinical course of alcoholic liver disease. Chemical analysis and immunofluorescence localization have now releaved the excess deposition of most of the known connective tissue components. Extensive reviews are available (Hahn et al. 1980a; Rojkind and Kershenobich 1981a, 1983; Orrego et al. 1981b; Rauterberg et al. 1981; Hahn and Schuppan 1983) and, in this section, our aim is to focus on new data relevant to the pathogenesis and clinical assessment of alcoholic fibrogenesis discussed in the sections on the mechanisms and clinical assessment of fibrogenesis.

Interstitial Matrix

Immunofluorescent Localization

It is now clear that all four major groups of connective tissue substances are increased in the fibrotic tissue of alcoholic liver disease (collagens, structural glycoproteins, proteoglycans, elastin). Though of major functional significance, little is known about proteoglycans and elastin. Native proteoglycans have never been isolated from human liver, but glycosaminoglycans (proteoglycans minus protein core) may be increased 4–5 times in cirrhosis. All types of glycosaminoglycans are involved, but the increase in chondroitin sulfate and dermatan sulfate appears to be prominent (Galambos and Shapira 1973; Kojima et al. 1975). Elastin is quite inconspicuous in normal liver, where it is visualized by elastica stains as scanty fine fibrillar deposits in portal tracts and vessel walls (Scheuer 1980). Its true amount and distribution is not known since we lack specific markers of elastin. Due to its high insolubility and resistance to proteolysis, no quantitative data are available, but judging by conventional staining methods, large amounts of elastin may accumulate in central parts of fibrotic septa. It may be of more significance than usually suspected (Scheuer and Maggi 1980) but has not been evaluated properly at the present time.

Because of their unusual biological properties, collagens and structural glycoproteins have recently attracted increasing attention. Investigations on liver tissue have been greatly facilitated by the use of affinity purified antibodies (Gay et al. 1975; Ott et al. 1977; Wick et al. 1978; Hahn et al. 1980b; Biempica et al. 1980; Grimaud et al. 1980) for immunofluorescence and immunoelectron microscopy. Excess deposition of virtually all of the known collagens and glycoproteins in the fibrotic matrix of the liver in alcoholic liver disease was demonstrated as summarized in Table 1. This includes fibrillar collagens types I and III, which are found in the portal tract, in Disse's space, and in all fibrotic areas including septa in approximately equal amounts (Rojkind et al. 1979). Basement membrane collagens types IV and V are found not only in basement membranes of bile ducts, vessels, and nerves, but also in Disse's space and in large amounts in all fibrotic areas and in the septum (Table 1, Figs. 2, 3). Type VI collagen is restricted to the interstitial connective tissue, sparing basement membranes (Von der Mark et al. 1984; Schulz et al. 1984). It is a constituent of microfibrillar components of the portal tract and the Disse space and, by immunofluorescence, is codistributed with types I and III collagens. Like these, type VI collagen is abundant in fibrotic areas and septa of alcoholic liver disease (R. Schulz, D. Schuppan, E. O. Riecken, E. G. Hahn 1984, unpublished results). Figure 2 shows, schematically, the distribution of these collagen types in normal liver as revealed by immunofluorescence studies, and Fig. 3 presents the distribution in alcoholic cirrhotic liver.

The distribution of two noncollagenous glycoproteins of the extracellular matrix in the liver is fairly well known: laminin (Timpl et al. 1979; Hahn et al. 1980b) and fibronectin (Stenman and Vaheri 1978; Rouslahti et al. 1981). Laminin is mainly located in basement membranes of the portal tract and originally could not be detected in the Disse space (Hahn et al. 1980b) using antibodies to mouse laminin. Recently, by employing antibodies to human laminin, it was possible to visualize a distinct staining along the Disse space, and the portal space showed dots of laminin

Table 1. Designation, molecular weight of precursor molecules, and location of tissue form of major connective tissue proteins in the normal and fibrotic liver. One should keep in mind that the molecular elements interact in many ways by chemical and physicochemical bonds and thereby build up the highly insoluble three-dimensional interstitial matrix. See also Figs. 2 and 3

	Fibrillar collagens			Microfibrillar collagen	Basement membrane (pericellular) collagen	
Designation (type)	I	III	I-Trimer	VI	IV	V
Molecular weight (kilodalton) of precursor molecule	450	450	450	510	540	660
Location	Portal tract, Disse's space fibrosis and septum; I-trimer not known				Basement membranes of bile ducts, vessels, nerves; Disse's space, pericellular in portal tract. Fibrosis and septum	

	Structural glycoproteins		Elastin	Proteoglycans (glycosaminoglycan plus protein core)
Designation (type)	Laminin	Fibronectin	Elastin monomer	Heparan sulfate Dermatan sulfate Chondroitin sulfate Keratan sulfate
Molecular weight (kilodalton) of precursor molecule	1,000	440	72	70–1,000
Location	Laminin like BMC Fibronectin: microfibrillar in portal tract and Disse's space; pericellular like BMC. Fibrosis and septum		Delicate fibrils in portal tract, vessel walls. Fibrosis and septum	"Ground substance"

BMC, basement membrane collagen

deposits (R. Schulz, D. Schuppan, E. G. Hahn 1984, unpublished results). In fibrotic areas and in septa of cirrhosis in alcoholic liver disease, laminin appears to be moderately but definitely increased (Hahn et al. 1980b) and is particularly conspicuous in compact basement membranes of sinusoids. An increase in sinusoidal laminin has already been observed in the alcoholic fatty liver before fibrosis occurs (Hahn et al. 1980b).

Fibronectin is prominent in the interstitial space including the Disse space, but basement membranes also stain, albeit rather weakly. An early focal increase in zone III of Rappaport's acinus was shown in alcoholic fatty liver, and a diffuse, partly fibrillar staining of hepatic fibrosis is evident (Hahn et al. 1980b; Kojima et al. 1981). The rim of septa is accentuated by immunofluorescence like for all the other

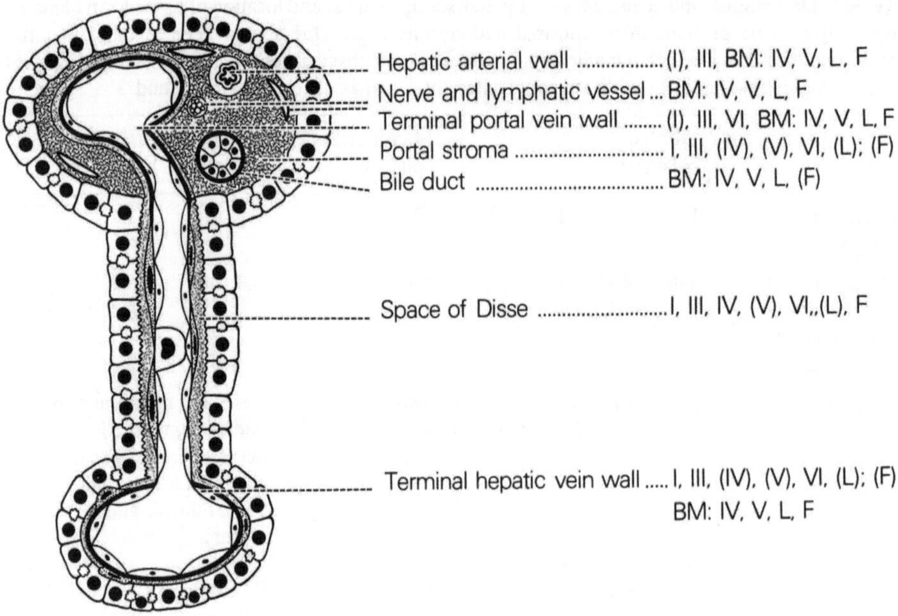

Fig. 2. Schematic drawing of a terminal portal tract and hepatic venule. The location of collagen types, laminin, and fibronectin is indicated. *BM*, basement membrane; *L*, laminin; *F*, fibronectin. See also Table 1

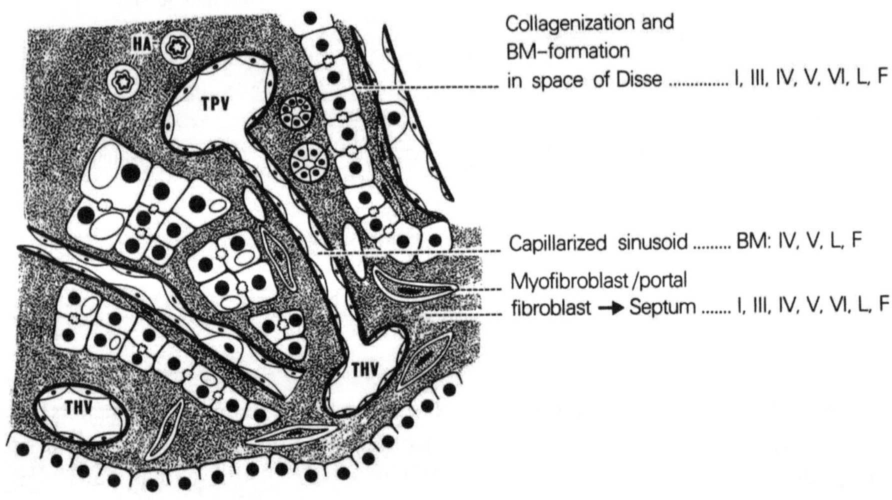

Fig. 3. Schematic drawing of a severely cirrhotic piece of liver. The location of collagen types, laminin, and fibronectin is indicated. *TPV*, terminal portal vein; *THV*, terminal hepatic vein; *HA*, hepatic arteriole; *BM*, basement membrane; *L*, laminin; *F*, fibroncetin. See also Table 1 and Fig. 1

connective tissue proteins, indicating the appositional area of fibrosis. Frequently, fibronectin is seen between hepatocytes along the cell membrane and within the cytoplasm (Hahn et al. 1980a).

Quantitative Analysis

Due to the highly insoluble character of most connective tissue proteins, their quantitative analysis poses difficult technical problems. Total collagen as measured by hydroxyproline content of liver tissue is 2–8 mg/g wet weight or 20–80 mg/g protein. In normal liver, types I and III collagens each comprise approximately one-third of total collagen, and the remaining third was thought to consist of basement membrane collagens (Rojkind et al. 1979). We now know that types IV, V, and VI collagen are only minor constituents of the normal liver but may increase 8–10-fold, even more so than total collagen does (6–7-fold). Using a newly developed technique of pepsin digestion of small pieces of liver and measuring defined fragments of types IV and VI collagen in the supernatant, type IV is less than 1% and type VI is less than 1‰ of total collagen in normal liver (D. Schuppan, C. Birkefeld, E. G. Hahn 1984, unpublished). These relations are maintained during the development of cirrhosis, though subtle shifts have been observed during alcoholic liver disease. In livers with more than 200 mg of total collagen per gram protein, type I collagen prevails (ratio I/III approximately 1.2). Type IV collagen increases much earlier in alcoholic hepatitis than type VI collagen, which increases later but then even more strongly than type IV collagen (D. Schuppan, C. Birkefeld, E. G. Hahn 1984, unpublished). Little is known about type V collagen (form A_2B) but one investigator found it to account for 7%–10% of total collagen in normal livers (Rojkind et al. 1979).

Laminin, when quantitated in normal liver, equals type VI collagen but increases only 3-fold in full-blown cirrhosis (D. Schuppan, C. Birkefeld, E. Baumer, E. G. Hahn 1984, unpublished). Nothing is known about the quantity of fibronectin and elastin, and limited data exist on proteoglycans as mentioned above.

Even though approximative, these data have established a qualitative and quantitative pattern of alcoholic liver fibrosis and cirrhosis that must be accounted for on the cellular and molecular level and may help to understand the underlying pathomechanism. It may also help to select quantitative approaches to monitor the process of fibrosis in alcoholic liver disease.

Cells

The production of an extracellular matrix is a fundamental characteristic of virtually all cells in the human body including those in the liver (Table 2). Cell differentiation and regeneration is accompanied by well-defined changes in the composition and quantity of connective tissue, and scarring of tissue after loss of parenchymal cell mass is considered to be a physiological repair mechanism ubiquitous in the body. Liver fibrosis, like wound healing, may therefore present an interesting model to study the type(s) of cells involved in the scarring of parenchymal organs. No single cell has been unequivocally identified, isolated, and grown for the study of its

Table 2. Cells of the normal and fibrotic human liver with a potential to contribute directly or indirectly to fibrogenesis in alcoholic liver disease. See also Fig. 4.

Parenchymal cells	Nonparenchymal cells		Cells of inflammatory infiltrate
	Sinusoid lining	Portal tract	
Hepatocytes	Kupffer cells	Portal fibroblasts	Monocyte-histiocyte
	Fat-storing cells	Smooth muscle cells of portal vessels	Neutrophil
	Endothelial cells of sinusoids	Bile duct epithelia	Lymphocytes T-lymphocytes B-lymphocytes
		Endothelial cells of portal blood and lymphatic vessels	Plasma cells
			Eosinophil

regulation from hepatic fibrotic tissue, but this approach is being used in several groups with interesting results, all pointing to a contractile modified fibroblast, operationally called myofibroblast (Gabbiani 1981). Electron-microscopically, some features of this cell type were first described by Bhathal in 1972. Up to 60% of cells in septa of alcoholic cirrhotic liver were myofibroblasts containing the classical features of 60–80-Å microfilament bundels with electron-dense bodies and prominent microtubules (Rudolph et al. 1979). In 3 noncirrhotic livers, such cells were not found, but Nakano et al. (1982) demonstrated perivenular myofibroblasts in normal livers and an increase in the number of these cells in early and late stages of perivenular sclerosis. Moreover, these authors also showed that myofibroblasts are not limited to the perivenular area. Thus, beyond any doubt, contractile fibroblasts (myofibroblasts) represent one of the fibroplastic cells in early and later alcoholic liver disease. Recently, the transformation of fat-storing cells into transitional cells was described in patients with alcoholic liver disease (Minato 1983; Okanoue et al. 1983) and baboons (Mak et al. 1984), and this was related to Disse's space fibrosis. Transitional cells were defined as perisinusoidal cells with a volume density of lipid droplets less than 20% of the cell volume (Mak et al. 1984). Defined in this way by these authors, there was an increased percentage of transitional cells in association with the progression of hepatic fibrosis in baboons fed alcohol. Fat-storing cells, transitional cells, and myofibroblasts described above share some common features and may belong to one family of cells responsible for the production and deposition of the fibrotic matrix in alcoholic liver disease.

The unusual biochemical composition of the fibrotic matrix in alcoholic liver disease (see the section "Interstitial Matrix") has also suggested the origin from modified fibroblasts (Hahn et al. 1980b). The over-abundant expression of basement membrane proteins (types IV and V collagen, laminin) in particular suggests a close relation to smooth muscle cells, which have been shown to elaborate all these proteins both in vivo and in cell culture (Hahn et al. 1980b; Kao et al. 1980; Grotendorst et al. 1981; von der Mark et al. 1984). Cell culture studies further support the myofibroblast concept: smooth muscle-like cells have been grown from liver biopsies (Voss et al. 1982a), and a cell resembling smooth muscle cell was isolated from human fibrotic liver, grown in primary culture, and shown to produce types I, III, IV and V collagens (Krauledat et al. 1982).

The question arises about the precursor cell of myofibroblasts in alcoholic liver fibrosis. This question is still unsolved, but some recent observations suggest that the fat-storing (Ito) cell may be the best candidate. Transitional cells between Ito cells and myofibroblasts have been observed by several authors for human livers (Bhathal 1972; Minato et al. 1983; Okanoue et al. 1983). Midzonal Ito cells appeared to decrease in number of alcoholic fatty liver and alcoholic hepatitis and were replaced by activated Ito cells containing less fat and more endoplasmic reticulum. This happened in close association with inflammatory changes and hepatic necrosis in small foci of fibrosis (Okanoue et al. 1983) and in association with increased Disse's space collagen (Minato et al. 1983). No such changes were seen in endothelial cells or hepatocytes.

Now that methods have been developed to isolate and culture Ito cells as well as other liver-derived cell types (Knook et al. 1982; De Leeuw et al. 1983, 1984), the precise role and regulation of Ito cells in alcoholic liver fibrosis will soon be clarified.

Hepatocytes have also been implicated in fibrogenesis of alcoholic liver disease as a direct source of collagen and fibronectin. This possibility is based on animal studies and the studies of primary cell culture of isolated rat hepatocytes (Voss et al. 1979; Guzelian et al. 1981; Tseng et al. 1982, 1983; Diegelmann et al. 1983). In man, there is no evidence for the production of scar tissue by hepatocytes in vivo. Cooperation of separate sets of cells may be important to maintain the physiology of human hepatocytes in culture, including the elaboration of an extracellular matrix (Clement et al. 1984). Hepatocytes, endothelial sinusoidal cells, Kupffer cells (Voss et al. 1982), portal fibroblasts, bile duct cells, and even nonsessile cells derived from the blood stream (Alitalo et al. 1980) may all have their own spectrum of connective tissue protein production and influence fibrogenesis in alcoholic liver disease, but this is not known.

Ethanol-Induced Mechanisms of Fibrogenesis

General pathogenic aspects of hepatic fibrogenesis have been discussed in several recent reviews (Hahn et al. 1980a; Rojkind and Kershenobich 1981a, b, 1982, 1983; Rauterberg et al. 1981; Rojkind 1982, Hahn and Schuppan 1983). The specific role of alcohol in the generation of liver fibrosis has also been addressed in excellent reviews (Chap. 2; Orrego et al. 1981b; Lieber and Leo 1982; Lieber 1983).

Any concept of the pathogenesis of liver fibrosis as being induced by alcohol will have to account for the characteristic changes on the cellular and molecular level described above. This focuses the attention on the Disse spaces, particularly around the terminal venules, on fat-storing cells and their proliferation and transition to fibroblasts (myofibroblasts), on the increase in number and activation of Kupffer cells, and last but not least on the inflammatory infiltrate in alcoholic hepatitis. This last condition has not been reproduced in an animal model, yet is the most active of all conditions in the production of liver fibrosis and cirrhosis in patients. This reminds us of another, genetic, aspect of the pathogenesis of alcoholic liver fibrosis (Jenkins and Thomas, 1981). Only a minority of alcoholics develop alcoholic hepatitis, and the pericellular and periportal fibrosis developing without overt necrosis and/or inflammation is not observed in all patients either, though the development of cirrhosis appears to be a function of the amount of alcohol consumed and of time (see Chap. 1). This has been challenged recently (Sørensen et al. 1984).

Direct Effects of Ethanol Metabolism

A widely discussed possibility has been the enhanced formation of lactate from pyruvate secondary to NADH production by alcohol metabolism (see Chap. 2). Lactate inhibits proline oxidase activity (Kowaloff et al. 1977), and in this capacity increases the hepatic proline pool. Patients with alcoholic liver cirrhosis have been shown to have greatly increased blood lactic acid levels, suggesting they are unable to compensate for excess formation of NADH (Kershenobich et al. 1981). These authors also found hyperprolinemia in these patients which correlated with blood lactate, and they postulated that hyperprolinemia is secondary to elevated blood lactate. Since the rate of collagen synthesis in humans and in experimental cirrhosis is highly dependent on liver proline concentration (Rojkind 1982) and lactate activates proline hydroxylase activity (Cardinale and Udenfriend 1974), Kershenobich et al. (1981) proposed that serum proline and blood lactate may be markers of liver fibrogenesis.

Contrary to these results, Shaw et al. (1984) found hyperprolinemia and hyperlactacidemia to occur infrequently in patients with alcoholic cirrhosis, which, of course, does not exclude hepatic elevations. But a high hepatic proline concentration does not induce collagen synthesis in rat liver (Forsander et al. 1983), and mice with congenital hyperprolinemia exhibit no change in liver collagen. To establish a role of elevated liver and/or serum proline through enhanced lactate formation in alcoholic fibrogenesis it would appear necessary to have more direct evidence in patients or for cells isolated from liver. This is encouraged by results in the alcoholic baboon model; lactate increased collagen synthesis in myofibroblasts isolated from the liver of these animals (Savolainen et al. 1983).

Increased Intrahepatic Pressure

Orrego et al. (1981 a) found a correlation between intrahepatic pressure and collagen deposition in the Disse space in alcoholic patients. This is part of an interesting concept, postulating the induction of hepatic collagen formation by increased pressure in the sinusoids. Hepatocytes have been shown to increase in volume mainly by intracellular water and, to a lesser degree, by protein and fat accumulation, so that increased luminal pressure would already be present in prefibrotic alcoholic liver disease. The authors (Orrego et al. 1981a) pointed out that increased collagen synthesis had been described as a reaction to pressure in many tubular organs like arteries, bile ducts, and ureters.

An alternate view is voiced by those who consider perivenular sclerosis and associated outflow obstruction to be a determinant and cause of portal hypertension (Lieber and Leo 1982; Goodman and Ishak 1982; Miyakawa et al. 1983). Both views could well be combined; increased intraluminal pressure alone does not explain the pattern of early alcoholic fibrosis (see the section "Morphologic Pattern of Fibrosis").

Perivenular Low Oxygen Tension and Ethanol-Induced Stimulation of Hepatic Oxygen Consumption

This aspect of ethanol metabolism is discussed in more detail by Lieber in Chap. 2 of this book. It is tempting to expect to determine a role for enhanced collagen formation by perivenular hypoxia; this is supported by the effect of oxygen tension on synthesis and degradation of bone collagen in tissue culture (Stern et al. 1966). The pattern of fibrosis in chronic right-sided heart failure would also suggest a role for hypoxia in the stimulation of (nonalcoholic) perivenular fibrosis, but some degree of necrosis may also be involved. In patients with severe generalized hypoxia not related to alcohol, liver fibrosis is not detected without additional necrosis of hepatocytes. Since perivenular low oxygen tension may also simply aggravate the toxic effects of alcohol, the true role of natural hypoxia is difficult to evaluate and by itself should probably not be considered to be of major importance for direct stimulation of fibrogenesis.

Hepatocyte Injury, Immunology, and Inflammation

Hepatocyte damage by alcohol ingestion may implicate a delicate alteration of the cell membrane not detected morphologically, and a loss of the cell by necrosis. Within this spectrum of injury to hepatocytes, a number of metabolic alterations occur (see Chap. 2) which may also influence other cells in the Disse space. For the alcoholic pattern described above to trigger fibrosis, overt alcoholic hepatitis is by no means necessary, and the fascinating discussion in the literature whether cirrhosis may develop in alcoholic patients with or without alcoholic hepatitis as a precursor (Lieber and Leo 1982; Thaler 1982) will not be reiterated here. Instead, by applying old (Zetterman and Sorrell 1981) and new (Anthony et al. 1983; Neuberger et al. 1984; Poralla et al. 1984; Wiedmann et al. 1984; Fallon et al. 1984) evidence for immunological reactions and by applying current concepts of the pathogenesis of fibrosis in other tissues (Rennard et al. 1984; Wahl 1984), we would like to propose a mechanism for alcohol-activated hepatic fibrosis as depicted in Fig. 4. This places the liver macrophage (Kupffer cell) in a central position as a source for a variety of activating and modulating factors, supported by T-cells under certain conditions. Macrophage factors and lymphocyte factors recruit and activate fat-storing cells from the Disse space to proliferate and to produce connective tissue. Neutrophils from the sinusoidal blood will also be attracted. Since liver macrophages and the principal fibroblast of the Disse space, the fat-storing cell, coexist in close proximity, any exogenous or endogenous factor(s) directly stimulating (activating) Kupffer cells will

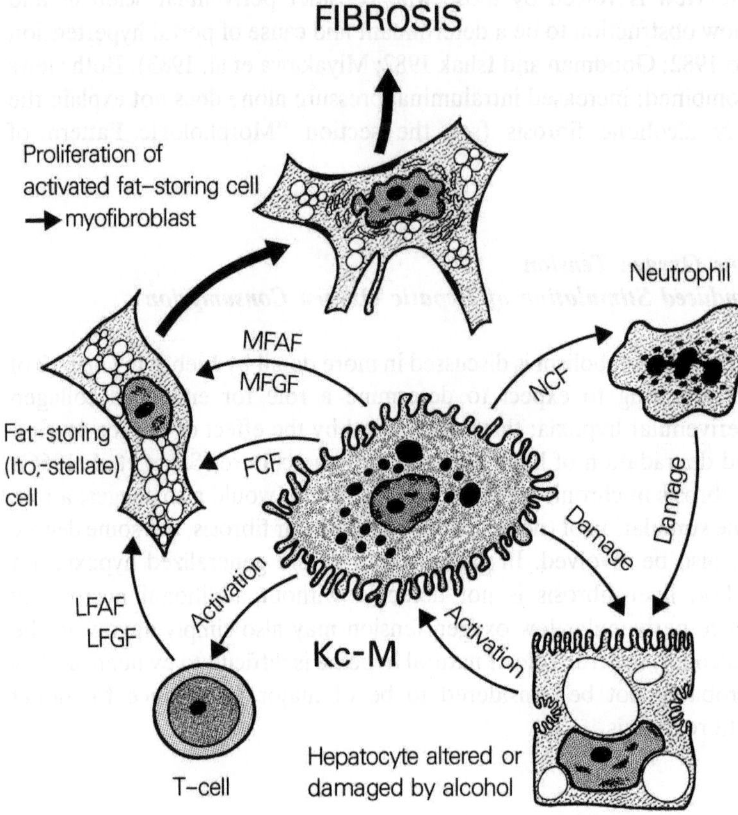

Fig. 4. Pathogenesis of alcohol-induced fibrogenesis. For explanation and references, see text. *Kc-M*, Kupffer cell or macrophage recruited from the circulation; *T-cell*, T lymphocyte; *FCF*, fibroblast chemotactic factor; *MFGF*, macrophage-derived fibroblast growth factor; *MFAF*, macrophage-derived fibroblast-activating factor; *NCF*, neutrophil chemotactic factor; *LFGF*, lymphocyte-derived fibroblast growth factor; *LFAF*, lymphocyte-derived fibroblast-activating factor

potentially lead to the development of Disse's space fibrosis. More extensive inflammation or necrosis will recruit increasing numbers of blood monocytes, who appear to be particularly sensitive to activation (Tanner et al. 1984).

Liver Macrophages

Liver macrophages may be local Kupffer cells or macrophages recently recruited from the blood (Van Furth and Blussé van oud Alblas 1982; Bouwens et al. 1984), the latter being the major source for the maintenance of the Kupffer cell pool and for its expansion. Established Kupffer cells appear to be relatively resistent to stimulation by endotoxin, but macrophages recently recruited from the blood can be effectively activated (Tanner et al. 1984). On the other hand, it was possible to stimulate nonparenchymal rat liver cells enriched for Kupffer cells to a higher interleukin 1 production than peritoneal macrophages (Leser et al. 1982), and interleukin 1 has been implicated in the stimulation and regulation of fibroblasts

(Mizel et al. 1981; Schmidt et al. 1982). Macrophages have been shown to produce factors chemotactic for fibroblasts, mitogenic to fibroblasts, and thus promoting proliferation, and stimulating collagen production by fibroblasts (Wahl 1984). While some of these capacities have been directly shown for lung macrophages in idiopathic pulmonary fibrosis (Rennard et al. 1984), they are still circumstantial for liver macrophages, particularly in the human liver. Advances in the isolation and purification of Kupffer cells, their maintenance in cell culture (de Leeuw et al. 1983), and appropriate coculture techniques will allow the study of factors elaborated by these cells. They may need conditioning or more than one signal to be stimulated (Tanner et al. 1984). It is already clear that Kupffer cells produce fibronectin (Voss et al. 1982b) and that they synthesize chemoattractants for smooth muscle cells (McIvor et al. 1982). Their precise role in alcohol-induced liver fibrogenesis remains to be established.

Lymphocytes

Lymphocyte infiltration is not very conspicuous in alcoholic liver disease, but is at least prominent in alcoholic hepatitis (Christoffersen and Poulsen 1979) and in alcohol-associated chronic active hepatitis (Crapper et al. 1983). Lymphocytes have been shown to produce factors stimulating fibroblast proliferation and collagen production (Johnson and Ziff 1976; Wahl et al. 1978), and T-cells appear to be mainly involved (Neilson et al. 1980; Wahl and Gately 1983). The production of these factors are antigen specific and macrophage dependent (Wahl et al. 1978).

In alcoholic liver disease, several lines of evidence point to an involvement of cytotoxic T-cells (for review see Zetterman and Sorrell 1981; Actis et al. 1983; Izumi et al. 1983; Poralla et al. 1984), of antibodies to liver cell antigens (Burt et al. 1982; Wiedmann et al. 1984), and of antibodies to alcohol altered hepatocytes (MacSween et al. 1981; Anthony et al. 1983; Neuberger et al. 1984) in hepatocyte injury. It is not clear in what way these mechanisms will perpetuate alcoholic hepatic damage and fibrosis in particular. We suggest a role for lymphokines as depicted in Fig. 4, particularly in severe alcoholic hepatitis, where lymphocytes may serve as amplifiers of both macrophage and neutrophil recruitment and stimulate both lipocyte proliferation and collagen synthesis. This will also have to be born out for human liver by modern cell separation and culture techniques.

Fat-Storing Cells

Fat-storing cells (FSC) have been shown to play a significant role in alcoholic liver fibrosis as described in the section on "Cells". FSC after isolation and culture synthesize collagens (de Leeuw 1984) as found in the Disse space and in fibrotic liver tissue (Hahn and Schuppan 1983). Their transformation into transitional cells in Disse's space (Minato et a. 1983; Okanue et al. 1983; Mak et al. 1984) and the appearance of myofibroblasts in cirrhotic septa (Bhathal 1972; Rudolph et al. 1979) and in perivenular fibrosis in patients and alcohol-fed baboons (Nakano et al. 1982); Nakano and Lieber 1982) must be stimulated by alcohol-induced mechanisms. This may involve a conditioning factor as has been shown in rats (Leo and Lieber 1983), and the local machinery of macrophage factors may then be operative as depicted in

Fig. 4. The conditioning factors in patients with alcoholic liver damage have yet to be defined and may be genetic, gut derived, or exogenous. The close similarity of hepatic fibrosis in jejunoileal bypass (O'Leary 1983) or Indian childhood cirrhosis (Nayak 1979) should offer models to uncover some of these factors.

Hepatocytes

Hepatocytes have been implicated in alcoholic fibrogenesis in various ways. Alcohol may alter the cell membrane in vivo, produce necrosis, or stimulate T-cells and membrane specific antibody production (see above, schematically shown in Fig. 4). Substances derived from intact or necrotic hepatocytes involved in the stimulation of lymphocytes and/or macrophages have not been unequivocally defined, though Mallory bodies are still under consideration (Zetterman and Sorrel 1981, Lieber and Leo 1982). Hepatocytes cannot be excluded to make some of the fibrotic matrix themselves (Tseng et al. 1983; Diegelmann et al. 1983) though they could never be stained for extracellular proteins or their precursors other than fibronectin (Hahn et al. 1980a; Hahn et al. 1981). Immunoelectron microscopy (Fleischmajer et al. 1983) of the liver and in situ hybridization using collagen-specific DNA probes (Saber et al. 1983) should soon settle this question. Increased mRNA for type I procollagen has been found in the livers of alcohol-fed baboons (Zern et al. 1984), particularly in cirrhotic livers, and appropriate measurements in distinct cell populations also appear feasible.

Advances in the Clinical Assessment of Alcoholic Fibrogenesis

There are two major determinents for the outcome of alcoholic liver disease: hepatic failure and the development of cirrhosis. Thus, assessment of fibrogenesis is of clinical interest before cirrhosis develops, and it is common knowledge that early changes can only be defined by a liver biopsy and/ or laparoscopy. In this section, the prognostic value of fibrogenesis will be reviewed first and then advances in the assessment of alcoholic fibrosis by blood tests will be discussed.

Alcoholic Fibrosis as a Precursor Lesion of Cirrhosis

It has long been recognized that the long-term prognosis of alcoholic hepatitis is unfavorably influenced by severe histologic inflammation and fibrosis in the initial liver biopsies (Galambos 1974). We have defined and discussed alcoholic liver fibrosis at the beginning of this chapter. Fibrosis is predominantly pericellular and accentuated around terminal venules (see Fig. 1f) (Christoffersen and Poulsen 1979). This pattern of fibrosis also occurs at the fatty liver stage in the absence of hepatitis (Edmondson et al. 1967; van Waes and Lieber 1977; Nasrallah et al. 1980; Nakano et al. 1982) and may then be considered to be one end of a spectrum, the other end being the same fibrotic pattern in conjunction with severe inflammation

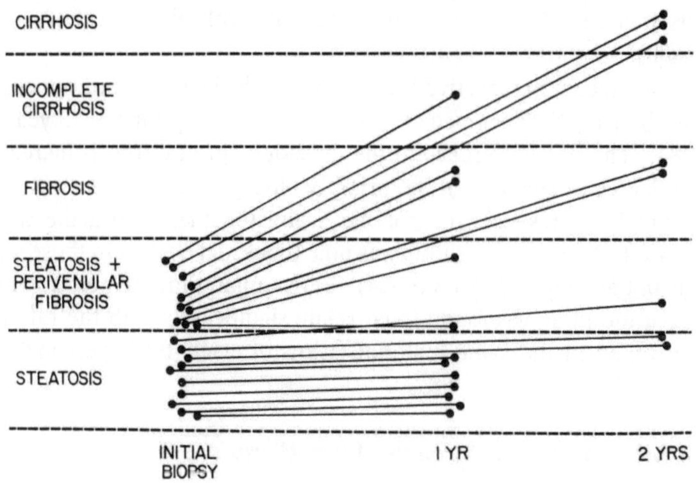

Fig. 5. Perivenular fibrosis as a predictor of progression of alcoholic fibrosis. Of 20 alcoholics selected on the basis of absence of alcoholic hepatitis in the liver biopsy and who were subsequently rebiopsied during follow-up, 10 showed simple fatty liver; 9 of them did not progress after 1–2 years. In contrast, patients with perivenular fibrosis who continued to drink showed progression (from Nakano et al. 1982)

and necrosis with Mallory bodies. One author even extends the histologic term "alcoholic hepatitis" to a lesion with perivenular/pericellular fibrosis without overt necrosis/inflammation (Thaler 1982). A mononuclear infiltrate with a few granulocytes has also been observed in perivenular fibrosis of alcoholic patients (Nakano et al. 1982) and in alcohol-fed baboons (Popper and Lieber 1980; Nakano and Lieber 1982). Nakano et al. (1982) pooled the material of 21 venules and found 43% mononuclear cells among the cellular population of perivenular fibrosis. This qualifies as a mild degree of inflammation and would be compatible with the concept of alcoholic fibrogenesis depicted in Fig. 4.

Perivenular fibrosis (Fig. 1b) is a predictor of the progression of fibrogenesis in patients who continue to drink, even in the absence of full blown alcoholic hepatitis. Of 20 alcoholics who were selected on the basis of absence of alcoholic hepatitis on liver biopsy and who were subsequently rebiopsied during follow-up, 10 showed simple fatty liver initially and of these, 9 did not progress after 1–2 years. By contrast, patients with perivenular fibrosis who continued to drink showed progression (Fig. 5; Nakano et al. 1982). Three developed cirrhosis. Similar observations have been published by Nasrallah et al. (1980), who have also pointed out that pericellular (Disse's space) fibrosis may be more predictive for the development of cirrhosis than perivenular fibrosis. Pericellular and pervenular fibrosis around terminal hepatic veins may be the same lesion, as discussed at the begin of this chapter. The importance of zone III (perivenular) fibrosis is supported by observations in alcohol-fed baboons where this lesion progressed to septum formation and cirrhosis (Van Waes and Lieber 1977; Popper and Lieber 1980). The earliest change in the alcoholic fatty liver was focal deposition of new connective tissue in zone III around terminal veins as detected by immunofluorescence (Hahn et al. 1980a, 1983), and this was also described in three patients with alcoholic fatty liver (Hahn et al. 1980b).

In a recent prospective evaluation of 258 alcoholics without cirrhosis on primary liver biopsy, 38 developed cirrhosis after follow-up for 10–13 years (Sørensen et al. 1984). The preponderance to the development of cirrhosis increased stepwise with the degree of steatosis in the primary biopsy, and, in those with alcoholic hepatitis (defined as presence of alcoholic hyalin together with acute intralobular inflammation), the rate of cirrhosis was nine times higher than in those with no steatosis in the initial biopsy (5/10 vs 3/54). Perivenular fibrosis was not mentioned in this publication but was found to correlate significantly with the rate at which cirrhosis developed in the same material (Christoffersen 1984, personal communication).

Blood Analysis for Alcoholic Liver Fibrosis

Liver tests are presently tailored to detect and quantitate hepatobiliary function, and hepatic fibrosis is nolens volens excluded from everyday decision-making. A recent issue of *Seminars in Liver Diseases* (1983) reflects this tradition and testifies to the limitations of even the more advanced assessments of liver function (McIntyre 1983; Bircher 1983). Since hepatic fibrosis is one of the most important factors of disturbed liver function, one may predict that more information about altered connective tissue metabolism in the liver of an individual patient with alcoholic liver disease will allow better interpretation of liver function tests, improve prognosis, and support new therapeutic approaches. To prevent or even reverse progressive liver fibrosis may be the ultimate goal if the cause of chronic liver disease cannot be eliminated.

Liver biopsy is precise in defining the presence and pattern of excess connective tissue at a given point of time, and histopathology is and will be the mainstay for the diagnosis of liver fibrosis and cirrhosis. But diagnosis is only one essential part of defining liver fibrogenesis. We also need information about the *activity* (connective tissue build-up in time) and the *reversibility* (connective tissue degraded in time) of liver fibrosis, and we need it at short intervals. Only then will we be effective in monitoring the natural course and therapy of a fibrotic liver disease. Ideally, probes of fibrotic activity and reversibility should be noninvasive, easy to perform and to control, widely available, and independent of the patient's general condition. They should also reflect the liver as a whole and be cost-effective. We believe that only laboratory tests can fulfill these requirements, and therefore, in this section, we will discuss blood methods for the analysis of liver fibrosis in patients with alcoholic fibrotic liver disease.

Against the background of the known composition of hepatic connective tissue (see above) various approaches appear feasible. Enzymes and metabolic products of collagen and other proteins have been used for some time to quantitate altered connective tissue metabolism in liver disease, and this is summarized in Table 3. Many of these methods require either biopsy material or have not attracted routine application for various, usually technical, reasons. When used in patients with liver disease, they have given clear evidence of enhanced hepatic collagen synthesis (i.e., prolyl hydroxylase activity) and degradation (i.e., urine hydroxyproline, tissue collagenase activity), and this has been discussed in many reviews (Hahn and Martini

Table 3. Enzymes and metabolic products that have been used (+) to quantitate altered connective tissue metabolism in liver disease

	Biopsy	Serum/plasma	Urine
General			
Proline	+	+	+
Hydroxyproline	+[a]	+[b]	+
(^{14}C) or (^{3}H) proline transformation into hydroxyproline	+	–	–
Prolyl hydroxylase			
(a) Activity	+	+	–
(b) Immunoreactive	+	+	–
Galactosylhydroxylysyl glycosyl transferase	+	+	–
Lysyl oxidase	+	+	–
Collagenase	+[b]	–	–
β-N-acetylglucosaminidase	+	+	+
Specific proteins			
Procollagen peptides			
(a) Type I	–	+	–
(b) Type III	+	+	+
7-S collagen[c]	+	+	+
NCl peptide[c]	+	+	+
Type VI collagen	+	+	+
Laminin	+	+	+

[a] Measures collagen content
[b] Measures collagen degradation
[c] Fragment of type IV collagen

1980; Rojkind and Kershenobich 1983; Hahn and Schuppan 1983; Hahn et al. 1984). Only one of the parameters mentioned in Table 1 has been avidly accepted by many clinicians and researchers as a blood test for collagen metabolism, particularly in liver disease; this is the aminoterminal procollagen type III peptide (PIIIP). The first full report about a radioimmunometric assay for this precursor peptide of collagen type III, its quantitation in human serum, and application to liver disease appeared in 1979 (Rohde et al. 1979), and a commercial kit has been available since November 1983 (Behring Werke, Marburg, F.R.G.).

Procollagen peptides are globular terminal structures on both sides of the procollagen molecule and particularly well characterized for types I and III collagen (Timpl and Glanville 1981). To produce a collagen type III fibril in the extracellular space, these procollagen peptides have to be cleaved off the helical moiety of the procollagen molecule. This appears to happen on the surface of thin fibrils (Fleischmajer et al. 1983), and special procollagen endopeptidases exist which remove the procollagen en bloc, so that the fibril can grow further. The PIIIP has been isolated from fetal calf skin and used to develop specific antibodies for immunofluorescence and for a sensitive radioimmunoassay (PIIIP-RIA). These antibodies showed high cross-reactivity with human material, so that quantiative measurements in serum, urine, and ascites of patients were possible (Rohde et al. 1979). The exact pathway of procollagen peptides liberated from the precursor molecule are now under active study. They may be partly further degraded locally, but some of them appear in the blood. The intact peptide (mol. wt. 40,000), a smaller fragment thereof (Col 1, mol. wt. 10,000), and less defined higher molecular material were detected by molecular sieve chromatography of serum and ascites using the PIIIP-RIA (Rohde et al. 1979; Niemelä et al. 1982, 1983).

Certain salient features of the PIIIP-RIA became clarified when used for human serum. The inhibition curve was not quite parallel with the standard curve, particularly when serum of healthy persons was used. This is due to degradation products of the peptide in serum, which generate a flatter inhibition curve (Rohde et al. 1979; Niemelä et al. 1982, 1983). The parallelism increased in serum from patients with ongoing liver fibrosis (like alcoholic hepatitis); these sera also have high or very high concentrations of the peptide. In ascites, the values were extremely high and the inhibition curves were always exactly parallel. This situation necessitates that at least two serum dilutions be used to generate the inhibition curve crossing the 50% inhibition level and the 50% inhibition point as a basis for all calculations. When this was done, an intra-assay variation of 1%–3% and an interassay variation of 10%–12% of normal and pathological sera resulted. Careful quality control must also be performed. The reference range (95% confidence limits) in healthy controls was found to be 3–13 mg/ml (Hahn and Schuppan 1983; Hochweiss et al. 1983), and values up to 20-fold the upper limit were found in sera of patients with alcoholic hepatitis.

The inconvenience of measuring two or more dilutions of a sample can be circumvented by a recently published modification of the original PIIIP assay using Fab-fragments of the antibodies (Rohde et al. 1983). Fab-fragments are inhibited equally well by the intact peptide and by smaller degradation products, and all sera, urine, and ascites give rise to inhibition curves exactly parallel to the standard curve. Hence, any point along these curves between 30% and 70% inhibition can be used for calculation of the concentration. Using a research version of this new assay (generously provided by Hoechst AG, Frankfurt, F.R.G.), we found reference values of 9–67 ng/ml in 54 healthy persons aged 19–92. Higher values with this assay are to be expected since smaller fragments of the native PIIIP are now recognized, which may be products of unphysiological degradation. This opens the fascinating possibility of measuring new synthesis and "physiological" scission of PIIIP with the one version and additional degradation with the other (Fab). Indeed, our results in sera covering the range between normal and highly pathological values using both versions of the PIIIP assay in patients with alcoholic liver disease support this possibility, and further clinical studies are in progress.

Using the PIIIP-RIA, elevated procollagen peptide serum levels have been found in carefully conducted retrospective and prospective studies of several fibrotic conditions of specific organs like lung (Low et al. 1983), bone, or bone marrow (Hochweiss et al. 1983; Simon et al. 1984) and of various liver diseases (Hahn 1984). These studies also show that interpretation of changes in serum PIIIP levels must incorporate clinical information about the organ affected, as is true for most biochemical tests. Although many tissues may contribute to the serum pool of PIIIP, calcified bone is excluded since it contains only type I collagen; this renders a marker of type III collagen particularly useful as a marker of fibrosis of an internal organ like the liver.

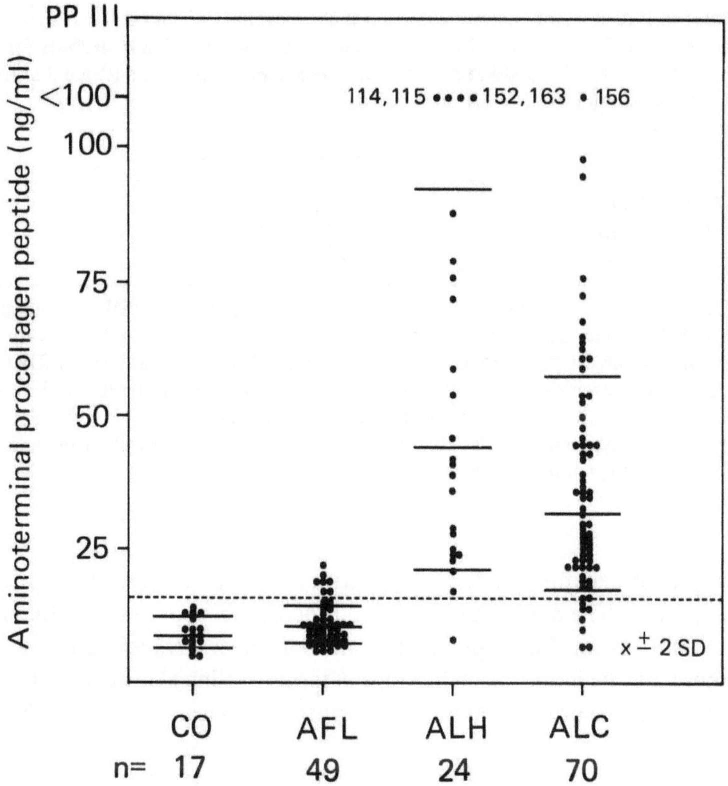

Fig. 6. Serum concentration of aminoterminal type III procollagen peptide (PIIIP) in 143 patients with alcoholic liver disease compared with 17 controls. These data are part of a prospective clinical study designed to determine the clinical significance of serum PIIIP. A serum was required at the time of biopsy, the histological diagnosis and grading was done independently by two pathologists (Dr. Kalbfleisch, Marburg, F.R.G.; Dr. Bruguera, Barcelona, Spain); extrahepatic disease was an exclusion criterion. Clinical investigation and diagnosis were done as described in Rohde et al. (1979), discriminant analysis as outlined in Pencev et al. (1982). The *dotted line* marks $\bar{x} \pm 2$ SD of normal controls. *Horizontal bars* indicate mean ± 1 SD. PIIIP was not normally distributed in patients, and statistical calculations were done after logarithmic transformation. *AFL*, alcoholic fatty liver; *ALH*, alcoholic hepatitis; *ALC*, alcoholic cirrhosis; *CO*, controls

Several papers have recently appeared studying alcoholic liver disease with the PIIIP-RIA; they all conclude that serum procollagen concentration may be helpful in monitoring alcoholic liver fibrogenesis (Pencev et al. 1982; Hahn and Schuppan 1983; Niemelä et al. 1983; Nakano et al. 1983; Savolainen et al. 1984; Maruyama et al. 1984; Mendenhall et al. 1984). The activity ("speed") of ongoing fibrosis appears to be an important factor for producing elevated PIIIP values in serum; in alcoholic hepatitis they are very high (Rohde et al. 1979; Pencev et al. 1982; Niemelä et al. 1983) and decline when the disease subsides (Hahn and Martini 1980). PIIIP distinguishes between alcoholic fatty liver and alcoholic hepatitis with a very high efficacy (Fig. 6, Table 4; Pencev et al. 1982; Hahn and Pencev 1983; Hahn and Schuppan 1983; Niemelä 1983). There is some overlap between alcoholic fatty liver,

Table 4. Discriminant analysis with clinical chemical data of patients with alcoholic liver disease as described in Pencev et al. (1982) and Hahn and Schuppan (1983). Results for a subgroup of 49 patients with fatty liver and 15 patients with alcoholic hepatitis demonstrate the effect of serum PIIIP on the separation by clinical chemical investigations

	Rank of parameters	Effectivity
PIIIP excluded	Bil, ESR	0.81
PIIIP included	PIIIP, ALAT	0.92
PIIIP alone	PIIIP	0.89

Calculations were done with all laboratory data excluding serum PIIIP, including serum PIIIP and with PIIIP alone. Only parameters with a significant contribution to diagnosis ($p < 0.01$) are listed, and a rank was established. The effectivity of separation shows that serum PIIIP used alone is better than all conventional blood tests taken together to distinguish between alcoholic fatty liver and alcoholic hepatitis. This conclusion can also be drawn from Fig. 6. PIIIP, aminoterminal procollagen type III peptide; Bil, serum bilirubin; ESR erythrocyte sedimentation rate; ALAT, SGOT, alanineaminotransferase

alcoholic hepatitis, and alcoholic cirrhosis, and an initial biopsy will definitely be necessary. The meaning of elevated procollagen peptides in alcoholic fatty liver and of normal procollagen peptides in alcoholic liver cirrhosis remains to be elucidated; these may be important subgroups in terms of clinical prognosis, and this question is presently being evaluated in prospective clinical studies (Hahn and Schuppan 1983).

From all these studies one may conclude that serum procollagen peptides help to distinguish between fibrotic and nonfibrotic alcoholic liver disease and that they are closely related to the activity of collagen type III metabolism in the liver regardless of serum transaminases and other laboratory tests (Rohde et al. 1979; Pencev et al. 1982; Niemelä et al. 1983; Maruyama et al. 1984). And yet, to define the clinical value of serum PIIIP as a blood test of liver fibrosis in alcoholic liver disease much more needs to be done. The activity of prolyl hydroxylase in liver specimens has been shown to correlate with PIIIP levels in serum of the same patient at the time of biopsy (Hahn and Schuppan 1983), but more direct evidence is needed on the relation between serum procollagen peptides and collagen type III metabolism in the liver. We also eagerly await the results of prospective clinical studies designed to evaluate the long-term outcome of patients with elevated serum PIIIP. Such studies are pesently being conducted by several clinical groups. Do the patients develop fibrosis and/or cirrhosis faster? Does the fibrosis arrest when PIIIP declines spontaneously or as a result of therapy? The question whether PIIIP reflects more synthesis or more degradation will also need more molecular studies of serum (Niemelä et al. 1982, 1983). The new Fab-PIIIP-RIA will undoubtedly help to elucidate this question. Degradation detected by these methods (Niemelä et al. 1982; Rohde et al. 1983) is certainly more related to type III collagen metabolism and remodelling than to net degradation of the fibrotic matrix. For instance, in babies, infants, and growing children, the PIIIP serum level is elevated above adult levels (Trivedi et al. 1984). It is elevated in the blood of pregnant women but declines immediately after delivery when there is maximal degradation in the involuting uterus containing large amounts of type III collagen at this time (E. G. Hahn,

U. Hahn, D. Schuppan 1984, unpublished results; L. Risteli and J. Risteli 1984, personal communication). Net degradation of the fibrotic matrix would of course also have clinical importance because it may be more closely related to the reversibility of liver fibrosis. Stable degradation products of type IV collagen or laminin may prove to be suitable serum parameters in this context (Table 4).

To date, fibrogenesis in alcoholic liver disease is always a potential surprise. We will now be able to assess the metabolism of liver fibrogenesis as we can measure the hepatobiliary function, and we should include this possibility in future clinical studies of alcoholic liver disease.

References

Actis G, Mieli-Vergani G, Portmann B, Eddleston ALWF, Davis M, Williams R (1983) Lymphocyte cytotoxicity to autologous hepatocytes in alcoholic liver disease. Liver 3: 8–12

Alitalo K, Hovi T, Vaheri A (1980) Fibronectin is produced by human macrophages. J Exp Med 151: 602–613

Anthony PP, Ishak KG, Nayak NC, Poulsen HE, Scheuer PJ, Sobin LH (1978) The morphology of cirrhosis. J Clin Pathol 31: 395–414

Anthony RS, Farquharson M, MacSween RNM (1983) Liver membrane antibodies in alcoholic liver disease II. Antibodies to ethanol-altered hepatocytes. J Clin Pathol 36: 1302–1308

Balázs M, Várkonyi S, Pintér A (1977) Electron microscopic study of alcoholic liver disease with special attention to the changes of mesenchymal cells of the liver. Exp Path 14: 340–350

Bhathal PS (1972) Presence of modified fibroblasts in cirrhotic liver in man. Pathology 4: 139–144

Biempica L, Morecki R, Wu CH, Giambrone M-A, Rojkind M (1980) Immunocytochemical localization of type B collagen. Am J Path 98: 591–602

Bircher J (1983) Quantitative assessment of deranged hepatic function: a missed opportunity? Sem Liver Dis 3: 275–284

Bloch EH, McCuskey RS (1977) Biodynamics of phagocytosis: an analysis of the dynamics of phagocytosis in the liver by in vivo microscopy. In: Wisse E, Knook DL (eds) Kupffer and other liver sinusoidal cells. Elsevier, Amsterdam, pp 21–32

Bouwens L, Baekeland M, Wisse E (1984) Importance of local proliferation in the expanding Kupffer cell population of rat liver after zymosan stimulation and partial hepatectomy. Hepatology 4: 213–219

Brunt PW, Kew MC, Scheuer PJ, Sherlock S (1974) Studies in alcoholic liver disease in Britain. 1. Clinical and pathologic patterns related to natural history. Gut 15: 52–58

Burt AD, Anthony RS, Hislop WS, Bouchier IAD, MacSween RNM (1982) Liver membrane antibodies in alcoholic liver disease: I. Prevalence and immunoglobulin class. Gut 23: 221–225

Cardinale GJ, Udenfriend S (1974) Prolyl hydroxylase. Adv Enzymol 41: 245–300

Carlisle R, Galambos JT, Warren WD (1979) The relationship between conventional liver tests, quantitative function tests, and histopathology in cirrhosis. Dig Dis Sci 24: 358–362

Christoffersen P, Poulsen H (1979) Alcoholic liver disease. In: MacSween RNM, Anthony PP, Scheuer PJ (eds) Pathology of the liver. Churchill Livingstone, Edinburgh, pp 232–247

Clement B, Guguen-Guilluzo C, Campion J-P, Glaise D, Bourel M, Guilluzo A (1984) Long-term co-cultures of adult human hepatocytes with rat liver epithelial cells: modulation of albumin secretion and accumulation of extracellular material. Hepatology 4: 373–380

Conn HO (1982) Cirrhosis. In: Schiff L, Schiff ER (eds) Diseases of the liver. Lippincott, Philadelphia, pp 847–977

Crapper RM, Bhathal PS, Mackay IR (1983) Chronic active hepatitis in alcoholic patients. Liver 3: 327–337

De Leeuw AM, Browner A, Barelds RJ, Knook DL (1983) Maintenance cultures of Kupffer cells isolated from rats of various ages: ultrastructure, enzyme cytochemistry and endocytosis. Hepatology 3: 497–506

De Leeuw AM, McCarthy SP, Geerts A, Knook DL (1984) Purified rat liver fat-storing cells in culture devide and contain collagen. Hepatology 4: 392–403
Diegelmann RF, Guzelian PS, Gay R, Gay S (1983) Collagen formation by the hepatocyte in primary monolayer culture and in vivo. Science 219: 1343–1345
DiLucio NR (1979) The role of Kupffer cells in homeostasis. In: Davidson CS (ed) Problems in liver diseases. Stratton Intercontinental, New York, pp 113–120
Edmondson HA, Peters RL, Frankel HH, Borowsky S (1967) The early stage of liver injury in the alcoholic. Medicine 46: 119–129
Fallon A, Bradley JF, Burns J, McGee J O'D (1984) Collagen stimulating factors in hepatic fibrogenesis. J Clin Pathol 37: 542–548
Fleischmajer R, Olsen BR, Timpl R, Perlish JS, Lovelace O (1983) Collagen fibril formation during embryogenesis. Proc Natl Acad Sci USA 80: 3354–3358
Forsander OA, Pikkarainen JAJ, Salaspuro MP (1983) A high hepatic concentration of free proline does not induce collagen synthesis in rat liver. Hepato-gastroenterol 30: 6–8
Gabbiani G (1981) The myofibroblast: a key cell for wound healing and fibrocontractive diseases. In: Adam M (ed) Connective tissue research: chemistry, biology, and physiology. Liss, New York, pp 183–194
Galambos JT (1972) Natural history of alcoholic hepatitis. III. Histological changes. Gastroenterology 63: 1026–1035
Galambos JT (1974) Alcoholic hepatitis. In: Schaffner F, Sherlock S, Leevy CM (eds) The liver and its diseases. Intercontinental Medical, New York, pp 255–267
Galambos JT, Shapira R (1973) Natural history of alcoholic hepatitis IV: Glycosaminoglycans and collagen in hepatic connective tissue. J Clin Invest 60: 107–111
Gay S, Fietzek PP, Remberger K, Eder M, Kühn K (1975) Liver cirrhosis: immunofluorescence and biochemical studies demonstrate two types of collagen. Klin Wochenschr 53: 205–208
Goldberg SJ, Mendenhall CL, Connell AM, Chedid A (1977) "Non-alcoholic" chronic hepatitis in the alcoholic. Gastroenterology 72: 598–604
Goodman ZD, Ishak KG (1982) Occlusive venous lesions in alcoholic liver disease. A study of 200 cases. Gastroenterology 83: 786–796
Goresky CA (1980) Uptake in the liver: the nature of the process. Int Rev Physiol 21: 65–101
Grimaud J-A, Druguet M, Peyrol S, Chevalier O, Herbage D, El Badrawy N (1980) Collagen immunotyping in human liver – light and electron microscopy study. J Histochem Cytochem 28: 1145–1156
Grotendorst GR, Seppä HEJ, Kleinmann HK, Martin GR (1981) Attachment of smooth muscle cells to collagen and their migration toward platelet-derived growth factor. Proc Natl Acad Sci USA 78: 3669–3672
Guzelian PS, Quereshi GD, Diegelmann RF (1981) Collagen synthesis by the hepatocyte. Studies in primary culture of parenchymal cells from adult rat liver. Coll Rel Res 1: 83–93
Hahn EG (1985) Blood analysis for liver fibrosis. J Hepatology 1: 67–73
Hahn EG, Martini GA (1980) Clinical parameters of fibroplasia. Ital J Gastroenterol 12: 41–46
Hahn EG, Pencev D (1983) Serum procollagen concentration as a new tool in the study of alcoholic liver disease. In: Lieber CS (ed) Biological approach to alcoholism. Research monograph No 11, U.S. Department of Health and Human Services, Washington, pp 399–402
Hahn EG, Schuppan D (1983) Collagen metabolism in liver disease. In: Bianchi L, Gerok W, Landmann, L, Sickinger K, Stalder GA (eds) Liver in metabolic diseases. MTP, Lancaster, pp 309–323
Hahn EG, Ott U, Martini GA (1980a) Die Leberfibrose. Z Gastroenterol 18: 453–469, 507–523
Hahn EG, Wick G, Pencev D, Timpl R (1980b) Distribution of basement membrane proteins in normal and fibrotic human liver: collagen type IV, laminin and fibronectin. Gut 21: 63–71
Hahn EG, Ott U, Martini GA (1981) Fibronectin and the liver. In: Berk PD, Chalmers TC (eds) Frontiers in liver diseases. Thieme-Stratton, New York, pp 27–43
Hahn EG, Timpl R, Lieber CS (1983) Distribution of hepatic collagens, elastin, and structural glycoproteins during the development of alcoholic liver injury in baboons. In: Lieber CS (ed) Biological approach to alcoholism. Research monograph No 11, U.S. Department of health and human services, Washington, pp 403–408

Hahn EG, Hochweiss S, Berk PD (1984) Diagnostic parameters of altered collagen turnover. In: Berk PD, Castro-Malaspina H, Wasserman LR (eds) Myelofibrosis and the biology of connective tissue. Liss, New York, pp 335–343

Hochweiss S, Fruchtman S, Hahn EG, Gilbert H, Donovan PB, Johnson J, Goldberg JD, Berk PD (1983) Increased serum procollagen III aminoterminal peptide in myelofibrosis. Am J Hematol 15: 343–351

International group (1981) Alcoholic liver disease: morphologic manifestations. Lancet I: 707–711

Izumi N, Hasumura Y, Takeuchi J (1983) Lymphocyte cytotoxicity for autologous human hepatocytes in alcoholic liver disease. Clin Exp Immunol 53: 219–224

Jakobsen KR, Ranek L, Tygstrup N (1969) Liver function and blood flow in normal man during infusion of vasopressin. Scand J Clin Lab Invest 24: 279

Jenkins WJ, Thomas HC (1981) Genetic factors in determining susceptibility to alcohol dependence and development of alcohol induced liver disease. Clinics in Gastroenterology 10: 307–314

Johnson RL, Ziff M (1976) Lymphokine stimulation of collagen accumulation. J Clin Invest 58: 246–252

Kao WW-Y, Gay S, Bressan GM, Prockop DJ (1980) Demonstration by immunofluorescence that the same cell from chick embryo aortas synthesize elastin and collagen types I and III. Artery 7: 176–190

Kershenobich D, García-Tsao G, Alvarez Saldaña S, Rojkind M (1981) Relationship between blood lactic acid and serum proline in alcoholic liver cirrhosis. Gastroenterology 80: 1012–1015

Knook DL, Seffelaar AM, de Leeuw AM (1982) Fat-storing cells of the rat liver. Their isolation and purification. Exp Cell Res 139: 468–471

Kojima J, Nakamura N, Kanatani M, Ohmori K (1975) The glycosaminoglycans in human hepatic cancer. Cancer Res 35: 542–547

Kojima N, Isemura M, Yosizawa Z, Ono T, Shinada S, Soga K, Aoyagi Y, Ichida F (1981) Distribution of fibronectin in fibrotic human livers at various states. Tohoku J Exp Med 135: 403–412

Kowaloff EM, Phang JM, Granger AS, Downing SJ (1977) Regulation of proline oxidase activity by lactate. Proc Natl Acad Sci USA 74: 5368–5371

Krauledat PB, Rohde H, Hahn EG, Martin GA (1982) Isolation and characterization of distinct cell types from normal and fibrotic liver of man and monkey and their production of connective tissue protein. In: Gerlach U, Pott G, Rauterberg J, Voss B (eds) Connective tissue of the normal and fibrotic human liver. Thieme, Stuttgart, pp 91–92

Leevy CM, Popper H, Sherlock S (criteria committee) (1977) Diseases of the liver and biliary tract. Fogarty Intern Center Proc No 22. US Government Printing Office, Washington, DC 20402, Stock No. 017-053-00049-3

Leo MA, Lieber CS (1983) Hepatic fibrosis after long-term administration of ethanol and moderate vitamin A supplementation in the rat. Hepatology 3: 1–11

Leser H-G, Debatin K-M, Northoff H, Gemsa D (1982) Kupffer cell-enriched nonparenchymal liver cells and peritoneal macrophages differ in prostaglandin release, production of interleukin-1 and tumor cytotoxicity. In: Knook DL, Wisse E (eds) Sinusoidal liver cells. Elsevier, Amsterdam, pp 369–376

Levin DM, Baker AL, Riddell RH, Rochman H, Boyer JL (1979) Nonalcoholic liver disease. Overlooked causes of liver injury in patients with heavy alcohol consumption. Am J Med 66: 429–434

Lieber CS (1983) Alcohol and the liver. In: Arias IM, Frenkel M, Wilson JHP (eds) The liver annual 3. Elsevier, Amsterdam, pp 106–148

Lieber CS, Leo MA (1982) Alcohol and the liver. In: Lieber CS (ed) Medical disorders of alcoholism. Saunders, Philadelphia, pp 259–312

Low RB, Cutroneo KR, Davis GS, Giancola MS (1983) Lavage type III procollagen N-terminal peptides in human pulmonary fibrosis and sarcoidosis. Lab Invest 48: 755–759

MacSween RNM, Anthony RS, Farquharson M (1981) Antibodies to alcohol-altered hepatocytes in patients with alcoholic liver disease. Lancet II: 803–804

Mak KM, Leo MA, Lieber CS (1984) Alcoholic liver injury in baboons: transformation of lipocytes to transitional cells. Gastroenterology 87: 188–200

Maruyama K, Okazaki I, Kobayashi T, Okuno F, Kamegaya K, Takagi S, Oda M, Ishii H, Tsuchiya M (1984) Measurement of serum type III procollagen aminopeptide as a parameter of hepatic fibrosis. Acta Hepatol Jpn 25: 24–32

McCuskey RS (1966) A dynamic and static study of hepatic arterioles and hepatic sphincters. Am J Anat 119: 455–478

McIntyre N (1983) The limitations of conventional liver function tests. Sem Liver Dis 3: 265–274

McIvor CA, Cooksley WGE, Grotendorst GR, Jones EA (1982) Synthesis of chemoattractants for smooth muscle cells by Kupffer cells and monocytes – potential implications for hepatic fibrogenesis. Hepatology 2: 725A

Mendenhall C, Chedid A, Morgan D, Kromme C (1983) Serum N-terminal procollagen III (Pro III) peptide (P) in alcoholic hepatitis (AH). Hepatology 3: 851A

Minato Y, Hasumura Y, Takeuchi J (1983) The role of fat-storing cells in Disse space fibrogenesis in alcoholic liver disease. Hepatology 3: 559–566

Miyakawa H, Iida S, Leo MA, Greenstein R, Lieber CS (1983) Pathogenesis of portal hypertension in precirrhotic alcoholic liver injury. Gastroenterology 84: 1385A

Mizel SB, Dayer JM, Krane SM, Mergenhagen SE (1981) Stimulation of rheumatoid synovial cell collagenase and prostaglandin production by partially purified lymphocyte activating factor (Interleukin 1). Proc Natl Acad Sci USA 78: 2474–2477

Morgan MY, Sherlock S, Scheuer PJ (1978) Portal fibrosis in the livers of alcoholic patients, Gut 19: 1015–1021

Nakano M, Lieber CS (1982) Ultrastructure of initial stages of perivenular fibrosis in alcohol-fed baboons. Am J Pathol 106: 145–155

Nakano M, Worner TM, Lieber CS (1982) Perivenular fibrosis in alcoholic liver injury: ultrastructure and histologic progression. Gastroenterology 83: 777–785

Nakano H, Kawasaki T, Miyamura M, Fukuda Y, Imura H (1983) Serum levels of N-terminal type III procollagen peptide in normal subjects and comparison to hepatic fibrosis. Acta Hepatol Jpn 24: 1230–1234

Nasrallah SM, Nassar VH, Galambos JT (1980) Importance of hepatic venule thickening. Arch Pathol Lab Med 104: 84–86

Nayak NC (1979) Indian childhood cirrhosis. In: MacSween RNM, Anthony PP, Scheuer PJ (eds) Pathology of the liver. Churchill Livingstone, Edinburgh, pp 268–269

Neilson EG, Jimenez SA, Phillips SM (1980) Cell mediated immunity in interstitial nephritis 3. T-lymphocyte mediated fibroblast proliferation and collagen synthesis, an immune mechanism for renal fibrogenesis. J Immunol 125: 1708–1714

Neuberger J, Crossley IR, Saunders JB, Davis M, Portmann B, Eddleston ALWF, Williams R (1984) Antibodies to alcohol altered liver cell determinants in patients with liver disease. Gut 25: 300–304

Neufeld EF, Ashwell G (1980) Carbohydrate recognition systems for receptor-mediated pinocytosis. In: Lennarz WJ (ed) The biochemistry of glycoproteins and proteoglycans. Plenum, New York, pp 241–266

Niemelä O, Risteli L, Sotaniemi EA, Risteli J (1982) Heterogeneity of the antigens related to the aminoterminal propeptide of type III procollagen in human serum. Clin Chim Acta 124: 39–44

Niemelä O, Risteli L, Sotaniemi EA, Risteli J (1983) Aminoterminal propeptide of type III procollagen in serum in alcoholic liver disease. Gastroenterology 85: 254–259

Ohnishi K, Iida S, Iwama S, Goto N, Nomura F, Takashi M, Mishima A, Kono K, Kimura K, Musha H, Kotata K, Okuda K (1982) The effect of chronic habitual alcohol intake on the development of liver cirrhosis and hepatocellular carcinoma. Cancer 49: 672–677

Okanoue T, Burbige EJ, French SW (1983) The role of the Ito cell in perivenular and intralobular fibrosis in alcoholic hepatitis. Arch Pathol Lab Med 107: 459–463

O'Leary JP (1983) Hepatic complications of jejunoidal bypass. Sem Liver Dis 3: 203–215

Orrego H, Medline A, Blendis LM, Rankin JG, Kreaden DA (1979) Collagenisation of the Disse space in alcoholic liver disease. Gut 20: 673–679

Orrego H, Blendis LM, Crossley IR, Medline A, Macdonald A, Ritchie S, Israel Y (1981a) Correlation of the intrahepatic pressure with collagen in the Disse space and hepatomegaly in humans and rats. Gastroenterology 80: 546–556

Orrego H, Israel Y, Blendis LM (1981b) Alcoholic liver disease: information in search of knowledge? Hepatology 1: 267–283

Ott U, Hahn EG, Moshudis E, Bode JCh, Martini GA (1977) Immunhistologischer Nachweis von Typ I- und Typ III-Kollagen in Leberbiopsien: frühe und späte Veränderungen bei alkoholischer Lebererkrankung. Verh Dtsch Ges Inn Med 83: 537–540

Pencev D, Pittner P, Hahn EG, Timpl R, Martini GA (1982) Discriminant analysis of laboratory, histological and serum procollagen peptide data in patients with acute and chronic liver disease. In: Gerlach U, Pott G, Rauterberg J, Voss B (eds) Connective tissue of the normal and fibrotic human liver. Thieme, Stuttgart, pp 212–214

Popper H, Lieber CS (1980) Histogenesis of alcoholic fibrosis and cirrhosis in the Baboon. Am J Pathol 98: 695–715

Poralla T, Hütteroth TH, Meyer zum Büschenfelde K-H (1984) Cellular cytotoxicity against autologous hepatocytes in alcoholic liver disease. Liver 4: 117–121

Preisig R, Bircher J, Paumgartner G (1972) Physiologic and pathophysiologic aspects of the hepatic hemodynamics. In: Popper H, Schaffner F (eds) Progress in liver diseases, Vol. 4, Grune and Stratton, New York, pp 201

Rappaport AM (1980) Hepatic blood flow: morphologic aspects and physiologic regulation. In: Javitt NB (ed) Liver and biliary tract physiology I. University Park Press, Baltimore, pp 1–63, (Int Rev Physiol, vol 21)

Rappaport AM, MacPhee PJ, Fisher MM, Phillips MJ (1983) The scarring of the liver acini (cirrhosis) - tridimensional and microcirculatory considerations. Virchows Arch A 402: 107–137

Rauterberg J, Voss B, Pott G, Gerlach U (1981) Connective tissue components of the normal and fibrotic human liver. Klin Wschr 59: 767–779

Rennard SI, Bitterman PB, Crystal RG (1984) Current concepts of the pathogenesis of fibrosis: lessons from pulmonary fibrosis. In: Berk PD, Castro-Malaspina H, Wasserman LR (eds) Myelofibrosis and the biology of connective tissue. Liss, New York, pp 359–377

Reynolds TB, Hidemura R, Michel H, Peters R (1969) Portal hypertension without cirrhosis in alcoholic liver disease. Ann Intern Med 70: 497–506

Rohde H, Vargas L, Hahn EG, Kalbfleisch H, Bruguera M, Timpl R (1979) Radioimmunoassay for type III procollagen peptide and its application to human liver disease. Europ J Clin Invest 9: 451–459

Rohde H, Langer I, Krieg T, Timpl R (1983) Serum and urine analysis of the aminoterminal procollagen peptide type III by radioimmunoassay with antibody Fab fragments. Collagen Rel Res 3: 371–379

Rojkind M (1982) Fibrogenesis. In: Arias I, Popper H, Schachter D, Shafritz DA (eds) The liver: biology and pathobiology. Raven, New York, pp 801–809

Rojkind M, Kershenobich D (1981a) Hepatic fibrosis. Clin Gastroenterol 10: 737–754

Rojkind M, Kershenobich D (1981b) Fibrosis and cirrhosis. In: Arias IM, Frenkel M, Wilson JHP (eds) The liver annual 1. Elsevier, Amsterdam, pp 126–148

Rojkind M, Kershenobich D (1982) Fibrosis and cirrhosis. In: Arias IM, Frenkel M, Wilson JHP (eds) The liver annual 2. Elsevier, Amsterdam

Rojkind M, Kershenobich D (1983) The extracellular matrix: fibrosis and cirrhosis. In: Arias IM, Frenkel M, Wilson JHP (eds) The liver annual 3. Elsevier, Amsterdam, pp 189–205

Rojkind M, Giambrone M-A, Biempica L (1979) Collagen types in normal and cirrhotic liver. Gastroenterology 76: 710–719

Rudolph R, McClure WJ, Woodward M (1979) Contractile fibroblasts in chronic alcoholic cirrhosis. Gastroenterology 76: 704–709

Ruoslahti E, Engvall E, Hayman EG (1981) Fibronectin: current concepts of its structure and functions. Coll Res 1: 95–128

Saber MA, Shafritz DA, Zern MA (1983) Changes in collagen and albumin mRNA in liver tissue of mice infected with Schistosoma mansoni as determined by in situ hybridisation. J Cell Biol 97: 986–992

Savolainen E-R, Goldberg B, Leo MA, Velez M, Lieber CS (1984) Diagnostic value of serum procollagen peptide measurements in alcoholic liver disease. Alcoholism Clin Exp Res 8: 384–389

Savolainen E-R, Leo MA, Timpl R, Lieber CS (1983) Lactate and acetaldehyde stimulate collagen synthesis of cultured baboon liver myofibroblasts. Hepatology 3: 801 A

Schaffner F (1970) The structural basis of altered hepatic function in viral hepatitis. Am J Med 49: 658–668

Scheuer PJ (1980) Liver biopsy interpretation. Baillière Tindall, London

Scheuer PJ, Maggi G (1980) Hepatic fibrosis and collapse: histological distinction by orcein staining. Histopathology 4: 487–490

Schmidt JA, Mizel SB, Cohen D, Green I (1982) Interleukin 1, a potential regulator of fibroblast proliferation. J Immunol 128: 2177–2182

Schneider JF, Baker AL, Haines NW, Hatfield G, Boyer JL (1980) Aminopyrine N-demethylation: a prognostic test of liver function in patients with alcoholic liver disease. Gastroenterology 79: 1145–1150

Schulz R, Hahn U, Schuppan D, Hahn EG, Riecken EO (1984) Expression neuer Kollagentypen durch portale Fibroblasten bei der obstruktiven Gallengangserkrankung. Verh Dtsch Ges Inn Med 90: 1499–1501

Shaw S, Worner TM, Lieber CS (1984) Frequency of hyperprolinemia in alcoholic liver cirrhosis: relationship to blood lactate. Hepatology 4: 295–299

Simon LS, Krane SM, Wortman PD, Krane IM, Kovitz KL (1984) Serum levels of type I and III procollagen fragments in Paget's disease of bone. J Clin Endocrinol Metab 5: 110–120

Sørensen TIA, Orholm M, Bentsen KD, Høybye G, Eghøje K, Christoffersen P (1984) Prospective evaluation of alcohol abuse and alcoholic liver injury in men as predictors of development of cirrhosis. Lancet II: 241–244

Starzl TE, Porter KA, Terblanche (1980) Inter-organ communications: with particular reference to hepatotrophic factors and intrinsic liver growth factors. In: Popper H, Bianchi L, Gudat F, Reutter W (eds) Communications of liver cells. MTP, Lancaster, pp 93–97

Stenman S, Vaheri A (1978) Distribution of a major connective tissue protein, fibronectin, in normal human tissues. J Exp Med 147: 1054–1064

Stern B, Glimcher MJ, Goldhaber P (1966) The effect of various oxygen tensions on the synthesis and degradation of bone collagen in tissue culture. Proc Soc Exp Biol Med 212: 869–872

Tanner AR, Arthur MJP, Wright R (1984) Macrophage activation, chronic inflammation and gastrointestinal disease. Gut 25: 760–783

Thaler H (1982) Leberkrankheiten. Springer, Berlin Heidelberg New York, pp 333–359

Timpl R, Glanville RW (1981) The aminopropeptide of collagen. Clin Orthopaed 158: 224–242

Timpl R, Rohde H, Gehron-Robey P, Rennard SR, Foidart JM, Martin GR (1979) Laminin, a glycoprotein from basement membranes. J Biol Chem 254: 9933–9937

Trivedi P, Cheeseman P, Portmann B, Hegarty J, Mowat AP (1985) Variation in serum type III procollagen peptide with age in healthy subjects and its comparative value in the assessment of disease activity in children and adult with chronic active hepatitis. Europ J Clin Invest, in press

Tseng SCG, Lee PC, Ells PF, Bissell DM, Smuckler EA, Stern R (1982) Collagen production by rat hepatocytes and sinusoidal cells in primary monolayer culture. Hepatology 2: 13–18

Tseng SCG, Smuckler EA, Stern R (1983) Types of collagen synthesized by normal rat liver hepatocytes in primary culture. Hepatology 3: 955–963

Van Furth R, Blussé van oud Alblas D (1982) New aspects on the origin of Kupffer cells. In: Knook DL, Wisse E (eds) Sinusoidal liver cells. Elsevier, Amsterdam, pp 173–183

Van Waes LV, Lieber CS (1977) Early perivenular sclerosis in alcoholic fatty liver, an index of progressive liver injury. Gastroenterology 73: 646–650

Von der Mark H, Aumailley M, Wick G, Fleischmajer R, Timpl R (1984) Immunochemistry, genuine size and tissue localization of collagen VI. Europ J Biochem 142: 493–502

Voss B, Allam S, Rauterberg J, Ullrich K, Gieselmann V, von Figura K (1979) Primary culture of rat hepatocytes synthesize fibronectin. Biochem Biophys Res Comm 90: 1238–1354

Voss B, Rauterberg J, Pott G, Brehmer U, Allam S, Lehmann R, von Bassewitz DB (1982a) Nonparenchymal cells cultivated from explants of fibrotic liver resemble endothelial and smooth muscle cells from blood vessel walls. Hepatology 2: 19–28

Voss B, Rauterberg J, Brehmer U, Pott G (1982b) Investigations on the biosynthesis of connective tissue components by cultured mouse liver macrophages and mouse peritoneal macrophages. In: Knook DL, Wisse E (eds) Sinusoidal liver cells. Elsevier, Amsterdam, pp 201–208

Wahl SM (1984) Lymphocyte and macrophage derived fibroblast growth factors. In: Berk PD, Castro-Malaspina H, Wasserman LR (eds) Myelofibrosis and the biology of connective tissue. Liss, New York, pp 143–154

Wahl SM, Gately CL (1983) Modulation of fibroblast growth by a lymphokine of human T cell and continuous T cell line origin. J Immunol 130: 1226–1230

Wahl SM, Wahl LM, McCarthy JB (1978) Lymphocyte mediated activation of fibroblast proliferation and collagen production. J Immunol 121: 942–946

Wick, G, Brunner H, Penner E, Timpl R (1978) The diagnostic application of specific anti-procollagen sera II. Analysis of liver biopsies. Int Arch Allerg Appl Immunol 56: 316–324

Wiedmann KH, Bartholomew TC, Brown DJC, Thomas HC (1984) Liver membrane antibodies detected by immunoradiometric assay in acute and chronic virus-induced and autoimmune liver disease. Hepatology 4: 199–204

Wisse E, Knook DL (1979) The investigation of sinusoidal cells: a new approach to the study of liver function. In: Popper H, Schaffner F (eds) Progress in liver diseases. Grune and Stratton, New York, pp 153–171

Zern MA, Leo MA, Lieber CS (1984) Increased type I procollagen mRNA levels in the baboon model of chronic alcoholic liver disease. Gastroenterology (5.2) 86: 1082A

Zetterman RK, Sorrell MF (1981) Immunologic aspects of alcoholic liver disease. Gastroenterology 81: 616–624

8 Ethanol, Mallory Bodies, and the Microtubular System*

H. Denk

Institute of Pathology, University of Graz, School of Medicine, 8036 Graz, Austria

Alcoholic hepatitis is a disease of still increasing importance in the Western societies and, by producing liver cell necroses, it is an important link between reversible fatty liver and irreversible cirrhosis of the alcoholic (for review see French 1981, 1983; French and Burbige 1979; Denk et al. 1979a). Despite considerable efforts toward elucidation of its pathogenesis the questions still open by far outnumber those resolved. Alcoholic hepatitis has a complex morphological pattern with Mallory bodies (MBs, alcoholic hyalin) as the most prominent features (Mallory 1911). Correlated clinical and pathologic studies have revealed that MBs are quantitatively related to the severity of hepatic parenchymal injury and duration and extent of alcohol abuse (Gregory and Levi 1972; Harinasuta and Zimmerman 1971; Birschbach et al. 1974; Christoffersen et al. 1973; see also Denk et al. 1979a for review and further references). MBs, however, are not alcohol specific, and a diversity of different disorders has been described which are also associated with MB formation, including hepatocellular neoplasia and hyperplasia, primary biliary cirrhosis, Wilson's disease, a-β-lipoproteinemia, Indian childhood cirrhosis, diabetes, fatty liver of various origin, intestinal bypass operation for morbid obesity, Weber-Christian disease, glucocorticoid administration, toxic liver injury, but also asbestosis, and radiation damage in pneumocytes (for review see Denk et al. 1981b; French 1983; Capron et al. 1982; see also Popper et al. 1979). Consequently, the advent of MBs may not only be characteristic of metabolic disturbances related to chronic ethanol intoxication but may also indicate a specific pathway of liver cell injury. In a series of studies MBs have been linked by biochemical as well as ultrastructural and immunohistochemical analyses to the cytokeratin type of intermediate-sized filaments (Franke et al. 1979a; French 1981; Denk et al. 1981a), a filament system which, together with the actin-containing microfilaments and the microtubules, forms a cytoskeleton with mutual interaction as well as interaction with the cell membrane and cytoplasmic organelles (see Lazarides 1980 and Denk and Franke 1982 for review). Disturbance of the cytoskeleton may profoundly influence cellular functions, cell stability, and mobility as well as cell-to-cell interaction (see Rungger-Brändle and Gabbiani 1983 for review). MBs represent altered and aggregated cytokeratin filaments, eventually also associated with other products of hepatocytes (Franke et al. 1979a; see Denk et al. 1979a for review). Animal models

* The original work presented in this review was partly supported by *Fonds zur Förderung der wissenschaftlichen Forschung,* grants nos. P 3580 and 4708

Alcohol Related Diseases in Gastroenterology
Edited by H. K. Seitz and B. Kommerell

in which MBs are produced under standardized conditions, for example chronic griseofulvin (GF), 3,5-diethoxycarbonyl-1,4-dihydrocollidine (DDC), or dieldrin intoxication of mice, greatly facilitate systematic studies of MB formation, confirming, in addition, the lack of alcohol specificity of these inclusions (Denk et al. 1975; Yokoo et al. 1982; Meierhenry et al. 1983). It is an interesting facet of the griseofulvin model that MBs are produced by prolonged feeding of a drug, i.e., griseofulvin, with antimicrotubular properties (Weber et al. 1976; Grisham et al. 1973). This makes a connection between microtubular failure and disturbances of the intermediate filament cytoskeleton possible. In the present review various aspects of structure and pathogenesis of MBs will be covered, and, in addition, arguments for and against a relationship of ethanol intoxication, MB formation, and microtubular disturbances will be discussed.

The Cytoskeleton of Hepatocytes

An elaborate system of cytoskeletal structures exists in most, if not all, eukaryotic cells, including hepatocytes (Lazarides 1980; Osborn et al. 1982; Denk and Franke 1982; see there for further references). Microfilaments, microtubules and intermediate-sized filaments are prominent components of this system and, in mutual interaction and interaction with membranes, may play a role in the modulation of cell shape, cellular movements, secretion, and cell division. Actin-containing microfilaments form a three-dimensional meshwork in hepatocytes, which is concentrated around bile canaliculi as pericanalicular web, and are present in more parallel arrays in the microvillar cores (Oda et al. 1974). The filaments consist of β- and γ-non-muscle actin which, in cooperation with myosin, acts as a contractile system (see Denk and Franke 1982 for review). The pericanalicular microfilament web plays a role in the regulation of bile flow through cycles of contractions and relaxations and microvillar movements (Phillips et al. 1983). Drugs which affect actin filament assembly or disassembly, e.g., phalloidin and cytochalasins, profoundly interfere with bile flow (leading to cholestasis), supporting the role of these cytoskeletal structures in bile secretion and flow (Phillips et al. 1975; Gabbiani et al. 1975; Oda and Phillips 1977).

Microtubules are straight tubular structures with an outer diameter of 20–26 nm and are composed of tubulin subunits. In hepatocytes only about 40% of tubulin is assembled into microtubules (see Denk and Franke 1982 for review). Antimicrotubular agents, e.g., colchicine, disturb intracellular translocation of proteins, phospholipids, and triglycerides as well as the export of albumin and other plasma proteins and lipoproteins (Gregory et al. 1978; Redman et al. 1975; Stein et al. 1974; Reaven and Reaven 1980; see there for further references).

Intermediate-sized filaments are unbranched tubular filamentous structures with diameters between 6 and 11 nm and pronounced resistance toward solubilization in nondenaturing buffers and detergents over a broad range of ionic strength at neutral and alkaline pH (Osborn et al. 1982; see there for further information). Although their biologic significance in all details is not yet elucidated, their features suggest

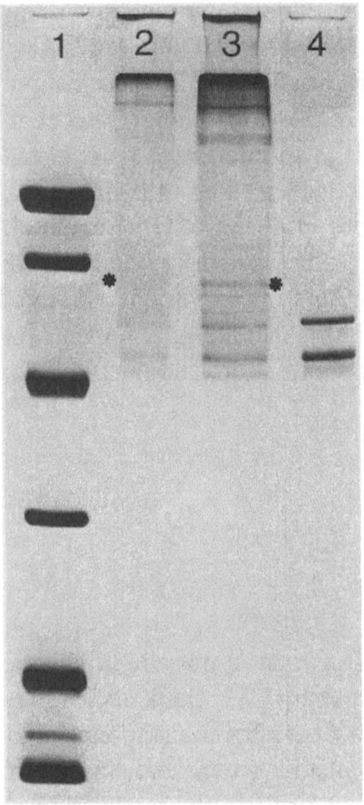

Fig. 1. SDS-polyacrylamide gel electrophoresis of the high salt-detergent-resistant intermediate filament cytoskeleton from mouse hepatocytes (slot 4) in comparison to the polypeptide composition of MB material isolated from the livers of griseofulvin-fed mice (slots 2 and 3). Major polypeptides of mouse liver cytokeratin migrate corresponding to apparent molecular weights of 55,000 (component A) and 48,000 (component D). MBs (slots 2 and 3; slot 3 contains double the amount of material of slot 2) contain additional polypeptides at higher and lower molecular weights with a major one around 65,000 (*asterisk;* an additional polypeptide at approximately 62,000 M_r is also present in this preparation). Slot 1 contains reference proteins (*from top to bottom:* phosphorylase B, bovine serum albumin, ovalbumin, carbonic anhydrase, soy bean trypsin inhibitor, lysozyme)

cytoskeletal functions of more static nature, including maintenance of cell shape and stability, and they may provide support to microfilaments and microtubules. Despite morphological similarities, intermediate filaments associated with different cell types differ in chemical composition and antigenicity: cytokeratin filaments are associated with epithelial cells; vimentin filaments with mesenchymal cells, including vascular smooth muscle cells; desmin filaments with muscle and myogenic cells; and neurofilaments and glial filaments with cells of the central nervous system. Moreover, cytokeratin filaments associated with different types of epithelial cells differ in chemical composition and 19 different polypeptides have so far been detected (Moll et al. 1982). In contrast to cells of stratified epithelia with a rather complex cytokeratin polypeptide composition, hepatocytes of various animal species, for example, contain cytokeratins of a rather simple composition with two major polypeptides (e.g., with molecular weights of 48,000 and 55,000 in mice; Fig. 1; Denk et al. 1981a; Moll et al. 1982). This characteristic and simple chemical composition is rapidly changed in vitro. Under these conditions not only are several novel cytokeratin polypeptides with higher and lower molecular weights expressed but also the intermediate filament protein specific for mesenchymal cells, i.e., vimentin, appears. This indicates that despite considerable structural stability the system of intermediate filaments is prone to rapid modulation depending, e.g., on environmental influences (Franke et al. 1981).

Ultrastructure and Chemical Composition of MBs

Ultrastructure

Mallory bodies consist of irregularly arranged filamentous rods covered by fuzzy material (Smuckler 1968; Yokoo et al. 1972; Franke et al. 1979a; Fig. 2). Most authors agree that MB filaments are tubular (see, however, Petersen 1977, who did not observe tubular structures). They are unbranched as shown by Franke et al. (1979a), but branched with lateral connections according to Wiggers et al. (1973). The diameters of MB filaments reported in the literature seem to depend on the disorder with which MBs are associated, on the mode of measurement in situ or after isolation, and particularly on the amount of nonfilamentous material which may be attached to variable degrees to the filament proper (Wiggers et al. 1973; see Denk et al. 1979a and French 1981, 1983 for review). Treatment of isolated MB filaments with deoxycholate (Wiggers et al. 1973) and particularly with trypsin (Denk 1984) considerably reduces the average diameters of MB filaments apparently by removal of nonfilamentous material (Fig. 3). The diameters of MB filaments reach from 3 to 23 nm according to different reports in the literature (Denk et al. 1979a). Yokoo et al. (1972) distinguished three types of arrangements of MB filaments in human material which can also be observed in the griseofulvin mouse model (Franke et al. 1978): in type 1 MB filaments are aligned in parallel arrays; the most common type in which filaments are randomly oriented is designated type 2. Type 3 MBs are characterized by a considerable increase in electron density, progressive loss of filamentous structure, and replacement by a more granular to amorphous material. Type 3 MBs are most often seen in the centers of type 2 MBs. Filaments of smaller diameters and more regular arrangement, resembling typical tono- or cytokeratin filaments, are often associated with the peripheral portions of MBs (Petersen 1977; French 1981; Kimoff and Huang 1981; Irie et al. 1982). Chemical composition and pathogenesis of the ultrastructurally different MB components is presently unknown. Inclusion of various other cellular constituents (e.g., membranous material) within MB filament accumulations has been observed and may account for the heterogeneous histochemical reactions published by several authors (see Denk et al. 1979a for further information).

Chemical Composition

If isolated and purified MBs of human and murine origin are analyzed on one- and two-dimensional sodium dodecyl sulfate (SDS-)polyacrylamide gel electrophoresis, a complex polypeptide composition is found. In MBs of human origin five major constituent polypeptides ranging in molecular weight from 30,000 to 56,000 (30,000, 43,000, 47,000, 50,000, 56,000) have been resolved (Tinberg et al. 1978). MBs experimentally produced in mice by prolonged griseofulvin administration consist of major polypeptide constituents with molecular weights of 48,000 (component III), 55,000 (component II), and 65,000 (component I) (Franke et al. 1979a; Denk et al. 1982; Tinberg 1981; minor polypeptides show migration distances corresponding to

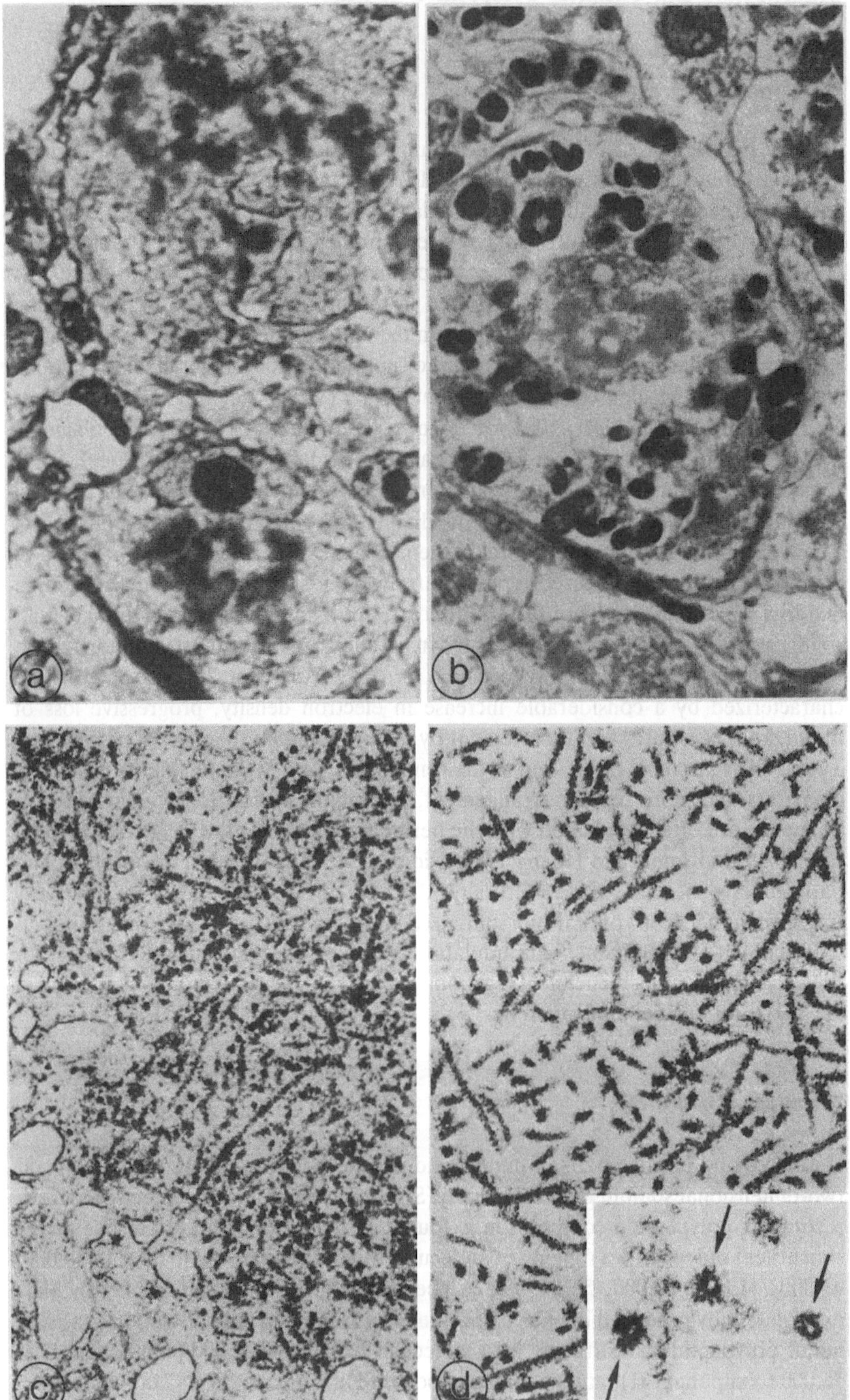
a
b
c
d

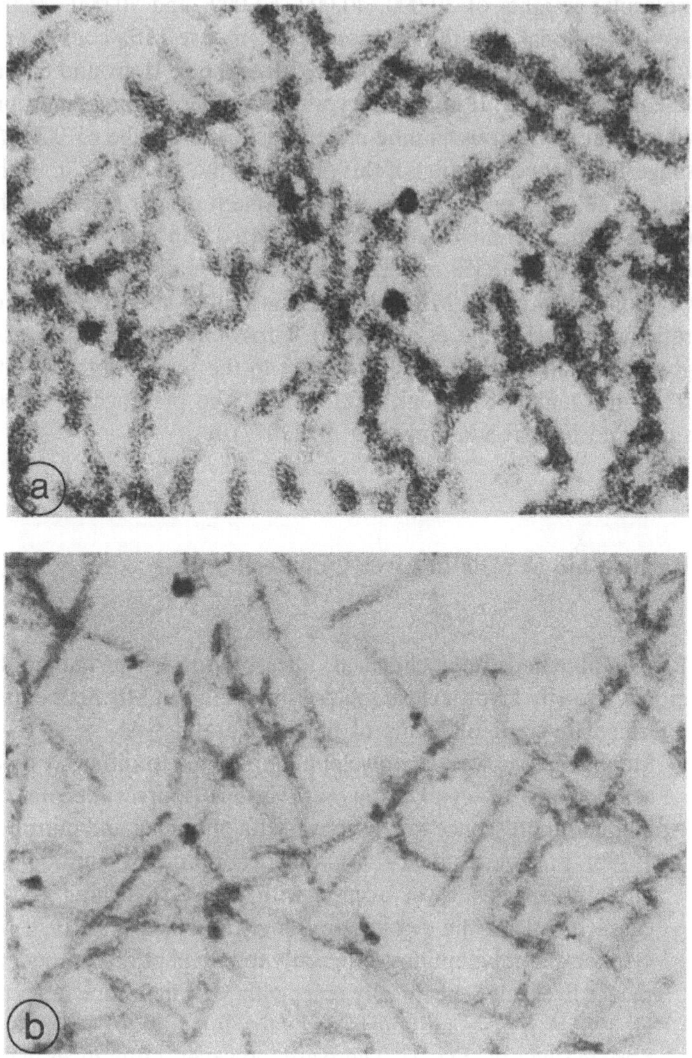

Fig. 3a, b. Treatment of MBs after isolation from griseofulvin-intoxicated mouse liver with trypsin (crystallized trypsin, type IX, from porcine pancreas; Sigma Chemical Co., St. Louis; enzyme-to-protein ratio 1 : 50; incubation for 10 min at 37° C) conspicuously alters the ultrastructure of MB filaments, apparently by removing attached material **(b).** Digested MB filaments have reduced diameters and display a smooth surface ("naked filaments"). Compare **(a)** untreated MB filaments with **(b)** trypsinized MBs. *a,b,* × 180,000

◄

Fig. 2. a Mallory bodies (MBs) appear as irregularly shaped cytoplasmic inclusions in hepatocytes, which are **b** often enlarged, damaged, and surrounded and penetrated by polymorphonuclear leukocytes. At the ultrastructural level, MBs consist of filamentous rods, mostly **c** in irregular (type 2) arrangement in situ as well as **d** after isolation. The filaments display a fuzzy outline apparently due to the presence of material attached to the filament proper *(d)*. In cross sections they appear as tubular structures *(d, inset)*. *a,b,* × 1,200; *c,* × 58,000; *d,* 64,000; *d inset,* 260,000

molecular weights of 30,000, 40,000, 62,000, and 70,000–71,000; see Fig. 1). In two-dimensional gels the components of mouse MBs consist of several isoelectric variants, component I with pI values around 5.4, II around 6.4, and III around 5.4 (Denk et al. 1982). If MBs are trypsinized under limiting digestion conditions under which nonfilamentous coating material is removed, the 65,000-M_r band disappears, suggesting a concentration of this protein in the associated non-filamentous material and of the other polypeptides in the filament proper (Denk, to be published). The nature of this protein has still to be elucidated. Since under nondenaturing conditions only about 80%–85% of total MB material is dissolved, there may be additional components in native MBs which are insoluble and, therefore, escape electrophoretic analysis (Franke et al. 1979a). Cross-links, other than disulfide bonds, e.g., glutamyl-lysine bonds, may contribute to the remarkable stability of a certain MB fraction (Denk 1984; Denk et al. 1984). The presence of carbohydrate has been suggested by Luisada-Opper et al. (1977).

Relationship of MBs to Liver Cytoskeleton

Results obtained by biochemical, ultrastructural, and immunohistochemical techniques strongly favor a close relationship between MB filaments and hepatocellular intermediate-sized filaments of the cytokeratin type:

1. Monoclonal as well as polyclonal antibodies to different types and polypeptide components of cytokeratins of various sources decorate (e.g., in indirect immunofluorescence microscopy) MBs of human and murine origin (Denk et al. 1979b; Franke et al. 1979a); the decoration by antibodies to cytokeratin of different species is in accord with the far-reaching cross-species reactivity characteristic of the cytokeratin family (Franke et al. 1978). Antibodies to liver cytokeratin polypeptides, especially to polypeptide component D (M_r 48,000) of mouse liver, also specifically react with MBs in murine and human liver (Denk et al. 1981a). These antibodies decorate, in addition, a hepatocellular cytokeratin fibril meshwork (Fig. 4). There are, however, differences in antigenicity between MBs and normal hepatocellular cytokeratins, since antibodies to mouse liver cytokeratin component A (M_r 55,000) and antibodies to bovine hoof prekeratin and prekeratin components (Denk et al. 1979b, 1981a) only stained MBs in frozen tissue sections, but not to a significant degree normal liver cytokeratins, indicating the existence of novel antigenic determinants in MBs or conformational changes with exposure of new antigenic determinants in situ.

Fig. 4a–c. Indirect immunofluorescence microscopy using antibodies to mouse liver cytokeratin component D. **a** A complex cytoplasmic fibril meshwork is stained normal mouse hepatocytes. **b, c** This meshwork becomes irregular in livers of mice fed griseofulvin and after prolonged feeding MBs appear. In the early stages of MB development granular MBs appear in association with the cytoplasmic cytokeratin meshwork, apparently at the intersections of the fibrils (**b**). The cytokeratin meshwork staining then gradually disappears in MB-containing cells and the typical MB inclusions develop and enlarge by confluence of the granules (**c**), *a–c*, × 720

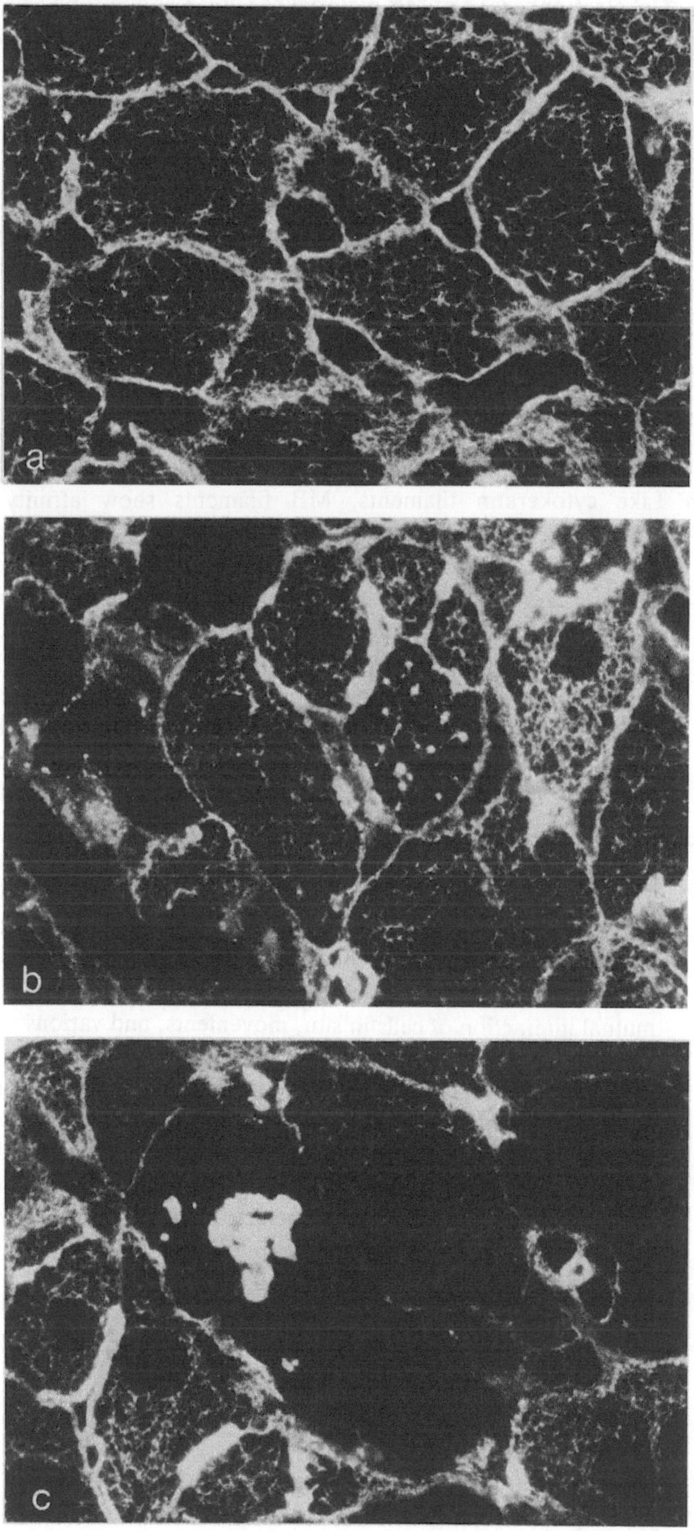
a
b
c

2. Gel electrophoresis of MB material of human and murine origin resolves several constituent polypeptides, the major ones in the molecular weight range from 40,000 to 65,000, a pattern typical of cytokeratins (Franke et al. 1979a; Tinberg et al. 1978). Moreover, human and murine MBs contain polypeptides equivalent in electrophoretic coordinates to those present in high salt-detergent-resistant cytoskeletons of hepatocytes (Denk et al. 1982). This is particularly obvious if MB proteins of mouse liver are compared to liver cytokeratins in two-dimensional gel electrophoresis, where MB proteins II and III closely resemble the major liver cytokeratin polypeptides A and D. However, MB component I is different and not detectable in significant amounts in normal hepatocytes, suggesting the pathologic nature of this protein. Antibodies prepared against this polypeptide and absorbed with liver cytokeratins react on frozen liver sections only with MBs but neither with the hepatocytic cytokeratin fibril meshwork, as revealed by immunofluorescence microscopy, nor with the respective polypeptides A and D of mouse liver cytoskeleton after transfer to nitrocellulose paper by blotting.
3. Like cytokeratin filaments, MB filaments show affinity to desmosomes: particularly during MB involution in the recovery period from intoxication MB filaments attached to desmosomes can be seen (Denk and Franke 1981).
4. The appearance of MBs is associated with derangement and diminution of the cytoplasmic cytokeratin fibril meshwork as shown by Denk et al. (1981 a), but also as shown by Morton et al. (1981) and Kimoff and Huang (1981), by immunofluorescence microscopy using polyclonal and monoclonal antibodies to different types of cytokeratins, including those derived from MBs, which suggests that aggregation to MBs occurs at the expense of normally arranged cytokeratin filaments. The validity of this finding has, however, been challenged by Irie et al. (1982), who, by electron microscopy, could not confirm the diminution of intermediate-sized filaments in hepatocytes containing heaps of MB filaments. The discrepancy may, however, be due to the fact that Irie et al. (1982) apparently have studied early lesions in which the disturbance of cytoskeleton may be less pronounced. Although the function of the intermediate filament system is as yet unclear, the derangement of its architecture may adversely affect cell stability, mutual interaction of cells in situ, movements, and various metabolic processes, including secretion, and may thus contribute to cell death.
5. Antibodies to MB proteins stain a reticular cytoplasmic network in hepatocytes in addition to epidermal cells and bile ducts (Kimoff and Huang 1981).
6. Intermediate filaments are often closely associated with MBs (Petersen 1977; Kimoff and Huang 1981), suggesting a histogenetic relationship. Moreover, immunofluorescence and immunoelectronmicroscopy revealed a hyperplasia of intermediate-sized filament bundles in GF-treated mice prior to and concomitant with MB development (Kimoff and Huang 1981).

The Hepatocytic Cytoskeleton in Ethanol Intoxication

An increase in the number of intermediate-sized filaments in hepatocytes of chronically ethanol-intoxicated individuals with alcoholic hepatitis has been found in

MB-free hepatocytes (Petersen 1977). On the contrary, MB-containing cells have been shown, by indirect immunofluorescence microscopy using various types of antibodies to MB proteins and cytokeratin polypeptides, to contain a deranged and often considerably diminished cytoskeleton meshwork, most of the cytoskeleton apparently being aggregated into MBs (Denk et al. 1981a; Morton et al. 1981; Kimoff and Huang 1981). Lieber and co-workers (see Baraona and Lieber 1982, for review) have demonstrated that liver and liver cell enlargement in alcohol intoxication is due to intracellular accumulation of fat, water, and export proteins (albumin, transferrin), resulting in swelling and "ballooning" of hepatocytes. The Golgi complex increases in size with numerous secretory vesicles loaded with lipoprotein material. Chronic as well as acute ethanol administration is associated with a decrease in hepatic content of polymerized and concomitant rise in free tubulin, in acutely treated animals less so than in those chronically fed alcohol-containing diets, which suggests a colchicine-like effect of ethanol or its metabolic products (for review see Baraona and Lieber 1982). Microtubules are rapidly restored after recovery from ethanol intoxication. Moreover, according to Matsuda et al. (1979), hepatic microtubules of ethanol-fed rats were not only decreased in number but showed morphological aberrations by being shorter and thicker. The functional consequences were found to be similar to those produced by colchicine and *Vinca* alkaloids, namely inhibition of protein and low-density lipoprotein secretion (Reaven and Reaven 1980). The antimicrotubular activity seems to reside in acetaldehyde, since it is partially prevented by 4-methylpyrazole, an inhibitor of alcohol dehydrogenase, and aggravated by disulfiram-induced inhibition of aldehyde dehydrogenase. It has been shown that acetaldehyde binds to tubulin in vitro in a way similar to colchicine. At concentrations higher than 1 m*M*, acetaldehyde has been shown to inhibit in vitro polymerization of brain neurotubuli, with maximum effect at 10 m*M* (Jennett et al. 1980; for further information see also Baraona and Lieber 1982). Yet acetaldehyde concentrations sufficient to produce this effect were well above those found in chronically intoxicated animal livers. However, microtubules seem to be more sensitive to acetaldehyde in rats chronically fed alcohol-containing diets (Baraona and Lieber 1982). Even at five times the concentrations of acetaldehyde, ethanol neither affected tubulin polymerization nor potentiated the acetaldehyde effect. Sodium acetate led to concentration-dependent increase in tubulin polymerization similarly to sodium chloride. In comparison to colchicine, the antimicrotubular effect of acetaldehyde was relatively weak, since colchicine was, on a molar basis, about 1,000–2,000 times more potent. Podophyllotoxin-induced tubulin depolymerization was not affected by ethanol or its oxidation products. However, these data are in conflict with morphometric studies recently published by Berman et al. (1983). In order to obtain data more relevant to the in vivo situation, several hepatocellular organelles, including microtubules, were quantitated by morphometry in chronically ethanol-fed rats. No differences in number and morphology of microtubules between the ethanol-fed and the control group could be detected. Furthermore, Berman et al. (1983) were also unable to find any significant effect of ethanol or its metabolites on the morphology of bovine brain microtubule polymerization in vitro. The increase in cell protein attributed by most workers to secretory defects could, at least in part, also be explained by an increase in endomembranes. These data did, therefore, not provide a firm morphological basis of microtubular alterations involved in alcoholic liver cell injury, although, of course, subtle alterations of microtubular structure and function may still escape detection by morphometric methods.

Relationship of MBs to Microtubular Dysfunction or Damage?

The association of MBs with chronic ethanol intoxication (Mallory 1911), the experimental induction of MBs in mouse livers by prolonged feeding of the antimicrotubular agent griseofulvin (Denk et al. 1975, 1976), and the rapid reinduction of MBs in mice recovered from griseofulvin intoxication by colchicine (Denk and Eckerstorfer 1977) render a role of microtubular failure in MB development feasible. Since MBs reflect abnormal hepatocytic cytokeratin filaments, their development could result from a disturbance of the functional and structural interaction of the three major cytoskeletal systems, microtubules, microfilaments, and intermediate filaments. Vimentin filament organization has been shown to be severely affected by disruption of microtubules by antimicrotubular agents and, in cultured cells, filaments collapse and form perinuclear whorls (Goldman 1971). Under comparable conditions cytokeratin filaments remain unaffected. Recently, however, Knapp et al. (1983) have shown that treatment of epithelial cells, e.g., Hela cells and fetal mouse epidermal cells, by a combination of antimicrotubular and antimicrofilament drugs, such as colchicine and cytochalasins, also results in keratin filament disorganization, i.e., in a rapid transition from a uniform distribution to an open lattice of cytokeratin fibrils attached to and stabilized by membrane-associated focal centers. Any of these drugs alone was ineffective. These findings at least suggest that antimicrotubular and antimicrofilament effects in combination could alter cytokeratin arrangement and eventually lead to MB aggregation. Griseofulvin is an antimicrotubular agent (Weber et al. 1976), at least under in vitro conditions, and has also been shown to lead to derangement of the hepatocyte pericanalicular microfilament web in vivo (Franke et al. 1979b; Yokoo et al. 1979). However, several arguments can be raised against the significance of antimicrotubular action in the induction of MBs:

1. In alcoholic hepatitis as well as in griseofulvin-treated mouse livers, MB development is not associated with significant morphological changes of microtubules or their arrangement (Petersen 1977; Yokoo et al. 1979; Irie et al. 1982).
2. Treatment of mice with colchicine alone did not result in MB formation. Treatment of mice recovering from chronic griseofulvin intoxication with other antimicrotubular agents, e.g., *Vinca* alkaloids and podophyllotoxin, did not reinduce MBs (Denk, unpublished observations).
3. Griseofulvin treatment of rats resulted neither in protoporphyria nor in MB formation, although antitubulin action should be effective in this animal species (Denk, unpublished observations).

Despite the "microtubular failure" hypothesis just discussed, several other means of MB development can be imagined, although none has been definitely proven to date. In view of the prekeratin nature of at least a major portion of MB filaments and the considerable amount of material aggregated as MBs, one has to postulate an initial increase in cytokeratin due either to increased synthesis or decreased degradation. Indeed, an accumulation of intermediate filaments in MB-free hepatocytes in alcoholic hepatitis has been observed (Petersen 1977; Kimoff and Huang 1981). But no data have so far been obtained on the turnover of cytokeratin polypeptides under conditions leading to MB formation. Pathologic accumulation of keratin material in cells is associated with vitamin A deficiency and, consequently, hypovitaminosis A may also play a role in MB development, particularly since vitamin A content may be diminished in patients with alcoholic liver disease as well as in chronically ethanol-intoxicated rats and baboons (Leo and Lieber 1982; Denk et al. 1981b). Leo and Lieber (1982) have shown that alcoholic liver disease is associated with severely decreased hepatic vitamin A levels. Moreover, a diversity of nonalcoholic disorders in which MBs appear are associated with vitamin A deficiency, such as chronic cholestasis, intestinal bypass for morbid obesity, juvenile cirrhosis, and chronic griseofulvin intoxication of mice (see Denk et al. 1979a, 1981b for further information). However, no difference existed in liver vitamin A levels between alcoholics with and without MBs. Moreover, the degree of liver damage did not correlate with vitamin A levels (Leo and Lieber 1982). Vitamin A supplementation of the experimental animals in amounts which increased hepatic vitamin A levels even above normal did not prevent MB formation by griseofulvin (Denk, unpublished observations). This does not support an etiological connection between vitamin A deficiency and MB formation, although a potentiating effect of the vitamin A deficiency state, at least on ethanol toxicity, cannot be ruled out.

Mallory bodies have been regarded as "preneoplastic" structures by Borenfreund and Bendich (1978) and French and co-workers (Meierhenry et al. 1983) on the basis of studies of cultured diethylnitrosamine-induced rat hepatocellular carcinoma cells and griseofulvin- or dieldrin-fed mice in which MBs appear in parallel with an increase in gamma-glutamyltranspeptidase. In cultured hepatoma cells, however, intermediate filament aggregates do not resemble MBs in their ultrastructure and chemical composition. Under these circumstances cytokeratin filaments of normal morphology accumulate in juxtanuclear position intermingled with filaments of the vimentin type (Borenfreund et al. 1980). Moreover, several pathologic conditions associated with MBs are certainly not preneoplastic, which casts some doubt on the validity of this hypothesis.

An additional hypothesis concerning the pathogenesis of MBs rests on electrophoretic analysis of MB material which revealed the presence of a polypeptide with electrophoretic coordinates (M_r 65,000–66,000; pI between 5.2 and 5.4) similar to that of stress proteins produced by heat shock, other stress conditions, and chemicals, including chelating agents and ethanol, in eukaryotic cells (stress proteins, heat shock proteins). These proteins are produced at the expense of other cellular protein constituents, and at least some of them show affinity to intermediate filaments (Wang et al. 1981; Li and Werb 1982; Pruss et al. 1981; Anderton 1981; see there for further information). It is possible that, under certain conditions, these proteins interfere with normal intermediate filament assembly and arrangement and, therefore, lead to MB formation. Moreover, it remains to be investigated whether proteins comparable to filaggrin of epidermis (Steinert et al. 1981), which specifically interact with intermediate filaments effecting aggregation and macrofibril formation, may also play a role in MB formation. Additional stabilization of cytokeratin aggregates could be expected from transglutaminase-induced cross-linking by ε-(γ-glutamyl)lysine bonds which confer considerable structural stability upon proteins. This type of stabilization would be consistent with the fact that a certain percentage of MB material (15%–20%) resists solubilization in high salt-detergent buffers containing disulfide-reducing agents (Franke et al. 1979a). Recent studies have revealed that transglutaminase activity is considerably increased in mouse liver in the course of griseofulvin feeding and that liver cytokeratins can act as substrates of this enzyme (Denk et al. 1984, unpublished work).

Fate of MBs and Relationship to Liver Cell Death

Mallory bodies in their earliest stages of development appear as tiny heaps of filaments, resembling granules in immunofluorescence microscopy (Denk et al. 1981a), and enlarge by confluence (Fig. 4). The development of MBs is a potentially reversible phenomenon and not necessarily associated with cell death (see, e.g., Denk and Franke 1981). However, MBs are very durable structures and persist, in man and mice, for several months following discontinuation of intoxication (see Denk et al. 1976; Denk et al. 1799a for further references). In the early recovery period a progressive transition from filamentous (mostly type 2) to electron-dense amorphous (type 3) MBs with concomitant changes of histologic staining pattern has been noted, eventually resulting from proteolysis (Denk et al. 1976). Degradation of MBs within lysosomes is controversial (Denk et al. 1979a). Dispersal of MBs into small granular subunits which accumulate at the cell periphery, often but not exclusively in association with desmosomes, is another way of involution, predominantly occurring in the recovery period. The apparent affinity of MB filaments to desmosomes supports the relationship between cytokeratin (tono)filaments and MB filaments and may suggest utilization of MB material for the reformation of tonofilament-desmosome complexes (Denk and Franke 1981).

The relationship of MBs to liver cell death is still hypothetical. On one hand, MBs certainly arise as a consequence of severe metabolic alteration of hepatocytes and are, therefore, indicators of a specific and severe type of liver cell damage. On the other, one can expect that severe disturbance of the cytokeratin skeleton, reflected by MB formation, adversely affects a variety of vital cell function, including cell stability, and may therefore contribute to cell death. However, several studies with tissue culture cells have shown that an orderly arranged intermediate filament cytoskeleton is not an absolute requirement for cell viability, at least under tissue culture conditions. Injection of cytokeratin antibodies led to disappearance of cytoskeleton without cell death (Klymkowsky et al. 1983). Neither microtubule nor microfilament organization was affected by antibody-induced keratin filament disruption in PtK cells. Moreover, interference with cytokeratin arrangement had no effect on cell shape and mitotic rate of cultured cells. These experiments show that the integrity of the intermediate filament system may not be absolutely required at least by cells cultured on a rigid substrate, but do not allow definitive conclusions with regard to cells in situ. Furthermore, in some mitotic cells, the cytoskeleton temporarily disappears (Horwitz et al. 1981; Franke et al. 1982, 1983; Lane et al. 1982). In the early embryonic development as well as in germ cells, normal cytoskeletal structures are also undetectable (Jackson et al. 1980). These findings may, however, not be entirely relevant to most cells in situ in the adult organ, and hepatocytes expressing large MB inclusions may, at least from a certain point of cytokeratin derangement on, be severely and irreversibly damaged.

About 30%–35% of chronic alcoholics develop alcoholic hepatitis, often with progression to cirrhosis in relation to alcohol consumption, extent and duration of alcohol abuse being important. Occasionally, however, cirrhosis develops despite cessation of alcohol intake. It is now established that genetic factors considerably modulate susceptibility to alcohol intoxication, but in addition immunological mechanisms could possibly be involved in the initiation and perpetuation of alcoholic liver injury. As far as MBs are concerned, a humoral and cell-mediated response to MB proteins has been found by some authors but not by others (see MacSween and Anthony 1982 for further information). However, cytokeratin antigens have a cytoplasmic position in intact cells and are not exposed at the cell surface. An immunologic reaction directed against them is, therefore, only plausible after cell death and liberation of cellular components, but may then aggravate the lesion. Hence it is unlikely that immunologic reactions against MBs play an initial role in cell damage. Still, immunologic mechanisms may play a role in the disposal and degradation of MBs after release from the damaged cells and may also contribute to some morphologic features of alcoholic hepatitis, e.g., leukotaxis. It is interesting in this context that Linder (1981) observed antibody-independent binding of complement component C1q to intermediate-sized filaments followed by complement activation.

Concluding Remarks

Mallory bodies remain an important criterion for severe alcoholic liver disease despite their association with a wide diversity of other pathologic conditions. The data so far obtained allow the conclusions that MBs are the morphologic expression of dysregulation and derangement of the intermediate filament cytoskeleton. Their pathogenesis is still obscure, but further studies may not only offer clues to the mechanisms operative in alcoholic cell injury in particular and to principles of cell damage in general, but may also provide insight into as yet unresolved questions of regulation and function of the intermediate filament system.

References

Anderton BH (1981) Intermediate filaments. A family of homologous structures. J Muscle Res Cell Motil 2: 141–166

Baraona E, Lieber CS (1982) Effects of alcohol on hepatic transport of proteins. Annu Rev Med 33: 281–292

Berman WJ, Gil J, Jennett RB, Tuma D, Sorrell MF, Rubin E (1983) Ethanol, hepatocellular organelles and microtubules. A morphometric study in vivo and in vitro. Lab Invest 48: 760–767

Birschbach HR, Harinasuta U, Zimmerman HJ (1974) Alcoholic steatonecrosis II. Prospective study of prevalence of Mallory bodies in biopsy specimens and comparison of severity of hepatic disease in patients with and without this histological feature. Gastroenterology 66: 1195–1202

Borenfreund E, Bendich A (1978) In vitro demonstration of Mallory body formation in liver cells from rats fed diethylnitrosamine. Lab Invest 38: 295–303

Borenfreund E, Schmid E, Bendich A, Franke WW (1980) Constitutive aggregates of intermediate-sized filaments of the vimentin and cytokeratin type in cultured hepatoma cells and their dispersal by butyrate. Exp Cell Res 127: 215–235

Capron JP, Delamarre J, Dupas JL, Braillon A, Degott C, Quenum C (1982) Fasting in obesity. Another cause of liver injury with alcoholic hyalin? Dig Dis Sci 27: 265–268

Christoffersen P, Eghöje K, Juhl E (1973) Mallory bodies in liver biopsies from chronic alcoholics. A comparative morphological, biochemical, and clinical study of two groups of chronic alcoholics with and without Mallory bodies. Scand J Gastroenterol 8: 341–346

Denk H (1984) The intermediate filament cytoskeleton in neoplastic and non-neoplastic liver disorders. J. Submicr Cytol 16: 141–145

Denk H, Eckerstorfer R (1977) Colchicine-induced Mallory body formation in the mouse. Lab Invest 36: 563–565

Denk H, Franke WW (1981) Rearrangement of the hepatocyte cytoskeleton after toxic damage: involution, dispersal and peripheral accumulation of Mallory body material after drug withdrawal. Eur J Cell Biol 23: 241–249

Denk H, Franke WW (1982) Cytoskeletal filaments. In: Arias IM, Popper H, Schachter D, Shafritz DA (eds) Liver biology and pathology. Raven, New York, pp 55–71

Denk H, Gschnait F, Wolff K (1975) Hepatocellular hyalin (Mallory bodies) in long-term griseofulvin-treated mice: a new experimental model for the study of hyalin formation. Lab Invest 32: 773–776

Denk H, Eckerstorfer R, Gschnait F, Konrad K, Wolff K (1976) Experimental induction of hepatocellular hyalin (Mallory bodies) in mice by griseofulvin treatment. 1. Light microscopic observations. Lab Invest 35: 377–382

Denk H, Franke WW, Kerjaschki D, Eckerstorfer R (1979a) Mallory bodies in experimental animals and man. Int Rev Exp Pathol 20: 77–121

Denk H, Franke WW, Eckerstorfer R, Schmid E, Kerjaschki D (1979b) Formation and involution of Mallory bodies ("alcoholic hyalin") in murine and human liver revealed by immunofluorescence microscopy with antibodies to prekeratin. Proc Natl Acad Sci USA 76: 4112–4116

Denk H, Franke WW, Dragosics B, Zeiler I (1981a) Pathology of cytoskeleton of liver cells: demonstration of Mallory bodies (alcoholic hyalin) in murine and human hepatocytes by immunofluorescence microscopy using antibodies to cytokeratin polypeptides from hepatocytes. Hepatology 1: 9–20

Denk H, Franke WW, Kerjaschki D (1981b) Mallory bodies: new facts and findings. In: Berk PD, Chalmers TC (eds) Frontiers in liver disease. Thieme-Stratton, New York, pp 93–105

Denk H, Krepler R, Lackinger E, Artlieb U, Franke WW (1982) Immunological and biochemical characterization of the keratin-related component of Mallory bodies: a pathological pattern of hepatocytic cytokeratins. Liver 2: 165–175

Denk H, Bernklau G, Krepler R (1984) Effect of griseofulvin treatment and neoplastic transformation on transglutaminase activity in mouse liver. Liver 4: 208–213

Franke WW, Weber K, Osborn M, Schmid E, Freudenstein C (1978) Antibody to prekeratin: decoration of tonofilament-like arrays in various cells of epithelial character. Exp Cell Res 116: 429–445

Franke WW, Denk H, Schmid E, Osborn M, Weber K (1979a) Ultrastructural, biochemical, and immunologic characterization of Mallory bodies in griseofulvin-treated mice. Fimbriated rods of filaments containing prekeratin-like polypeptides. Lab Invest 40: 207–220

Franke WW, Schmid E, Kartenbeck J, Mayer D, Hacker HJ, Bannasch P, Osborn M, Weber K, Denk H, Wanson JC, Drochmans P (1979b) Characterization of the intermediate-sized filaments in liver cells by immunofluorescence and electron microscopy. Biol Cell 34: 99–110

Franke WW, Mayer D, Schmid E, Denk H, Borenfreund E (1981) Differences of expression of cytoskeletal proteins in cultured rat hepatocytes and hepatoma cells. Exp Cell Res 134: 343–359

Franke WW, Schmid E, Grund C, Geiger B (1982) Intermediate filament proteins in non-filamentous structures: transient disintegration and inclusion of subunit proteins in granular aggregates. Cell 30: 103–113

Franke WW, Schmid E, Wellsteed J, Grund C, Gigi O, Geiger B (1983) Change of cytokeratin filament organization during the cell cycle: selective masking of an immunologic determinant in interphase PtK_2 cells. J Cell Biol 97: 1255–1260

French SW (1981) The Mallory body: structure, composition and pathogenesis. Hepatology 1: 76–83

French SW (1983) Present understanding of the development of Mallory's body. Arch Pathol Lab Med 107: 445–450

French SW, Burbige EJ (1979) Alcoholic hepatitis: clinical, morphologic, pathogenic, and therapeutic aspects. In: Popper H, Schaffner F (eds) Progress in liver diseases. Grune and Stratton, New York, pp 557–579

Gabbiani G, Montesano R, Tuchweber B, Salas M, Orci L (1975) Phalloidin induced hyperplasia of actin filaments in rat hepatocytes. Lab Invest 33: 562–569

Goldman R (1971) The role of three cytoplasmic fibers in BHK-21 cell motility. I. Microtubules and the effects of colchicine. J Cell Biol 51: 752–762

Gregory DH, Levi DF (1972) The clinical-pathologic spectrum of alcoholic hepatitis. Am J Dig Dis 17: 479–488

Gregory DH, Vlahcevic ZR, Prugh MF, Swell L (1978) Mechanism of secretion of biliary lipids: role of a microtubular system in hepatocellular transport of biliary lipids in the rat. Gastroenterology 74: 93–100

Grisham LM, Wilson L, Beusch KG (1973) Antimitotic action of griseofulvin does not involve disruption of microtubules. Nature 244: 294–296

Harinasuta U, Zimmerman HJ (1971) Alcoholoc steatonecrosis. I. Relationship between severity of hepatic disease and presence of Mallory bodies in the liver. Gastroenterology 60: 1036–1046

Horwitz B, Kupfer H, Eshar Z, Geiger B (1981) Reorganization of arrays of prekeratin filaments during mitosis. Exp Cell Res 134: 281–290

Irie T, Benson NC, French SW (1982) Relationship of Mallory bodies to the cytoskeleton of hepatocytes in griseofulvin-treated mice. Lab Invest 47: 336–345

Jackson BW, Grund C, Schmid E, Burki K, Franke WW, Illmensee K (1980) Formation of cytoskeletal elements during mouse embryogenesis. I. Intermediate filaments of the cytokeratin type and desmosomes in preimplantation embryos. Differentiation 17: 161–179

Jennett RB, Tuma DJ, Sorrell MF (1980) Effect of ethanol and its metabolites on microtubule formation. Pharmacology 21: 363–368

Kimoff AJ, Huang S (1981) Immunocytochemical and immunoelectron microscopic studies on Mallory bodies. Lab Invest 45: 491–503

Klymkowsky MW, Miller RH, Lane EB (1983) Morphology, behavior, and interaction of cultured epithelial cells after the antibody-induced disruption of keratin filament organisation. J Cell Biol 96: 494–509

Knapp LW, O'Guin WM, Sawyer RH (1983) Rearrangement of the keratin cytoskeleton after combined treatment with microtubule and microfilament inhibitors. J Cell Biol 97: 1788–1794

Lane EB, Goodman SL, Trejdosiewicz LK (1982) Disruption of the keratin filament network during epithelial cell division. EMBO J 1: 1365–1372

Lazarides E (1980) Intermediate filaments as mechanical integrators of cellular space. Nature 283: 249–256

Leo MA, Lieber CS (1982) Hepatic vitamin A depletion in alcoholic liver injury. N Engl J Med 307:597–601
Li GC, Werb Z (1982) Correlation between synthesis of heat shock proteins and development of thermotolerance in Chinese hamster fibroblasts. Proc Natl Acad Sci USA 79:3218–3222
Linder E (1981) Binding of C1q and complement activation by vascular endothelium. J Immunol 126:648–658
Luisada-Opper AV, Kanagasundaram N, Leevy CM (1977) Chemical nature of alcoholic hyalin. Gastroenterology 73:1374–1376
Mac Sween RNM, Anthony RS (1982) Immune mechanisms in alcoholic liver disease. J Clin Lab Immunol 9:1–5
Mallory FB (1911) Cirrhosis of the liver. Five different types of lesions from which it may arise. Bull Johns Hopkins Hosp 22:69–75
Matsuda Y, Baraona E, Salaspuro M, Lieber CS (1979) Effects of ethanol on liver microtubules and Golgi apparatus. Possible role in altered hepatic secretion of plasma protein. Lab Invest 41:455–463
Meierhenry EF, Ruebner BH, Gershwin ME, Hsieh LS, French SW (1981) Mallory body formation in hepatic nodules of mice ingesting dieldrin. Lab Invest 44:392–996
Meierhenry EF, Ruebner BH, Gershwin ME, Hsich LS, French SW (1983) Dieldrin-induced Mallory bodies in hepatic tumors of mice of different strains. Hepatology 3:90–95
Moll R, Franke WW, Schiller DL, Geiger B, Krepler R (1982) The catalog of human cytokeratin in normal epithelia, tumors and cultured cells. Cell 31:11–24
Morton JA, Bastin J, Fleming KA, McMichael A, Burns J, McGee JOD (1981) Mallory bodies in alcoholic liver disease: identification of cytoplasmic filament/cell membrane and unique antigenic determinants by monoclonal antibodies. Gut 22:1–7
Oda M, Phillips MJ (1977) Bile canalicular membrane pathology in cytochalasin B-induced cholestasis. Lab Invest 37:350–356
Oda M, Price VM, Fischer MM, Phillips MJ (1974) Ultrastructure of bile canaliculi with special reference to the surface coat and the pericanalicular web. Lab Invest 31:314–323
Osborn M, Geisler N, Shaw G, Sharp G, Weber K (1982) Intermediate filaments. Cold Spring Harbor Symp Quant Biol 46:413–429
Petersen P (1977) Alcoholic hyalin, microfilaments and microtubules in alcoholic hepatitis. Acta Pathol Microbiol Scand [A] 85:384–394
Phillips MJ, Oda M, Mak E, Fischer MM, Jeejeebhoy KN (1975) Microfilament dysfunction as a possible cause of intrahepatic cholestasis. Gastroenterology 69:48–58
Phillips MJ, Oshio C, Miyairi M, Smith CR (1983) Intrahepatic cholestasis as a canalicular motility disorder. Evidence using cytochalasin. Lab Invest 48:205–211
Popper H, Goldfischer S, Sternlieb I, Nayak NC, Madhavan TV (1979) Cytoplasmic copper and its toxic effects. Studies in Indian childhood cirrhosis. Lancet I:1205–1208
Pruss RM, Mirsky R, Raff MC, Thorpe R, Dowding AJ, Anderton BH (1981) Cell classes of intermediate filaments have a common antigenic determinant defined by a monoclonal antibody. Cell 27:419–428
Reaven EP, Reaven GM (1980) Evidence that microtubules play a permissive role in hepatocyte very low density lipoprotein secretion. J Cell Biol 84:28–29
Redman CM, Banerjee D, Howell K, Palade GE (1975) Colchicine inhibition of plasma protein release from rat hepatocytes. J Cell Biol 66:42–59
Rungger-Brändle D, Gabbiani G (1983) The role of cytoskeletal and cytocontractile elements in pathologic processes. Am J Pathol 110:361–392
Smuckler EA (1968) The ultrastructure of human alcoholic hyalin. Am J Clin Pathol 49:790–797
Stein O, Sanger L, Stein Y (1974) Colchicine-induced inhibition of lipoprotein and protein secretion of biliary phospholipids and cholesterol by rat liver in vivo. J Cell Biol 62:90–103
Steinert PM, Cantieri JS, Teller DC, Lonsdale-Eccles JD, Dale BA (1981) Characterization of a class of cationic proteins that specifically interact with intermediate filaments. Proc Natl Acad Sci USA 78:4097–4101
Tinberg HM (1981) Intermediate filaments: analysis of filamentous aggregates induced by griseofulvin, an antitubulin agent. Biochem Biophys Res Commun 99:458–465

Tinberg HM, Regan RJ, Geier EA, Peterson GE, French SW (1978) Mallory bodies. Isolation of hepatocellular hyalin and electrophoretic resolution of polypeptide components. Lab Invest 39: 483–490

Wang C, Orner RH, Lazarides E (1981) Heat-shock proteins are methylated in avian and mammalian cells. Proc Natl Acad Sci USA 78: 3531–3535

Weber K, Wehland J, Herzog W (1976) Griseofulvin interacts with microtubules both in vivo and in vitro. J Mol Biol 102: 827–829

Wiggers KD, French SW, French BA, Carr BN (1973) The ultrastructure of Mallory body filaments. Lab Invest 29: 652–658

Yokoo H, Minick OT, Batti F, Kent G (1972) Morphologic variants of alcoholic hyalin. Am J Pathol 69: 25–40

Yokoo H, Craig RM, Harwood TR, Cochrane C (1979) Griseofulvin-induced cholestasis in Swiss albino mice. Gastroenterology 77: 1082–1087

Yokoo H, Harwood TR, Racker D, Arak S (1982) Experimental production of Mallory bodies in mice by diet containing 3,5-diethoxycarbonyl-1,4-dihydrocollidine. Gastroenterology 83: 109–113

9 Interaction of Ethanol with Drugs and Xenobiotics

C. Sato, Y. Hasumura, and J. Takeuchi

The Second Department of Internal Medicine, Tokyo Medical and Dental University,
5-45, Yushima 1-chome, Bunkyo-ku, Tokyo 113, Japan

Introduction

Interaction of ethanol with drugs and xenobiotics has drawn much attention in clinical medicine. It is well known that alcoholic patients are tolerant to a variety of drugs, whereas they are susceptible to various hepatotoxins. Furthermore, in the presence of ethanol, drug action has been shown to be potentiated. However, the mechanisms of the observed effects of ethanol were not well understood until Lieber and his colleagues first showed that ethanol is metabolized not only via an alcohol dehydrogenase (ADH)-dependent pathway but also via a cytochrome-*P*-450-mediated pathway which is now known as the "microsomal ethanol-oxidizing system" (MEOS).

In the last 15 years, the effects of ethanol on drug metabolism have been studied extensively. Ethanol can affect any of the following steps: absorption, plasma protein binding, hepatic blood flow, distribution, hepatic uptake, and phase I and phase II hepatic metabolism. In this chapter, the effects of ethanol on these steps are reviewed briefly and the clinical relevance of the observed changes is discussed. The interaction of ethanol with mixed function oxidases is discussed in other chapters in relation to ethanol metabolism.

Absorption

Ethanol has been shown to affect the morphology and functions of the stomach and small intestine (Beck and Dinda 1981). Therefore, it is theoretically possible that ethanol may influence drug absorption. That question, however, remains to be evaluated. A study by Hayes et al. (1977) which showed increased diazepam absorption has been criticized (Hoyumpa and Schenker 1982). Other studies by Magnussen (1968), Hayton (1975), and Koysooko et al. (1975) have suggested that ethanol does not alter drug absorption significantly.

Alcohol Related Diseases in Gastroenterology
Edited by H. K. Seitz and B. Kommerell

Plasma Protein Binding

Although it is not clear whether drug binding to plasma proteins is affected by the presence of ethanol, decreased binding of several drugs in chronic alcoholics has been reported (Affrime and Reidenberg 1975; Thiessen et al. 1976). This change, however, appears to be due to liver injury rather than to ethanol itself (Brodie and Boobis 1978). It is important to note that in patients with severe alcoholic liver disease drug toxicity may develop, even when the serum concentration of the drug is within the normal therapeutic range (Lieber 1982).

Hepatic Blood Flow

Hepatic blood flow plays an important role in the hepatic uptake of drugs which undergo high first-pass elimination, whereas drugs which show low first-pass elimination are little affected. The former category includes lidocaine, meperidine, and propranolol.

There are conflicting reports concerning the effect of ethanol on hepatic blood flow, depending on the dose and the route of administration. At small doses, ethanol did not change (Castenfors et al. 1960) or reduced hepatic blood flow (Lundquist) et al. 1962), whereas at larger doses increased hepatic blood flow has been reported (Childs et al. 1963; Villeneuve et al. 1981). Cohn et al. (1972) have reported that hepatic blood flow is increased in patients with alcoholic hepatitis. To what extent these changes affect drug metabolism is not known. It should also be noted that drugs which undergo high first-pass elimination in healthy individuals may be poorly eliminated in patients with reduced hepatic function; thus hepatic blood flow may no longer be the rate-limiting step in drug metabolism.

Distribution

Sellman et al. (1975a) have shown that intravenous diazepam induces lower peak serum concentrations in alcoholic patients than in healthy volunteers, and they suggested that chronic intake of high doses of ethanol may reduce the bioavailability of diazepam (Sellman et al. 1975b). Although the studies have been criticized because of the difference in the ages of the two groups, the increase in the volume of distribution of diazepam could be an explanation for the clinically observed cross-tolerance in alcoholics to the sedative effect of diazepam.

Hepatic Uptake

Ethanol administration has been shown to alter cell membranes. Therefore, it is conceivable that ethanol may affect drug transport across cell membranes and thus change drug uptake by the liver. However, a recent study by Kreek et al. (1981) in perfused rabbit liver showed that ethanol did not affect the hepatic uptake of narcotics. The effect of ethanol on Kupffer cells, which play an important role in uptake of endotoxin (Ruiter et al. 1981), is unknown.

Mixed Function Oxidation

Ethanol has been shown to be metabolized to acetaldehyde by microsomes in the presence of reduced nicotinamide-adenosine dinucleotide phosphate (NADPH) and O_2 (Lieber and DeCarli 1970). This enzyme system is now called the "microsomal ethanol-oxidizing system". Recent studies have demonstrated that the MEOS can be reconstituted with cytochrome *P*-450, cytochrome *c* (*P*-450) reductase, and phospholipids (Ohnishi and Lieber 1977). A precise discussion of this issue is found in another chapter of this book. As shown in the reconstitution studies, the MEOS shares common components with microsomal mixed function oxidases. Interaction of ethanol with this system, especially with cytochrome *P*-450, appears to explain a large part of the complicated effects of ethanol on drug metabolism.

Acute Effects of Ethanol

In the presence of ethanol, drug metabolism by mixed function oxidation is generally inhibited both in vivo and in vitro. In vivo, acute ethanol administration resulted in a prolongation of the half-life of various drugs such as meprobamate and pentobarbital (Rubin et al. 1970). This inhibitory effect of ethanol has been implicated in the synergistic effect of ethanol and central nervous system (CNS) depressants. In fact, chlordiazepoxide metabolism is impaired in the alcohol-intoxicated man (Whiting et al. 1979; Desmond et al. 1980), and Thomas et al. (1972) have found elevated brain levels of pentobarbital in ethanol-treated rats. The metabolism of other drugs which are known to be metabolized by mixed function oxidation, such as methadone (Borowsky and Lieber 1978), phenothiazine (Milner and Landauer 1971), and caffeine (Mitchell et al. 1983), is also impaired when they are given with ethanol. Ethanol has also been shown to increase the systemic availability of drugs which undergo high first-pass elimination, such as mephenytoin (Zysset et al. 1980) and propoxyphene (Oguma and Levy 1981), when the drugs are taken orally.

In vitro, ethanol inhibits drug oxidation by microsomes such as aniline hydroxylation, aminopyrine *N*-demethylation, ethylmorphine *N*-demethylation, and phenobarbital hydroxylation (Rubin et al. 1970). This inhibitory effect of ethanol has been explained by its interaction with cytochrome *P*-450. Ethanol produces a reverse type I binding spectrum when added to microsomes (Rubin et al. 1971), suggesting the binding of ethanol to cytochrome *P*-450 or the dissociation of drugs from cytochrome *P*-450. Ethanol inhibits aniline hydroxylation at low concentrations, whereas high ethanol concentrations are required to inhibit aminopyrine *N*-demethylation (Rubin et al. 1970). In the latter case, low doses of ethanol are rather stimulatory (Cinti et al. 1973). It is suggested that reduced nicotinamide-adenine dinucleotide (NADH) produced by ADH-dependent ethanol oxidation may serve hydrogen equivalent to cytochrome-*P*-450-mediated drug oxidation as well as NADPH (Lieber 1982). In general, ethanol is a stronger inhibitor of metabolism by microsomes of type-II- than type-I-binding drugs. Recent findings that ethanol-inducible cytochrome *P*-450 has a high affinity for aniline hydroxylation as well as ethanol oxidation (Morgan et al. 1982) may be a possible explanation. Thi inhibitory effect of ethanol on drug metabolism in vivo has been explained by the interaction of ethanol with cytochrome *P*-450 as shown in vitro experiments. On the other hand, Reinke et al. (1980) have suggested decreased production of NADPH as a cause of the inhibitory action of ethanol. Chung and Brown (1976) claimed the increased release of corticosteroid to be the cause for the inhibition of hexobarbital metabolism by acute ethanol administration. To what extent these mechanisms play a role is still to be determined.

Chronic Effects of Ethanol

Repeated administration of ethanol in rats and man has been shown to increase the rate of plasma clearance of meprobamate and phenobarbital (Misra et al. 1971). A similar increase has also been observed in tolbutamide metabolism (Calluri et al. 1971). This increased rate persists for 4–9 weeks after cessation of alcohol and is reproduced with intake of large doses of ethanol (Iber 1977). Cushman et al. (1982) have reported that alcoholic patients showed enhanced metabolic disposal of antipyrine. Following 2 weeks' abstinence, the plasma half-life of antipyrine was extended.

Enhanced drug metabolism has also been observed in vitro. After chronic ethanol administration, accelerated metabolism of a number of drugs such as aniline, pentobarbital (Rubin and Lieber 1968), and methadone (Borowsky and Lieber 1978) has been demonstrated. Ethanol increases the content of microsomal cytochrome *P*-450 and NADPH-cytochrome *c* (*P*-450) reductase (Joly et al. 1973). Thus the adaptive increase in these components after chronic ethanol administration is attributable to enhanced drug metabolism in chronic alcoholics. Of importance are recent observations that chronic ethanol treatment increases a cytochrome *P*-450 isozyme which is distinct from known inducible forms. Koop et al. (1982) have demonstrated that there is a unique isozyme of cytochrome *P*-450 in liver microsomes from ethanol-treated rabbits. This isozyme shows unusually high activity in aniline hydroxylation as well as ethanol oxidation (Morgan et al. 1982). These findings may explain the heterogeneity of the induction of drug metabolism by ethanol.

Conjugation Reactions

Glucuronidation

The effect of ethanol on glucuronidation is generally inhibitory. Whether glucuronyltransferase activity is affected by ethanol is not clear. In general, the formation of glucuronide conjugates depends mainly on the continuous production of UDP-glucuronic acid (Moldeus et al. 1979). Synthesis of uridine dinucleotide phosphate-glucuronic acid depends on the availability of nicotinamide-adenine dinucleotide (NAD) in one of the steps and is inhibited by any shift in the NADH/NAD ratio toward the redox state. Since the shift in redox state is observed during ethanol oxidation, this has been implicated in the inhibitory effect of ethanol on the rate of glucuronidation of various drugs (Moldeus et al. 1978). In keeping with this hypothesis is the observation that this effect of ethanol was abolished in the presence of 4-methylpyrazole, an inhibitor of ADH.

In vivo, however, this effect of ethanol is not clear. Hoyumpa et al. (1981) have observed that short-term treatment with ethanol inhibits the first-pass elimination of lorazepam when the drug is given orally. The synergistic effect of ethanol with sedatives such as lorazepam and oxazepam may be due to this inhibition of glucuronidation. Long et al. (1983), however, found no effect of ethanol on the disposition of propylthiouracil, a drug metabolized by conjugation reaction.

The chronic effect of ethanol on glucuronidation seems to be stimulatory. Ideo et al. (1971) reported marked increase in the activity of UDP-glucuronyltransferase after chronic ethanol treatment in the rat. In cultured chick embryo hepatocytes, Sinclair et al. (1981) observed increased enzyme activity after ethanol exposure. On the other hand, Koivusaari et al. (1981) have shown no effect of ethanol. The difference may be due to the existence of isozymes in UDP-glucuronyltransferase, and to the duration and dosage of ethanol treatment. It should also be noted that the acute effect of ethanol on glucuronidation in chronic alcoholics may be different from that observed in nonalcoholics, since chronic ethanol administration has been shown to attenuate the redox shift caused by acute ethanol treatment (Domschke et al. 1974; Salaspuro et al. 1981). This remains to be elucidated.

Sulfation

The effect of ethanol on sulfate conjugation is not fully understood. A study by Moldeus et al. (1978) in isolated hepatocytes suggests that sulfation is not altered in the presence of ethanol.

Glutathione Conjugation

Conjugation with glutathione offers a major protective mechanism against reactive species of drug metabolites. Acute ethanol administration in large doses produces depletion of hepatic reduced glutathione (Macdonald et al. 1977; Videla et al. 1980).

In isolated hepatocytes, ethanol decreases the concentration of glutathione, and acetaldehyde appears to be responsible for this effect of ethanol (Vina et al. 1980).

The effect of chronic ethanol administration on hepatic glutathione level has not been confirmed. There have been conflicting reports of increases (Hetu et al. 1982), decreases (Guerri and Grisolia 1978), or no change (Sato et al. 1981a). These differences may be due to differences in the experimental conditions. More recently, decreased hepatic glutathione has been reported in patients with alcoholic liver disease (Shaw et al. 1983). Increased biliary excretion may be partially responsible.

Acute ethanol administration does not affect glutathione peroxidase or glutathione reductase (Valenzuela et al. 1980), whereas chronic ethanol treatment increases these enzymes (Macdonald 1973) as well as glutathione *S*-transferase (Hetu et al. 1982; David and Nerland 1983).

These effects of ethanol have been discussed in relation to the lipid peroxidative process in alcoholic liver injury, but their participation in drug metabolism has not been studied; it is known only that drug-induced hepatotoxicity may be affected by these changes.

Acetylate Conjugation

The effect of ethanol on the acetylation of drugs is stimulatory. Olsen and Mørland (1978) have shown that sulfadimidine acetylation is enhanced by coadministration of ethanol in both slow and rapid acetylators. This finding was further supported in isolated hepatocytes. A shortened half-life of isoniazid was observed in the presence of ethanol in man (Lester 1964), and the excretion of *N*-acetylniazid was increased in the presence of ethanol in the rat (Thomas and Solomonraj 1977). Recently, however, Olsen (1982) reported that ethanol enhances sulfanilamide acetylation, whereas the acetylation of procainamide is unchanged in isolated hepatocytes. The same author reported increased procainamide acetylation in the presence of ethanol in man (Olsen and Mørland 1982). The effect of chronic ethanol treatment on this pathway has not been well documented. From these observations, one could speculate that ethanol may increase drug toxicity, which is linked to the rate of acetylation.

Drug-Induced Hepatotoxicity

Clinical observations have suggested that alcoholics are more susceptible to some toxic agents. Ethanol pretreatment enhances the hepatotoxicity of carbon tetrachloride (Cornish and Adefuin 1967; Traiger and Plaa 1971). Chronic ethanol treatment enhances microsomal drug-oxidizing enzyme activities as mentioned before, and it has been suggested that the increased hepatotoxicity of carbon tetrachloride in animals fed ethanol chronically is due to enhanced microsomal

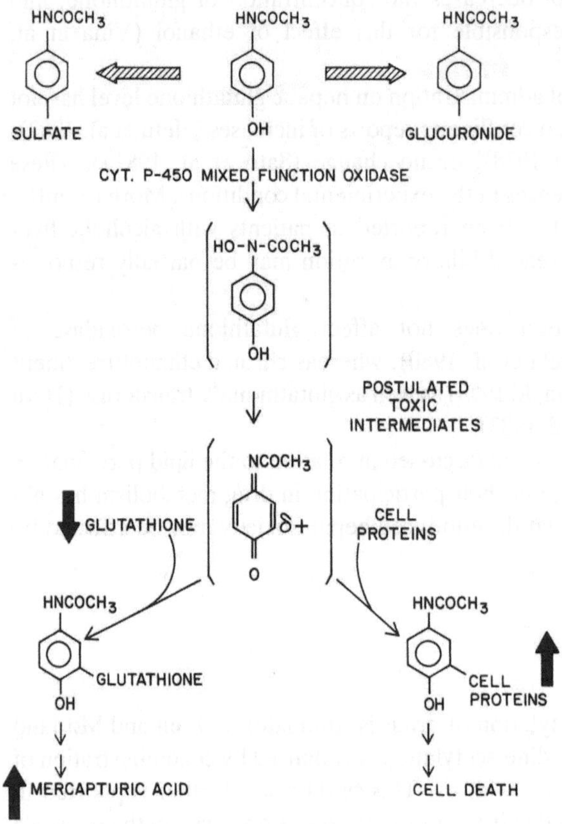

Fig. 1. Summary of the effects of chronic ethanol consumption on acetaminophen disposition

activation and biotransformation of carbon tetrachloride (Hasumura et al. 1974). A similar effect of ethanol can also be observed in the hepatotoxicity of acetaminophen.

Acetaminophen (paracetamol) is widely used as an analgesic and antipyretic, and is generally safe at recommended doses. Acetaminophen at high doses, however, has been shown to produce hepatic injury and is one of the most common causes of fulminant hepatic failure in Western countries. A similar lesion can be demonstrated in experimental animals (Boyd and Bereczky 1966). Clinical observations have suggested increased susceptibility of alcoholics to acetaminophen-induced hepatotoxicity (Emby and Fraser 1977), and increased hepatotoxicity of acetaminophen after chronic ethanol treatment has been shown in the rat (Teschke et al. 1979; Sato et al. 1981a).

Acetaminophen metabolism occurs mostly in the liver. The main route of metabolism involves conjugation with glucuronide and sulfate, and the metabolites are excreted in the urine. A small proportion of the acetaminophen is metabolized by a mixed function oxidase, and the reactive metabolites produced are detoxified by binding to glutathione, followed by excretion in the urine as mercapturic acid. When the production of the reactive metabolites is sufficiently large to deplete hepatic glutathione, the reactive metabolites can no longer be detoxified by this pathway and bind to cellular proteins. This convalent binding has been postulated to produce hepatic injury (Jollow et al. 1973; Mitchell and Jollow 1975). In keeping with this hypothesis is the observation that the increased hepatotoxicity in the rat fed ethanol chronically is associated with enhanced production of reactive metabolites of acetaminophen, both in vivo (Fig. 1) and in vitro (Sato et al. 1981a).

Similar effects of ethanol have also been observed in the hepatotoxicity of thioacetamide, dimethylnitrosamine (Maling et al. 1975), cocaine (Smith et al. 1981), and halothane (Ishii et al. 1983). Thus induction of mixed function oxidases by chronic ethanol consumption could result in enhanced hepatotoxicity of drugs and xenobiotics which fit the "reactive metabolite hypothesis."

On the other hand, acute ethanol administration inhibits drug oxidation by microsomes. Therefore, it is theoretically possible that ethanol prevents drug-induced hepatotoxicity. Indeed, acute ethanol administration has been demonstrated to prevent acetaminophen-induced hepatotoxicity in the rat (Sato et al. 1981b). This effect is associated with decreased production of reactive metabolites in vivo and in vitro (Sato and Lieber 1981). Clinical observations support this finding (Rumack et al. 1981; Critchley et al. 1983).

Liver Disease and Other Factors

The effect of ethanol consumption on drug metabolism in patients with liver disease is not predictable and, in some cases, is different from the observations in experimental animals. The discrepancies may be due partially to liver dysfunction caused by liver damage. It is well known that drug metabolism is impaired in patients with liver disease.

Harman et al. (1979) reported impaired antipyrine metabolism in alcoholics. These clinical observation appear to contrast with the findings in experimental animals that chronic ethanol consumption increases a variety of drug-metabolizing enzymes. Impaired drug metabolism in alcoholics was correlated with various biochemical or histological parameters of liver injury (Hepner and Vesell 1975; Bircher et al. 1976; Farrell et al. 1978). Lewis et al. (1977) observed that in patients with alcoholic cirrhosis the aminopyrine breath test was impaired, whereas that in alcoholics without liver cirrhosis showed higher values than normal. Saunders et al. (1980) have further supported this finding. Thus, although drug metabolism appears to be enhanced in alcoholics, the presence of liver injury has a great effect on drug metabolism in them.

Chronic alcoholics frequently exhibit a poor nutritional state. This poor nutritional state may additively influence drug metabolism in chronic alcoholics (Campbell et al. 1979). For instance, vitamin A deficiency, which may occur after chronic ethanol consumption (Sato and Lieber 1982), decreases hepatic cytochrome *P*-450 levels as well as UDP-glucuronyltransferase (Siddik et al. 1980). The poor nutritional state may further potentiate drug-induced hepatotoxicity by decreasing hepatic glutathione (Pessayre et al. 1979, 1980).

Conclusions

The interaction of ethanol with drugs and xenobiotics is complex because ethanol can affect any of the following steps: absorption, plasma protein binding, hepatic blood flow, distribution, hepatic uptake, and phase I and phase II hepatic metabolism. In general, in the presence of ethanol drug metabolism is impaired, due mainly to the inhibitory effect of ethanol on mixed function oxidation and glucuronidation. This effect is more exaggerated when the drugs are taken orally. The synergistic effects of ethanol and CNS depressants can be explained by this mechanism. In contrast, chronic ethanol consumption increases mixed function oxidation and accelerates drug metabolism. The cross-tolerance between ethanol and sedatives in chronic alcoholics may be due to this effect of ethanol. Enhanced production of reactive species by mixed function oxidation could result in susceptibility to hepatotoxins in alcoholics. The presence of liver disease has a great effect on drug metabolism in alcoholics.

References

Affrime M, Reidenberg MM (1975) The protein binding of some drugs in plasma from patients with alcoholic liver disease. Eur J Clin Pharmacol 8: 267–269

Beck IT, Dinda PK (1981) Acute exposure of small intestine to ethanol: effects on morphology and function. Dig Dis Sci 26: 817–838

Bircher J, Küpfer A, Gikalov I, Preisig R (1976) Aminopyrine demethylation measured by breath analysis in cirrhosis. Clin Pharmacol Ther 20: 484–492

Borowsky SA, Lieber CS (1978) Interaction of methadone and ethanol metabolism. J Pharmacol Exp Ther 207: 123–129

Boyd EM, Bereczky GM (1966) Liver necrosis from paracetamol. Br J Pharmacol 26: 606–614

Brodie MJ, Boobis S (1978) The effects of chronic alcohol ingestion and alcoholic liver disease on binding of drugs to serum proteins. Eur J Clin Pharmacol 13: 435–438

Calluri N, Manenti I, Gallo M, Salvioli GF (1971) Alcohol-drugs interaction in man: alcohol and tolbutamide. Eur J Clin Invest 1: 421–424

Campbell TC, Hayes JR, Merrill AH, Maso M, Goetchius M (1979) The influence of dietary factors on drug metabolism in animals. Drug Metab Rev 9: 173–184

Castenfors H, Hultman E, Josephson B (1960) Effect of intravenous infusions of ethyl alcohol on estimated hepatic blood flow in man. J Clin Invest 39: 776–781

Childs AW, Kivel RM, Lieberman A (1963) Effect of ethyl alcohol on hepatic circulation, sulfobromophthalein clearance, and hepatic glutamic-oxalacetic transaminase production in man. J Clin Invest 45: 176–181

Chung H, Brown DR (1976) Mechanism of the effect of acute ethanol on hexobarbital metabolism. Biochem Pharmacol 25: 1613–1616

Cinti DL, Grundin R, Orrenius S (1973) The effect of ethanol on drug oxidation in vitro and the significance of ethanol-cytochrome P-450 interaction. Biochem J 134: 367–375

Cohn JN, Khatri IM, Groszmann RJ, Kotelanski B (1972) Hepatic blood flow in alcoholic liver disease measured by an indicator dilution technic. Am J Med 53: 704–714

Cornish HH, Adefuin J (1967) Potentiation of carbon tetrachloride toxicity by aliphatic alcohols. Arch Environ Health 14: 447–449

Critchley JAJA, Dyson EH, Scott AW, Jarvie DR, Prescott LF (1983) Is there a place for cimetidine or ethanol in the treatment of paracetamol poisoning? Lancet I: 1375–1376

Cushman P, Barboriak JJ, Liao A, Hoffman NE (1982) Association between plasma high density lipoprotein cholesterol and antipyrine metabolism in alcoholics. Life Sci 30: 1721–1724

David RM, Nerland DE (1983) Induction of mouse glutathione *S*-transferase by ethanol. Biochem Pharmacol 32: 2809–2812

Desmond PV, Patwardhan RV, Schenker S, Hoyumpa AM (1980) Short-term ethanol administration impairs the elimination of chlordiazepoxide (Librium) in man. Eur J Clin Pharmacol 18: 275–278

Domschke S, Domschke W, Lieber CS (1974) Hepatic redox state: attenuation of the acute effects of ethanol induced by chronic ethanol consumption. Life Sci 15: 1327–1334

Emby DJ, Fraser BN (1977) Hepatotoxicity of paracetamol enhanced by ingestion of alcohol: report of two cases. S Afr Med J 51: 208–209

Farrel GC, Cooksley WGE, Hart P, Powell LW (1978) Drug metabolism in liver disease: identification of patients with impaired hepatic drug metabolism. Gastroenterology 75: 580–588

Guerri C, Grisolia S (1978) Decreased glutathione in rats due to chronic intake of ethanol. Fed Proc 37: 1541

Harman AW, Frewin DB, Priestly BG, Alexander CBJ (1979) Impairment of hepatic drug metabolism in alcoholics. Br J Clin Pharmacol 7: 45–48

Hasumura Y, Teschke R, Lieber CS (1974) Increased carbon tetrachloride hepatotoxicity and its mechanism, after chronic ethanol consumption. Gastroenterology 66: 415–422

Hayes SL, Pablo G, Radomski T, Palmer RF (1977) Ethanol and oral diazepam absorption. N Engl J Med 296: 186–189

Hayton WL (1975) Effects of normal alcohols on intestinal absorption of salicylic acid, sulfapyridine, and prednisolone in rats. J Pharm Sci 64: 1450–1456

Hepner GW, Vesell ES (1975) Quantitative assessment of hepatic function by breath analysis after oral administration of ^{14}C-aminopyrine. Am J Med 83: 632–638

Hetu C, Yelle L, Joly JG (1982) Influence of ethanol on hepatic glutathione content and on the activity of glutathione *S*-transferase and epoxide hydrase in the rat. Drug Metab Dispos 10: 246–250

Hoyumpa AM, Schenker S (1982) Major drug interactions: effect of liver disease, alcohol, and malnutrition. Annu Rev Med 33: 113–149

Hoyumpa AM, Patwardhan R, Maples M, Desmond PV, Johnson RF, Sinclair AP, Schenker S (1981) Effect of short-term ethanol administration on lorazepam clearance. Hepatology 1: 47–53

Iber FL (1977) Drug metabolism in heavy consumers of ethyl alcohol. Clin Pharmacol Ther 22: 735–742

Ideo G, De Franchis R, Del Ninno E, Dioguardi N (1971) Ethanol increases liver uridine-diphosphate-glucuronyltransferase. Experientia 27: 24–25

Ishii H, Takagi T, Okuno F, Ebihara Y, Tashiro M, Tsuchiya M (1983) Halothane-induced hepatic necrosis in ethanol pretreated rats. In: Lieber CS (ed) Biological approach to alcoholism. US Department of Health and Human Services, NIAAA, Maryland, pp 152–159 (Research monograph, vol 11)

Jollow DJ, Mitchell JR, Potter WZ, Davis DC, Gilette JR, Brodie BB (1973) Acetaminophen-induced hepatic necrosis. II. Role of covalent binding in vivo. J Pharmacol Exp Ther 187: 195–202

Joly JG, Ishii H, Teschke R, Hasumura Y, Lieber CS (1973) Effect of chronic ethanol feeding on the activities and submicrosomal distribution of reduced nicotinamide adenine dinucleotide phosphate (NADPH)-cytochrome P-450 reductase and the demethylase for aminopyrine and ethylmorphine. Biochem Pharmacol 22: 1532–1535

Koivusaari U, Norling A, Lang M, Hietanen E (1981) Structural and biotransformational membrane changes in the liver and intestine during chronic ethanol administration. Toxicology 20: 173–183

Koop DR, Morgan ET, Tarr GE, Coon MJ (1982) Purification and characterization of a unique isozyme of cytochrome P-450 from liver microsomes of ethanol-treated rabbits. J Biol Chem 257: 8472–8480

Koysooko R, Ellis EF, Levy G (1975) Effect of ethanol on theophylline absorption in humans. J Pharm Sci 64: 299–301

Kreek MJ, Rothschild MA, Oratz M, Mongelli J, Handley AC (1981) Acute effects of ethanol on hepatic uptake and distribution of narcotics in the isolated perfused rabbit liver. Hepatology 1: 419–423

Lester D (1964) The acetylation of isoniazid in alcoholics. Q J Stud Alcohol 25: 541–543
Lewis KO, Nicholson G, Lance P, Paton A (1977) Aminopyrine breath test in alcoholic liver disease and in patients on enzyme-inducing drugs. J Clin Pathol 30: 1040–1043
Lieber CS (1982) Medical disorders of alcoholism: pathogenesis and treatment. Saunders, Philadelphia (Major problems in internal medicine, vol XXII)
Lieber CS, DeCarli LM (1970) Hepatic microsomal ethanol-oxidizing system: in vitro characteristics and adaptive properties in vivo. J Biol Chem 245: 2505–2512
Long JP, Giles HG, Bennett J, Dorian P, Thiessen JJ, Orrego H, Sellers EM (1983) Effect of alcohol on propylthiouracil disposition. Clin Pharmacol Ther 33: 663–667
Lundquist F, Tygstrup N, Winkler K, Mellemgaard K, Munck-Petersen S (1962) Ethanol metabolism and production of free acetate in the human liver. J Clin Invest 41: 955–961
Macdonald CM (1973) The effects of ethanol on hepatic lipid peroxidation and the activities of glutathione reductase and peroxidase. FEBS Lett 35: 227–230
Macdonald CM, Dow J, Moore MR (1977) A possible protective role for sulphydryl compounds in acute alcoholic liver injury. Biochem Pharmacol 26: 1529–1531
Magnussen MP (1968) The effect of ethanol on the gastrointestinal absorption of drugs in the rat. Acta Pharmacol Toxicol 26: 130–144
Maling HM, Stripp B, Sipes IG, Highman B, Saul W, Williams MA (1975) Enhanced hepatotoxicity of carbon tetrachloride, thioacetamide, and dimethylnitrosamine by pretreatment of rats with ethanol and some comparisons with potentiation by isopropanol. Toxicol Appl Pharmacol 33: 291–308
Milner G, Landauer AA (1971) Alcohol thioridazine and chlorpromazine effects on skills related to driving behavior. Br J Psychiatry 118: 351–352
Misra PS, Lefevre A, Ishii H, Rubin E, Lieber CS (1971) Increase of ethanol, meprobamate and pentobarbital metabolism after chronic ethanol administration in man and in rats. Am J Med 51: 346–351
Mitchell JR, Jollow DJ (1975) Metabolic activation of drugs to toxic substances. Gastroenterology 68: 392–410
Mitchell MC, Hoyumpa AM, Schenker S, Johnson RF, Nichols S, Patwardhan RV (1983) Inhibition of caffeine elimination by short-term ethanol administration. J Lab Clin Med 101: 826–834
Moldeus P, Andersson B, Norling A (1978) Interaction of ethanol oxidation with glucuronidation in isolated hepatocytes. Biochem Pharmacol 27: 2583–2588
Moldeus P, Andersson B, Gergely V (1979) Regulation of glucuronidation and sulfate conjugation in isolated hepatocytes. Drug Metab Dispos 7: 416–419
Morgan ET, Koop DR, Coon MJ (1982) Catalytic activity of cytochrome P-450 isozyme 3a isolated from liver microsomes of ethanol-treated rabbits. J Biol Chem 257: 13951–13957
Oguma T, Levy G (1981) Acute effect of ethanol on hepatic first-pass elimination of propoxyphene in rats. J Pharmacol Exp Ther 219: 7–13
Ohnishi K, Lieber CS (1977) Reconstitution of the microsomal ethanol-oxidizing system: qualitative and quantitative changes of cytochrome P-450 after chronic ethanol consumption. J Biol Chem 252: 7124–7131
Olsen H (1982) Interaction between drug acetylation and ethanol, acetate, pyruvate, citrate, and L(−)carnitine in isolated rat liver parenchymal cells. Acta Pharmacol Toxicol 50: 67–74
Olsen H, Mørland J (1978) Ethanol-induced increase in drug acetylation in man and isolated rat liver cells. Br Med J 2: 1260–1262
Olsen H, Mørland J (1982) Ethanol-induced increases in procainamide acetylation in man. Br J Clin Pharmacol 13: 203–208
Pessayre D, Dolder A, Artigou JY, Wandscheer JC, Descatoire V, Degott C, Benhamou JP (1979) Effect of fasting on metabolite-mediated hepatotoxicity in the rat. Gastroenterology 77: 264–271
Pessayre D, Wandscheer JC, Cobert B, Level R. Degott C, Batt AM, Martin N, Benhamou JP (1980) Additive effects of inducers and fasting on acetaminophen hepatotoxicity. Biochem Pharmacol 29: 2219–2223
Reinke LA, Kauffman FC, Belinsky SA, Thurman RG (1980) Interactions between ethanol metabolism and mixed-function oxidation in perfused rat liver: inhibition of *p*-nitroanisole *O*-demethylation. J Pharmacol Exp Ther 213: 70–78

Rubin E, Lieber CS (1968) Hepatic microsomal enzymes in man and rat: induction and inhibition by ethanol. Science 162: 690–691

Rubin E, Gang H, Misra PS, Lieber CS (1970) Inhibition of drug metabolism by acute ethanol intoxication: a hepatic microsomal mechanism. Am J Med 49: 801–806

Rubin E, Lieber CS, Alvares AP, Levin W, Kuntzman R (1971) Ethanol binding to hepatic microsomes – its increase by ethanol consumption. Biochem Pharmacol 20: 229–231

Ruiter BJ, Van Der Meulen J, Brouwer A, Hummel MJR, Mauw BJ, Van Der Ploeg JCM, Wisse E (1981) Uptake by liver cells of endotoxin following its intravenous injection. Lab Invest 45: 38–45

Rumack BH, Peterson RC, Koch GC, Amara IA (1981) Acetaminophen overdose: 662 cases with evaluation of oral acetylcysteine treatment. Arch Intern Med 141: 380–385

Salaspuro MP, Shaw S, Jayatilleke E, Ross WA, Lieber CS (1981) Attenuation of the ethanol-induced hepatic redox change after chronic alcohol consumption in baboons: metabolic consequences in vivo and in vitro. Hepatology 1: 33–38

Sato C, Lieber CS (1981) Mechanism of the preventive effect of ethanol on acetaminophen-induced hepatotoxicity. J Pharmacol Exp Ther 218: 811–815

Sato C, Matsuda Y, Lieber CS (1981a) Increased hepatotoxicity of acetaminophen after chronic ethanol consumption in the rat. Gastroenterology 80: 140–148

Sato C, Nakano M, Lieber CS (1981b) Prevention of acetaminophen induced hepatotoxicity by acute ethanol administration in the rat: comparison with carbon tetrachloride-induced hepatotoxicity. J Pharmacol Exp Ther 218: 805–810

Sato M, Lieber CS (1982) Hepatic vitamin A depletion after chronic ethanol consumption in baboons and rats. J Nutr 111: 2015–2023

Saunders JB, Lewis KO, Paton A (1980) Early diagnosis of alcoholic cirrhosis by the aminopyrine breath test. Gastroenterology 79: 112–114

Sellman R, Pekkarinen A, Kangas L, Raijola E (1975a) Reduced concentrations of plasma diazepam in chronic alcoholic patients following an oral administration of diazepam. Acta Pharmacol Toxicol 36: 25–32

Sellman R, Kanto J, Raijola E, Pekkarinen A (1975b) Human and animal study on elimination from plasma and metabolism of diazepam after chronic alcohol intake. Acta Pharmacol Toxicol 36: 33–38

Shaw S, Rubin K, Lieber CS (1983) Depressed hepatic glutathione and increased diene conjugates in alcoholic liver disease: evidence of lipid peroxidation. Dig Dis Sci 28: 585–589

Siddik ZH, Drew R, Litterst CL, Mimnaugh EG, Sikic BI, Gram TE (1980) Hepatic cytochrome P-450-dependent metabolism and enzymatic conjugation of foreign compounds in vitamin A-deficient rats. Pharmacology 21: 383–390

Sinclair JF, Sinclair PR, Smith EL, Bement WJ, Pomeroy J (1981) Ethanol-mediated increase in cytochrome P-450 in cultured hepatocytes. Biochem Pharmacol 30: 2805–2809

Smith AC, Freemen RW, Harbison RD (1981) Ethanol enhancement of cocaine-induced hepatotoxicity. Biochem Pharmacol 30: 453–458

Teschke R, Stutz G, Strohmeyer G (1979) Increased paracetamol-induced hepatotoxicity after chronic alcohol consumption. Biochem Biophys Res Commun 91: 368–374

Thiessen JJ, Sellers EM, Denbeigh P, Dolman L (1976) Plasma protein binding of diazepam and tolbutamide in chronic alcoholics. J Clin Pharmacol 16: 345–351

Thomas BH, Solomonraj G (1977) Drug interactions with isoniazid metabolism in rats. J Pharm Sci 66: 1322–1326

Thomas BH, Coldwell BB, Solomonraj G, Zeitz W, Trenholm HL (1972) Effect of ethanol on the fate of pentobarbital in the rat. Biochem Pharmacol 21: 2605–2614

Traiger GJ, Plaa GL (1971) Differences in the potentiation of carbon tetrachloride in rats by ethanol and isopropanol pretreatment. Toxicol Appl Pharmacol 20: 105–112

Valenzuela A, Fernandez N, Fernandez V, Ugarte G, Videla LA (1980) Effect of acute ethanol ingestion on lipoperoxidation and on the activity of the enzymes related to peroxide metabolism in rat liver. FEBS Lett 111: 11–13

Videla LA, Fernandez V, Ugarte G, Valenzuela A (1980) Effect of acute ethanol intoxication on the content of reduced glutathione of the liver in relation to its lipoperoxidative capacity in the rat. FEBS Lett 111: 6–10

Villeneuve JP, Pomier G, Huet PM (1981) Effect of ethanol on hepatic blood flow in unanesthetized dogs with chronic portal and hepatic vein catheterization. Can J Physiol Pharmacol 59: 598–603
Vina J, Estrela JM, Guerri C, Romero FJ (1980) Effect of ethanol on glutathione concentration in isolated hepatocytes. Biochem J 188: 549–552
Whiting B, Lawrence JR, Skellern GC, Meier J (1979) Effect of acute alcohol intoxication on the metabolism and plasma kinetics of chlordiazepoxide. Br J Clin Pharmacol 7: 95–100
Zysset T, Preisig R, Bircher J (1980) Increased systemic availability of drugs during acute ethanol intoxication: studies with mephenytoin in the dog. J Pharmacol Exp Ther 213: 173–178

10 Cytochrome *P*-450: Its Involvement in the Microsomal Ethanol Oxidation and Quantitative and Qualitative Changes After Chronic Alcohol Consumption

K. Ohnishi and F. Nomura

The First Department of Medicine, Chiba University School of Medicine, 1-8-1 Inohana, Chiba City, Chiba 280, Japan

Involvement of Cytochrome *P*-450 in Ethanol Metabolism

In 1968, Lieber and DeCarli described a microsomal ethanol-oxidizing system (MEOS) which could be differentiated from the catalase-dependent system by its lesser sensitivity to either azide or cyanide. The reaction required reduced nicotinamide-adenine dinucleotide phosphate (NADPH) and molecular oxygen and was partially inhibited by carbon monoxide (Lieber and DeCarli 1970a). Moreover, it was found that the activity of MEOS was enhanced by the increase in cytochrome *P*-450, NADPH-cytochrome *c* reductase, and phospholipid after chronic alcohol consumption (Ishii et al. 1973). Ethanol was also shown to bind to cytochrome *P*-450, producing a modified type II spectrum (Rubin et al. 1971). All these results suggested that the mechanism of the MEOS is similar to that of the microsomal cytochrome-*P*-450-dependent monooxygenase system that metabolizes many xenobiotics. Subsequently, MEOS was solubilized and isolated from alcohol dehydrogenase and catalase activities by DEAE-cellulose column chromatography (Teschke et al. 1974). The column fractions rich in MEOS activity were consistently accompanied by cytochrome *P*-450, indicating the obligatory role of cytochrome *P*-450 in MEOS activity.

The most straightforward way to demonstrate the involvement of cytochrome *P*-450 in a reaction is to reconstitute the reaction with cytochrome *P*-450, NADPH-cytochrome *c* reductase, and phospholipid. Ohnishi and Lieber (1977) were able to reconstitute MEOS, which appears to be the most convincing evidence that cytochrome *P*-450 is involved in the microsomal ethanol oxidation. Cytochrome *P*-450 was partially purified by the method of Imai and Sato (1974) from liver microsomes of ethanol-treated rats (Lieber and DeCarli 1970b), and NADPH-cytochrome *c* reductase was purified by the method of Yasukochi and Masters (Yasukochi and Masters 1976) using phenobarbital-treated rats. The activity of the reconstituted MEOS showed a dependency on cytochrome *P*-450 and reductase and required synthetic phospholipid such as lecithin for its maximal activity (Fig. 1). The K_m of the reconstituted MEOS was 10 m*M*, which is similar to the K_m measured in the crude microsomes. This reconstituted system required NADPH as a cofactor, did not react to a H_2O_2-generating system, and was insensitive to the catalase inhibitor azide. These characteristics were also similar to those observed in the crude microsomes (Lieber and DeCarli 1970a). This reconstitution of MEOS was

Alcohol Related Diseases in Gastroenterology
Edited by H. K. Seitz and B. Kommerell

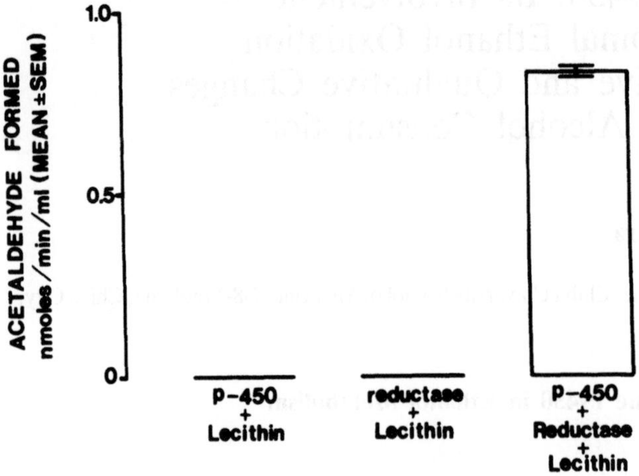

Fig. 1. Ethanol oxidation by the reconstituted system containing cytochrome *P*-450 (0.045 nmol/ml), NADPH-cytochrome *c* reductase (200 units/ml), and dilauroyl L-3-phosphatidylcholine (5 μg/ml)

confirmed by Miwa and co-workers, who used cytochrome *P*-450 purified from phenobarbital-treated rats (Miwa et al. 1978).

Although the issue concerning the involvement of cytochrome *P*-450 in MEOS activity has been thus settled, the biochemical steps involved in this pathway are not yet completely resolved. Superoxide has been implicated as a form of active oxygen in the mixed function oxidases (Strobel and Coon). Hydroxyl radical has been reported to be generated by the microsomes in the presence of oxygen and NADPH (Cohen and Cederbaum 1979), and it has been implicated in the activity of the MEOS in naive rats (Cederbaum et al. 1977). We investigated whether superoxide and hydroxyl radicals are required for the reconstituted MEOS (Ohnishi and Lieber 1978). Cytochrome *P*-450 and NADPH-cytochrome *c* reductase were partially purified as described above. Using the reconstituted system consisting of cytochrome *P*-450 (0.036 nmol/ml), NADPH-cytochrome *c* reductase (200 units/ml), and dilauroyl L-3-phosphatidylcholine (5 μg/ml), we tested the effect of superoxide dismutase (scavenger of O_2^-.) on the activity of the reconstituted MEOS. It was found that superoxide dismutase did not inhibit the activity of the reconstitnted MEOS. By contrast, at the concentration used, the superoxide dismutase completely abolished the xanthine-xanthine oxidase supported ethanol oxidation. Recently, it was reported that superoxide anions are necessary for the cytochrome-*P*-450-dependent oxidation of ethanol in reconstituted membrane vesicles (Ingelman-Sundberg and Johanson 1981). In their system, a phenobarbital-inducible form of rabbit cytochrome *P*-450 was used instead of the ethanol-inducible form, and the assay system for the activity of the reconstituted MEOS was less sensitive and might not be accurate compared to our method using gas chromatography. Possible differences between the phenobarbital-inducible and ethanol-inducible forms of cytochrome *P*-450 in terms of ethanol oxidation remain to be investigated.

To determine whether hydroxyl radical is required for the activity of the reconstituted MEOS, we tested the effect of known hydroxyl radical scavengers.

Table 1. Effect of hydroxyl radical scavengers on ethanol oxidation catalyzed by the reconstituted system

	Concentration (*M*)	Ethanol oxidation (percent of control)
Control		100
Mannitol	1×10^{-1}	33.1
Benzoate	1×10^{-1}	52.3
Dimethyl sulfoxide	5×10^{-2}	24.3
Thiourea	2×10^{-2}	34
Urea	1×10^{-1}	86.7

The reconstituted systems consisted of cytochrome *P*-450 (0.036 nmol/ml), NADPH-cytochrome *c* reductase (200 unit/ml), and dilauroyl L-3-phosphatidylcholine (5 μg/ml). The incubations were carried out at 37° C in 1 ml of medium containing 50 μmol ethanol, 6 μmol $MgCl_2$, 1 μmol Na_2-ethylenediaminetetraacetate, 0.4 μmol NADPH, and 100 μmol potassium phosphate buffer (pH 7.4) in a 25-ml glass bottle designed for use in a Perkin Elmer F-40 gas-liquid chromatograph. The reaction was initiated by adding ethanol and was stopped by adding 0.2 ml 35% perchloric acid containing (per milliliter) 100 μmol thiourea. Values represent means of two experiments (in duplicate)

These hydroxyl radical scavengers (mannitol 100 m*M*, benzoate 100 m*M*, dimethyl sulfoxide 50 m*M*, and thiourea 20 m*M*) reduced the activity of the reconstituted MEOS by 67%, 48%, 76%, and 66% respectively (Table 1). It is very likely that the hydroxyl radical is, at least in part, involved in the activity of the reconstituted MEOS. Alternatively, scavengers could simply act as substrates for cytochrome *P*-450 and compete with ethanol for active oxygen generated. This role of the hydroxyl radical in the reconstituted MEOS was more recently confirmed by Winston and Cederbaum (1983). It has also been shown that microsomes actively generate the corresponding hydrocarbon gas from each hydroxyl radical scavengers (3-thiomethylpropanol, 2-keto-4-thiomethyl butyric acid, and dimethyl sulfoxide) in the presence of NADPH and that the production of hydrocarbon gas is inhibited by the addition of ethanol (Cohen and Cederbaum 1980).

Quantitative and Qualitative Changes of Cytochrome *P*-450 After Chronic Alcohol Consumption

Following chronic alcohol consumption, MEOS activity significantly increases (Lieber and DeCarli 1970a). This is associated with an increase in various constituents of the smooth fraction of the endoplasmic reticulum, such as phospholipids, cytochrome *P*-450 reductase, and cytochrome *P*-450 (Ishii et al. 1973). In addition to this quantitative change of cytochrome *P*-450, a number of studies have shown that chronic ethanol consumption can induce selectively some forms of cytochrome *P*-450 (ethanol-inducible forms of cytochrome *P*-450). A cytochrome *P*-450 species showing high affinity for cyanide has been reported as preferentially induced by ethanol (Joly et al. 1972). Evidence in favor of an increase

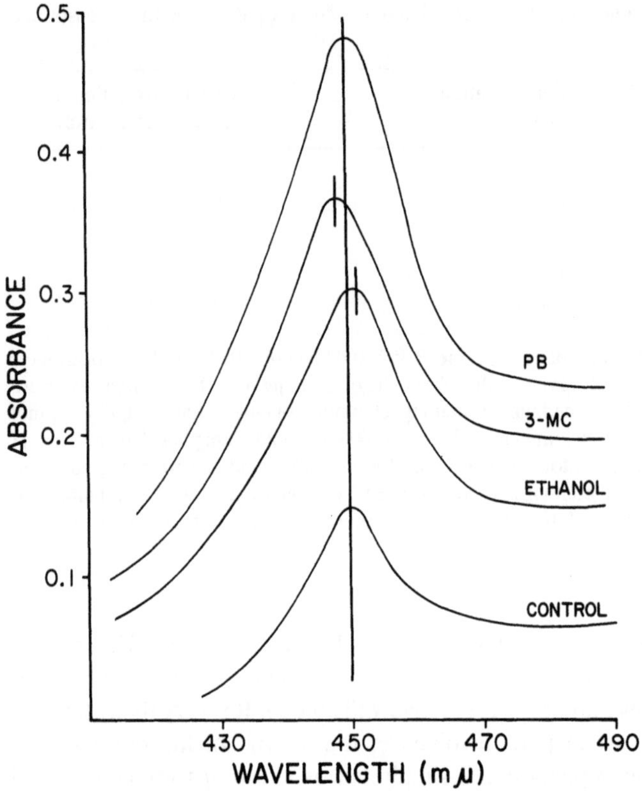

Fig. 2. Effect of various inducers [phenobarbital *(PB)*, 3-methylcholanthrene (3-MC), and ethanol] on the CO-reduced difference spectra of hepatic microsomal cytochrome *P*-450. Male rats were treated with phenobarbital (80 mg/kg intraperitoneally, 4 days), 3-methylcholanthrene (20 mg/kg intraperitoneally, 2 days), or ethanol liquid diet (8 weeks). The protein concentration of the microsomal suspension was 1.0 mg/ml

in a special species of cytochrome *P*-450 after ethanol treatment was also derived from inhibitor studies. Ullich et al. (1975) studied the inhibitory effect of tetrahydrofuran on *O*-dealkylation activity for 7-ethoxycoumarine in liver microsomes from ethanol-treated, phenobarbital-treated, benzopyrene-treated, and control rats. They found that tetrahydrofuran inhibited specifically the *O*-dealkylation in microsomes from ethanol-treated rats.

More direct proof of the induction of specific forms of cytochrome *P*-450 after chronic alcohol consumption was obtained by the following study by Ohnishi and Lieber (1977). Liver microsomes were prepared from ethanol-treated, control, phenobarbital-treated, and 3-methylcholanthrene-treated rats. The CO-reduced difference spectrum maxima of cytochrome *P*-450 shifted toward 451 nm (Fig. 2). Furthermore, when liver microsomes prepared from ethanol-treated rats were subjected to protein electrophoresis on sodium dodecyl sulfate polyacrylamide gel with continuous buffer system, 3,3′,5,5′-tetramethyl benzidine staining of these gels in the presence of H_2O_2 showed three distinct hemoproteins whose molecular weights ranged between 50,000 and 60,000. Chronic ethanol feeding increased two

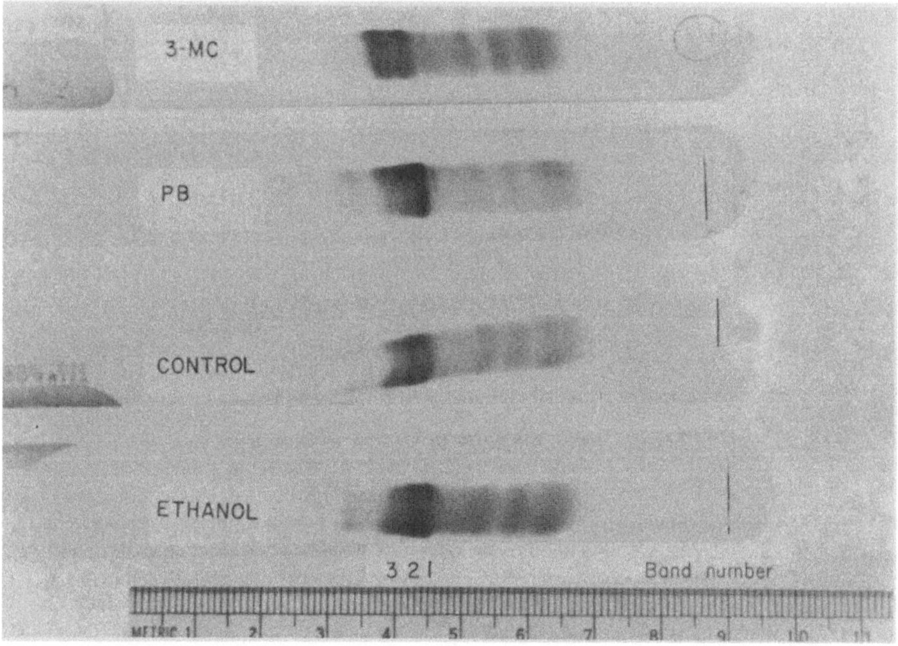

Fig. 3. Microsomal hemoprotein profiles of control, ethanol-fed, phenobarbital-treated, and 3-methylcholanthrene-treated rats. Treatments of rats with inducers were carried out as described in Fig. 2. Microsomal protein (120 μg) was applied to each 5.0% polyacrylamide gel. After electrophoresis, gels were stained by 3,3′,5,5-tetramethylbenzidine plus H_2O_2. The apparent molecular weights of band numbers 1, 2, and 3 are 47,000, 50,000, and 56,000 respectively. *PB,* phenobarbital; *3-MC,* 3-methylcholanthrene

hemoproteins (nos. 1 and 2). One hemoprotein (no. 2) is distinct from hemoprotein no. 1 induced by phenobarbital treatment and hemoprotein no. 3 induced by 3-methylcholanthrene treatment (Fig. 3). There were also changes in microsomal protcins. Liver microsomes prepared from ethanol-treated and control rats were subjected to protein electrophoresis on sodium dodecyl sulfate polyacrylamide gel with discontinuous buffer system. These gels were scanned (Figs. 4 and 5). Major differences were in the distribution of the protein constituents with an apparent molecular weight of approximately 50,000, which is the cytochrome *P*-450 region. After chronic alcohol feeding, a protein with an apparent molecular weight of 53,400 (band no. 3) strikingly increased and a protein with an apparent molecular weight of 52,000 (band no. 2) also increased. Judging from the relative location of the protein and hemoprotein by different treatments, the increase of proteins with apparent molecular weight of 52,000 and 53,400 may reflect the increase in apoproteins of hemoproteins no. 1 and no. 2 respectively after chronic ethanol feeding.

As described above, one form of cytochrome *P*-450 induced by chronic ethanol feeding is similar to a phenobarbital-inducible form as judged by its behavior on gel clcctrophorcsis. Morcover, it is well known that the enhancement of hepatotoxicity of xenobiotics such as carbon tetrachloride, acetaminophen, and halothane by phenobarbital pretreatment is also observed after chronic alcohol consumption in

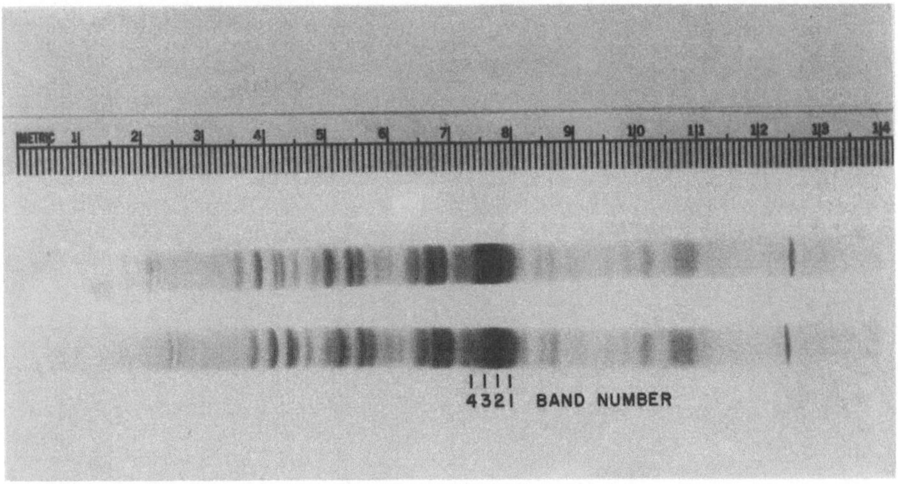

Fig. 4. Coomassie blue protein profiles obtained by electrophoresis of liver microsomes from control and ethanol-fed rats in 0.1% sodium dodecyl sulfate polyacrylamide gels. Microsomal protein was electrophoresed by the method of Laemmli (1970) with 7.5% separating gel after applying 40 µg protein to each gel. Protein profiles are from hepatic microsomes of ethanol-fed *(above)* and control *(below)* rats. Band number 3 indicates the position to which a protein of molecular weight 53,400 would migrate. The area of bands 1–4 was scanned (Fig. 5)

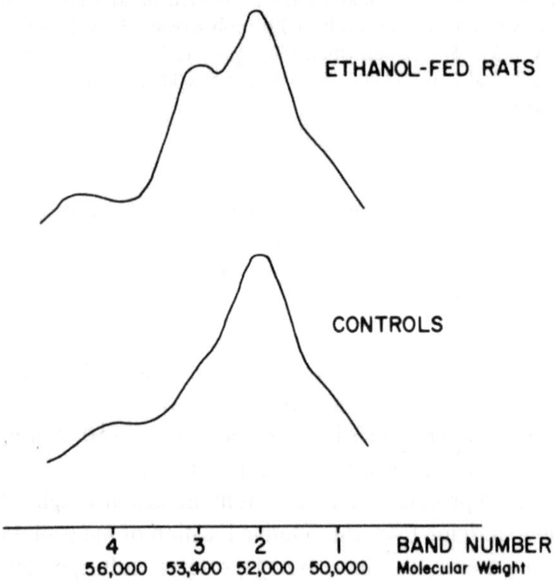

Fig. 5. Scans of the 50,000 molecular weight region of the gels shown in Fig. 4. Relative density of each band at 550 nm is plotted on the *ordinate*. Peak heights of band number 2 in both gels are automatically set to the same height. Migration distance is on the *abscissa*. The *upper scan* was obtained with microsomes from ethanol-fed and the *lower* from control rats. Bands 1, 2, 3, and 4 have the apparent molecular weights shown underneath the band number

rats (Hasumura et al. 1974; Sato et al. 1981; Takagi et al. 1982). Therefore, we intended to quantitate the phenobarbital-inducible form of cytochrome *P*-450 (PB-P450) and to study its localization in the hepatic lobule. To this end, we purified the phenobarbital-inducible form of cytochrome *P*-450 to electrophoretic homogeneity (16 nmol/mg protein) as described by West et al. (1979) and prepared rabbit

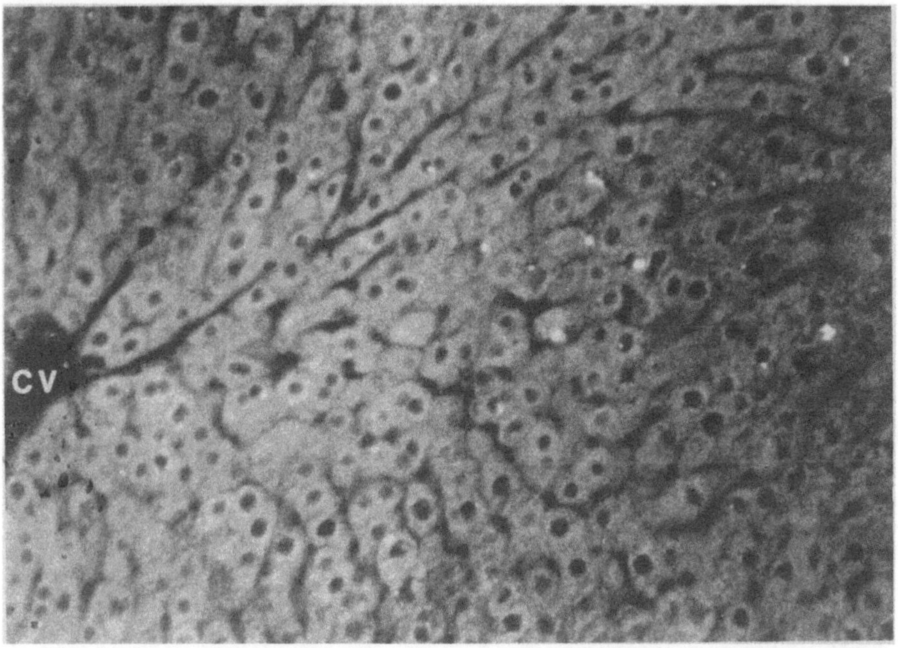

Fig. 6. Immunohistochemical localization of phenobarbital-inducible form of cytochrome *P*-450 in the hepatic lobule of normal rats. Section of liver tissue (4 μm) from normal rat was exposed to fluorescein-isothiocyanate-conjugated rabbit antibody to the phenobarbital-inducible form of cytochrome *P*-450. A central vein *(cv)* is indicated

IgG-antibody (Ab) to PB-P450. Fluorescein isocyanate was conjugated to Ab to PB-P450 and nonimmune rabbit IgG as described previously (Ohnishi et al. 1982a). Chronic ethanol feeding increased total microsomal cytochrome *P*-450 (0.71 ± 0.09 nmol/mg protein vs 0.56 ± 0.03 in controls, $n = 6, p < 0.05$). Using single radial immunodiffusion with specific Ab to PB-P450, microsomal PB-P450 was 0.043 ± 0.014 nmol/mg protein in ethanol-fed rats (control 0.008 ± 0.011, $n = 6, p < 0.005$). Specific fluorescence was seen in the hepatocyte cytoplasm throughout the liver. In control rats, fluorescence was barely discerned but was slightly more intense in the centrilobular than perilobular zones. In phenobarbital-treated rats, fluorescence was much more intense, with a similar difference between the centrilobular and perilobular zones (Figs. 6 and 7). Similar enhancement of fluorescence was also observed, but was less striking, in ethanol-fed rats (Ohnishi et al. 1982b).

Once the induction of a specific form of cytochrome *P*-450 after chronic alcohol consumption is established, the next concern is its catalytic properties. Villeneuve et al. (1976) reported that cytochrome *P*-450 partially purified from liver microsomes obtained from ethanol-fed rats has more capacity to hydroxylate aniline than those isolated from liver microsomes obtained from phenobarbital-treated rats and 3-methylcholanthrene-treated rats. One of the major questions in terms of the properties of ethanol-inducible forms of cytochrome *P*-450 is whether the qualitative changes in cytochrome *P*-450 could play a role in the adaptive increase in the activity of MEOS after chronic alcohol consumption. To delineate a role of these qualitative

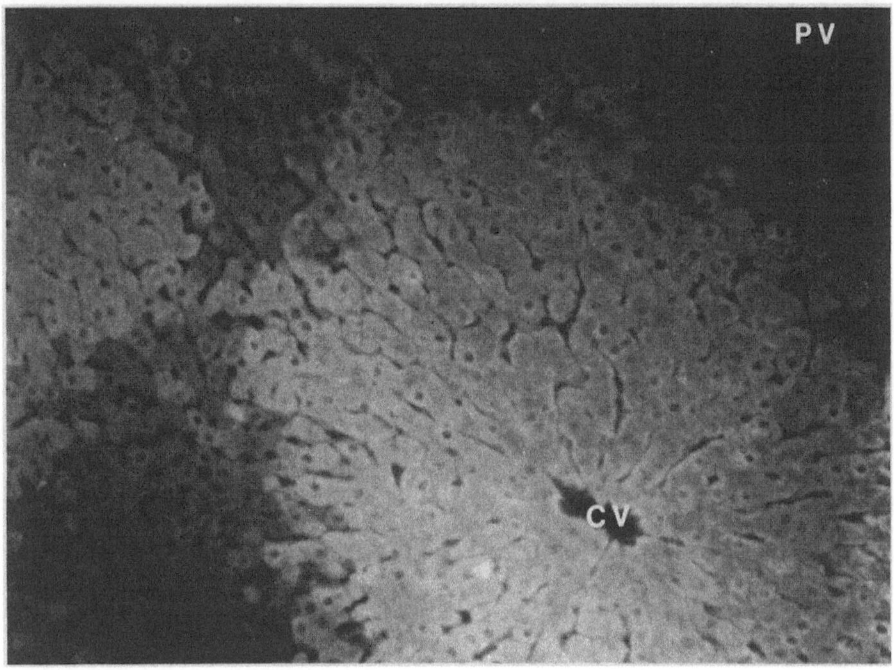

Fig. 7. Immunohistochemical localization of phenobarbital-inducible form of cytochrome *P*-450 in the hepatic lobule of phenobarbital-treated rat. Section of liver tissue (4 μm) was exposed to fluorescein-isothiocyanate-conjugated rabbit antibody to the phenobarbital-inducible form of cytochrome *P*-450. A central vein *(CV)* and a portal vein *(PV)* are indicated

changes in the adaptive increase in the activity of MEOS, the capacity of cytochrome *P*-450 from ethanol-treated rats or controls to promote ethanol oxidation was compared in the reconstituted system, in the presence of excess NADPH-cytochrome *c* reductase and phospholipid. The partially purified cytochrome *P*-450 from ethanol-fed rats was found to be about twice as active for ethanol oxidation as for the control preparation (Fig. 8) (Ohnishi et al. 1983). This increased capacity of ethanol-inducible forms of cytochrome *P*-450 to oxidize ethanol was recently confirmed by Koop et al. (1982). They have purified a new isozyme of cytochrome *P*-450 to electrophoretic homogeneity from hepatic microsomes of rabbits treated chronically with ethanol and found that this form of cytochrome *P*-450 displays the highest activity of all of the rabbits isozyme of cytochrome *P*-450 in the oxidation of ethanol to acetaldehyde and the *p*-hydroxylation of aniline when reconstituted with NADPH-cytochrome *P*-450 reductase and phospholipid in the presence of NADPH and oxygen.

Next one may wonder whether the hydroxyl radical is also involved in the adaptive increase of MEOS after chronic alcohol consumption. To that effect, microsomes were prepared from ethanol-fed rats and their pair-fed controls and the effect of the hydroxyl radical scavengers on MEOS activity in these two groups was tested. As shown in Table 2, chronic ethanol feeding resulted in a striking increase in MEOS activity and it was found that hydroxyl radical scavengers inhibited the

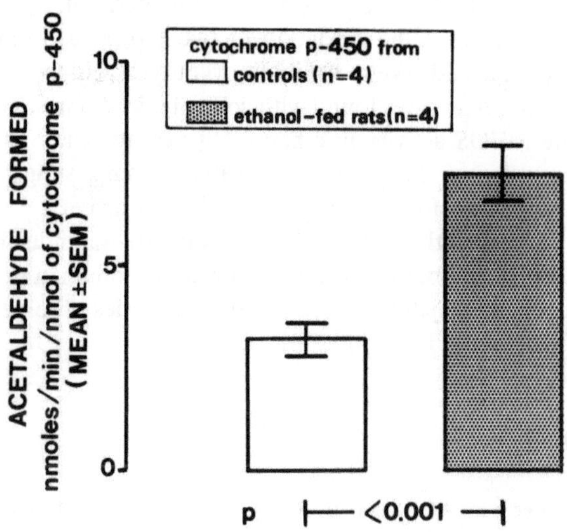

Fig. 8. Ethanol oxidation by the microsomal ethanol-oxidizing system reconstituted with cytochrome *P*-450 from ethanol-fed rats or control rats. The reconstituted system consisted of cytochrome *P*-450 (0.045 nmol/ml) from ethanol-fed or control rats, NADPH-cytochrome *c* reductase (200 units/ml), and dilauroyl L-3-phosphatidylcholine (5 μg/ml)

Table 2. Effect of hydroxyl radical scavengers on ethanol oxidation by microsomal preparations from ethanol-fed rats and their pair-fed controls

Additions to reaction mixture	Microsomal preparations			
	Ethanol-fed		Controls	
	Specific activity (nmol acetal-dehyde/min/mg protein)	Relative activity (%)	Specific activity (nmol acetal-dehyde/min/mg protein)	Relative activity (%)
None	15.9 ± 0.97	100	5.8 ± 0.64	100
Dimethyl sulfoxide (50 m*M*)	4.5 ± 0.13	28	1.4 ± 0.05	24
Mannitol (100 m*M*)	5.7 ± 0.30	36	1.8 ± 0.04	31
Benzoate (100 m*M*)	7.6 ± 0.46	48	1.5 ± 0.05	26
Thiourea (20 m*M*)	5.9 ± 0.23	37	1.6 ± 0.19	28

The reaction mixtures (in a final volume of 1.0 ml) were incubated at 37° C and contained 1 mg microsomal protein, 1 μmol of sodium azide, 100 μmol potassium phosphate buffer (pH 7.4), 6 μmol $MgCl_2$, 1 μmol Na_2-ethylenediaminetetraacetate, 0.4 μmol NADPH, and 50 μmol ethanol. The reaction was initated by adding ethanol and stopped at 0, 5, 10 min by adding 0.2 ml 35% percholic acid containing (per mililiter) 100 μmol thiourea. When the effect of hydroxyl radical scavengers was studied, the respective compound was included in the incubation medium. Values represent means ± SE of four experiments

activity of MEOS in both ethanol-fed rats and their pair fed controls. Indeed, the amount of MEOS activity in ethanol-fed rats inhibited by scavengers was more than the baseline levels found in controls. This may indicate that hydroxyl radicals are related to the enhancement of MEOS activity after chronic alcohol consumption (Ohnishi et al. 1983). This partial involvement of hydroxyl radicals in the enhanced MEOS activity after chronic alcohol consumption was also supported by data reported by Klein et al. (1983). They found that liver microsomes from ethanol-fed rats catalyzed the oxidation of two typical hydroxyl radical scavengers, dimethyl sulfoxide and 2-keto-4-thiomethyl butyric acid, at rates which were two- to threefold greater than rates found with control microsomes. However, the difference between the MEOS activity in ethanol-fed rats and their pair-fed controls remained in the presence of the scavengers (Table 2), suggesting that other mechanisms are also responsible for the increase in MEOS activity.

Finally, the enhanced capacity of ethanol-inducible form of cytochrome *P*-450 to oxidize ethanol was recently shown in subhuman primates, baboons (Nomura et al. 1983), in which a variety of hepatic injuries observed in alcoholics can be produced (Lieber and DeCarli 1976).

References

Cederbaum AI, Dicker E, Rubin E, Cohen G (1977) The effect of dimethylsulfoxide and other hydroxyl radical scavengers on the oxidation of ethanol by rat liver microsomes. Biochem Biophys Res Commun 78: 1254–1262

Cohen G, Cederbaum AI (1979) Chemical evidence for production of hydroxyl radicals during microsomal electron transfer. Science 204: 66–67

Cohen G, Cederbaum AI (1980) Microsomal metabolism of hydroxyl radical scavenging agents: relationship to the microsomal oxidation of alcohols. Arch Biochem Biophys 199: 438–447

Hasumura Y, Teschke R, Lieber CS (1974) Increased carbon tetrachloride hepatotoxicity, and its mechanism, after chronic ethanol consumption. Gastroenterology 66: 415–422

Imai Y, Sato R (1974) An affinity column method for partial purification of cytochrome P-450 from phenobarbital-induced rabbit liver microsomes. J Biochem (Tokyo) 75: 689–697

Ingelman-Sundberg M, Johanson I (1981) The mechanism of cytochrome P-450 dependent oxidation of ethanol in reconstituted membrane vesicles. J Biol Chem 256: 6321–6326

Ishii H, Joly JG, Lieber CS (1973) Effect of ethanol on the amount and enzyme activities of hepatic rough and smooth microsomal membranes. Biochim Biophys Acta 291: 411–420

Joly JG, Ishii H, Lieber CS (1972) Microsomal cyanide binding cytochromes: its role in hepatic ethanol oxidation. Gastroenterology 62: 174

Klein SM, Cohen G, Lieber CS, Cederbaum AI (1983) Increased microsomal oxidation of hydroxyl radical scavenging agents and ethanol after chronic consumption of ethanol. Arch Biochem Biophys 223: 425–432

Koop DR, Morgan ET, Tarr GE, Coon MJ (1982) Purification and characterization of a unique isozyme of cytochrome P-450 from liver microsomes of ethanol-treated rabbits. J Biol Chem 257: 8472–8480

Laemli UK (1970) Cleavage of structural proteins during the assembly of the head of bacteriophage T 4. Nature 227: 680–685

Lieber CS, DeCarli LM (1968) Ethanol oxidation by hepatic microsomes: adaptive increase after ethanol feeding. Science 162: 917–918

Lieber CS, DeCarli LM (1970a) Hepatic microsomal ethanol oxidizing system: in vitro characteristics and adaptive properties in vivo. J Biol Chem 245: 2505–2512

Lieber CS, DeCarli LM (1970b) Quantitative relationship between amount of dietary fat and the severity of the alcoholic fatty liver. Am J Clin Nutr 23: 474–478

Lieber CS, DeCarli LM (1976) Animal models of ethanol dependence and liver injury in rats and baboons. Fed Proc 35: 1232–1236

Miwa G, Levin W, Thomas PE, Lu AYH (1978) The direct oxidation of ethanol by a catalase- and alcohol dehydrogenase-free reconstituted system containing cytochrome P-450. Arch Biochem Biophys 187: 464–475

Nomura F, Pikkarainen PH, Jauhonen P, Arai M, Gordon ER, Baraona E, Lieber CS (1983) The effect of ethanol administration on the metabolism of ethanol in baboons. J Pharmacol Exp Ther 227: 77–83

Ohnishi K, Lieber CS (1977) Reconstitution of the microsomal ethanol oxidizing system (MEOS): qualitative and quantitative changes of cytochrome P-450 after chronic ethanol consumption. J Biol Chem 252: 7124–7131

Ohnishi K, Lieber CS (1978) Respective role sof superoxide and hydroxyl radicals in the activity of the reconstituted microsomal ethanol oxidizing system. Arch Biochem Biophys 191: 798–803

Ohnishi K, Mishima A, Okuda K (1982a) Immunofluorescence of phenobarbital inducible cytochrome P-450 in the hepatic lobule of normal and phenobarbital treated rats. Hepatology 2: 849–855

Ohnishi K, Mishima A, Nomura F, Nakayama T, Saito M, Hatano H, Okuda K (1982b) Effect of chronic ethanol feeding on the amount of phenobarbital inducible cytochrome P-450 and its localization. Hepatology 2: 685

Ohnishi K, Nomura F, Lieber CS (1983) Reconstitution of the microsomal ethanol-oxidizing system: qualitative and quantitative changes of cytochrome P-450 after chronic ethanol consumption. In: Lieber CS (ed) Biological approach to alcoholism. National Institute on Alcohol Abuse and Alcoholism, Washington DC, pp 279–290 (Research Monograph no 11)

Rubin E, Lieber CS, Alvares AP, Levin W, Kuntzman R (1971) Ethanol binding hepatic microsomes – its increase by ethanol consumption. Biochem Pharmacol 20: 229–231

Sato C, Matsuda Y, Lieber CS (1981) Increased hepatotoxicity of acetaminophen after chronic ethanol consumption in the rat. Gastroenterology 80: 140–148

Strobel HW, Coon MJ (1977) Effect of superoxide generation and dismutation on hydroxylation reactions catalyzed by liver microsomal cytochrome P-450. J Biol Chem 246: 7826–7829

Takagi T, Ishii H, Okuno F, Ebihara H, Kato S, Tsuchiya M (1982) Halothane hepatotoxicity: its potentiation by chronic ethanol administration in rats. In: Sato R, Kato R (eds) Microsomes, drug oxidations and drug toxicity. Japan Scientific Societies, Tokyo, pp 509–512

Teschke R, Hasumura Y, Lieber CS (1974) Hepatic microsomal ethanol oxidizing system: solubilization, isolation and characterization. Arch Biochem Biophys 163: 404–415

Ullich V, Weber P, Wollenberg P (1975) Tetrahydrofurane – an inhibitor for ethanol-induced liver microsomal cytochrome P-450. Biochem Biophys Res Commun 64: 808–813

Villeneuve JP, Mavier P, Joly JG (1976) Ethanol-induced cytochrome P-450: catalytic activity after partial purification. Biochem Biophys Res Commun 70: 723–728

West SB, Huang MT, Miwa G, Lu AYH (1979) A simple and rapid procedure for the purification of phenobarbital-inducible cytochrome P-450 from rat liver microsomes. Arch Biochem Biophys 193: 42–50

Winston GW, Cederbaum AI (1983) NADPH-dependent production of oxy radicals by purified component of the rat liver mixed function oxidase system II. Role in microsomal oxidation of ethanol. J Biol Chem 258: 1514–1519

Yasukochi Y, Masters BSS (1976) Some properties of a detergent solubilized NADPH-cytochrome C (cytochrome P-450) reductase purified by biospecific affinity chromatography. J Biol Chem 251: 5337–5344

11 Ethanol and Carcinogenesis

H. K. Seitz*

Medizinische Klinik, Abteilung Innere Medizin IV, Universität Heidelberg, Bergheimer Strasse 58, 6900 Heidelberg, Federal Republic of Germany

Epidemiology

Clinically, a link between alcoholism and certain types of cancer has been observed for many years. A great number of epidemiologic studies has shown an association between excessive drinking of alcoholic beverages and cancer of the oropharynx (Wynder et al. 1957; Keller 1967; Rothman and Keller 1972; Alcohol and Health 1975; Feldmann et al. 1975; Fortier 1975; Bross and Coombs 1976; Graham et al. 1977; McCoy 1978), the larynx (Vincent and Marchetta 1963; Gregoriades 1974; Alcohol and Health 1975; McMichael 1978; Tuyns 1978; Wynder et al. 1976; Tuyns 1979), and the esophagus (Wynder et al. 1957; Wynder and Bross 1961; Tuyns 1970; Schoenberg et al. 1971; Tuyns and Masse 1973; Wynder and Mabuchi 1973; Schottenfeldt et al. 1974). In a series of studies (Wynder et al. 1956a, b, 1957; Wynder and Bross 1957) heavy drinkers were found to have roughly a tenfold increased risk of developing cancer of the mouth. Subjects who drink heavily often also smoke heavily. This later fact was first taken into account by Flamant et al. (1964), who assessed both factors and reported that there was a "strong" association of alcohol intake with cancer of these sites that come most directly in contact with alcohol (tongue, hypopharynx, larynx, esophagus). It was calculated that 76% of these cancers in males could be eliminated if exposure to alcohol and tobacco were avoided (Rothman and Keller 1972). The risk of developing oral cancer for a heavy drinker who smokes was 6–15 times higher than for nondrinkers and nonsmokers (Fortier 1975; Feldman et al. 1975). Also, women who drink and smoke heavily develop cancer of the tongue and buccal cavity 15 years earlier than do women who abstain from both alcohol and tobacco (Bross and Coombs 1976). It appears that alcohol plays a more important role than smoking with respect to cancer of the esophagus, whereas smoking seems to be more strongly associated with cancer of the mouth and pharynx (Flamant et al. 1964). Recently, additional sites of cancer associated with alcohol were detected in the liver (Flamant et al. 1964; Lee 1966; Hakulinnen et al. 1974; Kissin and Kaley 1974; Alcohol and Health 1975; Jensen 1979; Lowenfels 1979; McCoy and Wynder 1979; Schottenfeld 1979; Lieber et al. 1981; Tamburro and Hua-Ming 1981; Ohnishi et al. 1982), the pancreas (Burch and

* The author wishes to thank Mrs. S. Veith for preparation of the manuscript and Mr. P. Froelich for the expert photographic work

Alcohol Related Diseases in Gastroenterology
Edited by H. K. Seitz and B. Kommerell

Ansari 1968; Ishii et al. 1968; Durbec et al. 1980; Okuda and Ohnishi 1981), the cardia of the stomach (MacDonald 1972), the lungs (Klatsky et al. 1981; Pollack et al. 1984), and the rectum (Wynder and Shigematsu 1967; Breslow and Enstrom 1974; Enstrom 1977; Williams and Horm 1977; Dean et al. 1979; Kono and Ikeda 1979; Pollack et al. 1984).

Hepatocellular carcinoma in alcoholics is commonly thought to occur in association with cirrhosis of the liver. Indeed, the incidence of cirrhosis in patients with hepatocellular carcinoma has been reported to vary between 16% and 80% (Chan 1975), with most reports indicating a 55%–80% association. Numerous etiologies have been proposed for carcinoma in alcoholic patients, especially in conjunction with a hyperregenerative state in cirrhosis. Both Brechot et al. (1982) and Ohnishi et al. (1982) reported an increased incidence of liver cancer in hepatitis B surface antigen carriers due to heavy alcohol intake. However, as shown by Lieber et al. (1981), alcoholism also stimulates the development of cancer in the noncirrhotic liver.

Further evidence for the cocarcinogenic role of alcohol in hepatocarcinogenesis may derive from case reports: Gottfried et al. (1982) reported a case of histologically proven hepatocellular carcinoma in an alcoholic. The tumor vanished spontaneously following abstention from alcohol. In addition, an unusual paired human accident demonstrated the cocarcinogenic role of alcohol in two chemical workers working with vinyl chloride. One worker consumed large quantities of alcohol in addition to his exposure to vinyl chloride and developed both angiosarcoma and hepatocellular cancer, while his colleague, a nondrinker, developed only angiosarcoma (Tamburro and Hua-Ming 1981).

More recently, another target organ of alcohol with respect to carcinogenesis has been focused on, namely the rectum. First reports of an association between alcohol consumption and cancer of the rectum were published by Breslow and Enstrom (1974). These authors found a good correlation between beer drinking and rectal cancer. Meanwhile, other epidemiologic studies in Ireland (Dean et al. 1979), the United States (Wynder and Shigematsu 1967; Williams and Horm 1977), and Japan (Kono and Ikeda 1979) have confirmed these data. However, Jensen (1979) in Denmark failed to show such an association. More recently, Pollack et al. (1984) found a highly significant ($p < 0.001$) positive correlation between beer consumption and rectal cancer in a prospective study of 8,006 subjects.

Animal Studies

Ethanol per se is not a carcinogen (Ketcham et al. 1963). However, when administered in combination with a chemical carcinogen, ethanol enhances carcinogenesis in some organs under certain experimental conditions. Thus ethanol is a cocarcinogen. Table 1 summarizes results on the effect of ethanol on chemically induced carcinogenesis in the animal model. Ethanol enhances the local (Horie et al. 1965; Elzay 1966; Henefer 1966; Stenback 1969) and the systemic (Capel et al. 1978) carcinogenic effect of some polycyclic hydrocarbons. In addition, Radike et al.

Table 1. Effect of ethanol on chemically induced carcinogenesis

Authors	Species	Carcinogen[a]	Ethanol administration	Target organ	Ethanol effect
Horie et al. (1965)	Mouse	BP, orally	As solvent	Esophagus	Stimulation
Henefer (1966)	Hamster	DMBA, locally	As solvent	Pouch	Stimulation
Elzay (1966)	Hamster	DMBA, locally	As solvent	Pouch	Stimulation
Stenback (1969)	Mouse	DMBA, locally	As solvent	Skin	Stimulation
Capel et al. (1978)	Mouse	BP, i.p.	Prior to carcinogen	Muscle	Stimulation
Gibel (1967)	Rat	DENA, i.g.	30%, i.g. with carcinogen	Esophagus, liver	Stimulation/ no effect
Schmähl (1976)	Rat	MPNA, s.c.	25%, in drinking water, continuously	Esophagus	No effect
Teschke et al. (1983)	Rat	DMNA, i.p.	6%, liquid diet prior to carcinogen	Liver, kidney	No effect/ inhibition[b]
Habs and Schmähl (1981)	Rat	DMNA, orally	25%, in drinking water, after carcinogen	Liver	Inhibition
Griciute et al. (1981)	Mouse	DMNA, i.g.	40%, i.g. with carcinogen	Liver, olfactory nerve	Inhibition/ Stimulation
McCoy et al. (1981)	Hamster	NPYR, i.p. NNN, i.p.	5%, liquid diet, prior to and with carcinogen	Nasal cavitytrachea	Stimulation/ no effect
Radike et al. (1977)	Rat	Vinyl chloride in the air	5%, in drinking water prior to and with carcinogen	Liver	Stimulation
Mendenhall and Chedid (1980)	Rat	Aflatoxin B1, i.p.	6%, liquid diet, continuously	Liver	Stimulation of hepatis Peliosis
Weisburger et al. (1964)	Rat	N-OH-2-AAF, orally	10%, in drinking water, with carcinogen	Liver	No effect
Seitz et al. (1984)	Rat	1,2-DMH, s.c.	6%, liquid diet prior to and intermittent with carcinogen	Rectum	Stimulation
Hamilton et al. (1983)	Rat	Azoxymethane s. c.	Liquid diet prior to and with carcinogen, various ethanol concentrations	Distal colon, proximal colon	Stimulation[c]/ no effect[c]/ inhibition

[a] BP, benzo(a)pyrene; DMBA, dimethylbenzanthracene; MPNA, methylphenylnitrosamine; DENA, diethylnitrosamine; DMNA, dimethylnitrosamine; N-OH-2-AAF, *n*-hydroxy-2-acetaminofluorene; 1,2-DMH, 1,2-dimethylhydrazine dihydrochloride; NPYR, *N*-nitrosopyrrolidine; NNN, *N*-nitrosonornicotine; i.p., intraperitoneally; s.c., subcutaneously; i.g., intragastrically

[b] Similar tumor yield, but prolonged latency period after alcohol

[c] Dependent on alcohol content of the diet

(1977) found a fourfold increase in vinyl-chloride-induced hepatocellular carcinoma after ethanol administration in the rat, and such an ethanol-induced effect of vinyl-chloride-induced hepatocarcinogenesis has been noted in man, as already discussed (Tamburro and Hua-Ming 1981). Ethanol also changed the cell type of the hepatic tumors induced by vinyl chloride. Mendenhall and Chedid (1980) observed a significantly increased appearance of peliosis hepatis induced by aflatoxin B1 following chronic ethanol ingestion in the rat.

Different results exist with respect to nitrosamine-induced carcinogenesis. Ethanol stimulates diethylnitrosamine- (Gibel 1967) but not methylphenylnitrosamine- (Schmähl 1976) induced esophageal carcinogenesis, while hepatic carcinogenesis seems not to be influenced (Gibel 1967; Teschke et al. 1983) or even to be inhibited (Habs and Schmähl 1981; Griciute et al. 1981; Teschke et al. 1983) by alcohol. Griciute et al. (1981) observed the development of olfactoric neuroepithelioma following alcohol and dimethylnitrosamine application, but such tumors were not noted when the carcinogen was given alone. The seemingly contradictory results on the effect of alcohol in nitrosamine-induced hepatocarcinogenesis can be partly explained by different experimental conditions (Lester 1979). Factors which influence the effect of ethanol on carcinogenesis in animal experiments are:

- Type of chemical carcinogen (organ specificity, metabolism).
- Dose of carcinogen and duration of its application.
- Means of carcinogen application (local, oral, parenteral).
- Amount, concentration, and duration of ethanol administration.
- Means of ethanol administration (in drinking water, as liquid diet, intragastrically, parenteral injection).
- Combination of carcinogen and ethanol (ethanol application prior to, with, or after carcinogen application).
- Species, strain, sex.

One of the most important factors seems to be the means of alcohol administration. When ethanol is given in the drinking water as in some studies, an inadequate intake of ethanol by the animals results. This is not the case when animals are pair-fed using liquid diets as described by Lieber and DeCarli (1970a). Since ethanol, like many procarcinogens, is metabolized via a hepatic microsomal cytochrome-*P*-450-dependent biotransformation system (Lieber and DeCarli 1970a), an interaction in the metabolism of the two compounds cannot be excluded. After chronic ethanol consumption, drug and xenobiotic metabolism is enhanced when ethanol is absent from the organism (Mezey 1976; Lieber and DeCarli 1977). However, in the presence of alcohol, microsomal drug metabolism is inhibited (Mezey 1976; Lieber and DeCarli 1977). A similar ethanol effect may also play a role in the activation of procarcinogens. When ethanol and procarcinogens are given simultaneously, alcohol may inhibit the activation of the procarcinogen and the initiation of the carcinogenic process may not occur. When ethanol is administered prior to the procarcinogen the microsomal cytochrome-*P*-450-dependent enzyme system is induced and the activation of the procarcinogen may be enhanced, leading to a stimulation of carcinogenesis.

We have investigated the effect of chronic ethanol consumption on 1,2-dimethylhydrazine (DMH)-induced large-intestinal carcinogenesis in the rat (Seitz et al.,

1984). Ethanol was given as part of a liquid diet of which it comprised 36% of total calories [controls received isocaloric carbohydrates (Lieber and DeCarli 1970a)] 4 weeks prior to the first of 16 weekly injections of DMH. During the application of the procarcinogen, liquid diets were omitted and replaced by standard laboratory diets. After 32 weeks, animals were killed and the intestine was histologically examined. The results are shown in Fig. 1. Ethanol administration increased the tumor yield only in the rectum but not in other segments of the intestinal tract. This is in accordance with the data of Hamilton et al. (1983). These authors found a cocarcinogenic effect of chronic ethanol ingestion in the distal colon after tumor induction by azoxymethane. However, this stimulation of tumor development was only seen when ethanol was given as 11% of total calories. When ethanol was administered as 22% of total calories an inhibitory effect of alcohol on the proximal colon was noted and alcohol had no effect on the distal colon.

The occurence of spontaneous tumors of the mammary gland is also affected by alcohol. Schrauzer et al. (1979) showed that chronic ethanol consumption can decrease the latent period and increase the tumor volume of spontaneously occurring mammary adenocarcinoma in female C3H/St mice.

Possible Mechanisms Responsible for the Cocarcinogenicity of Alcohol

General Systemic Mechanisms

As shown in Fig. 2, ethanol is capable of affecting carcinogenesis at different stages during initiation and promotion. More than 90% of environmental carcinogens exist in their procarcinogenic form and require further activation by microsomal cytochrome-*P*-450-dependent enzymes (Miller and Miller 1959; Weisburger 1975). The activated procarcinogens exhibit a high capacity to bind to macromolecules such as DNA, RNA, or proteins and thus lead to initiation of the carcinogenic process. Activation of many procarcinogens by microsomal enzymes is an obligatory event for mutagenesis in microbiological assays and cell transformation in tissue cultures (Weisburger 1975). Induction of microsomal enzyme activity increases the mutagenic effect of many compounds in the Ames *Salmonella* mutagenesis assay (Popper et al. 1973; Ames et al. 1975; Guttenplan and Garro 1977). Since the extent of metabolic activation of various secondary carcinogens can be correlated with microsomal enzyme activity (Conney et al. 1971; Czygan et al. 1974; Felton and Nebert 1975; Guttenplan et al. 1976; Conney 1982), factors such as environmental pollutants, drugs, and diet which can influence the activity of this enzyme system are also expected to affect tumor formation in animals exposed to carcinogens (Marugami et al. 1967; McLean and Marshall 1971; Peraino et al. 1971; Rogers and Newburne 1971; Wattenberg 1975b; Kouri et al. 1978; Stohs and Wu 1982; Moore et al. 1983). In the light of this fact it seems important that alcohol is a well-known microsomal enzyme inducer (Mezey 1976; Lieber and DeCarli 1977; Lieber et al. 1979). The effect of alcohol on microsomal enzymes is discussed intensively by Lieber and Sato et al. in other chapters of this book (see Chaps. 2 and 10). It has

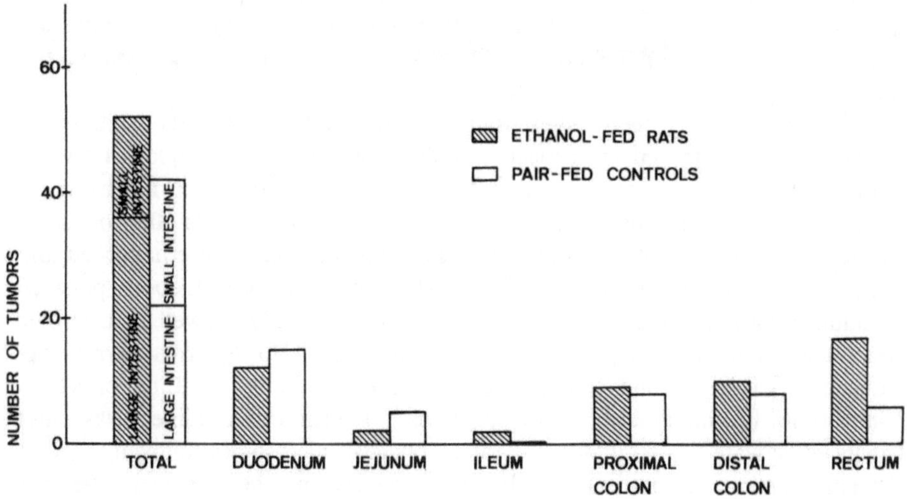

Fig. 1. Effect of chronic ethanol consumption (36% of total calories) on 1,2-dimethylhydrazine (DMH)-induced intestinal carcinogenesis in 26 pair-fed rats. Carcinogenesis was induced by 16 weekly subcutaneous injections of DMH (30 mg/kg body weight). Tumor yield was determined after 32 weeks. Tumor development in the rectum (last 5 cm of large intestine) was significantly enhanced ($p < 0.02$) following chronic ethanol ingestion (Seitz et al., 1984)

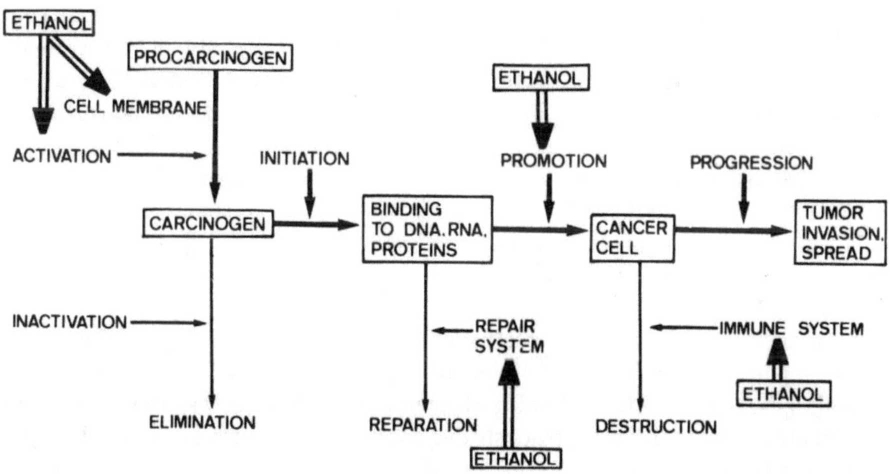

Fig. 2. Simplified scheme of two-step carcinogenesis and possible sites of action of ethanol

been demonstrated that chronic ethanol consumption increases microsomal procarcinogen-activating enzyme activities in the liver (Lieber et al. 1979; McCoy et al. 1979; Schwarz et al. 1980; Garro et al. 1981; Seitz et al. 1981a) and in the gastrointestinal tract (Seitz et al. 1979; Stohs and Wu 1982). In addition, the capacity of hepatic and intestinal microsomes to activate a variety of chemical procarcinogens to mutagens, including benzo(a)pyrene (Seitz et al. 1978; Seitz et al. 1981a),

2-aminofluorene (Seitz et al. 1981b), tryptophane pyrolysate (Seitz et al. 1981b), nitrosamines (Garro et al. 1981; Glatt et al. 1981), and aflatoxine B1 (Obidoa and Okolo 1979), has been enhanced following chronic ethanol ingestion in the animal.

Furthermore, lung microsomes from chronically ethanol-fed rats exhibited an increased capacity to activate procarcinogens present in tobacco pyrolysate when compared to microsomes from controls (Lieber et al. 1979; Seitz et al. 1981b). It has been shown that the promutagenic compound responsible for the observed difference in mutagenicity between pulmonary microsomes from ethanol-fed and control rats was *N*-nitrosopyrrolidine (AJ Garro and CS Lieber, personal communication). In addition, McCoy et al. (1981) recently showed that chronic ethanol consumption was cocarcinogenic with respect to *N*-nitrosopyrrolidine-induced nasal cavity and tracheal tumors in hamsters but had no effect on *N*-nitrosonornicotine-induced carcinogenesis; *N*-nitrosopyrrolidine was also enhanced activated to a mutagen by rat esophageal microsomes following chronic ethanol ingestion (Farinati et al. 1984b). These data seem to be of relevance because of the synergistic effect of alcohol drinking and smoking with respect to tumor development in the upper aerodigestive tract.

Although the microsomal cytochrome-*P*-450-dependent biotransformation system is essential for the activation of most chemical carcinogens, induction of this enzyme system does not necessarily entail an increased cancer risk (Peraino et al. 1971; Capel et al. 1978). This is probably related to the fact that microsomal metabolism of some compounds like benzo(a)pyrene gives rise to multiple products and that the components of the microsomal enzyme system and associated enzymes such as epoxide hydratase and glutathione transferase are also involved in the detoxification of many of the same chemicals that require activation. Ultimately, what is probably most critical in the relationship between enzyme activities and carcinogenesis is the relative effect of inducers, such as ethanol, on the steady state levels of activated carcinogens. In this context it must be mentioned that, although in vitro activation of dimethylnitrosamine by hepatic microsomes was enhanced following ethanol administration in the rat, in vivo activation using a host-mediated assay was not affected by a short-term alcohol pretreatment (Glatt et al. 1981).

Another mechanism by which ethanol may affect carcinogenesis is inhibition of DNA repair. A significant inhibition of repair of dimethylnitrosamine-induced hepatic DNA methylation following chronic ethanol feeding was found in the rat (Mufti et al. 1981). This was obviously due to an inhibitory effect of alcohol on liver O_6methylguaninetransferase (Farinati et al. 1984).

Furthermore, less specific factors involved in carcinogenesis may be influenced by alcohol administration. Malnutrition, deficiencies in vitamins (A, B_1, B_2, B_6, E, folic acid) and trace elements (Mg, Zn, Fe), hormonal disturbances, and immune system alterations may all contribute to an alcohol-mediated stimulation of cancer formation (Table 2) (Lieber et al. 1979).

Alcoholic beverages may contain a variety of procarcinogens. In parts of Africa and Puerto Rico increased concentrations of branched alcohols, aldehydes, and fusel oils have been found in home-brewed alcoholic beverages. In addition, nitrosamines, polycyclic hydrocarbons, aflatoxins, and asbestos fibers have been detected in a great

Table 2. Possible mechanisms of the cocarcinogenic effect of ethanol

1. Initiation
   - Carcinogens in alcoholic beverages
   - Enhanced membrane transport and absorption of carcinogens
   - Favoured gastric nitrosamine production
   - Enhanced activation of procarcinogens
   - Increased cell injury and cell regeneration in the presence of carcinogens
2. Promotion
   - Direct cell injury and increased cell regeneration
   - Inhibition of DNA repair
   - Change in secretion, motility, and microflora of the gastrointestinal tract
   - Change in hormone metabolism
   - Alteration of the immune system
   - Malnutrition

variety of alcoholic beverages, including certain types of beer, whisky, sake, vermouth, and sherry (Matsuda et al. 1966; Biles and Emerson 1968; Tuyns 1978; Marshall 1979; Spiegelhalder et al. 1979).

Local, Tissue-Specific Mechanisms

Upper Alimentary Tract

The upper digestive tract is exposed to ethanol concentrations several times higher than other tissues. Thus alcohol leads to direct cell alterations in these tissues. Such cellular injury is usually answered by cellular hyperregenerativity. However, tissues in an increased regenerative state exhibit an enhanced sensitivity toward chemical carcinogens (Craddock 1978; Wright 1983). In the animal experiment, ethanol administration results in a shortening of the villi and in a reduction of the cell number in the jejunum (Baraona et al. 1974). On the other hand, the number of crypt cells increases in association with an increased mitotic index reflecting enhanced DNA synthesis in the jejunum and in the ileum following alcohol administration (Baraona et al. 1974; Seitz et al. 1982b, 1983). To what extent such alterations occur in the esophagus is not known. However, DNA synthesis is stimulated in reflux esophagitis, which has frequently been observed in the alcoholic (Wienbeck and Berges 1981). Recently, Kouros et al. (1983) reported an increased methylation of DNA in the esophagus of female Wistar rats by *N*-nitrosomethylbenzylamine, a potent esophageal carcinogen, following ethanol ingestion. Morphological and functional abnormalities in the esophagus due to ethanol are discussed in detail by Wienbeck in another chapter of this book.

In addition, ethanol-induced changes in gastrointestinal secretion may influence carcinogenesis. Acute alcohol administration decreases the production and secretion of saliva and is associated with an increased viscosity (Kissin and Kaley 1974). This might result both in higher local concentrations of procarcinogens and lesser rinsing of the mucosal surface. Alcohol has also been observed to facilitate the gastric production of nitrosamines (Pignatelli et al. 1976) and to increase absorption of carcinogens from the gastrointestinal tract (Capel et al. 1981).

The importance of gastrointestinal bacteria in the pathogenesis of gastrointestinal cancer has been emphasized by various investigators (Hill 1975). However, with respect to the effect of ethanol on the gastrointestinal and especially on the colonic flora, only limited results exist. Bode et al. (1980) found a significant increase in anaerobics in the small intestine of alcoholics in comparison to a control population. It is also known that the colonic flora, when exposed to ethanol, is capable to produce a variety of toxic compounds (Levitt et al. 1982).

Liver

In man, chronic ethanol consumption results in the appearance of fatty liver, alcoholic hepatitis, and finally alcoholic cirrhosis as described in detail by Lieber and Mendenhall in other chapters of this book (see Chaps. 2 and 17). Thus cellular injury and cellular hyperregeneration may favour hepatocarcinogenesis. Schwarz et al. (1981) found an increased hepatic DNA synthesis following chronic ethanol ingestion in the rat. However, these authors could not find a difference in the levels of hepatic DNA alkylations between ethanol-fed and control rats following dimethylnitrosamine application. It must be noted that in these experiments ethanol was given in the drinking water. On the other hand, when ethanol is given as part of a liquid diet hepatic regeneration is found to be inhibited (Wands et al. 1979) or unchanged (Orrego et al. 1981), depending on the duration of ethanol application. Thus the effect of alcohol on hepatic regeneration may depend on the experimental conditions used and remains controversial (see Chap. 14).

As already mentioned, the effect of ethanol on hepatic microsomal enzymes and on the hepatic DNA repair system may play an additional role in hepatocarcinogenesis. Furthermore, liver function seems to be important also in extrahepatic carcinogenesis. Mice which received either ethanol or carbon tetrachloride and therefore developed liver damage more frequently exhibited benzo(a)pyrene-induced buccal tumors than controls (Protzel et al. 1964).

Large Intestine

Epidemiologic data suggest a significant correlation between rectal cancer and heavy alcohol consumption (Breslow and Enstrom 1974; Williams and Horm 1977; Kono and Ikeda 1979; Dean et al. 1979; Pollack et al. 1984). Such a cocarcinogenic effect of ethanol on the rectum was also observed in our animal study and by Hamilton et al. (1983), depending on the amount of alcohol in the diet.

As important pathophysiologic mechanisms in colorectal carcinogenesis, fecal excretion of bile acids and neutral steroids (Reddy et al. 1977) and their metabolism by colonic bacteria (Hill 1975), as well as colonic metabolism of procarcinogens (Fang and Strobel 1978), have been implicated. We have investigated fecal excretion of bile acids after chronic ethanol consumption in the rat. However, fecal bile acids were not found to be increased after alcohol ingestion when measured in the presence of ethanol in the blood (2.04 ± 1.35 vs 1.09 ± 0.25 mg/24 h, $p < 0.05$). This is supported by the data of other authors who found a decreased biliary secretion of bile acids due to ethanol in the rat (Maddrey and Boyer 1973) and an increased ileal

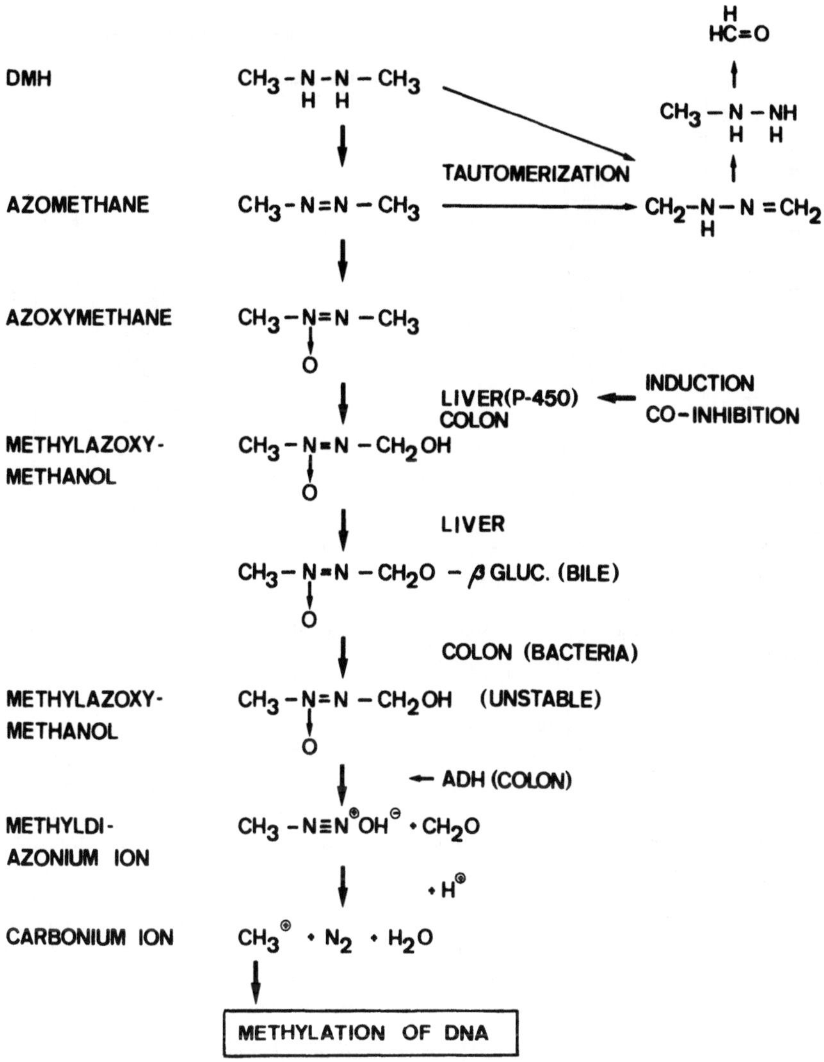

Fig. 3. Metabolism of 1,2-dimethylhydrazine *(DMH)*

absorption of bile acids in the presence of ethanol in the cat (Goebell et al. 1980). In addition, Cohen and Raicht (1981) found a similar fecal output of acidic steroids between ethanol fed and control Fisher as well as Sprague Dawley rats.

To elucidate the mechanism of the cocarcinogenic effect of ethanol in DMH-induced colorectal carcinogenesis, one also has to focus on the complex metabolism of the procarcinogen, DMH (Fig. 3). Two enzymatic steps in the activation of DMH seem to be of considerable importance:

1. The conversion of azoxymethane to methylazoxymethanol takes place in the liver (Fiala 1977) and in the colon (Wargovich and Felkner 1982; Glauert and Bennink 1983). This conversion is catalyzed by a microsomal cytochrome-*P*-450-dependent *N*-hydroxylase (Fiala 1977). Pretreatment of animals with microsomal enzyme inducers such as phenobarbital or chrysene leads to an increased metabolism of azoxymethane to carbon monoxide, probably through an induction of the

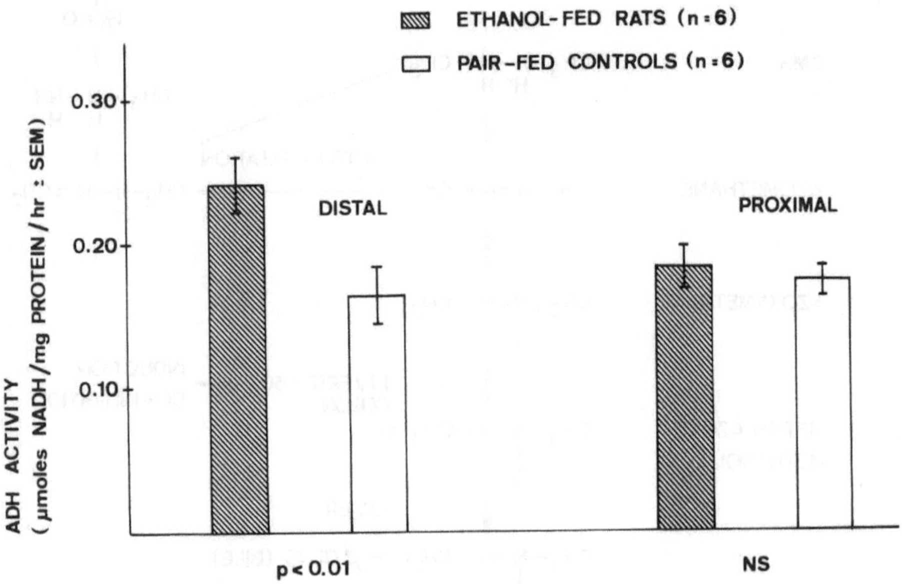

Fig. 4. Effect of chronic ethanol consumption on large-intestinal alcohol dehydrogenase *(ADH)* activity

microsomal enzyme (Fiala 1977). On the other hand, administration of disulfiram and related compounds inhibits DMH metabolism (Fiala et al. 1977) and colorectal carcinogenesis in vivo (Wattenberg 1975 a). Since ethanol is also a microsomal enzyme inducer in a variety of tissues, an increased activity of azoxymethane *N*-hydroxylase in liver and/or colon due to alcohol could theoretically at least in part explain the cocarcinogenic effect of ethanol. Since a gradient exists in the microsomal cytochrome-*P*-450-dependent activation of DMH along the large intestine (Newaz et al. 1983), chronic ethanol consumption may affect this microsomal enzyme differently in various segments of the colon, although we could not find an inducing effect of chronic ethanol feeding on colonic microsomal enzymes when mucosal microsomes were used from the total large intestine (Seitz et al. 1982 a). An enhanced hepatic microsomal enzyme induction due to ethanol, however, does not explain the stimulation of rectal tumor development compared to the remaining large intestine.

2. The last step in the activation of DMH is the conversion of methylazoxymethanol to methyldiazonium ion. It has been suggested by Schoenthal (1973) that this conversion is enzymatically catalyzed by alcohol dehydrogenase (ADH). Indeed, several observations support this concept:
   a) In the isolated perfused rat liver, methylazoxymethanol is metabolized much faster than would be expected from the spontaneous decomposition rate (Wolter and Frank 1982).
   b) The incidence of DMH- or methylazoxymethanol-induced intestinal tumors is paralleled by the activity of intestinal mucosal ADH (Grab and Zedeck 1977). Both tumor yield and ADH activity are highest in the large intestine and in the duodenum but low in the jejunum or ileum (Grab and Zedeck 1977; Mezey 1975).
   c) Butylated hydroxyanisole inhibits both colonic carcinogenesis and colonic ADH activity (Wattenberg and Sparnens 1979).
   d) Pyrazole, a potent ADH inhibitor, reveals a protective effect on methylazoxymethanolinduced colonic carcinogenesis (Zedeck and Tan 1978; Zedeck 1980).

   However, it must be mentioned that both butylated hydroxyanisole (Stohs and Wu 1982) and pyrazole (Fiala et al. 1978) also inhibit microsomal cytochrome-*P*-450-dependent enzyme activities.

Because of these facts we have measured ADH activity in the cytoplasm of the colonic mucosa. Chronic ethanol consumption increases ADH activity significantly in the distal colorectum but not in the proximal part of the large intestine (Fig. 4). It remains unclear why chronic ethanol ingestion affects colonic mucosal ADH activity differently along the large intestine. Recent immunohistochemical studies by Pestalozzi et al. (1983) show a correlation between the amount of mucosal ADH and the number of mucosal cells coming into contact with the intestinal lumen. However, whether chronic ethanol administration influences cell regeneration in the rectum, leading to an increase in mucosal ADH and also to an increased sensitivity toward chemical carcinogens, or whether ethanol increases microsomal activation of DMH in the rectum appears purely speculative at present.

Addendum: Since this manuscript was prepared, additional information on the effect of ethanol on rectal cell regeneration has accumulated. Baier et al. (1984) demonstrated a significant enhancement of rectal cell regeneration and an extension of the proliferative compartment of the rectal crypt following chronic ethanol ingestion in the rat.

References

Alcohol and health. Second special report to the U.S. Congress (1975) National Institute of Mental Health, National Institute on Alcohol Abuse and Alcoholism, Department of Health, Education and Welfare Publication no. (ADM) 75-212. United States Government Printing Office, Washington DC

Ames BN, McCann J, Yamasaki E (1975) Methods for detecting carcinogens and mutagens with the *Salmonella*/mammalian-microsomes mutagenicity test. Mutat Res 31: 347–364

Baier B, Simanowski U, Czygan P, Veith S, Sulger B, Seitz HK (1984) Erhöhte Zellregeneration im Rektum der Ratte nach chronischer Alkoholgabe: Ursache der Cokarzinogenität von Alkohol? Z Gastroenterol 22: 469 (abstract)

Baraona E, Pirola RC, Lieber CS (1974) Small intestinal damage and changes in cell population produced by ethanol ingestion in the rat. Gastroenterology 66: 226–234

Biles B, Emerson TR (1968) Examination of fibers in beer. Nature 219: 93–94

Bode JCh, Heidelbach R, Bode Ch, Mannheim W, Dürr HK, Martini GA (1980) Bacterial microflora in the jejunum of chronic alcoholics. In: Stock C, Sarles H (eds) Alcohol and gastrointestinal tract. INSERM, Paris, pp 451–456

Brechot C, Nalpas B, Courouce AM, Duhamel G, Callard P (1982) Evidence that hepatitis B virus has a role in liver cell carcinoma in alcohol disease. N Engl J Med 306: 1384–1387

Breslow NE, Enstrom JE (1974) Geographic correlations between mortality rates and alcohol, tobacco consumption in the United States. JNCI 53: 631–639

Bross IJ, Coombs J (1976) Early onset of oral cancer among women who drink and smoke. Oncology 33: 136–139

Burch CE, Ansari A (1968) Chronic alcoholism and carcinoma of the pancreas, a correlative hypothesis. Arch Intern Med 122: 273–275

Capel ID, Turner M, Pinnock MH, Williams DC (1978) The effect of chronic alcohol intake upon the hepatic microsomal carcinogen activation system. Oncology 35: 168–170

Capel ID, Cosier RS, Pinnock MH, Williams DC (1981) Use of inverted intestinal sacs to assess the effect of gastrointestinal insult on carcinogen absorption. Oncology 38: 243–248

Chan CH (1975) Primary carcinoma of the liver. Med Clin North Am 59: 989–994

Cohen BI, Raicht RF (1981) Sterol metabolism in the rat: effect of alcohol on sterol metabolism in two strains of rats. Alcoholism: Clin Exp Res 5: 225–229

Conney AH (1982) Induction of microsomal enzymes by foreign chemicals and carcinogenesis by polycyclic hydrocarbons. Cancer Res 42: 4875–4917

Conney AH, Welch R, Kuntzman R, Poland R, Poppers PJ, Finster M, Wolff JA, Munro-Faure AD, Peck AW, Bye A, Chang R, Jacobsen M (1971) Effects of environmental chemicals on the metabolism of drugs, carcinogens and normal body constituents in man. Ann NY Acad Sci 179: 155–172

Craddock VM (1978) Cell proliferation and induction of liver cancer. In: Remmer H, Bolt HM,

Bannasch P, Popper H (eds) Primary liver tumors. MTP Press, lancaster, pp 377–383

Czygan P, Greim H, Garro A, Schaffner F, Popper H (1974) The effect of dietary protein deficiency on the ability of isolated hepatic microsomes to alter the mutagenicity of a primary and a secondary carcinogen. Cancer Res 34: 119–123

Dean G, MacLennan R, McLoughlin H, Shelley E (1979) The cause of death of blue collar workers at a Dublin brewery 1954–1973. Br J Cancer 40: 581–589

Durbec JP, Bidart JM, Sarles H (1980) Interaction between alcohol and other foodstuffs. Epidemiological aspects. In: Stock C, Sarles H (eds) Alcohol and the gastrointestinal tract. INSERM, Paris, pp 33–51

Elzay RP (1966) Local effect of alcohol in combination with DMBA on hamster cheek pouch. J Dent Res 45: 1788–1795

Enstrom JE (1977) Colorectal cancer and beer drinking. Br J Cancer 35: 674–683

Fang WF, Strobel HW (1978) Activation of carcinogens and mutagens by rat colon mucosa. Cancer Res 38: 2939–2944

Farinati S, Espina N, Lieber CS, Garro AJ (1984a) Ethanol mediated inhibition of DNA repair: a specific effect on liver O_6 methyl guanine transferase. Gastroenterology 86: (part 2) 1317 (abstract)

Farinati F, Zhou ZC, Lieber CS, Garro AJ (1984b) Chronic ethanol consumption increases microsomal cytochrome P-450 and the activation of nitrosopyrrolidine to a mutagen in rat esophagus. Gastroenterology 86: (part 2) 1073 (abstract)

Feldmann JG, Hazan M, Nagaranjan M, Kissin B (1975) A case control investigation of alcohol, tobacco and diet in head and neck cancer. Prev Med 4: 444–463

Felton JS, Nebert DW (1975) Mutagenesis of certain activated carcinogens in vitro associated with genetically mediated increases in monooxygenase activity and cytochrome P-450. J Biol Chem 250: 6769–6778

Fiala ES (1977) Investigations into the metabolism and mode of action of the colon carcinogen 1,2 dimethylhydrazine and azoxymethane. Cancer 40: 2436–2445

Fiala ES, Bobota G, Kulakis C, Wattenberg W, Weisburger JH (1977) Effects of disulfiram and related compounds on the metabolism in vivo of the colon carcinogen 1,2 dimethylhydrazine. Biochem Pharmacol 26: 1763–1768

Fiala ES, Kulakis C, Christiansen G, Weisburger JH (1978) Inhibition of the metabolism of the colon carcinogen azomethane by pyrazole. Cancer Res 38: 4515–4521

Flamant R, Lassarre O, Lazar P, Leguerinais J, Denoix P, Schwartz D (1964) Differences in sex ratio according to cancer site and possible relationship with use of tobacco and alcohol: review of 65,000 cases. JNCI 32: 1309–1316

Fortier RA (1975) Étude sur les facteurs étiologiques des cancers buccauxpharynges dans le province de Quebec. J Can Dent Assoc 41: 235–241

Garro AJ, Seitz HK, Lieber CS (1981) Enhancement of dimethylnitrosamine metabolism and activation to a mutagen following chronic ethanol consumption. Cancer Res 41: 120–124

Gibel W (1967) Experimentelle Untersuchungen zur Synkarzinogenese beim Ösophaguskarzinom. Arch Geschwulstforsch 30: 181–189

Glatt H, de Balle L, Oesch F (1981) Ethanol or acetone pretreatment of mice strongly enhances the bacterial mutagenicity of dimethylnitrosamine in assays mediated by liver subcellular fractions, but not in host mediated assays. Carcinogenesis 2: 1057–1061

Glauert HP, Bennink MR (1983) Metabolism of 1,2 dimethylhydrazine by cultured rat colon epithelial cells. Nutr Cancer 5: 78–86

Goebell H, Hartmann W, Schulenburg F, Hotz J (1980) Enhanced absorption of bile acids from the gut in the presence of ethanol in the blood. In: Stock C, Sarles H (eds) Alcohol and the gastrointestinal tract. INSERM, Paris, pp 463–468

Gottfried EB, Steller R, Paronetto F, Lieber CS (1982) Spontaneous regression of hepatocellular carcinoma. Gastroenterology 82: 770–774

Grab DJ, Zedeck MS (1977) Organ specific effects of the carcinogen methylazoxymethanol related to metabolism by nicotinamide adenosine dinucleotide dependent dehydrogenases. Cancer Res 37: 4182–4190

Graham S, Dayal H, Rohrer T, Swanson M, Sultz H, Shedd D, Fishman S (1977) Dentition, diet, tobacco and alcohol in the epidemiology of oral cancer, JNCI 59: 1611–1616

Gregoriades G (1974) Cancer of the endolarynx: analysis of 415 cases. J Laryngol Otol 88: 749–755

Griciute L, Castegnaro M, Bereziat JC (1981) Influence of ethyl alcohol on carcinogenesis with *N*-nitroso-dimethylamine. Cancer Lett 13: 345–352

Guttenplan JB, Garro AJ (1977) Factors affecting the induction of dimethylnitrosamine demethylase by Aroclor 1254. Cancer Res 37: 329–330

Guttenplan JB, Hutterer F, Garro AJ (1976) Effect of cytochrome P-448 and P-450 inducers on microsomal dimethylnitrosamine demethylase activity and the capacity of isolated microsomes to activate dimethylnitrosamine to a mutagen. Mutat Res 35: 415–422

Habs M, Schmähl D (1981) Inhibition of the hepatocarcinogenic activity of diethylnitrosamine (DENA) by ethanol in rats. Acta Gastroenterol 28: 242–244

Hakulinnen T, Lehtimäki L, Lehtonen M, Teppo L (1974) Cancer morbidity among two male cohorts with increased alcohol consumption in Finland. JNCI 52: 1711–1714

Hamilton SR, Hyland J, McAvinchey D, Boitnott JK (1983) Effect of dietary beer or ethanol on experimental colonic carcinogenesis. Gastroenterology 84: (part 2) 1180 (abstract)

Henefer EP (1966) Ethanol, 30 per cent and hamster pouch carcinogenesis. J Dent Res 45: 838–844

Hill MJ (1975) The role of colon anaerobes in the metabolism of bile acids and steroids and its relation to colon cancer. Cancer 36: 2387–2400

Horie A, Kohchi S, Kuratsune M (1965) Carcinogenesis in the esophagus II: experimental production of esophageal cancer by administration of ethanolic solutions of carcinogens. Gann 56: 429–441

Ishii K, Nakumara K, Ozaki H, Yamada N, Takeuchi T (1968) Epidemiological problems of pancreatic cancer. Jp J Clin Med 26: 1839–1842

Jensen OM (1979) Cancer morbidity and cause of death among Danish brewery workers. Int J Cancer 23: 454–463

Keller AZ (1967) Cirrhosis of the liver, alcoholism and heavy smoking associated with cancer of the mouth and pharynx. Cancer 20: 1015–1022

Ketcham AS, Wexler H, Mantel N (1963) Effects of alcohol in mouse neoplasia. Cancer Res 23: 667–670

Kissin B, Kaley MM (1974) Alcohol and cancer. In: Kissin B, Begleiter H (eds) The biology of alcoholism, vol 3. Plenum, New York, pp 481–511

Klatsky AL, Friedmann GD, Siegelaub AB (1981) Alcohol and mortality. A ten year Kaiser-permanente experience. Ann Intern Med 95: 139–145

Kono S, Ikeda M (1979) Correlation between cancer mortality and alcoholic beverages in Japan. Br J Cancer 40: 449–455

Kouri RE, Rude TH, Joglekar R, Dansette DM, Jerina DM, Atlas SA, Owens LS, Nebert DW (1978) 2,3,7,8-Tetrachlorodibenzo-*p*-dioxin as cocarcinogen causing 3-methylcholanthrene-initiated subcutaneous tumors in mice genetically "nonresponsive" at Ah locus. Cancer Res 38: 1777–2783

Kouros M, Mönch W, Reiffer J, Dehnen W (1983) The influence of various factors on the methylation of DNA by the esophageal carcinogen *N*-nitrosomethylbenzylamine I. The importance of alcohol. Carcinogenesis 4: 1081–1084

Lee FI (1966) Cirrhosis and hepatoma in alcoholics. Gut 7: 77–85

Lester D (1979) Modes of alcohol administration appropriate for the study of the role of alcohol in carcinogenesis. Cancer Res 39: 2891–2893

Levitt MD, Doizaki W, Levine AS (1982) Hypothesis: metabolic activity of the colonic bacteria influences organ injury from ethanol. Hepatology 2: 598–600

Lieber CS, DeCarli LM (1970a) Quantitative relationship between the amount of dietary fat and the severity of the alcoholic fatty liver. Am J Clin Nutr 23: 474–478

Lieber CS, DeCarli LM (1970b) Hepatic microsomal ethanol oxidizing system. In vitro characteristics and adaptive properties in vivo. J Biol Chem 245: 2505–2512

Lieber CS, DeCarli LM (1977) Metabolic effects of alcohol on the liver. In: Lieber CS (ed) Metabolic aspects of alcoholism. MTP Press, Lancaster, pp 31–79

Lieber CS, Seitz HK, Garro AJ, Worner TM (1979) Alcohol related diseases and carcinogenesis. Cancer Res 39: 2863–2886

Lieber CS, Seitz HK, Garro AJ, Worner TM (1981) Alcohol as a cocarcinogen. In: Berk PD, Chalmers TC (eds) Frontiers of liver disease. Thieme, Stuttgart, pp 320–335

Lowenfels AB (1979) Alcohol and cancer. A review and update. Br J Alcohol Alcoholism 14: 148–153
MacDonald WC (1972) Chemical and pathological features of adenocarcinoma of the gastric cardia. Cancer 29: 724–732
Maddrey WC, Boyer JC (1973) The acute and chronic effect of ethanol administration on bile secretion in the rat. J Lab Clin Med 82: 215–225
Marshall E (1979) Carcinogens in Scotch. Science 205: 768–769
Marugami M, Ito N, Konishi Y, Hiaso Y, Faber E (1967) Influence of 3-methylcholanthrene on liver carcinogenesis in rats ingested DL-ethionine, 3-methyl-4-dimethyl-aminoazobenzene and *N*-2-fluorene-acetamide. Cancer Res 27: 2011–2019
Matsuda Y, Mori K, Hirohata T, Kuratsune M (1966) Carcinogenesis in the esophagus III: polycyclic hydrocarbons and phenols in whiskey. Gan 57: 549–557
McCoy GD (1978) A biochemical approach to the etiology of alcohol related cancers of the head and neck. Laryngoscope [Suppl 8] 88: 59–62 (part 2)
McCoy GD, Wynder EL (1979) Etiological and preventive implications in alcohol carcinogenesis. Cancer Res 39: 2844–2850
McCoy GD, Chen CB, Hecht SS (1979) Enhanced metabolism and mutagenesis of nitrosopyrrolidine in liver fractions isolated from chronic ethanol consuming hamsters. Cancer Res 39: 793–796
McCoy GD, Hecht SS, Katayama S, Wynder EL (1981) Differential effect of chronic ethanol consumption on the carcinogenicity of *N*-nitrosopyrrolidine and *N'*-nitrosonornicotine in male Syrian golden hamsters. Cancer Res 41: 2849–2854
McLean AEM, Marshall A (1971) Reduced carcinogenic effect of aflatoxine in rats given phenobarbitone. Br J Exp Pathol 52: 322–329
McMichael AJ (1978) Increases in laryngeal cancer in Britain and Australia in relation to alcohol and tobacco consumption trends. Lancet 1: 1244–1247
Mendenhall CL, Chedid LA (1980) Peliosis hepatis: its relationship to chronic alcoholism, aflatoxin B1 and carcinogenesis in male Holtzman rats. Dig Dis Sci 25: 587–592
Mezey E (1975) Intestinal function in chronic alcoholism. Ann NY Acad Sci 252: 215–227
Mezey E (1976) Ethanol metabolism and ethanol drug interactions. Biochem Pharmacol 25: 869–875
Miller EC, Miller JA (1959) Biochemistry of carcinogenesis. Annu Rev Biochem 28: 291–300
Moore MA, Hacker HJ, Kunz HW, Bannasch P (1983) Enhancement of NNM-induced carcinogenesis in the rat liver by phenobarbital: a combined morphologic and enzyme histochemical approach. Carcinogenesis 4: 473–479
Mufti SJ, Lieber CS, Garro AJ (1981) Effect of chronic ethanol consumption on the extent and persistence of DMN induced DNA alkylation. Cancer Res 21: 58 (abstract)
Newaz SN, Fang WF, Strobel HW (1983) Metabolism of the carcinogen 1,2 dimethylhydrazine by isolated human colon microsomes and human colon tumor cells in culture. Cancer 52: 794–798
Obidoa O, Okolo TC (1979) Effect of ethanol administration on the metabolism of aflatoxin B1. Biochem Med 22: 145–148
Ohnishi K, Iida S, Iwana S (1982) The effect of chronic habitual alcohol intake on the development of liver cirrhosis and hepatocellular carcinoma-reaction to hepatitis B surface antigen. Cancer 49: 672–677
Okuda K, Ohnishi K (1981) Pancreatic cancer and alcohol. In: Leevy CM (ed) Alcohol and the gastrointestinal tract. Clinics in gastroenterology. Saunders, Philadelphia, pp 479–484
Orrego H, Crossley IR, Saldiria Y, Medline A, Varghese G, Israel Y (1981) Long term ethanol administration and short- and long-term liver regeneration after partial hepatectomy. J Lab Clin Med 97: 221–230
Peraino C, Frey RJM, Steffelt E (1971) Reduction and enhancement by phenobarbital of hepatocarcinogenesis induced in the rat by 2-acetataminofluorene. Cancer Res 31: 1506–1512
Pestalozzi DM, Bühler R, Von Warthburg JP, Hess M (1983) Immunohistochemical localisation of alcohol dehydrogenase in the human gastrointestinal tract. Gastroenterology 85: 1011–1016
Pignatelli B, Castegnaro M, Walker EA (1976) Effects of gallic acid and of ethanol on formation of nitrosodiethylamine. IARC Sci Publ 14: 173–178

Pollack ES, Nomura AMY, Heilbrun LK, Stemmermann GN, Green SB (1984) Prospective study of alcohol consumption and cancer. N Engl J Med 310: 617–621
Popper H, Czygan P, Greim H, Schaffner F, Garro AJ (1973) Mutagenicity of primary and secondary carcinogens altered by normal and induced hepatic microsomes. Proc Soc Exp Biol Med 142: 727–729
Protzel M, Giardina AC, Albano EH (1964) The effect of liver imbalance on the development of oral tumors in mice following the application of benzo(a)pyrene or tobacco tar. Oral Surg 18: 622–635
Radike MJ, Stemmer KL, Brown PB, Larson E, Bingham E (1977) Effect of ethanol and vinyl chloride on the induction of liver tumors. Environ Health Perspect 21: 153–155
Reddy BS, Magat S, Sheinfil A, Weisburger JH, Wynder EL (1977) Effect of type and amount of dietary fat and 1,2 dimethylhydrazine on biliary bile acids, fecal bile acids and neutral steroids in rats. Cancer Res 37: 2132–2137
Rogers AE, Newburne PM (1971) Nutrition and aflatoxin carcinogenesis. Nature 229: 62–63
Rothman KJ, Keller A (1972) The effect of joint exposure to alcohol and tobacco on risk of cancer of the mouth and the pharynx. J Chronic Dis 25: 711–716
Schmähl D (1976) Investigations on esophageal carcinogenicity by methyl-phenylnitrosamine and ethyl alcohol in rats. Cancer Lett 1: 215–218
Schoenberg BS, Bailar JC, Fraumeni JF (1971) Certain mortality pattern of esophageal cancer in the United States. JNCI 46: 63–73
Schoenthal R (1973) The mechanism of carcinogenic nitro- and related compounds. Br J Cancer 28: 436–439
Schottenfeld D (1979) Alcohol as a co-factor in the etiology of cancer. Cancer 43: 1962–1966
Schottenfeld D, Gantt RC, Wynder EL (1974) The role of alcohol and tobacco in multiple primary cancers of the upper digestive system, larynx and lung. A prospective study. Prev Med 3: 277–293
Schrauzer GN, McGinness JE, Ishmael D, Bell LJ (1979) Alcoholism and cancer: effects of long-term exposure to alcohol on spontaneous mammary adenocarcinoma and prolactin levels in C3H/St mice. J Stud Alcohol 40: 240–246
Schwarz M, Appel KE, Schrenk, Kunz W (1980) Effect of ethanol on microsomal metabolism of dimethylnitrosamine. J Cancer Res Clin Oncol 97: 233–240
Schwarz M, Wiesbeck G, Hummel J, Kunz W (1981) Effect of ethanol on dimethylnitrosamine activation and DNA synthesis in rat liver. Carcinogenesis 3: 1071–1075
Seitz HK, Garro AJ, Lieber CS (1978) Effect of chronic ethanol ingestion on intestinal metabolism and mutagenicity of benzo(a)pyrene. Biochem Biophys Res Commun 85: 1061–1066
Seitz HK, Korsten MA, Lieber CS (1979) Ethanol oxidation by intestinal microsomes: microsomal activity after chronic ethanol administration. Life Sci 25: 1443–1448
Seitz HK, Garro AJ, Lieber CS (1981a) Sex dependent effect of chronic ethanol consumption in rats on hepatic microsome mediated mutagenicity of benzo(a)pyrene. Cancer Lett 13: 97–102
Seitz HK, Garro AJ, Lieber CS (1981b) Enhanced pulmonary and intestinal activation of procarcinogens and mutagens after chronic ethanol consumption in the rat. Eur J Clin Invest 11: 33–38
Seitz HK, Bösche J, Czygan P, Veith S, Kommerell B (1982a) Microsomal ethanol oxidation in the colonic mucosa of the rat: effect of chronic ethanol ingestion. Arch Pharmacol 320: 81–84
Seitz HK, Czygan P, Kommerell B (1982b) Stimulation of thymidine incorporation in isolated rat intestinal mucosal cells by feeding an ethanol-containing liquid diet. Digestion 23: 65–67
Seitz HK, Czygan P, Kienapfel H, Veith S, Schmidt-Gayk H, Kommerell B (1983) Changes in gastrointestinal DNA synthesis produced by acute and chronic ethanol consumption in the rat: a biochemical study. Z Gastroenterol 21: 79–84
Seitz HK, Czygan P, Waldherr R, Veith S, Raedsch R, Kässmodel H, Kommerell B (1984) Enhancement of 1,2 dimethylhydrazine induced rectal carcinogenesis following chronic ethanol consumption in the rat. Gastroenterology 86: 886–891
Spiegelhalder B, Eisenbrand G, Preussmann R (1979) Contamination of beer with trace quantities of N-nitroso-dimethylamine. Food Cosmet Toxicol 17: 29–31
Stenback F (1969) The tumorigenic effect of ethanol. Acta Pathol Microbiol Scand 77: 325–326
Stohs SJ, Wu CLJ (1982) Effect of various xenobiotics and steroids on aryl hydrocarbon hydroxylase of intestinal and hepatic microsomes from male rats. Pharmacology 25: 237–249

Tamburro CH, Hua-Ming L (1981) Primary hepatic cancer in alcoholics. In: Leevy CM (ed) Alcohol and the gastrointestinal tract. Clinics in gastroenterology. Saunders, Philadelphia, pp 457–477

Teschke R, Minzlaff M, Oldiges H, Frenzel H (1983) Effect of chronic alcohol consumption on tumor incidence due to dimethylnitrosamine administration. J Cancer Res Clin Oncol 106: 58–64

Tuyns A (1970) Cancer of the esophagus. Further evidence of the relation to drinking habits in France. Int J Cancer 5: 152–162

Tuyns A (1978) Alcohol and cancer. Alcohol Health Res World 2: 20–31

Tuyns A (1979) Epidemiology of alcohol and cancer. Cancer Res 39: 2840–2843

Tuyns A, Masse LMF (1973) Mortality from cancer of the esophagus in Brittany. Int J Epidemiol 2: 241–245

Vincent RG, Marchetta F (1963) The relationship of the use of tobacco and alcohol to cancer of the oral cavity, pharynx or larynx. Am J Surg 106: 501–505

Wands JR, Carter EA, Bucher NLR, Isselbacher KJ (1979) Inhibition of hepatic regeneration in rats by acute and chronic ethanol intoxication. Gastroenterology 77: 528–531

Wargovich MJ, Felkner IC (1982) Metabolic activation of DMH by colonic microsomes: a process influenced by type of dietary fat. Nutr Cancer 4: 146–153

Wattenberg LW (1975a) Inhibition of dimethylhydrazine induced neoplasia of the large intestine by disulfiram. JNCI 54: 1005–1006

Wattenberg LW (1975b) Effect of dietary constituents on the metabolism of chemical carcinogens. Cancer Res 35: 3326–3331

Wattenberg LW, Sparnens VL (1979) Inhibitory effects of butylated hydroxyanisole on methylazoxymethanol acetate induced neoplasia of the large intestine and on nicotinamide adenine dinucleotide dependent alcohol dehydrogenase activity in mice. JNCI 63: 219–222

Weisburger JH (1975) Chemical carcinogens. In: Casarett LJ, Duoll J (eds) Toxicology, the basic science of poisons. Macmillan, New York, pp 333–378

Weisburger JH, Yamamoto RS, Pai SR (1964) Ethanol and the carcinogenicity of *N*-hydroxy-*N*-2-fluorenyl-acetamide in male and female rats. Toxicol Appl Pharmacol 6: 363 (abstract)

Wienbeck M, Berges W (1981) Esophageal lesions in the alcoholic. In: Leevy CM (ed) Alcohol and the gastrointestinal tract. Clinics in gastroenterology. Saunders, Philadelphia, pp 375–388

Williams RR, Horm JW (1977) Association of cancer sites with tobacco and alcohol consumption and socioeconomic status of patients. Interview study from the third National Cancer Survey. JNCI 58: 525–547

Wolter S, Frank N (1982) Metabolism of 1,2 dimethylhydrazine in isolated perfused rat liver. Chem Biol Interact 42: 335–344

Wright NA (1983) The histogenesis of gastrointestinal cancer. In: Hodgson HJF, Bloom SR (eds) Gastrointestinal and hepatobiliary cancer. Chapman Hall, London, pp 79–118

Wynder EL, Bross IJ (1957) Aetiological factors in mouth cancer: an approach to its prevention. Br Med J 1: 1137–1143

Wynder EL, Bross IJ (1961) A study of etiological factors in cancer of the esophagus. Cancer 14: 389–395

Wynder EL, Mabuchi K (1973) Etiological and environmental factors in esophageal cancer. JAMA 226: 1546–1548

Wynder EL, Shigematsu T (1967) Environmental factors of cancer of the colon and rectum. Cancer 20: 1520–1561

Wynder EL, Bross IJ, Day E (1965a) Epidemiological approach to the etiology of cancer of the larynx. JAMA 160: 1384–1391

Wynder EL, Bross IJ, Day E (1956b) A study of environmental factors in cancer of the larynx. Cancer 9: 86–110

Wynder EL, Bross IJ, Feldman RM (1957) A study of etiological factors in cancer of the mouth. Cancer 10: 1300–1323

Wynder EL, Covey LS, Mabuchi K, Muskinski M (1976) Environmental factors in cancer of the larynx. Second look. Cancer 38: 1591–1601

Zedeck MS (1980) Colon carcinogenesis and the role of dehydrogenase enzyme activity: inhibition of tumorgenesis by pyrazole. Prev Med 9: 346–351

Zedeck MS, Tan QH (1978) Effect of pyrazole on tumor induction by methylazoxymethanol (MAM) acetate: relationship to metabolism of MAM. Pharmacologist 20: 174–180

12 Ethanol and Biological Membranes: Experimental Studies and Theoretical Considerations*

T. F. Taraschi and E. Rubin

Department of Pathology and Laboratory Medicine, Hahnemann University School of Medicine, Philadelphia, PA 19102, USA

Introduction

An enigma facing investigators who seek to explain the pathogenesis of ethanol-induced diseases is the wide variety of tissues and organs affected by this compound. Moreover, the effects of acute and chronic ethanol intoxication are not only different, but often of an opposite nature. Whereas acute ethanol intoxication exerts only reversible functional changes, chronic abuse may lead to significant degenerative and inflammatory diseases of many organs, including the liver, pancreas, brain, heart, and skeletal muscle (Majchrowicz and Noble 1979). Oxidation of ethanol by the liver results in an early increase in the hepatic reduced nicotinamide-adenine dinucleotide/nicotinamide-adenine dinucleotide (NADH-NAD) ratio, but after chronic consumption this ratio returns to normal (Williamson et al. 1969). In extrahepatic organs, little metabolism of ethanol takes place (Erickson 1979). While acetaldehyde, the primary metabolite of ethanol oxidation, might react with cell constituents in the liver, the organ where it is produced, circulating levels of acetaldehyde are so low that it is unlikely to play a role in diseases of other organs, e.g., heart or brain (Erickson et al. 1982). For these and other reasons, it appears unlikely that the metabolism of ethanol can be incriminated as the cause of the widespread manifestations of ethanol-induced maladies. In recent years, we and others have been intrigued by the observations which indicate that ethanol acts on all biological membranes as a physical agent, similarly to other anesthetics. All general anesthetics intercalate in or bind to biological membranes. Evidence has been accumulating that as an adaptive response to this effect of ethanol, changes in the composition of many, if not all, membranes in the body take place. Such a universal effect might be involved, at least to some degree, in the ubiquitous effects of ethanol.

Ethanol and other anesthetic agents are molecules which may differ greatly in chemical structure, but which share the common properties of inhibiting the Na^+ channel in excitable membranes (Seeman 1972) and binding to or intercalating in biological membranes. The correlation between solubility in a polar environment, such as that encountered in a biological membrane, and anesthetic potency (Seeman 1972) suggests a specific type of membrane interaction. Seeman (1972), using

* Supported in part by USPHS Grants #AA3442, #HL26903, and #AA5662

Alcohol Related Diseases in Gastroenterology
Edited by H. K. Seitz and B. Kommerell

erythrocyte plasma membranes, demonstrated that binding of ethanol to the membrane is a nonsaturable phenomenon, thus dispelling the notion that ethanol binding may be a receptor-mediated process. Moreover, the acute action of ethanol on cell membranes is neither species specific [e.g., the same alterations are seen in prokaryotic and eukaryotic cells (Michaelis and Michaelis 1983)] nor organ specific (cells of all organs are affected). Thus it seems reasonable to assume that the mode of action of ethanol is similar for all cell membranes.

One of the main strategies which has been developed in trying to unlock the mystery concerning the mechanism by which ethanol affects biological membranes has been to examine, using physical and analytical techniques, alterations in membrane chemical composition (e.g., fatty acid composition) and molecular order caused by acute and chronic alcohol treatment, and to relate these to observed functional changes. Since many membrane-bound enzymes are thought to require specific "boundary" lipids for optimal function, alterations in the physical properties of membrane lipids caused by acute and chronic ethanol intake may play a role in the attenuation of membrane function.

The emphasis of this review will be on the interactions of ethanol and other anesthetic agents with biological membranes, with particular emphasis on the chemical and physical properties of membrane lipids and proteins.

Effects of Ethanol on Membrane Fluidity in Model and Biological Membranes

Membrane Fluidity

The term "membrane fluidity" has a variety of meanings, depending on the discipline in which the investigator has been trained. Singer and Nicolson (1972), in presenting their fluid mosaic model, introduced the concept of fluidity to describe the ability of membrane lipids and proteins to diffuse laterally through and rotate within the "fluid" membrane matrix. To the physical chemist or spectroscopist who utilizes physical techniques, such as nuclear magnetic resonance (NMR), electron paramagnetic resonance (EPR), fluorescence polarization, Raman spectroscopy, and differential scanning calorimetry (DSC), this term refers to the physical state of the fatty acyl chains which comprise the hydrophobic region of the membrane. The physical state, usually referred to as the degree of "order" of the membrane, reflects the rate of motion and the angular range of motion of the acyl chains. This order, which is usually arrived at by the spectroscopist using a mathematically derived "order parameter", is sensitive to the degree of unsaturation of the fatty acyl chains and the length of the chains. In general, longer acyl chains exhibit greater motion than shorter ones, and more unsaturated fatty acids chains display greater motion than saturated ones of similar length. Membrane fluidity, or order, is also influenced by other membrane components, such as cholesterol, proteins, and polar phospholipid head groups. These interact with both the hydrophilic and the hydrophobic regions of membrane lipids. Conversely, the physical state of membrane lipids and their associated fatty acyl chains can have an effect on the function of membrane proteins and enzymes. Temperature, ionic strength, ions such as Ca^{2+} and Mn^{2+}, and pH can also have pronounced effects on membrane fluidity. Thus the fluidity of the membrane is a resultant from the complex interaction of multiple membrane entities, and can be modulated by chemical and/or structural alterations in these components, as well as by external perturbants.

Techniques Used to Monitor Membrane Fluidity

Two of the most useful physical techniques which have been extensively used to study membrane fluidity in model and biological membranes are EPR and NMR. Since the results derived from investigations using these techniques have contributed considerably to our understanding of the effects that ethanol exerts on membrane structure, it is useful to summarize these and other methods.

Electron Paramagnetic Resonance

The incorporation of stable free radicals (spin probes) into biological membranes enables one to study specific environments within the membrane. The spin label has a three-line nitrogen hyperfine structure, the splitting of which varies with the orientation of the magnetic field relative to the nitroxide axes. It is this spectral anisotropy which has rendered spin label EPR a powerful tool for studying molecular motions which are the characteristic features of the highly dynamic structure of biological membranes. In studies dealing with model and biological membranes, the probe employed is usually a fatty acid, such as 5-12,- or 16-doxyl stearic acid; a cholesterol-like molecule, such as 3-doxyl cholestane; or a phospholipid molecule which has a stable nitroxide free radical attached to it, either in the hydrocarbon chain or the polar headgroup region. The nitroxide group can be attached in a stereospecific way to a fatty acid or phospholipid molecule, thus offering many different levels at which to probe the membrane of interest.

It is well established that phospholipids present in lipid bilayers undergo anisotropic motion in the form of side-to-side "wobbling" (e.g., trans-gauche isomerizations), as well as fast long axis rotation. Thus spectra arising from probes incorporated into biological membranes yield information which reflects this anisotropy. Using EPR spectroscopy, two characteristic hyperfine splittings, $A_{\parallel}$ and $A_{\perp}$, can be directly measured in systems sensitive to the rapid molecular motions of the EPR time scale (correlation time, $\tau\ c < 3\times10^{-9}$s). $A_{\parallel}$ is the experimentally observed hyperfine splitting when the external magnetic field H_o is applied along the main axis (z') of the probe molecule. Conversely, $A_{\perp}$ is the hyperfine splitting which corresponds to membrane areas where the static magnetic field is perpendicular to the main axis (z'). The measurement of these splittings is illustrated in Fig. 1. In the case of a natural biological membrane, all possible orientations of the bilayer normal between the two extremes of parallel and perpendicular are present and thus give rise to a spectral intensity in the intermediate regions between the parallel and perpendicular splittings (for extensive review see Marsh 1981).

The hyperfine splittings, $A_{\parallel}$ and $A_{\perp}$, describe the anisotropy of a certain environment within a membrane. This has led to the determination of a spin label

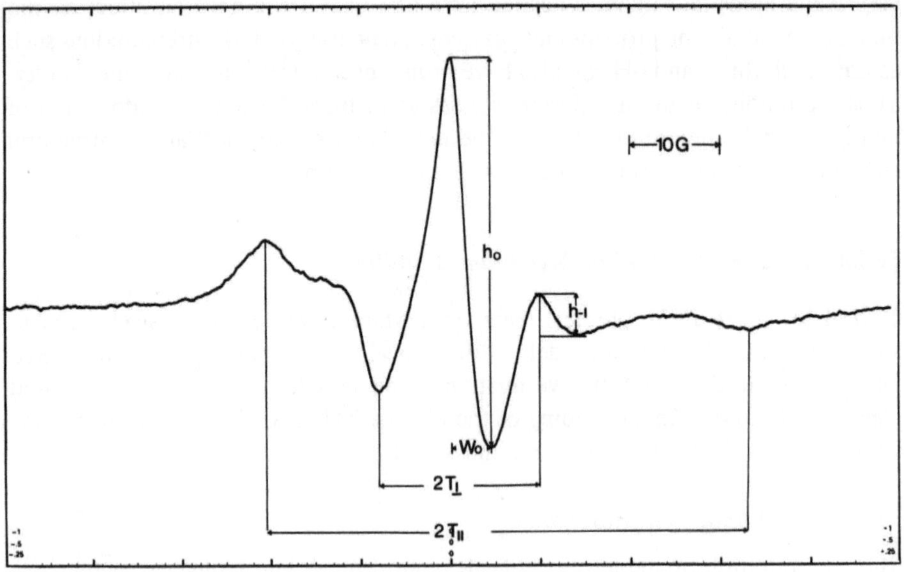

Fig. 1. Electron paramagnetic resonance spectra of mitochondrial membranes (40 mg protein/ml) labeled with 5-doxylstearic acid. Measurements of the outer hyperfine extrema ($2T_{\|}$) and inner hyperfine splittings ($2T_{\perp}$) and the time-dependent parameters of midline width (W_o), midline height (h_o), and high field height ($h_{-\perp}$) are indicated. Spectral scan, 100 G; midfield-line, 3,250 G; modulation amplitude, 0.63 G; microwave power, 5 mW, average of four scans

order parameter, S, by Hubbell and McConnell (1971). S is effectively the anisotropy ($A_{\|}-A_{\perp}$) normalized to the maximum possible anisotropy for a rigidly oriented (no motion) situation, such as that encountered in a single crystal. This order parameter is defined as:

$$S = \frac{A_{\|} - A_{\perp}}{A_{zz} - (1/2)\,(A_{xx} + A_{yy})} \cdot \frac{a_o'}{a_o}$$

where $a_o' = 1/3\,(A_{zz} + A_{yy} + A_{xx})$ and $a_o = 1/3\,(A_{\|} + 2A_{\perp})$ and A_{zz}, A_{yy}, and A_{xx} are the principal hyperfine splittings which correspond to the main (molecular) axis of the spin probe.

In situations where the spin label motion is too rapid for accurate measurement of the outer hyperfine splitting ($2A_{\|}$), the order parameter can be estimated from the inner hyperfine extreme ($2A_{\perp}$) using the equation developed by Gaffney (1974).

$$S = \left(\frac{43.7\,G - 3\,A_{\perp}}{46.1}\right) \times 1.723.$$

Nuclear Magnetic Resonance

^{2}H NMR

Deuterium (^{2}H) NMR has been widely used to characterize membrane structure in model and biological membranes (for review see Seelig 1977). Using chemical or biochemical means, it is possible selectively to replace ^{1}H by ^{2}H in lipid molecules. Replacement by ^{2}H, which is a very small atom and causes little, if any, perturbation, does not affect the molecular organization in the membrane. Since virtually all parts of a lipid molecule can be labeled, ^2H NMR has provided detailed insight into the conformational and motional properties of the polar headgroup region, glycerol backbone, and fatty acyl chains of membrane phospholipids. Anisotropic motion is readily detected by ^2H NMR, since for large molecules, such as phospholipids, which undergo anisotropic motion, the signal of each deuteron is a doublet, owing to the ^2H quadrupole splitting, $\Delta\nu Q$, which can be used to calculate the ^2H order parameter, S_{CD}, according to:

$$\Delta\widehat{\nu Q} = (3/4)\left(\frac{e^2 qQ}{h}\right) S_{CD}$$

where $e^2 qQ/h$ is the static quadrupole coupling constant.

Determination of the quadrupole splitting for CD_2 groups at a number of positions along a fatty acyl chain in a phospholipid has revealed that splitting is most pronounced where the degree of order is greatest. This occurs in the half of the chain closest to the phosphoryl glycerol head group. CD_2 groups near the end of the chain can undergo greater motion, and consequently the splitting and degree of order are much smaller (Davis 1983).

^{31}P NMR

^{31}P NMR has proven to be a powerful technique for studying membrane structure (for review see Seelig 1978). Since most membrane phospholipids contain a single phosphate in the polar head group, and since the ^{31}P isotope is 100% naturally abundant, ^{31}P NMR provides a nonperturbing probe of a known region of the membrane. ^{31}P NMR has also been shown to provide a convenient method of discriminating quantitatively between the most important phases (e.g., bilayer, hexagonal (H_{II}), isotropic, rhombic, cubic) found for membrane phospholipids. The ^{31}P NMR spectrum contains contributions from both structural and motional origins; thus its use to obtain structural information is extrapolative.

The lineshape of the ^{31}P NMR signal of phospholipids is determined both by the chemical shift anisotropy (CSA) of the lipid phosphate moiety and the $^1H-^{31}P$ dipolar interactions. Therefore, phospholipids present in different structural states (e.g., gel, liquid-crystalline, and intermediate phases) which experience different degrees of motion will give rise to different ^{31}P NMR lineshapes. Unlike EPR and ^2H NMR, analysis of data obtained by ^{31}P NMR does not involve the use of an order parameter. In ^{31}P NMR experiments, fluidity is often measured by plotting the value obtained for the chemical shift anistotropy, which is sensitive to the motion and therefore to the structural environment of the phospholipid headgroup region, versus the desired parameter (e.g., temperature, ethanol concentration). As a general rule, the smaller the CSA, the greater the motion.

Fluorescence Polarization

Fluorescent probe molecules have been extensively employed to study the physical properties of model and biological membranes (Shinitzky and Barenholz 1974, 1978). These probes yield information concerning both the microviscosity and polarity of their environment. Similarly to EPR probes and ^2H-labeled molecules, fluorescent probes allow specific membrane regions to be monitored.

The basic principle involved in obtaining information by fluorescence polarization experiments relates to the fact that plane-polarized light is used to excite a fluorescent system, such as a membrane containing 1,6-diphenyl-1,3,5-hexatriene (DPH), and the linearly polarized components of the emission are detected. In a typical experiment, light is incident along the x axis and is detected along the y axis. The incident light is polarized along the z axis. Two components of the emitted light are measured. $I_{\parallel}$ is polarized along the z axis and $I_{\perp}$ along the x axis. No emission propagates along the y axis that is also polarized along this axis, since light is a transverse wave. In practice, $I_{\parallel}$ and $I_{\perp}$ are measured and compared. Two convenient comparisons ($I_{\parallel}$ and $I_{\perp}$) are the polarization (P) and the anisotropy (A), which are defined as follows:

$$P = \frac{(I_{\parallel} - I_{\perp})}{(I_{\parallel} + I_{\perp})}\,; \quad A = \frac{(I_{\parallel} - I_{\perp})}{(I_{\parallel} + 2\,I_{\perp})}$$

In experiments involving biological membranes, the fluorescence polarization (P) is usually plotted versus temperature, or in the case of the experiments of interest to this review, the concentration of ethanol. P correlates, in general, with molecular order. Thus a decrease in P is interpreted as indicating decreased order (increased fluidity).

Effects of Ethanol on Model Membranes

It is well established that the addition of ethanol and other similar agents (e.g., anesthetics) increases the fluidity of model liposomal membranes. In such systems, this fluidization is reflected in a decrease in the temperature of the gel to liquid-crystalline phase transition (Jain and Wu 1977).

Most of the information which is presently available regarding the effect of ethanol on membrane structure has been obtained from studies which utilize electron spin resonance. Michaelis et al. (1980) have investigated the effect of ethanol on the membrane structure and phase transition of one-component (egg phosphatidylcholine [PC]) and multiple-component (bovine brain phospholipids) model membranes. The amplitude of motion of a series of nitroxide-bearing probes, labeled at different positions on the hydrocarbon chain, were incorporated into the vesicles, after which ethanol was added in increasing amounts. It was reported that ethanol, at concentrations $> 0.2\,M$, fluidized the membrane, and that this effect was more pronounced with increasing distance of the label from the phospholipid headgroup region. Using 5-doxyl stearic acid, these authors also demonstrated that at 23° C ethanol decreased fatty acid chain mobility near the surface of the bilayer. Similar effects were observed for egg PC and bovine brain phospholipids, when the intercalated spin probe was 3-doxyl cholestane. Interestingly, vesicles composed of dipalmitoyl phosphatidylcholine (DPPC), which unlike egg PC exist in the gel phase at room temperature, were not disordered when similar amounts of ethanol were added. On the basis of this result, these authors suggested that lipid packing may be an important consideration when assessing the mechanism by which ethanol fluidizes the membrane.

The presence of ethanol has also been shown to increase membrane fluidity by other physical techniques. Jain and Wu (1977) reported an ethanol-induced decrease in the transition temperature for DPPC liposomes by DSC. Longer-chain alcohols were shown not only to decrease the transition temperature, but also to broaden the transition heat capacity profile. Hill (1974) reported changes in light scattering, a finding indicative of a decrease in the transition temperature of DPPC liposomes following addition of increasing amounts of ethanol. Lee also found a decreased phase transition temperature by fluorescence techniques when ethanol was added to DPPC (1976).

In recent years Cullis and co-workers (Cullis et al. 1980) have examined the effects of ethanol and other agents on the polymorphic properties of membranes using NMR techniques. They investigated the effects of a series of alcohols and anesthetics on the phase organization of egg phosphatidylethanolamine (PE), a phospholipid which exists in a bilayer organization below 25° C and the hexagonal (H_{II}) phase above 25° C. Ethanol and butanol were found to stabilize the bilayer structure for egg PE, whereas longer-chain ($n > 6$) alcohols and alkanes strongly promoted hexagonal (H_{II}) phase formation (Hornby and Cullis 1981).

Dibucaine and chlorpromazine, two local anesthetics, have been shown to induce liquid-crystalline bilayer hexagonal (H_{II}) phase transitions in beef heart cardiolipin (Cullis et al. 1978) and dioleoylphosphatidic acid (Verkleij et al. 1982). Chlorpromazine, dibucaine, tetracaine, and procaine were reported to stabilize bilayer structures when added to aqueous dispersions of unsaturated PE, which otherwise would exist in a nonbilayer hexagonal (H_{II}) phase. On the basis of these results, the authors suggested that the ability of these molecules to affect the phase structure of PE was related to their dynamic shape.

Effects of Ethanol on Biological Membranes In Vitro

Incubation of ethanol in vitro with brain synaptosomes and erythrocyte ghosts has been shown by EPR to increase the fluidity of these membranes (Chin and Goldstein 1977). Using 5-doxyl stearic acid there was a progressive decrease in the order parameter with increasing concentrations (20–350 mM) of ethanol. In agreement with these results, the fluorescence polarization of DPH, and hence the molecular order, was decreased when synaptic plasma membranes were exposed in vitro to low concentrations (10–20 mM) of ethanol (Harris and Schroeder 1981). Since this technique reports on the environment neighboring the fluorescent probe, and DPH is thought to partition near the core of the bilayer, these results suggest that the interior acyl chains of the membrane underwent increased motion in the presence of ethanol. When a fluorescent probe, such as 1-aminopyrene, which partitions near the surface of the membrane, was added, low concentrations of ethanol did not affect the fluorescence polarization (Harris and Schroeder 1981).

Similar fluidizing effects of ethanol on mitochondrial membranes have been reported from this laboratory (Waring et al. 1981, 1982). Mitochondrial membranes labeled with 5-doxyl stearic acid exhibited a progressive decrease in the order parameter as a result of increasing the concentration of ethanol (0.05–1.0 M) (Fig. 2). These observations have been confirmed by Lenaz et al. (1976).

The disordering effects of ethanol on lipid fatty acyl chain motion, which have been described above for model systems and biological membranes, have also been

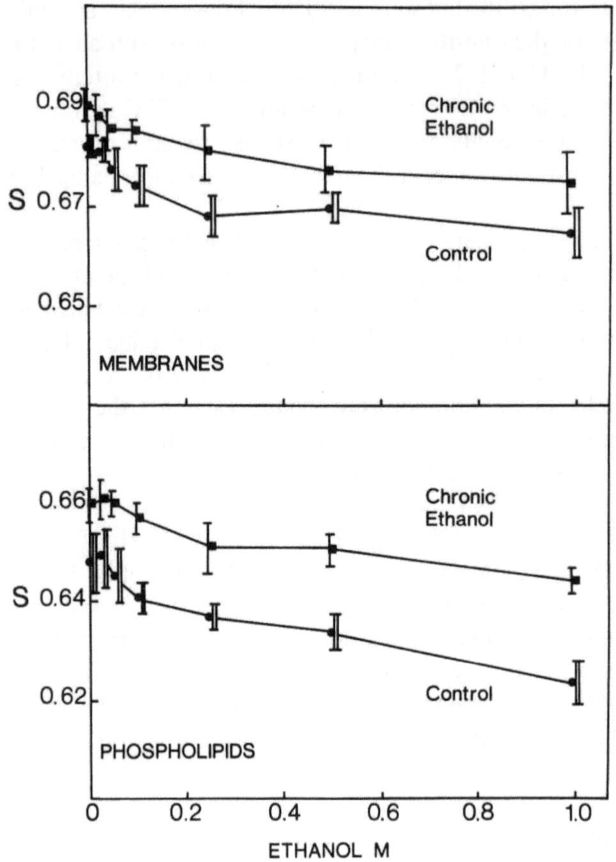

Fig. 2. The effect of ethanol, in vitro, on the order parameter of 5-doxyl stearic acid in mitochondrial membranes and phospholipids from control and ethanol-fed rats at 15° C. Points and bars represent the mean ± SD of four scans

observed in liposomes prepared from the total lipid extracts of biological membranes. Liposomes made of phospholipids isolated from the mitochondrial membranes of normal rat livers and labeled with 5-doxyl stearic acid exhibited the same sensitivity to ethanol as did intact mitochondrial membranes (27). Johnson et al. (1979) and Harris and Schroeder (1981) demonstrated that exposure of liposomes prepared from the total lipid extract of brain synaptic plasma membranes to 20–80 m*M* ethanol caused a significant decrease in the fluorescence polarization of membrane-bound DPH. However, higher concentrations of ethanol were needed to induce the changes in fluorescence polarization which were reported for intact membranes. The opposite, i.e., that the liposomes formed from the lipid extracts of natural membranes are more sensitive to the effects of ethanol than the intact membrane, has been reported for myelin membrane systems (Harris and Schroeder 1981). The interpretation of these two seemingly contrary results may well be related to cholesterol content and the interaction of differing phospholipids of the two membranes and their complex interaction with cholesterol. In this regard, Chin and Goldstein (1981) have reported that addition of increasing amounts of cholesterol to liposomes of egg PC attenuates the fluidizing effect of ethanol. Thus cholesterol, a molecule which in itself has differing effects on lipid structure, depending on the lipid(s) with which it is associated, may play a role in sensitizing the membrane to the action of ethanol.

Effect of Chronic Ethanol Ingestion of Membrane Fluidity

The effect of chronic ethanol consumption on biological membranes is opposite to that produced by the presence of ethanol, an effect which presumably reflects an adaptive response. Chronic ethanol intoxication leads to resistance to the membrane-disordering effects of acute exposure to ethanol. This phenomenon was first reported by Chin and Goldstein (1977), who studied erythrocyte and synaptic plasma membranes from mice chronically exposed to ethanol. In EPR studies, exposure of these membranes to ethanol in vitro (20–150 m*M*) produced a much smaller decrease in the order parameter than was observed when membranes from untreated animals were titrated with ethanol in the some fashion. Interestingly, the baseline order parameter derived from the 5-doxylstearic acid spectrum was comparable in membranes from control and alcoholic animals.

We have reported similar data in studies of rat liver mitochondria (Waring et al. 1981), and microsomes (Ponnappa et al. 1982) (Fig. 3) and pancreatic plasma membranes (B. C. Ponnappa, personal communication) obtained from animals chronically exposed to ethanol. Liposomes prepared from hepatic mitochondrial phospholipids extracted from alcoholic rats displayed the same smaller disordering effect when compared to liposomes of phospholipids obtained from control mitochondria. Thus, since mitochondria contain negligible amounts of cholesterol, it is not likely that in mitochondrial membranes cholesterol plays a significant role in this adaptive phenomenon.

The resistance to disordering by ethanol as a result of chronic ethanol treatment has also been demonstrated by fluorescence polarization of DPH in liposomes prepared from the total lipid extracts of brain synaptosomal membranes (Johnson et al. 1979). In these studies, cholesterol was suggested to be responsible for these observed differences. Recently, Demediuk et al. (1983) reported that spin-labeled aqueous dispersions of total lipid extracts from the whole brain of rats chronically consuming alcohol were more rigid than those obtained in control animals.

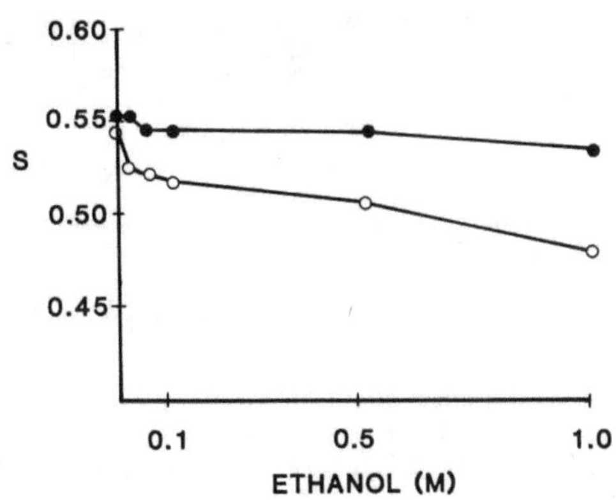

Fig. 3. The effect of in vitro ethanol titration on the molecular order parameter of 5-doxyl stearic acid in microsomal membranes from control (○) and ethanol-fed (●) groups. Experiments were carried out at 37° C. Points indicate means for eight scans

Effects of Ethanol on the Partition of Lipophilic Compounds into Biological Membranes

Despite the fact that ethanol and other anesthetic agents are chemically dissimilar, they share the common properties of their inhibitory action on the Na^+ channel and their high affinity for membranes. Their partitioning into the membrane bilayer has been shown to be affected by the molecular order of the membrane. This point has been demonstrated by Korten et al. (1980), who showed that the partitioning of these compounds into a model membrane system was reduced by increasing membrane cholesterol content. It seems, therefore, that membrane fluidity and partitioning of lipophilic compounds may be related, since increasing the cholesterol content renders the membrane more rigid. Our laboratory has studied the effects of acute and chronic ethanol exposure on the partition of lipophilic amounts into membranes, in view of the observations that (a) molecular order is inversely correlated with the partitioning of lipophilic agents (Waring et al. 1982); (b) erythrocyte and synaptosomal membranes obtained from animals chronically

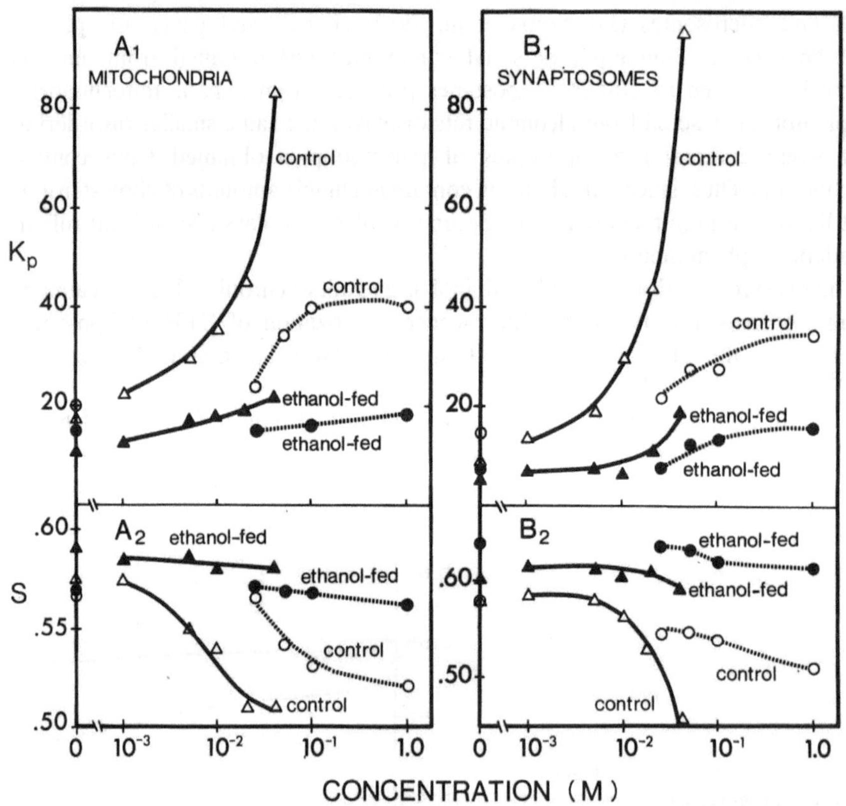

Fig. 4. The effect of ethanol (○, ●) and halothane (△, ▲) on the 5N10 partition coefficient K_p (panels A_1 and B_1) and the 5-doxyl stearic acid order parameter (panels A_2 and B_2) in liver mitochondrial membranes *(A)* and brain synaptosomal membranes *(B)* from ethanol-fed rats (△, ○) and their pair-fed controls (▲, ●) (Rottenberg et al. 1981)

Table 1. Partition coefficients of ethanol and anesthetics in brain synaptosomes and liver mitochondria from ethanol-fed rats and controls

		Ethanol K_p		Halothane K_p
Mitochondria				
Ethanol-fed[a]	(8)	1.17 ± 0.634	(5)	21.4 ± 2.71
Control		3.60 ± 0.740		28.6 ± 5.58
Control/ethanol-fed[b]		4.24		1.35
Synaptosomes				
Ethanol-fed[a]	(5)	0.33 ± 0.11	(7)	21.7 ± 10.2
Control		1.00 ± 0.42		27.5 ± 11.3
Control/ethanol-fed[b]		3.07		1.38

[a] The number of pairs is shown in parentheses. The values are mean ± SD
[b] The ratio of the partition coefficients is the average pair ratio. All the paired differences in the partition coefficients between ethanol-fed and control are significant ($p < 0.01$)

ingesting alcohol have been reported to contain increased amounts of cholesterol (Johnson et al. 1979; Chin et al. 1978; and (c) in the presence of disordering agents, the molecular order of membranes is increased in rats chronically treated with ethanol (Waring et al. 1981, 1982).

Recently, we have demonstrated by ESR spectroscopy that the presence of ethanol or the inhalation anesthetic halothane increased the binding of the partition spin probe 5-doxyl decane (5N10) into both rat liver mitochondria and brain synaptosomes (Fig. 4) (Rottenberg et al. 1981). By contrast, chronic ethanol ingestion reduced the binding of ethanol and halothane by both synaptosomal and mitochondrial membranes. The partition coefficients, K_p, for ethanol and halothane in control and alcohol-treated mitochondrial and synaptosomal membranes are shown in Table 1. Furthermore, the membranes derived from chronically treated animals were resistant to the increased partitioning of the spin probe which was produced by halothane and ethanol in vitro, a result which presumably reflects decreased partitioning of these compounds.

In a recent study, we examined the partitioning of lipophilic compounds into erythrocyte ghosts (Kelly-Murphy et al. 1984), using ^{14}C-halothane and the partitioning spin probe 5N10. Chronic ethanol ingestion caused a decreased partitioning of halothane and 5N10 into erythrocyte membranes. Both halothane and 5N10 are potent fluorescence quenchers. The binding of both of these molecules to the membrane can be ascertained by measuring the steady state fluorescence of membrane-bound DPH following addition of halothane or 5N10. Exposure to halothane, or halothane plus 5N10, caused significant fluorescence quenching of DPH in untreated membranes, indicative of partitioning of halothane/5N10 into the membrane. Partitioning was decreased in erythrocyte membranes from rats treated chronically with ethanol, as evidenced by reduced fluorescence quenching. Interestingly, when the effects of halothane were expressed as a function of the halothane in the membrane rather than the total concentration of halothane added, no differences were found between the control and treated rats. As a result of these two studies, three valuable pieces of information regarding the effect of alcohol on biological membranes were obtained:

1. Chronic ethanol ingestion, which results in resistance to membrane disordering, leads to reduced binding of ethanol and other compounds to the membrane.
2. Membrane cholesterol content is probably not solely responsible for the decreased partitioning of lipophilic compounds into the membranes obtained from alcoholic animals. While synaptosomes are rich in cholesterol, mitochondria contain minimal amounts.
3. The phenomena of physical dependence, tolerance, and cross-tolerance to ethanol may be related through a simple, unified model. Chronic ethanol consumption induces an adaptation of the membrane composition, causing decreased membrane fluidity (increased rigidity), leading to a reduction in the binding of alcohol to the membrane. Hence tolerance to alcohol is acquired. This increased rigidity, which impairs normal membrane function, becomes attenuated in the presence of moderate amounts of ethanol, such that the membranes become sufficiently fluid to resemble normal membranes, thus explaining alcohol dependence. Furthermore, since the membrane is more rigid, it binds fewer molecules of anesthetics (cross-tolerance). Other drugs which are dependent on membrane binding for activity would also be cross-tolerated by the alcoholic.

Alterations in Membrane Composition Induced by Chronic Ethanol Consumption

Lipids

Numerous physicochemical studies have demonstrated that plasma membranes and cell organelles from organisms which chronically consume ethanol have membranes with altered physical properties. An obvious starting place in studying the cause of these alterations is to determine chemical changes in the membrane, particularly in phospholipids.

Using prokaryotes, the effects of chronic ethanol treatment focused on changes in lipid composition in *Escherichia coli* (Ingram 1977; Ingram et al. 1980). When grown in the presence of ethanol (20–90 m*M*), a decrease in saturated fatty acids, especially palmitic acid (16 : 0), and an increase in unsaturated fatty acids, particularly oleic acid (18 : 1), were observed (Ingram et al. 1980). In tissue culture, hamster ovary cells grown with increasing concentrations of ethanol demonstrated a tendency to form short-chain (16 : 0) fatty acyl chains at the expense of longer-chain (18 : 0) fatty acids.

Goto and co-workers examined lipid composition in plasma membranes, mitochondria, and microsomes isolated from *Tetrahymena pyriformis* which had been exposed chronically to ethanol (Goto et al. 1983). All membranes exhibited a dramatic increase in PE, with little or no change in phosphatidylcholine. The fatty acyl composition of the membrane phospholipids was also found to be significantly modified. A pronounced increase in palmitic (16 : 0) and linoleic (18 : 2) acids was observed, as was a decrease in the amount of oleic (18 : 1) and gamma-linolenic (18 : 3) acids. Thus the membranes studied exhibited an overall decrease in the degree of unsaturation. These results are at variance with an earlier study in *Tetrahymena* of ethanol-induced changes by Nandini-Kishore et al. (1979), in which ethanol (10 m*M*) produced a shift in phospholipid fatty acid chains, with a decrease in 16 : 1 and 16 : 23 and an increase in 18 : 2 and 18 : 3. These changes would result in a membrane containing lipids of higher unsaturation, which might be expected to be more fluid than untreated *Tetrahymena*.

Herrero and co-workers have investigated the effects of adding ethanol to the growth medium of *Clostridium thermocellum* (Herrero et al. 1982). They observed a marked increase in normal and anteisobranched fatty acids at the expense of isobranched fatty acids, and an increase in short and unsaturated fatty acids. The adaptive response of *Clostridium* to ethanol leads to a membrane containing fatty acids with lower transition temperatures, suggesting a more fluid membrane. This apparently contrasts with changes found in mammalian cells.

Changes in fatty acid composition of plasma membranes and membranes of subcellular organelles chronically exposed to ethanol have also been investigated. An early study (Miceli and Ferrell 1973) of rat liver mitochondrial membranes obtained from animals chronically consuming alcohol reported a decrease in unsaturated (16 : 1, 18 : 1, 18 : 2, 20 : 4, and 22 : 6) fatty acids and a significant increase in 18 : 0. Cunningham et al. (1982), in a more recent study of rat liver mitochondria obtained from animals chronically exposed to ethanol, observed increases in 16 : 0 and 20 : 4, and decreases in 18 : 1 and 18 : 2, in phosphatidylcholine and PE. By contrast, 18 : 2 was significantly depressed in cardiolipin. No changes in total phospholipid or phospholipid phosphorus/protein ratio were apparent. Studies of changes in fatty acyl groups in liver mitochondria of chronically treated rats from this laboratory also found a significant decrease in 18 : 2 in cardiolipin, as well as an increase in 18 : 1 in PE. Mitochondria obtained from monkeys maintained for 1 year on an ethanol-containing diet were reported to show a marked increase in 16 : 1 and 18 : 1 and a decrease in 18 : 2 and 20 : 4 fatty acids.

Microsomal membranes isolated from livers of alcoholic rats were also scrutinized for changes in fatty acid composition. Miceli and Ferrell (1973) reported a moderate decrease in 16 : 0 and a small increase in 16 : 1. In another study, 16 : 0 was lowered in phosphatidylcholine (PC) and PE, and 18 : 1 was elevated in PE (Cunningham et al. 1982).

Littleton examined fatty acid composition in mice that were maintained for 10 days on ethanol inhalation (Littleton et al. 1980). In these membranes, an increase in 18 : 0 and a decrease in 20 : 4 fatty acids were reported. Chronic administration of ethanol also caused an increase in cholesterol content of synaptic plasma membranes (Chin et al. 1978). A recent study reported that chronic ethanol administration to guinea pigs elicited a specific increase in phosphatidylserine in synaptic plasma membranes (Sun and Sun 1983).

In summary, there has not emerged a consistent pattern of changes in fatty acid composition of membranes isolated from cells or animals chronically exposed to ethanol. In some cases, such as those of *Tetrahymena* and *Escherichia coli*, an increase in fatty acids of greater unsaturation was observed, whereas in mitochondria and microsomes obtained from rat livers a decrease in unsaturated fatty acyl chains was reported.

It is difficult at present to explain the differences in fluidity observed in ethanol-treated animals simply on the basis of adaptive changes in fatty acid composition. It should be noted, however, that the methodologies used to detect changes in fluidity, such as EPR and fluorescence polarization, utilize probes which monitor the average membrane molecular order. It may well be that only specific classes of membrane lipids are structurally altered by chronic ethanol treatment. Such changes, which might also account for altered enzyme function in these membranes, would go undetected by these techniques. The use of phospholipid EPR spin probes of differing classes (PC, PE, PS, PI, etc.) might shed considerable light on this matter.

Proteins

Content

The majority of studies involving changes in protein conformation, activity, and content induced by chronic ethanol treatment have been carried out on the liver, with particular focus on alterations in rat liver mitochondrial membranes. The earliest report of the modulation of membrane protein content in the mitochondrial membrane was a decrease in cytochrome oxidase content (Rubin et al. 1970). This finding has also been observed by a number of other investigators in recent years (Bernstein and Penniall 1978; Thayer and Rubin 1981). In a more recent study, in which rats were chronically fed ethanol (40 days), a 40%–50% decrease in cytochrome oxidase content was detected. Although a decreased content was observed, the functional properties of the protein appeared unchanged (Thayer and Rubin 1981). Chronic ethanol ingestion by rats has also been found to result in a 50% decrease in cytochrome *b* content of the mitochondrial membrane (Thayer and Rubin 1981). The contents of two other major membrane cytochromes, *c* and c_1, were unaffected by chronic ethanol consumption, as was that of ubiquinone (Thayer and Rubin 1981). The content of iron-sulfur clusters of NADH dehydrogenase have been demonstrated by Thayer et al. (1980) to be significantly reduced, as detected by EPR spectroscopy. By contrast, chronic ethanol ingestion does not alter the content of proteins associated with succinate dehydrogenase in the mitochondrial membrane. Finally, the mitochondrial membrane content of adenosine triphosphatase (ATPase) has been reported to be decreased as a result of chronic ethanol consumption in rats (Thayer and Rubin 1979).

Function

Intrinsic membrane protein complexes, many of which require specific phospholipids for maximal function, would be expected to have a change in their activity as a result of ethanol-induced changes in "boundary" lipids with which they are associated. Accordingly, numerous studies have been undertaken to assay enzyme function in membrane systems which have been reported to exhibit significant alterations in lipid content and molecular order. Those studies have previously been discussed at length in a review by Michaelis and Michaelis (1983) and will, therefore, not be repeated here. Rather, only the major functional changes which are influenced by ethanol will be mentioned.

The consensus among a number of investigators is that ethanol, in vitro, inhibits Na^+-K^+-ATPase activity in various tissues (Roach 1979). Two notable exceptions have recently been reported, however, showing that low concentrations of ethanol or other alcohols increase the activity of Na^+-K^+-ATPase in synaptosomal (Sun and Sun 1978) and liver membranes (Gordon et al. 1980). As discussed earlier, significant changes in fatty acid composition have been reported for these membranes.

The activities of mitochondrial NADH oxidase, succinoxidase, and ATP synthetase are decreased in ethanol-fed rats (Rubin and Lieber 1967; Svoboda and Manning 1964; French 1979). The fatty acid composition of mitochondrial membrane phospholipids, particularly that of cardiolipin, may play a role in the observed decrease in activities.

Mitochondrial Protein Synthesis

Studies from this laboratory have been performed in an attempt to explain the decreased content of certain components of the mitochondrial electron transport chain. Mitochondrial protein synthesis was assayed by following the incorporation of ^{14}C-leucine into membrane proteins of liver mitochondria from control and ethanol-fed rats. The effects of ethanol in vitro and in vivo were assayed (Burke and Rubin 1979). The incubation of mitochondria from untreated rats in vitro with ethanol inhibited protein synthesis. In vitro addition of ethanol to the mitochondria from chronically treated rats amplified this effect. Furthermore, hepatic mitochondria incubated with ethanol in vitro showed decreased ^{14}C-leucine incorporation into a polypeptide fraction of molecular weight 36,000–40,000. A similar decrease in this fraction was noted in liver mitochondria from rats chronically fed ethanol, incubated with ^{14}C-leucine in the absence of ethanol. Thus both acute and chronic ethanol consumption inhibits synthesis or incorporation of mitochondrial proteins in the molecular weight range of ATPase and certain cytochromes.

Summary

Increasing evidence suggests that the deleterious effects of chronic ethanol intoxication may be related to changes that occur at the cellular level. The interaction of ethanol with biological membranes, including the lipid and protein components, can cause significant changes in membrane biochemical function.

However, the interaction of ethanol with biological membranes is complex. It has not been proved that a single effect, such as an alteration in membrane fluidity or modulation of fatty acid composition, is solely responsible for all functional alterations. However, with the information currently available, it is possible to piece together a reasonable mode of action. The main points and problems are summarized below:

1. Acute administration of ethanol results in fluidization of model and biological membranes.
2. Chronic ethanol ingestion induces a resistance to the acute fluidizing effects of ethanol. This may result from adaptive alterations in fatty acid composition.
3. Physicochemical changes in the membrane alter the partition of ethanol into the membrane. The presence of ethanol increases the partition of lipophilic compounds, while chronic intoxication decreases it. These changes in partition may play a role in the increased sensitivity of acutely intoxicated individuals to drugs, and in chronic alcoholics to tolerance and cross-tolerance, physical dependence, and withdrawal.
4. The baseline membrane fluidity measured by EPR spectroscopy has not routinely been found to be different in membranes from untreated and chronically intoxicated animals. This may reflect a problem with the sensitivity of the technique, however, as the probes which have been used to date report on bulk membrane fluidity. Many of these studies have used 5-doxyl stearic acid, a compound which has been shown to cause disruption of the highly ordered structure of the membrane in the area of the bilayer closest to the surface (Taylor and Smith 1980). Taylor and Smith, using ^{2}H NMR, have also demonstrated that this probe, owing to its chemical nature, is inserted into the membrane in a somewhat abnormal way, due to the constraints imposed by the tightly packed lipids present at the surface of the bilayer (Taylor and Smith 1981). Thus, use of this probe may lead to erroneous results when assessing the effects of various agents on membrane phospholipid structure (Taylor and Smith 1981).

These problems can be overcome in two ways. First, studies using phospholipid spin probes of differing classes (PC, PE, PS, PI, cardiolipin, etc.) should help to localize the phospholipid structural alterations induced by chronic ethanol treatment. Second, the use of fatty acids and phospholipids selectively labeled at different positions in the hydrocarbon chains in model membranes, liposomes comprised of lipid extracts of membranes from alcoholic animals, and natural biological membranes could give valuable information derived by ^{2}H NMR regarding the site at which alcohol exerts its strongest effects. This method also has an advantage in that replacement of ^{1}H by ^{2}H does not perturb membrane structure.

5. The ability of certain membrane phospholipids, such as unsaturated PE, phosphatidylserine, phosphatidic acid, and cardiolipin, to undergo polymorphic phase transitions from the liquid crystalline bilayer to the hexagonal (H_{II}) phase is dependent on a number of factors. These include the degree of unsaturation of the fatty acyl chains, cholesterol content, ions such as Ca^{2+}, pH, proteins, and phospholipids. The ability of these lipids to adopt nonbilayer structures (e.g., inverted micelles) under relevant conditions has led to the suggestion that they

may be involved in membrane functional processes, such as transbilayer "flip-flop" of lipids, ion transport, fusion, and the insertion of membrane protein, (for review see Cullis and deKruiff 1979).

Cullis and co-workers have demonstrated that ethanol and other similar agents are capable of modulating the polymorphic properties of membrane lipids. Moreover, changes in fatty acid composition and the degree of unsaturation caused by chronic ethanol treatment may have profound effects on the polymorphic properties of membranes, particularly cardiolipin, which is significantly more saturated in the alcoholic membranes. In addition, ethanol may interfere with normal membrane function by altering the ability of certain membrane lipids to alter their physical structure to assume optimal functionality.

References

Bernstein JD, Penniall R (1978) Effect of chronic ethanol treatment upon rat liver mitochondria. Biochem Pharmacol 27: 2337–2342

Burke JP, Rubin E (1979) The effects of ethanol and acetaldehyde on the products of protein synthesis by liver mitochondria. Lab Invest 41: 393–400

Chin JH, Goldstein DB (1977) Effects of low concentrations of ethanol on the fluidity of spin-labelled erythrocyte and brain membranes. Mol Pharmacol 13: 435–441

Chin JH, Goldstein DB (1977) Drug tolerance in biomembranes: a spin label study of the effects of ethanol. Science 196: 684–685

Chin JH, Parson LM, Goldstein DB (1978) Increased cholesterol content of erythrocyte and brain membranes in ethanol-tolerant mice. Biochim Biophys Acta 513: 358–363

Chin JH, Goldstein DB (1981) Membrane disordering action of ethanol: variation with membrane cholesterol content and depth of the spin label probe. Mol Pharmacol 19: 425–431

Cullis PR, Vkrkleij AJ, Ververgaert PH (1978) Polymorphic phase behaviour of cardiolipin as detected by 31P NMR and freeze-fracture techniques. Effects of calcium, dibucaine and chlorpromazine. Biochim Biophys Acta 513: 11–20

Cullis PR, Hornby AP, Hope MJ (1980) Molecular mechanisms of anesthesia. In: Fisk BR (ed) Molecular mechanisms of anesthesia. Raven, New York, (Progress in anesthesia, vol 25)

Cunningham CC, Filus S, Bottenus RE, Spach PI (1982) Effect of ethanol consumption on the phospholipid composition of rat liver microsomes and mitochondria. Biochem Biophys Acta 712: 225–233

Davis JH (1983) The description of membrane lipid conformation, order and dynamics by ^{2}H-NMR. Biochim Biophys Acta 737: 117–171

Demediuk D, Dowan DL, Moscatelli EA (1983) Effects of plasmenylethonlamine on the dynamic properties of the hydrocarbon region of mixed phosphatidylcholine-phosphatidylethanolamine aqueous dispersions. Biochim Biophys Acta 730: 263–270

Erickson CK (1979) Factors affecting the distribution and measurement of ethanol in the body. In: Majchrowicz E, Noble EP (eds) Biochemistry and pharmacology of ethanol, Vol 1. Plenum, New York, pp 9–26

Erickson CJP, Mizoi Y, Funkunago T (1982) The determination of acetaldehyde in human blood by the perchloric acid precipitation method. Anal Biochem 125: 259–263

French SW (1979) Role of mitochondrial damage in alcoholic liver diseases. In: Majchrowicz E, Noble EP (eds) Biochemistry and pharmacology of ethanol, vol 1. Plenum, New York, pp 409

Gaffney BJ, McNamee CM (1974) Spin label measurements in EPR spectroscopy. Methods Enzymol 32B: 161–197

Gordon LM, Sauerheber RD, Esgate JA, Dipple I, Marchmont RJ, Houslay MD (1980) The increase in bilayer fluidity of rat liver plasma membranes achieved by the local anesthetic benzyl alcohol affects the activity of intrinsic membrane enzymes. J Biol Chem 255: 4519–4527

Goto M, Banno Y, Umeki S, Kameyama Y, Nozawa Y (1983) Effects of chronic ethanol exposure on composition and metabolism of *Tetrahymena* membrane lipids. Biochim Biophys Acta 751: 286–297

Harris RA, Schroeder F (1981) Ethanol and the physical properties of brain membranes. Mol Pharmacol 20: 128–137

Herrero AA, Gomez RF, Roberts MF (1982) Ethanol induced changes in the membrane lipid composition of *Clostridium thermocellum*. Biochim Biophys Acta 693: 195–204

Hill MW (1974) The effect of anesthetic-like molecules on the phase transition in smectic mesophases of dipalmitoyllecithin. Biochim Biophys Acta 356: 117–124

Hornby AP, Cullis PR (1981) Influence of local and neutral anesthetics on the polymorphic phase preferences of egg yolk phosphatidylethanolamine. Biochim Biophys Acta 647: 285–292

Hubbell WL, McConnell HM (1971) Molecular motion in phospholipids and spin-labelled membranes. J Am Chem Soc 93: 314–326

Ingram LO (1977) Preferential inhibition of phosphatidylethanolamine synthesis in *E. coli* by alcohols. Can J Microbiol 23: 779–789

Ingram LO, Ley KD, Hoffman EM (1978) Drug-induced changes in lipid composition of *E. coli* and of mammalian cells in culture: ethanol, pentobarbital and chlorpromazine. Life Sci 22: 489–494

Ingram LO, Dickens BF, Buttke TM (1980) Reversible effects of ethanol on *E. coli*. In: Begleiter H (ed) Biological effects of alcohol. Plenum, New York, pp 299–337

Jain MK, Wu NM (1977) Effect of small molecules on the dipalmitoyllecithin liposomal bilayer III phase transition in lipid bilayer. J Memb Biol 34: 157–201

Johnson PA, Lee NM, Cooke R, Loh HH (1979) Ethanol-induced fluidization of brain lipid bilayers: required presence of cholesterol in membranes for the expression of tolerance. Mol Pharmacol 15: 739–746

Kelly-Murphy S, Waring AJ, Rottenberg H, Rubin E (1984) Effects of chronic ethanol consumption on the partition of lipophilic compounds into erythrocyte membranes. Lab Invest 50: 174–183

Korten K, Sommer TJ, Miller KW (1980) Membrane composition modulates thiopental partitioning in bilayers and biomembranes. Biochim Biophys Acta 599: 271–279

Lee AG (1976) Interactions between anaesthetics and lipid mixtures. Amines Biochem Biophys Acta 448: 34–44

Lenaz G, Bertol E, Curatola G, Mazzanti L, Bigi A (1976) Lipid protein interactions in mitochondria: spin and fluorescence probe studies and the effect of *n*-alkanols on phospholipid vesicles and mitochondrial membranes. Arch Biochem Biophys 172: 278–288

Littleton JM, Grieve SJ, Griffiths PJ, John JR (1980) Ethanol-induced alteration in membrane phospholipid composition: possible relationship to development of cellular tolerance to ethanol In: Begleiter H (ed) Biological effects of alcohol. Plenum, New York, pp 7–19

Majchrowicz E, Noble EP (eds) (1979) Biochemistry and pharmacology of ethanol. Plenum, New York

Marsh D (1981) Electron spin resonance: spin labels. In: Grell E (ed) Membrane spectroscopy. Springer, Berlin Heidelberg New York, pp 51–142 (Molecular biology, biochemistry and biophysics, vol 31)

Miceli JN, Ferrell WJ (1973) Effects of ethanol on membrane lipids III: quantitative changes in lipid and fatty acid composition of nonpolar and polar lipids of mouse total liver mitochondria and microsomes following ethanol feeding. Lipids 8: 722–727

Michaelis EK, Zimbrick JD, McFaul JA, Lampe RA, Michaelis ML (1980) Ethanol effects on synaptic glutamate receptors and on liposomal membrane structure. Pharmacol Biochem Behav (Suppl 1) 13: 197–202

Michaelis EK, Michaelis ML (1983) Physico-chemical interactions between alcohol and biological membranes. In: Smart RG, Glaser H, Israel Y, Kalant JM, Popham A, Schmidt A (eds) Research advances in alcohol and drug problems, vol 7. Plenum, New York, pp 127–173

Nandini-Kishore SG, Mattox SM, Martin CE, Thompson GA (1979) Membrane changes during growth of *Tetrahymena* in the presence of ethanol. Biochim Biophys Acta 551: 315–327

Ponnappa BC, Waring AJ, Hoek JB, Rottenberg H, Rubin E (1982) Chronic ethanol ingestion increases calcium uptake and resistance to molecular disordering by ethanol in liver microsomes. J. Biol Chem 257: 10141–10146

Roach MK (1979) Changes in the activity of Na^+, K^+-ATPhase during acute and chronic administration of ethanol. In: Majchrowicz E, Noble EP (eds) Biochemistry and pharmacology of ethanol, vol 2. Plenum, New York, pp 67–80

Rottenberg H, Waring A, Rubin E (1981) Tolerance and cross-tolerance in chronic alcoholics: reduced membrane binding of ethanol and other drugs. Science 213: 583–585
Rubin E, Lieber CS (1967) Early fine structural changes in the human liver induced by alcohol. Gastroenterology 52: 1–13
Rubin E, Beattie DS, Lieber CS (1970) Effects of ethanol on the biogenesis of mitochondrial membranes and associated mitochondrial functions. Lab Invest 23: 620–627
Seelig J (1977) Deuterium magnetic resonance: theory and application to lipid membranes. Q Rev Biophys 10: 353–418
Seelig J (1978) Nuclear magnetic resonance and the head group structure of phospholipids in membranes. Biochim Biophys Acta 515: 105–140
Seeman P (1972) The membrane actions of anesthetics and tranquilizers. Pharmacol Rev 24: 583–655
Shinitzky M, Barenholz Y (1974) Dynamics of the hydrocarbon layer in liposomes of lecithin and sphingomyelin. J Biol Chem 249: 2652–2657
Shinitzky M, Barenholz Y (1978) Fluidity parameters of lipid regions determined by fluorescence polarization. Biochim Biophys Acta 515: 367–394
Singer SJ, Nicolson GL (1972) The fluid mosaic model of the structure of cell membranes. Science 175: 720–731
Sun GY, Sun AY (1978) The effects of chronic ethanol administration on acyl group composition of mitochondrial phospholipids from guinea pig adrenal. Res Commun Chem Pathol Pharmacol 21: 355–358
Sun GY, Sun AY (1983) Chronic ethanol administration induced an increase in phosphatidylserine in guinea pig synaptic plasma membranes. Biochem Biophys Res Comm 113: 262–268
Svoboda DJ, Manning RT (1964) Chronic alcoholism with fatty metamorphosis of the liver. Am J Pathol 44: 645–652
Thayer WS, Rubin E (1979) Effects of chronic ethanol intoxication on oxidative phosphorylation in rat liver submitochondrial particles. J Biol Chem 254: 7717–7723
Thayer WS, Ohnishi T, Rubin E (1980) Characterization of iron-sulfur clusters in rat liver submitochondrial particles by electron paramagnetic resonance spectroscopy. Alterations produced by chronic ethanol consumption. Biochem Biophys Acta 591: 22–36
Thayer WS, Rubin E (1981) Molecular alterations in the respiratory chain of rat liver after chronic ethanol consumption. J Biol Chem 256: 6090–6097
Taylor MG, Smith ICP (1980) The fidelity of response by nitroxide spin probes to changes in membrane organization. The condensing effect of cholesterol. Biochem Biophys Acta 599: 140–149
Taylor MG, Smith ICP (1981) Reliability of nitroxide spin probes in reporting membrane properties: a comparison of nitroxide and deuterium labeled steroids. Biochemistry 20: 5252–2255
Verkleij AJ, De Maagd R, Leunissen-Bijvelt J, de Kruijff B (1982) Divalent cations and chlorpromazine can induce non-bilayer structures in phosphatidic acid-containing model membranes. Biochim Biophys Acta 684: 255–262
Waring AJ, Rottenberg H, Ohnishi T, Rubin E (1981) Membranes and phospholipids of liver mitochondria from chronic alcoholic rats are resistant to membrane disordering by alcohol. Proc Natl Acad Sci USA 78: 2582–2586
Waring AJ, Rottenberg H, Ohnishi T, Rubin E (1982) The effect of chronic ethanol consumption on temperature-dependent physical properties of liver mitochondrial membranes. Arch Biochem Biophys 216: 51–61
Williamson JR, Scholz R, Thurman RG, Chance B (1969) In: Papa S, Tager JM, Quagliariello E, Slater EC (eds) The neergy level and metabolic control in mitochondria. Adriatica Editrice, Bari, pp 411–429

13 Alcohol and Porphyrin Metabolism

M. O. Doss*

Abteilung für Klinische Biochemie, Fachbereich Humanmedizin der Universität Marburg, Deutschhausstraße 17 1/2, 3550 Marburg, Federal Republic of Germany

Introduction

"Ethanol should be included in the list of drugs metabolized by hepatic microsomes" (Rubin and Lieber 1972). Such drugs stimulate their own metabolism, in which heme synthesis is involved, and affect porphyrin synthesis in various ways. Alteration of porphyrin synthesis in an organism without hereditary and acquired toxic disturbances of porphyrin metabolism is an adaptive response. Generally, these adaptive responses elicited by alcohol or other drugs in an organism without preexistent derangements in the heme biosynthesis sequences do not lead to clinical consequences. However, certain drugs, including alcohol, are capable of expressing not only the latent but also the clinical phase of different porphyrias. Especially acute hepatic porphyrias are triggered by drugs and/or alcohol and have therefore been designated as "pharmacogenetic diseases" (Kalow 1962). In addition, alcohol is the most frequent manifesting and aggravating factor in chronic hepatic porphyria, including porphyria cutanea tarda, which is genetically predetermined in some of the patients (Doss 1982; Kushner 1982). The clinical-biochemical interactions between alcohol and porphyrin/heme biosynthesis involve three main aspects:

1. Triggering the biochemical and clinical manifestation of acute and chronic hepatic porphyrias.
2. Development of symptomatic disturbances in porphyrin metabolism as secondary coproporphyrinuria and secondary protoporphyrinemia.
3. Inhibition and stimulation of certain enzymes in the heme biosynthetic chain.

These effects may be the main cause of the above-mentioned disorders, which are of diagnostic and clinical relevance.

Diseases and Disorders of Porphyrin Metabolism

Goldberg and Rimington (1962) recommended that the term "porphyria" be reserved "for those diseases of porphyrin metabolism caused by an 'inborn error of

* The author's work was supported by Grant Do 134 from the German Research Association

Alcohol Related Diseases in Gastroenterology
Edited by H. K. Seitz and B. Kommerell

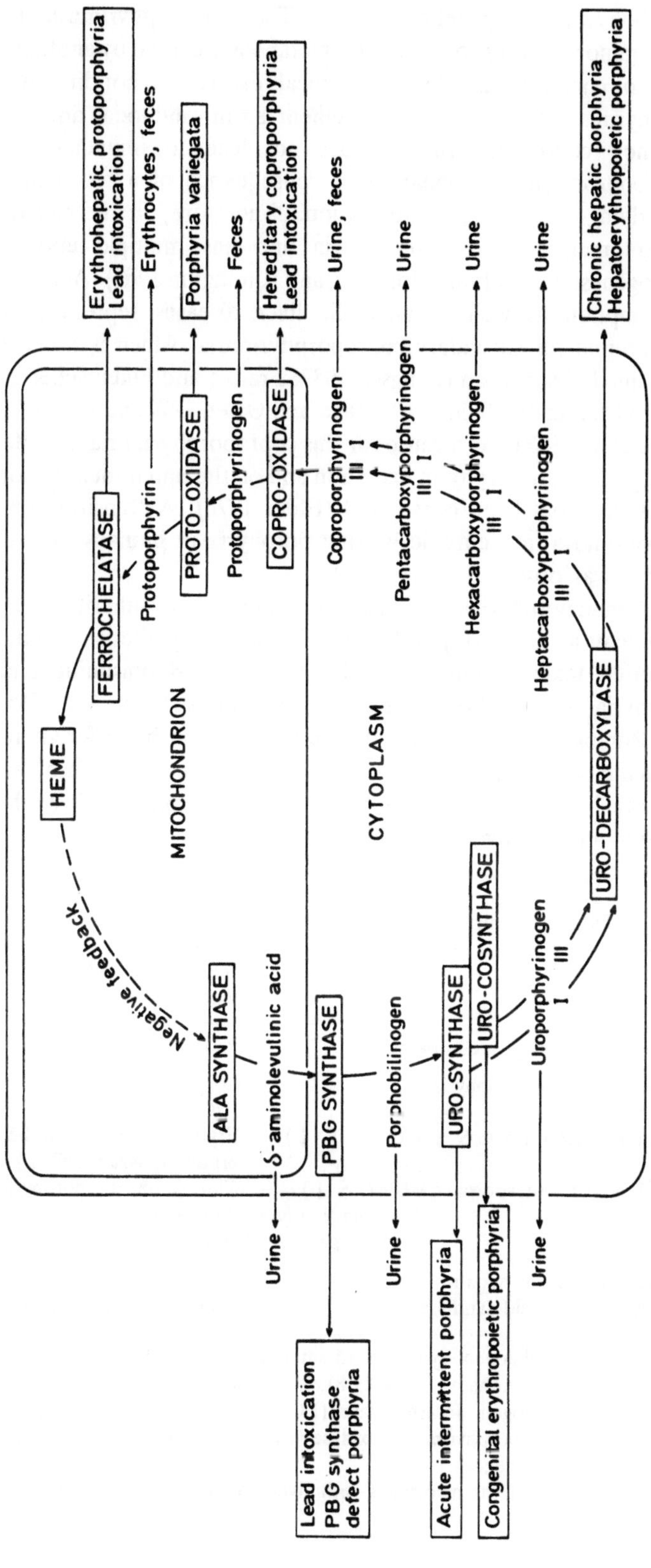

Fig. 1. Heme biosynthesis and localization of enzyme deficiencies in porphyrias and lead intoxication

metabolism', i.e., a genetic defect.... The term porphyrinuria should be confined to those minor disorders of porphyrin metabolism caused by another disease, or certain drugs or chemicals, in which the clinical features are not directly attributable to the porphyrin abnormality." This statement is still valid today and implies that the most common coproporphyrinuria does not lead to immanent clinical symptoms. Coproporphyrinuria appears as a consequence of a biochemical disorder in an underlying disease or intoxication. Therefore, it is called "secondary" or "symptomatic" coproporphyrinuria and has multiple etiologies and variable pathogenesis (Doss 1982; Goldberg and Rimington 1962). Acute and chronic alcohol consumption as well as alcoholic liver diseases represent the most frequent background for secondary coproporphyrinuria, which is a classic and repeatedly confirmed observation (Brugsch 1937; Franke and Fikentscher 1935; McColl et al. 1980; Orten et al. 1963; Shrivastava and Orten 1972; Sutherland and Watson 1951; Zieve et al. 1953). As a cause for this coproporphyrinuria the effect of alcohol both on hepatic coproporphyrin excretion and on the impairment of coproporphyrinogen metabolism has been considered (Elder 1976). A transition of secondary coproporphyrinuria into chronic hepatic porphyria in humans due to the influence of alcohol was observed (Doss 1980).

Human porphyrias are caused predominantly by inherited enzyme abnormalities (Brodie et al. 1977) of porphyrin and heme biosynthesis, as outlined in the Fig. 1. Acquired toxic porphyrias, such as acute lead poisoning and chronic hepatic porphyria, are possible under certain conditions (Doss 1982; Elder 1976; Goldberg and Rimington 1962). Generally, porphyrias can be differentiated into acute and nonacute porphyrias (Table 1) and, in the classical view, into "erythropoietic" and "hepatic" prophyrias, as the main sites of porphyrin biosynthesis are the bone marrow and the liver (Schmid et al. 1954).

Table 1. Different clinical expression of porphyrias

Acute porphyrias (acute syndromes)	Non acute porphyrias (cutaneous symptoms)
Abdominal colics, panneuropathy	Photosensitivity
Acute intermittent porphyria (US ↓)	Congenital erythropoietic porphyria (UCoS ↓)
Lead intoxication (PBG-S ↓ CO ↓ FC ↓)	Erythropoietic (erythrohepatic) protoporphyria[a] (FC ↓)
PBG-S defect acute porphyria (PBG-S ↓)	Chronic hepatic porphyria[a] (UD ↓)
Hereditary coproporphyria (CO ↓)	
Variegate porphyria (PO ↓)	
Strong induction of hepatic δ-aminolevulinic acid synthase	No or minimal induction of hepatic δ-aminolevulinic acid synthase

PBG-S, porphobilinogen synthase (δ-aminolevulinic acid dehydratase, EC 4.2.1.24); US, uroporphyrinogen synthase (EC 4.3.1.8); UCoS, uroporphyrinogen-III-cosynthase (unclassified); UD, uroporphyrinogen decarboxylase (EC 4.1.1.37); CO, coproporphyrinogen oxidase (EC 1.3.3.3); PO; protoporphyrinogen oxidase (EC 1.3.3.4); FC, ferrochelatase (heme synthase; EC 4.99.1.1); ↓ = deficiency

[a] Abdominal symptoms associated with liver and bile duct disease, not primarily caused by porphyria

In congenital erythropoietic porphyria (Nordmann and Deybach 1982), porphobilinogen synthase defect porphyria (Doss 1982), and hepatoerythropoietic porphyria (Elder et al. 1981), clinical manifestation will appear only in the homozygous state. These porphyrias are autosomal recessive disorders. In all the other inherited porphyrias outlined in Fig. 1 the clinical expression also affects the heterozygote conditions: the transmission of these porphyrias follows an autosomal dominant trait.

Effects of Alcohol on Porphyrin and Heme Biosynthesis

Fifty years ago, Franke and Fikentscher (1935) observed that alcohol consumption increases urinary porphyrin excretion, which was found to be approximately doubled after ingestion of 1 l beer or 90 ml cognac. They stated that "even minor fluctuations of porphyrin excretion may be important for research and clinical medicine". Later on the alcohol-mediated coproporphyrinuria was studied by Brugsch (1937), and has been subject to detailed investigations by several groups up to now (Doss 1980; Elder 1976; McColl et al. 1980; McEwin 1976; Orten et al. 1963; Shrivastava and Orten 1972; Sutherland and Watson 1951). Research on the interrelationships between alcohol and porphyrin metabolism has been enforced by clinical observations (Table 2): on the one hand, ingestion of alcohol, erruption of cutaneous lesions and increase of uroporphyrin excretion were closely correlated (Doss et al. 1971; Habermann et al. 1975; Shanley et al. 1969), and on the other hand alcohol was recognized as one of the "porphyrinogenic" drugs for the exacerbation of acute hepatic porphyrias (Bloomer 1976; Doss et al. 1982b; Eales 1979; Goldberg and Rimington 1962; Moore 1980; Welterberg 1976). Thus two important effects of alcohol on human hereditary hepatic porphyrias must be emphasized: the induction

Table 2. Forces of alcohol: biochemical and clinical expression of hepatic porphyrias

Phase	Acute/Chronic Hepatic porphyrias	Alcohol
A	Genetic diposition[a]	
↓		↓
B	Latency compensated	
↓		↓ ↓
C	Latency decompensated	
↓		↓ ↓ ↓
D	Clinical expression	

[a] For example: uroporphyrinogen synthase deficiency for acute intermittent porphyria ("regulatory disease" with induction of ALA synthase), and uroporphyrinogen decarboxylase deficiency for chronic hepatic porphyria or porphyria cutanea tarda ("membrane-disease" due to an associated liver damage, but without significant increase in ALA synthase)

of δ-aminolevulinic acid (ALA) synthase in view of the acute porphyrias and the inhibition of uroporphyrinogen decarboxylase for chronic hepatic porphyria, including its clinically overt phase, porphyria cutanea tarda.

Effects of Alcohol on the Enzymes of the Heme Biosynthetic Chain

Clinical and experimental studies in humans and animals have been extended to nearly all the enzymes involved in the sequences leading to heme (Fig. 1).

δ-Aminolevulinic Acid Synthase

Alcohol induces about threefold hepatic ALA synthase in starving rats (Beattie et al. 1973; Held 1977; Shanley et al. 1968) and cultured rat hepatocytes as well as in humans (Shanley et al. 1969; Lane and Stewart 1983). Consequently, porphyrin synthesis is stimulated (Held 1977). Pyrazole reduces the inducing effect of alcohol on ALA synthase (Beattie et al. 1973; Held 1977), whereas phenobarbital increases it. The activities of hepatic ALA dehydratase (porphobilinogen synthase), microsomal monooxygenase, and the concentration of cytochrome *P*-450 were not altered in the animal model (Held 1977), in contrast to a decrease in hepatocytes, where ethanol treatment also causes a transient decrease in intracellular heme (Lane and Stewart 1983). It was concluded that the porphyrinogenic effect is not caused by alcohol itself but either by its metabolites or by changes of the liver metabolism caused by the effect of alcohol on the reduced nicotinamide-adenine dinucleotide/nicotinamide-adenine dinucleotide (NADH/NAD) ratio (Beattie et al. 1973; Held 1977; Labbe 1967), because the hepatic metabolism of ethanol results in an increase in the redox potential. Although the mechanism of ALA synthase induction is not yet completely elucidated, this enzyme is under negative feedback control by heme (Granick 1966; Sassa and Granick 1970). As cytochrome *P*-450 plays a central role in heme utilization, drugs increasing *P*-450 often induce ALA synthase as well (Maxwell and Meyer 1976). Heme synthesis is regulated by the first and rate-limiting enzyme in the pathway, ALA synthase (Granick 1977). Alcohol and phenobarbital induce the drug-metabolizing enzymes in hepatic microsomes and increase their own metabolism (Rubin and Lieber 1971). Induction of ALA synthase is probably a secondary event due to changes in cellular demand for heme initiated by increased utilization, destruction (de Matteis 1978), or reduced formation. Inherited enzyme deficiencies will lead to a reduced formation of heme, which result in a labilization of the feedback control (Fig. 1). Under this condition, induction of ALA synthase in the liver reacts highly sensitively to various drugs, including alcohol. A decrease in "regulatory liver heme" will be compensated by the induction of ALA synthase (de Matteis 1978). This mechanism can be understood as a counterregulation in human acute hepatic porphyrias (Doss 1982). An increased hepatic ALA synthase activity was also observed in cirrhotic patients (Bonkowsky and Pomeroy 1977; Kodama et al. 1983; Levere 1967), but this increase does not seem to reach levels expected in acute intermittent porphyria and is not followed by an elevation of urinary ALA and porphobilinogen excretion in contrast to acute hepatic porphyrias (Bonkowsky and Pomeroy 1977; Doss et al. 1972).

Recent studies have revealed an age-dependent decrease in mitochondrial ALA synthase in normal and induced animals (Scotto et al. 1983), suggesting an age-related change in the process by which cytosolic ALA synthase is translocated into the mitochondria (Kikuchi and Hayshi 1981). Thus the basal levels of hepatic ALA synthase and the ethanol-induced activity decreased dramatically as a function of age, although the ratio of induced to basal activities did not change with age (Paterniti et al. 1978). In the heart ALA synthase activity also decreased with aging and fasting, but was not induced by ethanol. In contrast, kidney ALA synthase activity showed no age-related changes (Paterniti et al. 1978).

The effect of alcohol on ALA synthase is not limited to the hepatic enzyme. The activity of ALA synthase in leukocytes has been monitored after acute alcohol ingestion in healthy subjects as well as in chronic alcoholics (McColl et al. 1980, 1981). In both conditions the activity of leukocyte ALA synthase was significantly elevated. In the acute reversible sideroblastic anemia of alcoholics, increased bone marrow ALA synthase activity was noted (Fraser and Schacter 1980). In contrast, patients with idiopathic sideroblastic anemia showed diminished ALA synthase activity (Bottomley et al. 1973).

The induction of ALA synthase by alcohol may be mediated by changing the oxidation/reduction rate of the mitochondria (Shanley et al. 1968), because the NADH/NAD ratio in the cytosol as well as within the mitochondria is enhanced by ethanol (Lindros and Hillbom 1969). The alteration of the redox state may impair terminal oxidation and stimulate porphyrinogenesis in several unknown ways (Labbe 1967), which last not least lead to changes in the activity of the different heme pathway enzymes.

δ-Aminolevulinic Acid Dehydratase (Porphobilinogen Synthase)

It was first shown by Moore and co-workers (Krasner et al. 1974; Moore et al. 1971) that the ALA dehydratase activity was depressed by ethanol in human blood cells. In rats intoxicated with ethanol a significant depression of ALA dehydratase was found in blood, liver, and kidney, but not in the heart (Moore et al. 1971). Hepatic ALA dehydratase was also decreased in vitro both by ethanol and its metabolite acetaldehyde (Moore et al. 1971). In man the decrease in blood ALA dehydratase activity was closely related to the increase in blood alcohol. An increase of the intracellular redox potential by alcohol and its effects on the redox balance of other compounds within the cell has been regarded as a direct cause of the enzyme inhibition: the augmented redox potential may influence the equilibrium between oxidized and reduced forms of sulfhydryl compounds (Moore et al. 1971). On the other hand, the toxicity of acetaldehyde interacts with amino acids: binding of acetaldehyde with cystein and/or glutathione may decrease glutathione concentration in the liver (Lieber 1980). Glutathione depletion may contribute to the change in ALA dehydratase activity. In addition, it may promote peroxidative damage of cellular membranes, which is one pathogenetic factor for chronic hepatic porphyria (Koszo et al. 1982).

Several reports have confirmed the depression of ALA dehydratase activity in human erythrocytes after chronic and acute alcohol ingestion (Brodie et al. 1979; Hamlyn et al. 1979; McColl et al. 1980, 1981). The activation of ALA dehydratase by zinc counteracts the inhibitory effect of orally administered ethanol in rats (Abdulla et al. 1976). In porphyria cutanea tarda associated with alcoholic liver syndromes in most cases (Martini and Dölle 1965; Mezey 1982), erythrocyte ALA dehydratase activity was decreased (Simon and Kiss 1979). However, these findings remain controversial (Goertz et al. 1980). Furthermore, a decreased ALA dehydratase activity was demonstrated in the liver of porphyria cutanea tarda, in alcoholics (Kondo et al. 1983), and in cirrhosis of various origin (Kodama et al. 1983). In most of the patients with hepatic cirrhosis alcohol was a causative or aggravating factor. A decrease in enzyme activity to 34% of the control level (Kondo et al. 1983) must lead neither to a disturbance of heme biosynthesis nor to an elevated urinary ALA excretion, as the enzyme is excessively available and represents the most abundant enzyme along the sequence of heme biosynthesis (Meyer 1978). Addition of iron or uroporphyrinogen did not alter the activity of purified bovine liver ALA dehydratase (Kondo et al. 1983), in contrast to acute iron loading of rats. The decrease in activity of ALA dehydratase in liver homogenates caused by intraperitoneal administration of iron-dextran was mirrored by a reciprocal increase in ALA synthase (Bonkowsky et al. 1983). Thus the response of ALA dehydratase to iron seems to be different under in vitro and in vivo conditions. Generally, a reduction of hepatic ALA dehydratase by alcohol will be compensated by an increase in ALA synthase (Bonkowsky et al. 1983; Kodama et al. 1983) as is known to be the case with lead intoxication (Goldberg et al. 1978; Maxwell and Meyer 1976). Erythrocyte ALA dehydratase activity was assessed for outpatient detection of alcoholic liver disease and compared with gamma-glutamyltransferase and casual blood ethanol (Hamlyn et al. 1979): the highest specificity for alcoholism was achieved by ALA dehydratase; the best overall performance, with highest sensitivity and specificity, was, however, gamma-glutamyltransferase (Hamlyn et al. 1979). In another study on cirrhotic patients (Kodama et al. 1983), hepatic ALA dehydratase activity did not correlate with liver function tests and iron. Alcohol is oxidized to acetaldehyde by all three alcohol-metabolizing systems (alcohol dehydrogenase, microsomal ethanol-oxidizing system, and catalase) (Mezey 1976; Teschke et al. 1976). As shown from in vitro experiments (Moore et al. 1971), as well as from acute and chronic ethanol ingestion (McColl et al. 1980, 1981), both alcohol and acetaldehyde may decrease ALA dehydratase activity.

Table 3. Enzyme inhibitions (↓) by lead and alcohol and corresponding hereditary enzyme deficiencies in the porphyrin and heme pathway

Enzyme	Lead	Ethanol	Genetic defect
Porphobilinogen synthase (EC 4.2.1.24)	↓↓	↓	Heterozygotes without clinical symptoms but with increased sensitivity to lead Homozygotes with acute intermittent porphyria syndrome
Coproporphyrinogen oxidase (EC 1.3.3.3)	↓↓	↓	Hereditary coproporphyria
Ferrochelatase (EC 4.99.1.1)	↓↓	↓	Erythrohepatic (erythropoietic) protoporphyria

Alcohol interferes with lead in the suppression of ALA dehydratase activity in red blood cells (Table 3); alcohol can potentiate the rapid enzyme inhibition by lead as well as prolong the hepatic and erythropoietic phase of lead intoxication (Doss 1982). The frequent clinically reported synergism of lead toxicity and ethanol is not related to increased lead absorption or diminished lead excretion but rather to nutritional deficiencies and increased lead exposure among some alcoholics (Barton and Conrad 1978).

Uroporphyrinogen Synthase (Porphobilinogen Deaminase)

Acute alcohol ingestion in healthy subjects resulted in an increased activity of uroporphyrinogen synthase in erythrocytes (McColl et al. 1980). An increase was also observed in chronic alcoholics (McColl et al. 1981). Thus the two enzymes with rate-controlling function in the pathway, ALA synthase (Granick 1966; Sassa and Granick 1970) and uroporphyrinogen synthase (Doss 1982), are stimulated by alcohol.

Uroporphyrinogen Decarboxylase

Uroporphyrinogen decarboxylase is reduced in experimental chronic hepatic porphyria (Elder et al. 1976; Stonard 1978; Taljaard et al. 1971; von Tiepermann et al. 1980) and in human chronic hepatic porphyria or porphyria cutanea tarda (Doss 1982). Alcohol ingestion increases uroporphyrin excretion in humans (Doss et al. 1971; Shanley et al. 1969) and urinary porphyrins in animals (Doss et al. 1981; Koszo et al. 1982; Pearson and Malkinson 1965). Porphyrin excretion biochemograms reflecting the transition from secondary coproporphyrinuria to chronic hepatic porphyria support the concept that alcohol may disturb uroporphyrinogen decarboxylase in the liver, thereby initiating and aggravating the chronic hepatic porphyria process in alcoholic liver syndromes (Doss 1980). The metabolic effect of alcohol on hepatic porphyrin heme synthesis is very complex (Fig. 2). Its porphyrinogenic action not only consists in the induction of ALA synthase and the reduction of uroporphyrinogen decarboxylase and porphobilinogen synthase, but may also result from a disturbance of coproporphyrinogen oxidase (McColl et al. 1980, 1981), as well as changes in the mitochondrial NADH/NAD ratio (Lieber 1980). The redox changes are caused by the oxidation of alcohol through cytosolic alcohol dehydrogenase, since NAD is reduced to NADH in this reaction. Only part of the NADH produced can be reoxidized by hepatic mitochondria (Lieber 1980) and by the oxidation of alcohol through the microsomal ethanol-oxidizing system, a pathway which consumes reducing equivalents in the form of reduced nicotinamide-adenine dinucleotide phosphate (NADPH) (Teschke et al. 1975, 1976).

Alcohol, with respect to its toxic potential in the pathobiochemical development and clinical manifestation of a chronic hepatic porphyria, is less porphyrinogenic than hexachlorobenzene, if the findings in metabolite and enzymatical changes after chronic alcohol ingestion (Doss et al. 1981) are compared with those found in hexachlorobenzene porphyria in rats (Elder et al. 1976; Stonard 1978; von Tiepermann et al. 1980). After 25 days of alcohol feeding under the described conditions one cannot expect the development of experimental chronic hepatic porphyria, considering the fact that feeding of hexachlorobenzene for more than 30

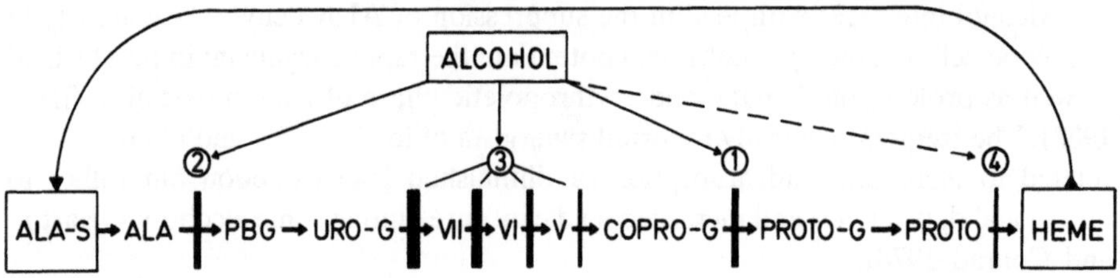

Fig. 2. Hypothesis on the effect of alcohol on heme synthesis: inhibition of coproporphyrinogen oxidase *(1)* and porphobilinogen synthase *(2)* is thought to be responsible for the development of secondary hepatic coproporphyrinuria and, in some cases, of δ-aminolevulinic-acid-uria too. Induction of δ-aminolevulinic acid synthase *(ALA-S)* is followed by heme depletion. Inhibition of uroporphyrinogen decarboxylase *(3)* may be an essential cause for triggering chronic hepatic porphyria under genetic and nongenetic conditions. Inhibition of ferrochelatase *(4)* in bone marrow may explain in the occurrence of a symptomatic protoporphyrinemia in patients with alcoholic liver syndromes. *PBG,* porphobilinogen; *URO-G,* uroporphyrinogen; *COPRO-G,* coproporphyrinogen; *PROTO-G,* protoporphyrinogen; *PROTO,* protoporphyrin; *VII,* heptacarboxyporphyrinogen; *VI,* hexacarboxyporphyrinogen; *V,* pentacarboxyporphyrinogen

days is necessary to simulate the state of latent chronic hepatic porphyria (Doss et al. 1981; von Tiepermann et al. 1980). The coproporphyrinuria following chronic alcohol consumption in rats with a coproporphyrin/uroporphyrin ratio of 4 : 1 associated with a decreased activity of hepatic uroporphyrin decarboxylase (Doss et al. 1981) most likely imitates the transition phase between secondary coproporphyrinuria and chronic hepatic porphyria found in human alcohol liver syndromes (Doss 1980). An alcohol-induced inhibition of hepatic uroporphyrinogen decarboxylase activity must be assumed (Doss 1980; Felsher et al. 1982) for an explanation of these transitional urinary porphyrin excretion constellations in alcoholic liver disease (Table 4).

In the hexachlorobenzene-induced porphyria in animals a concentration of 5% alcohol in the diet initially enhances the porphyrin excretion (Pearson and Malkinson 1965). Spin labeling investigations have shown that both hexachlorobenzene and alcohol increase the fluidity of the hepatocyte lipid membranes (Koszo et al. 1982). This mechanism and the augmentation of the hepatic microsomal ethanol-oxidizing system (Teschke et al. 1983) may be an important pathogenetic factor in human chronic hepatic porphyria, including porphyria cutanea tarda, which is regularly associated with liver injury and designated a "membrane disease" (Doss 1982). Furthermore, it appears convincing that redox changes and enhanced production of hepatotoxic acetaldehyde due to an increase of microsomal enzyme-oxidizing system (MEOS) activities may also contribute to the biochemical and clinical expression of chronic hepatic porphyria by alcohol.

Acute and chronic alcohol consumption inhibit uroporphyrinogen decarboxylase activity in erythrocytes (McColl et al. 1980, 1981). The activity of this enzyme is primarily decreased in the inherited type of porphyria cutanea tarda (Doss 1980; Elder et al. 1978, 1980; Kushner et al. 1976; Lehr and Doss 1981; von Tiepermann et al. 1978; Topi et al. 1976; Ventura et al. 1980; de Verneuil et al. 1978). One can assume that a defective uroporphyrinogen decarboxylase is more sensitive to alcohol than the normal enzyme. Therefore, alcohol can precipitate porphyria cutanea tarda in persons processing the genetic variant.

Table 4. Transition of secondary coproporphyrinuria to chronic hepatic porphyria (CHP) reflected by urinary porphyrin excretion: Progression (↓) induced by alcohol

Condition	Constellation of urinary porphyrins	Total porphyrins (μmol/24 h)	Uro (%)	Hepta (%)	Copro (%)	Porphyrin accumulation in the liver (U, 7)	Liver damage	Cutaneous symptoms
Normal	C >>> U >> 7 = 5 > 6 > 3	< 0.2	< 30	< 5	> 60	–	–	–
					↓			
Secondary copro-porphyrinuria	C >> U > 5 > 7 > 6 > 3	< 1.0	< 20	< 3	> 80	–	+	–
			– Transition –	↓	↓			
CHP A	C > U > 7 > 5 > 6 > 3	< 1.0	< 40	> 5	> 40	(+)	+	–
			↓	↓	↓			
CHP B	U > C > 7 > 5 > 6 > 3	< 1.5	> 40	< 20	< 40	+	+	–
				↓	↓			
CHP C	U > 7 > U > 5 > 6 > 3	> 2.0	> 50	> 20	> 10	++	+	–/(+)
					↓			
CHP D (PCT)	U > 7 >> C > 5 > 6 > 3	> 2.5	> 50	> 20	< 10	+++	+	+/++

U and C, uro- and coproporphyrin; 7, 6, 5, and 3, hepta-, hexa-, penta-, and tricarboxyporphyrin; PTC = prophyria cutanea tarda

Coproporphyrinogen Oxidase

The mitochondrial enzyme coproporphyrinogen oxidase is inhibited by alcohol as tested in leukocytes from patients with acute alcohol ingestion as well as in chronic alcoholics (McColl et al. 1980, 1981). It seems probable that the molecular injury to mitochondria produced by ethanol and acetaldehyde changes the activity of the hepatic enzyme, resulting in a pathologic coproporphyrinuria; alcohol interferes with functions of the inner mitochondrial membrane (Cederbaum and Rubin 1975). The change in the redox state of the cell with an increase in the hepatic NADH/NAD ratio therefore can not serve as an explanation of coproporphyrinuria, as NAD is not a cofactor of mammalian coproporphyrinogen oxidase (del Battle et al. 1965). On the other hand, coproporphyrinogen oxidase is highly sensitive to changes in oxygen concentrations (Keithly and Nadler 1983). Factors affecting the metabolism of coproporphyrinogen III are intriguing (Elder 1976): a decrease in the amount of the active enzyme as well as competitive inhibition has been discussed in connection with an increased oxidation and decreased uptake of coproporphyrinogen by the inner membrane of the mitochondrion.

Two main features of the pathobiochemical expression of coproporphyrinuria have been pointed out: impaired utilization of coproporphyrinogen III is associated with a compensatory production of coproporphyrinogen via an ethanol-induced augmentation of ALA synthesis (Fig. 2). Urinary ALA excretion is sometimes increased after alcohol abuse. In acute alcohol ingestion in man without preexisting liver damage the coproporphyrin III/I isomer ratio is nearly normal (Aziz et al. 1964; Orten et al. 1963). Urinary coproporphyrin I is relatively and absolutely elevated in alcoholic liver diseases, depending on the stage of intrahepatic cholestasis (Aziz et al. 1964; Orten et al. 1963; Doss 1980; Look et al. 1978; Sutherland and Watson 1951). The coproporphyrinuria in acute alcohol ingestion is mainly due to an inhibition of coproporphyrinogen oxidase (McColl et al. 1980) in the liver counterregulated by an increase of ALA synthase, whereas a stereospecific carrier disturbance of the symmetrical coproporphyrinogen I isomer via bile and its excretion via plasma and kidney leads to coproporphyrinuria in chronic alcohol liver diseases (Kaplowitz et al. 1972).

The coproporphyrinuria after acute alcohol ingestion is comparable to the hepatic reaction to lead (Doss 1982). An interference with lead is possible, which causes coproporphyrinuria besides ALAuria and protoporphyrinemia due to triple enzyme inhibition (Table 3). Thus the metabolic effects of lead and alcohol on the heme biosynthetic sequences catalyzed by ALA synthase, ALA dehydratase, coproporphyrinogen oxidase, and ferrochelatase are principally similar. The magnitude of alcohol-induced coproporphyrinuria in humans can be compared to that in subclinical lead intoxication. Generally, changes in coproporphyrinogen oxidase activity or impairment of coproporphyrinogen III utilization represent nonspecific reaction (Doss 1982): hepatic coproporphyrinogen metabolism reacts not only to alcohol and lead but also to drugs, other heavy metals, foreign chemicals, starvation, and fat diet. In experimental porphyria induced by hexachlorobenzene, coproporphyrinuria develops first (von Tiepermann et al. 1980) and corresponds to the alcohol-induced coproporphyrinuria in rats (Doss et al. 1981; Koszo et al. 1982).

Ferrochelatase and Hematopoiesis

Ferrochelatase is also inhibited by alcohol as studied in leukocytes (McColl et al. 1980, 1981). Protoporphyrinemia is of various origin (Labbe and Nielsen 1976). In alcoholics it can be explained by ferrochelatase inhibition if blood lead levels are normal, as the inference of lead intoxication in the development of protoporphyrinemia is evident (Table 3). Protoporphyrin is present as a zinc chelate both in alcohol- and lead-induced protoporphyrinemia. The alcohol-mediated inhibition of ferrochelatase will lead to alcohol-induced anemia in chronic alcoholism. Alcohol reduces concentrations of pyridoxal phosphate, which has been postulated as an essential cofactor for ferrochelatase (Labbe and Nielsen 1976). Probably, the depressed ferrochelatase (and coproporphyrinogen oxidase) activities in alcoholics are causally related to the increased bone marrow ALA synthase in patients with the acute reversible sideroblastic anemia of alcoholics (Fraser and Schacter 1980) in contrast to idiopathic sideroblastic anemia (Bottomley et al. 1973).

Alcohol is, like isoniazid and chloramphenicol, a hematopoietic toxin and inhibits in vitro rabbit reticulocyte heme and protein synthesis (Gruenspecht et al. 1979). As is known from alcohol-drug interactions in the liver (Mezey 1976; Rubin and Lieber 1972), human hematologic toxicity may result from additive interaction of chemically unrelated compounds, including alcohol and lead, which inhibit the heme pathway (Gruenspecht et al. 1979). Ethanol diminishes protein synthesis in human and rabbit reticulocytes in the presence of iron-transferrin for endogenous heme; alcohol is a direct toxin to developing red cell precursors via its effect on mitochondrial heme synthesis (Freedmann et al. 1975). In contrast to the alcohol-induced increase in bone marrow ALA synthase noted in man (Fraser and Schacter 1980), ethanol inhibits rabbit reticulocyte heme synthesis on the level of ALA synthase (Ibrahim et al. 1979). cAMP protects against ethanol toxicity by preventing inhibition of ALA synthase (Ibrahim et al. 1979).

Alcohol as Mediator of Inborn Errors of Porphyrin Metabolism

Acute Hepatic Porphyrias

Acute hepatic porphyrias are "regulatory diseases" with pharmacogenetic implications due to the enhanced inducibility of ALA synthase (Doss 1982; Kalow 1962; Maxwell and Meyer 1976). The clinical expression of acute porphyrias (Table 1) can be triggered by alcohol (Bloomer 1976; Moore 1980; Wetterberg 1976), which has been noted as the third commonest precipitating factor (Goldberg et al. 1981). Clinical observations have shown that fasting and/or drugs have a potentiating or additive effect on the porphyrinogenicity of alcohol (Doss et al. 1982a). In several cases the effect both of alcohol intake and alcohol withdrawal was monitored under metabolic control. Patients in the decompensated latent phase with high excretory profiles of heme precursors (Doss 1982) are especially endangered. Alcohol intake of > 60 g/day can increase urinary ALA, porphobilinogen, and porphyrin excretion up

Table 5. Response to alcohol intake in acute porphyrias

Patient no., sex, age	Day	Alcohol (g)	24-h urinary ratios of	
			ALA : PBG (both in μmol)	Uro : copro (both in nmol)
1, M, 24 (AIP)	1	~ 60	107 : 210	627 : 1,067
	2	~100	287 : 430	1,073 : 1,376
	3[a]		379 : 619	2,394 : 2,129
	2 weeks after alcohol withdrawal:			
	18	~ 80	94 : 187	748 : 232
	19[a]		418 : 443	1,760 : 1,553
2, F, 16 (AIP)	1	~ 20	260 : 416	903 : 689
	2[a]		397 : 720	1,233 : 1,166
3, M, 23 (PBG-S-DP)	1	~110	84 : 2	5 : 2,512
	2[a]		196 : 3	17 : 4,028
4, M, 25 (PBG-S-DP)	1	>100	150 : 2	73 : 2,577
	2[a]		277 : 7	80 : 3,874
5, F, 35 (AIP)	1		30 : 61	489 : 1,092
	After alcohol withdrawal:			
	9		23 : 23	484 : 511
6, M, 34 (VP)	1		12 : 9	102 : 1,092
	After alcohol withdrawal:			
	11		10 : 3	62 : 511[b]
Normal			<49 : <8	<30 : <120

AIP, acute intermittent porphyria; VP, variegate porphyria; PBC, porphobilinogen; PBG-S-DP, porphobilinogen synthase deficiency porphyria

[a] Expression of acute symptoms

[b] In addition, fecal protoporphyrin decreased from 970 to 360 nmol/g (normal < 150 nmol/g dry weight)

to three- to sixfold (Table 5), leading to an acute exacerbation. Recently, a case of variegate porphyria precipitated by the ingestion of heavy metals contained in illicit alcohol was reported (Hughes and Davis 1983).

In my own etiological study of 100 patients with acute hepatic porphyrias, alcohol was in second position as a causative factor (16%). Drugs, including estrogens, are responsible for about half of the clinical exacerbations. Fasting together with alcohol has led to manifestations of the acute phase in 5% of cases.

Alcohol increases the absorption of drugs like phenobarbital and inhibits the rats of disappearance of the porphyrinogenic drugs phenobarbital, tolbutamide, and meprobamate in vivo in man and animals (Mezey 1976). Inhibition of drug metabolism results from direct interference with alcohol. Acceleration of drug metabolism is due to induction of microsomal enzymes by ethanol and can increase toxicity (Mezey 1976). Inhibition and acceleration of drug metabolism by alcohol provide a synergistic basis for the expression of acute porphyrias as pharmacogenetic diseases. Alcohol increases the induction of ALA synthase in rats pretreated with phenobarbital (Held 1977). A sensitization of hepatic heme biosynthesis by the drug to the porphyrinogenic effect of alcohol is suggested. Drugs potentiate the porphyrinogenic effect of alcohol and vice versa in patients with genetic enzyme defects of acute hepatic porphyria (Fig. 1, Talbe 1).

Chronic Hepatic Porphyria (Porphyria Cutanea Tarda)

Porphyria cutanea tarda is the most frequent porphyria. It exists in latent as well as subclinical phases (phases A, B, and C) (Doss 1982) and is associated with liver lesions (Grossmann et al. 1979; Lefkowitch and Grossmann 1983; Topi et al. 1976). The urinary porphyrin constellation changes from mild to severe stages (Table 4). This progression may be induced or aggravated by alcohol. According to my own investigations, alcohol plays an important role in about two-thirds of all patients; its predominant influence in the expression of porphyria cutanea tarda has been reported repeatedly (Doss 1982; Grossmann et al. 1979; Habermann et al. 1975). Estrogens and oral contraceptives represent a second manifestation factor (Grossmann et al. 1979; Habermann et al. 1975; Sixel and Doss 1982). In half of the cases with estrogen-induced porphyria cutanea tarda in additive alcohol consumption was noted (Sixel and Doss 1982). Both alcohol and estrogens, including oral contraceptives, reduce the activity of uroporphyrinogen decarboxylase (Doss et al. 1981; McColl et al. 1980, 1981; Smith and Francis 1981). Women with estrogen-induced porphyria cutanea tarda have an inherited defect of the enzyme (Sixel and Doss 1982); this is also the case with so-called familial porphyria cutanea tards (Elder et al. 1980; Kushner 1982; Kushner et al. 1976; Lehr and Doss 1981; von Tiepermann et al. 1978; de Verneuil et al. 1978). A disturbance of uroporphyrinogen decarboxylase in the liver is a general prerequisite for the development of chronic hepatic porphyria; another premise is probably liver injury. A pathogenetic model is sketched out in Fig. 3.

The role of iron in the pathogenesis of porphyria cutanea tarda is controversial up to now; the conflicting aspects have been reviewed (Hines 1980; Pimstone 1982). It remains unclear how the multiple influence sof iron on the heme pathway (Bonkowsky et al. 1983; Elder et al. 1976; Kushner et al. 1976) coincide with a mutual action in evoking chronic hepatic porphyria. In experimental chronic hepatic porphyria the results on the function and importance of iron are also controversial (Elder et al. 1976; Shanley et al. 1970; Smith and Francis 1983; Schäfer et al. 1982). Elder and co-workers suggest that hexachlorobenzene or a metabolite acts directly to decrease the activity of hepatic uroporphyrinogen decarboxylase. In a study with purified enzyme preparations of human erythrocyte uroporphyrinogen decarboxylase, an inhibitory effect of iron could not be identified (de Verneuil et al. 1983). Clinical remission of porphyria cutanea tarda after phlebotomy without reversal of uroporphyrinogen decarboxylase deficiency suggests that iron may contribute to the clinical expression of the disease. However, a direct inhibition of uroporphyrinogen decarboxylase by iron does not seem to play a role in the pathogenesis of chronic hepatic porphyria (de Verneuil et al. 1983). The role of iron in the development of porphyria cutanea tarda has possibly also been overestimated (Hincs 1980). As to how far iron in combination with alcohol represents an especially porphyrinogenic factor for the development of porphyria cutanea tarda must be elucidated by further

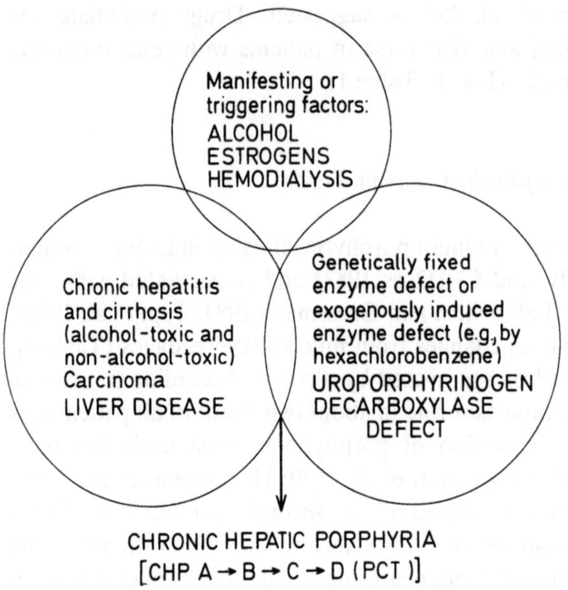

Fig. 3. Suggested pathobiochemical development of chronic hepatic porphyrias, including the coincidence of endogenous, especially genetic, and exogenous, especially toxic, factors. Exclusively exogenous factors may be responsible in chronic hepatic porphyria cases induced by environmental chemicals. Hepatic uroporphyrinogen decarboxylase activity is decreased in all cases; however, the activity of this enzyme in blood cells and in other tissues is decreased only in the genetically determined form of chronic hepatic porphyria *(CHP)* or porphyria cutanea tarda *(PCT)*

metabolic investigations concerning the interrelationship of chronic alcoholism and intestinal iron transport on the one hand, as well as the combined effect of alcohol and iron on the uroporphyrinogen decarboxylase on the other hand.

The concept of increasing the fluidity of hepatocyte membranes by ethanol or hexachlorobenzene (Koszo et al. 1982) appears to be an important model of the pathogenesis of porphyria cutanea tarda as a membrane disease (Doss 1982) in connection with the alcohol-induced and/or genetically predisposed reduction of uroporphyrinogen decarboxylase.

Erythrohepatic Protoporphyria

Hepatobiliary complications in erythropoietic protoporphyria are not caused by alcohol, but more likely by a toxic effect of protoporphyrin deposited in the liver and bile duct system (Bloomer 1982). As alcohol also inhibits hepatic ferrochelatase, which is genetically defect in this disease, and produces liver damage, alcohol ingestion will present a pathogenetic factor for the expression and/or aggravation of the potential hepatobiliary involvement.

Carrier of ALA Dehydratase Deficiency

In ALA dehydratase (porphobilinogen synthase)-deficient porphyria with an ALA dehydratase defect in the homozygous state (Doss 1982), the acute syndrome may be provoked by alcohol (Table 3). Heterozygous carriers are more sensitive to lead and also more sensitive to alcohol than healthy subjects (Doss et al. 1972).

Table 6. Effects of alcohol on the enzymes of the heme biosynthetic chain

Enzyme	Change in activity	Organ/cells studied	Reference
δ-Aminolevulinic acid synthase	Increase	Liver, bone marrow, leukocytes	Beattie et al. (1973); Held (1977); Kodama et al. (1983); Lane and Steward (1983); McColl et al. (1980, 1981); Paterniti et al. (1978); Shanley et al. (1968)
δ-Aminolevulinic acid dehydratase	Decrease	Erythrocytes, liver, kidney	Hamlyn et al. (1979); Kodama et al. (1983); Kondo et al. (1983); Krasner et al. (1974); McColl et al. (1980, 1981); Moore et al. (1971)
Uroporphyrinogen synthase	Increase	Erythrocytes	McColl et al. (1980)
Uroporphyrinogen cosynthase	Not studied		
Uroporphyrinogen decarboxylase	Decrease	Erythrocytes, liver, spleen	Doss et al. (1981); Elder et al. (1978); Felsher et al. (1982); McColl et al. (1980, 1981)
Coproporphyrinogen oxidase	Decrease	Leukocytes	McColl et al. (1980, 1981)
Protoporphyrinogen oxidase	Not studied		
Ferrochelatase	Decrease	Leukocytes	McColl et al. (1980, 1981)

Conclusion

Alcohol is an important precipitating and aggravating factor in human hepatic porphyrias: it is most important for porphyria cutanea tarda, including latent phases of chronic hepatic porphyria, but also important for acute hepatic porphyrias. The main reason for these effects consists in the inhibition of uroporphyrinogen decarboxylase and the induction of ALA synthase in the liver. Alcohol interferes with a preexisting uroporphyrinogen decarboxylase defect in chronic hepatic porphyria (membrane disease) and augments the inducibility of ALA synthase in the pharmacogenetic acute hepatic porphyrias (regulatory disease). Along this mechanism, synergistic forces between the primary enzyme deficiency, the depression of several heme biosynthetic enzymes by alcohol, and an altered drug metabolism may be responsible for the potentiation of the porphyrinogenic effect of alcohol. Therefore, alcohol consumption should be avoided in all hepatic porphyrias.

In healthy subjects and patients with alcoholic liver syndromes, alcohol can produce a secondary coproporphyrinuria which is not of clinical relevance and does not lead to clinical symptoms due to the porphyrin abnormality. In addition to the inhibition of uroporphyrinogen decarboxylase and induction of ALA synthase, alcohol inhibits ALA dehydratase, coproporphyrinogen oxidase, and ferrochelatase and increases uroporphyrinogen synthase as well (Table 6). The inhibition of ferrochelatase may produce secondary protoporphyrinemia in chronic alcoholism.

References

Abdulla M, Haeger-Aronsen B, Sevensson S (1976) Effect of ethanol and zinc on ALA-dehydratase activity in red blood cells. Enzyme 21: 248–252

Aziz MA, Schwartz S, Watson CJ (1964) Studies of coproporphyrin. VIII. Reinvestigation of the isomer distribution in jaundice and liver diseases. J Lab Clin Med 63: 586–604

Barton JC, Conrad ME (1978) Effects of ethanol on the absorption and retention of lead (40317). Proc Soc Exp Biol Med 159: 213–218

Beattie DS, Patton GM, Rubin E (1973) The control of δ-aminolevulinic acid synthase in rat liver mitochondria. Effect of pyrazole on the apparent 'induction'. Enzyme 16: 252–257

Bloomer JR (1976) The hepatic porphyrias. Pathogenesis, manifestations, and management. Gastroenterology 71: 689–701

Bloomer JR (1982) Protoporphyria. Semin Liv Dis 2: 143–153

Bonkowsky HG, Pomeroy JS (1977) Human hepatic δ-aminolaevulinate synthase: requirement of an exogenous system for succinyl-co-enzyme A generation to demonstrate increased activity in cirrhotic and anticonvulsant-treated subjects. Clin Sci Mol Med 52: 509–521

Bonkowsky HL, Healy JF, Sinclair PR, Sinclair JF, Shedlofsky SI, Elder GH (1983) Iron and the liver: acute effects of iron-loading on hepatic heme synthesis of rats. J Clin Invest 71: 1175–1182

Bottomley SS, Tanaka M, Self J (1973) δ-Aminolevulinic acid synthetase activity in normal human bone marrow and in patients with idiopathic sideroblastic anemia. Enzyme 16: 138–145

Brodie MJ, Moore MR, Goldberg A (1977) Enzyme abnormalities in the porphyrias. Lancet II: 699–701

Brodie MJ, Thompson GC, Moore MR, McColl KEL, Goldberg A, Hardie RA, Hunter JAA (1979) Haem biosynthesis in cutaneous hepatic porphyria: comparison with alcoholism and liver disease. Acta Hepatogastroenterol 26: 122–128

Brugsch JT (1937) Toxic porphyrinuria following chronic alcoholism. Proc Staff Meet Mayo Clin 12: 609–613

Cederbaum AI, Rubin E (1975) Molecular injury to mitochondria produced by ethanol and acetaldehyde. Fed Proc 34: 2045–2051

del Battle AMC, Benson A, Rimington C (1965) Purification and properties of coproporphyrinogenase. Biochem J 97: 731–740

de Matteis F (1978) Hepatic porphyrias caused by 2-allyl-2-isopropyl-acetamide, 3,5-diethoxycarbonyl-1,4-dihydrocollidine, griseofulvin and related compounds. In: de Matteis F, Aldringe WN (eds) Heme and hemoproteins. Springer, Berlin Heidelberg New York, pp 129–155

de Verneuil H, Aitken G, Nordmann Y (1978) Familial and sporadic porphyria cutanea. Two different diseases. Hum Genet 42: 145–151

de Verneuil H, Sassa S, Kappas A (1983) Purification and properties of uroporphyrinogen decarboxylase from human erythrocytes. A single enzyme catalyzing the four sequential decarboxylations of uroporphyrinogens I and III. J Biol Chem 258: 2454–2460

Doss M (1980) Pathobiochemical transition of secondary coproporphyrinuria to chronic hepatic porphyria in humans. Klin Wochenschr 58: 141–148

Doss M (1982) Hepatic porphyrias: pathobiochemical, diagnostic, and therapeutic implications. In: Popper H, Schaffner F (eds) Progress in liver diseases. Grune and Stratton, New York, pp 573–597

Doss M, Nawrocki P, Schmidt A, Strohmeyer G, Egbring R, Schimpff G, Dölle W, Korb G (1971) Einfluß von Diät, Glycin und Alkohol auf die Porphyrinurie bei chronischer hepatischer Porphyrie. Dtsch Med Wochenschr 96: 1229–1231

Doss M, Look D, Henning H, Nawrocki P, Schmidt A, Dölle W, Korb G, Lüders CJ, Strohmeyer G (1972) Hepatic porphyrins and urinary porphyrins and porphyrin precursors in liver cirrhosis. Klin Wochenschr 50: 1025–1032

Doss M, von Tiepermann R, Stutz G, Teschke R (1981) Alcohol-induced decrease in uroporphyrinogen decarboxylase activity in rat liver and spleen. Enzyme 26: 24–31

Doss M, Baumann H, Sixel F (1982a) Alcohol in acute porphyria. Lancet II: 1307–1307

Doss M, Becker U, Sixel F, Geisse S, Solcher H, Schneider J, Kufner G, Schlegel H, Stoeppler M (1982b) Persistent protoporphyrinemia in hereditary porphobilinogen synthase (δ-aminolevulinic acid dehydrase) deficiency under low lead exposure. A new molecular basis for the pathogenesis of lead intoxication. Klin Wochenschr 60: 599–606

Eales L (1979) Porphyria and the dangerous life-threatening drugs. S Afr Med J 56: 914–917

Elder GH (1976) Acquired disorders of haem synthesis. Essays Med Biochem 2: 75–114

Elder GH, Evans JO, Matlin SA (1976) The effect of the porphyrinogenic compound, hexachlorobenzene, on the activity of hepatic uroporphyrinogen decarboxylase in the rat. Clin Sci Mol Med 51: 71–80

Elder GH, Path MRC, Lee GB, Tovey JA (1978) Decreased activity of hepatic uroporphyrinogen decarboxylase in sporadic porphyria cutanea tarda. N Engl J Med 299: 274–278

Elder GH, Sheppard DM, de Salamanca RE, Olmos A (1980) Identification of two types of porphyria cutanea tarda by measurement of erythrocyte uroporphyrinogen decarboxylase. Clin Sci 58: 477–484

Elder GH, Smith SG, Herrero C, Lecha M, Mascaro JM, Muniesa AM, Czarnecki DB, Brenan J, Poulos V, de Salamanca RE (1982) Hepatoerythropoietic porphyria: a new uroporphyrinogen decarboxylase defect or homozygous porphyria cutanea tarda? Lancet I: 916–919

Felsher BF, Carpio NM, Engleking DW, Nunn AT (1982) Decreased hepatic uroporphyrinogen decarboxylase activity in porphyria cutanea tarda. N Engl J Med 306: 766–769

Franke K, Fikentscher R (1935) Die Bedeutung der quantitativen Porphyrinbestimmung mit der Lumineszenzmessung für die Prüfung der Leberfunktion und für Ernährungsfragen. MMW 82: 171–172

Fraser MB, Schacter BA (1980) Increased bone marrow delta-aminolevulinic acid synthetase activity in the acute reversible sideroblastic anemia of alcoholics. Am J Hematol 8: 149–156

Freedmann ML, Cohen HS, Rosman J, Forte FJ (1975) Ethanol inhibition of reticulocyte protein synthesis: the role of haem. Br J Haematol 30: 351–351

Goertz G, Jacobs A, Vizethum W (1980) Erythrocyte δ-aminolevulinic acid dehydratase activity in porphyria cutanea tarda. Arch Dermatol 268: 109–110

Goldberg A, Rimington C (1962) Diseases of porphyrin metabolism. Thomas, Springfield, pp 19–20

Goldberg A, Meredith PA, Miller S, Moore MR, Thompson GG (1978) Hepatic drug metabolism and haem biosynthesis in lead-poisoned rats. Br J Pharmacol 62: 529–536
Goldberg A, McColl KEL, Moore MR (1981) Alcohol and porphyria. Lancet II: 905–906
Granick S (1966) The induction of the synthesis of δ-aminolevulinic acid synthetase in chemical porphyria: a response to certain drugs, sex hormones and foreign chemicals. J Biol Chem 241: 1350–1375
Grossmann ME, Bickers DR, Poh-Fitzpatrick MB, Deleo VA, Harber LC (1979) Porphyria cutanea tarda. Clinical features and laboratory findings in 40 patients. Am J Med 67: 277–286
Gruenspecht NR, Torre J, Marcus DL, Freedman ML (1979) Additive inhibition of in vitro rabbit reticulocyte heme and protein synthesis with combinations of isoniazid, chloramphenicol and ethanol. Am J Med Sci 278(3): 207–215
Habermann HF, Rosenberg F, Menon IA (1975) Porphyria cutanea tarda: comparison of cases precipitated by alcohol and estrogens. Can Med J 113: 653–655
Hamlyn AN, Hopper J, Skillen AW (1979) Assessment of erythrocyte δ-aminolevulinate dehydratase for outpatient detection of alcoholic liver disease: comparison with γ-glutamyltransferase and casual blood ethanol. Clin Chim Acta 95: 453–459
Held H (1977) Effect of alcohol on the heme and porphyrin synthesis interaction with phenobarbital and pyrazole. Digestion 15: 136–146
Hines JD (1980) Effects of alcohol on inborn errors of metabolism: porphyria cutanea tarda and hemochromatosis. Semin Hematol 17: 113–118
Hughes GS, Davis L (1983) Variegate porphyria and heavy metal poisoning from ingestion of "moonshine". South Med J 76: 1027–1029
Ibrahim NG, Spieler PJ, Freedman ML (1979) Ethanol inhibition of rabbit reticulocyte haem synthesis at the level of δ-aminolaevulinic acid synthetase. Br J Haematol 41: 235–243
Kalow W (1962) Pharmacogenetics. Heredity and the response to drugs. Saunders, Philadelphia
Kaplowitz N, Javitt N, Kappas A (1972) Coproporphyrin I and III excretion in bile and urine. J Clin Invest 51: 2895–2899
Keithly JH, Nudler KD (1983) Protoporphyrin formation in *Rhizobium japonicum*. J Bacteriol 154: 838–845
Kikuchi G, Hayshi N (1981) Regulation by heme of synthesis and intracellular enzyme translocation of δ-aminolevulinate synthase in the liver. Mol Cell Biochem 37: 27–41
Kodama T, Kondo M, Urata G, Satoh H, Ohtake H, Iwasaki Y, Itakura H, Ohkubo A, Kosaka K (1983) Changes in aminolevulinate synthase and aminolevulinate dehydratase activity in cirrhotic liver. Gastroenterology 84: 236–241
Kondo M, Urata G, Shimizu Y (1983) Decreased liver δ-aminolaevulinate dehydratase activity in porphyria cutanea tarda and in alcoholism. Clin Sci 65: 423–428
Koszo F, Horvath LI, Simon N, Siklosi Cs, Kiss M (1982) The role of possible membrane damage in porphyria cutanea tarda: a spin label study of rat liver cell membranes. Biochem Pharmacol 31: 11–17
Krasner N, Moore MR, Thompson GG, McIntosh W, Goldberg A (1974) Depression of erythrocyte δ-aminolaevulinate acid dehydratase activity in alcoholics. Clin Sci Mol Med 46: 415–418
Kushner JP (1982) The enzymatic defect in porphyria cutanea tarda. N Engl J Med 306: 799–800
Kushner JP, Barbuto AJ, Lee GR (1976) An inherited enzymatic defect in porphyria cutanea tarda: decreased uroporphyrinogen decarboxylase activity. J Clin Invest 58: 1089–1097
Labbe RF (1967) Metabolic anomalie sin porphyria. The result of impaired biologicla oxidation. Lancet I: 1361–1364
Labbe RF, Nielsen L (1976) Clinical-biochemical interpretations of erythrocyte protoporphyrin. Ferrochelatase-pyridoxal phosphate studies. In: Doss M (ed) Porphyrins in human diseases. Karger, Basel, pp 141–147
Lane SE, Stewart ME (1983) Alcohol-mediated effects on the level of cytochrome P-450 and heme biosynthesis in cultured rat hepatocytes. Biochim Biophys Acta 755: 313–317
Lefkowitch JH, Grossman ME (1983) Hepatic pathology in porphyria cutanea tarda. Liver 3: 19–29
Lehr PA, Doss M (1981) Chronic hepatic porphyria with uroporphyrinogen decarboxylase defect in four generations. Dtsch Med Wochenschr 106: 241–245

Levere RD (1967) Porphyrin synthesis in hepatic cirrhosis: increase in δ-aminolevulinic acid synthetase. Biochem Med 1: 92–99

Lieber CS (1980) Metabolism and metabolic effects of alcohol. Semin Hematol 17: 85–99

Lindros KO, Hillbom ME (1969) Parallel changes in cytoplasmic and mitochondrial redox state of rat liver. Effects of ethanol, sorbitol and thyroid hormones. Life Sci 8: 1017–1022

Look D, Doss M, von Tiepermann R, Henning H (1978) Coproporphyrin isomer composition in urine in liver disease with intrahepatic cholestasis. J Clin Chem Clin Biochem 16: 49–50

Martini GA, Dölle W (1965) Krankheitsbilder bei Leberschädigung durch Alkohol. Dtsch Med Wochenschr 90: 793–799

Maxwell JD, Meyer UA (1976) Drug sensitivity in hereditary hepatic porphyria. In: Doss M (ed) Porphyrins in human diseases. Karger, Basel, pp 1–9

McColl KEL, ;Thompson GG, Moore MR, Goldberg A (1980) Acute ethanol ingestion and haem biosynthesis in healthy subjects. Eur J Clin Invest 10: 107–112

McColl KEL, Moore MR, Thompson GG, Goldberg A (1981) Abnormal haem biosynthesis in chronic alcoholics. Eur J Clin Invest 11: 461–468

McEwin R (1976) Porphyria in Australia. In: Doss M (ed) Porphyrins in human diseases. Karger, Basel, pp 233–244 (Proc I. int porphyrin meeting)

Meyer UA (1978) Regulation of heme biosynthesis in hepatic and erythroid cells: a brief introduction to the pathobiochemistry of human porphyrias. In: Doss M (ed) Diagnosis and therapy of porphyrias and lead intoxication. Springer, Berlin Heidelberg New York, pp 3–7

Mezey E (1976) Ethanol metabolism and ethanol-drug interactions. Biochem Pharmacol 25: 869–875

Mezey E (1982) Alcoholic liver disease. In: Popper H, Schaffner F (eds) Progress in liver diseases. Grune and Stratton, New York, pp 555–572

Moore MR (1980) International review of drugs in acute porphyria – 1980. Int J Biochem 12: 1089–1097

Moore MR, Beattie AD, Thompson GG, Goldberg A (1971) Depression of δ-aminolevulinic acid dehydrase activity by ethanol in man and rat. Clin Sci 40: 81–88

Nordmann Y, Deybach JC (1982) Congenital erythropoietic porphyria. Semin Liver Dis 2: 154–163

Orten JM, Doehr SA, Bond C, Johnson H, Pappas A (1963) Urinary excretion of porphyrins and porphyrin intermediates in human alcoholics. Q J Stud Alcohol 24: 598–609

Paterniti JR, Lin Ch-IP, Beattie DS (1978) δ-Aminolevulinic acid synthetase: regulation of activity in various tissues of the aging rat. Arch Biochem Biophys 191: 792–797

Pearson RW, Malkinson FD (1965) Some observations on hexachlorobenzene induced experimental porphyria. J Invest Dermatol 44: 420–432

Pimstone NR (1982) Porphyria cutanea tarda. Semin Liver Dis 2: 132–142

Rubin E, Lieber CS (1971) Alcoholism, alcohol, and drugs. Science 172: 1097–1102

Rubin E, Lieber CS (1972) Ethanol metabolism in the liver. In: Popper H, Schaffner F (eds) Progress in liver diseases. Grune and Stratton, New York, pp 549–566

Sassa S, Granick S (1970) Induction of δ-aminolevulinic acid synthetase in chick embryo liver cells in culture. Proc Natl Acad Sci USA 67: 517–522

Schäfer SG, Richter E, Meyer HD (1982) Iron absorption during the development of hexachlorobenzene-induced porphyria in rats. Arch Dermatol Res 274: 349–357

Schmid P, Schwartz S, Watson CJ (1954) Porphyrin content of bone marrow and liver in various forms of porphyria. Arch Intern Med 93: 167–190

Scotto AW, Rinehart RW, Beattie DS (1983) Aging-related decrease in hepatic mitochondrial and cytosolic δ-aminolevulinic acid synthase during experimental porphyria. Arch Biochem Biophys 222: 150–157

Shanley BC, Zail SS, Joubert SM (1968) Effect of ethanol on liver δ-aminolevulinate synthetase in rats. Lancet II: 70–71

Shanley BC, Zail SS, Joubert SM (1969) Effect of ethanol on liver δ-aminolevulinate synthase activity and urinary porphyrin excretion in symptomatic porphyria. Br J Haematol 17: 389–396

Shanley BC, Zail SS, Joubert SM (1970) Porphyrin metabolism in experimental hepatic siderosis in the rat. Br J Haematol 18: 79–87

Shrivastava KC, Orten JM (1972) Urinary excretion of coproporphyrin in alcoholics. Q J Stud Alcohol 33: 962–965

Simon N, Kiss M (1979) Study of erythrocyte δ-aminolevulinic acid dehydratase activity in porphyria cutanea tarda. Arch Dermatol 266: 121–126

Sixel F, Doss M (1982) Östrogeninduzierte chronische hepatische Porphyrie (Porphyria cutanea tarda). Verh Dtsch Ges Inn Med 88: 1066–1069

Smith AG, Francis JE (1981) Increased inhibition of hepatic uroporphyrinogen decarboxylase by hexachlorobenzene in male rats given the oestrogenic drugs diethylstilboestrol and chlorotrianisene. Biochem Pharmacol 30: 1849–1853

Smith AG, Francis JE (1983) Synergism of iron and hexachlorobenzene inhibits hepatic uroporphyrinogen decarboxylase in inbred mice. Biochem J 214: 909–913

Stonard MD (1978) Disturbances of porphyrin metabolism. In: Slater TF (ed) Biochemical mechanisms of liver injury. Academic Press, London, pp 443–468

Sutherland DA, Watson CJ (1951) Studies on coproporphyrin. VI. The effect of alcohol on the per diem excretion and isomer distribution of the urinary coproporphyrins. J Lab Clin Med 37: 29–39

Taljaard JJF, Shanley BC, Jourbert SM (1971) Decreased uroporphyrinogen decarboxylase activity in 'experimental symptomatic porphyria'. Life Sci 10: 887–893

Teschke R, Hasumura Y, Lieber CS (1975) Hepatic microsomal alcohol oxidizing system: affinity for methanol, ethanol, propanol and butanol. J Biol Chem 250: 7397–7404

Teschke R, Hasumura Y, Lieber CS (1976) Hepatic ethanol metabolism: the respective roles of alcohol dehydrogenase, the microsomal ethanol oxidizing system and catalase. Arch Biochem Biophys 175: 635–643

Teschke R, Bolsen K, Landmann H, Goertz G (1983) Effect of hexachlorobenzene on the activities of hepatic alcohol metabolizing enzymes. Biochem Pharmacol 32: 1745–1751

Topi GC, Gandolfo LD'Alessandro (1976) Liver in porphyria cutanea tarda. In: Doss M (ed) Porphyrins in human diseases. Karger, Basel, pp 312–324

Ventura E, Rocchi E, Gibertini P (1980) Hepatic and erythrocyte uroporphyrinogen synthetase and decarboxylase activities in porphyria cutanea tarda. Ital J Gastroenterol 12: 71–75

von Tiepermann R, Topi G, Gandolfo LD'Alessandro, Doss M (1978) Uroporphyrinogen-decarboxylase defect in erythrocytes in hereditary chronic hepatic porphyria. J Clin Chem Clin Biochem 16: 52–54

von Tiepermann R, Koss G, Doss M (1980) Uroporphyrinogen decarboxylase deficiency in experimental chronic hepatic porphyria. Hoppe Seylers Z Physiol Chem 361: 1217–1222

Wetterberg L (1976) Report on an international survey of safe and unsafe drugs in acute intermittent porphyria. In: Doss M, Nawrocki P (eds) Porphyrins in human diseases – report of the discussions. Falk, Freiburg, pp 191–202

Zieve L, Hill E, Schwartz S, Watson CJ (1953) Normal limits of coproporphyrin excretions determined by an improved method. J Lab Clin Med 41: 663–669

14 Ethanol and Hepatic Cell Regeneration

J.-G. Joly and L. Duguay

Centre de Recherches Cliniques, Hôpital Saint-Luc,
1058, rue Saint-Denis, Montréal, Québec H2X 3J4, Canada

Introduction

Regeneration of the liver is a remarkable phenomenon of fundamental biological significance. The normal adult hepatocyte is a stable cell with a life span similar to the life of the organism. Destruction of part of the liver sets in motion an explosive burst of mitotic activity that will restore the liver mass in a matter of days or weeks, depending on the species. Cell replication will stop when the deficit has been restored. Liver regeneration has been demonstrated in every species in which it was investigated, including man, but most studies have been done in rats. Following a 67% hepatectomy in the rat, DNA synthesis requires 5–8 h to begin and 24 h to peak, and declines progressively thereafter. Mitotic activity follows the same course with a time lag of 8–12 h. The initial response involves only the hepatocytes but, beginning a day later, the bile ducts and littoral cells share in the response (Bucher and Malt 1971). The magnitude of the response is proportional to the amount of liver mass excised. Hepatocytes throughout the liver participate in the process, not just those at the margin of resection. In rats, after a 67% hepatectomy the organ is restored in 7–10 days (Karran et al. 1974). In man it is well advanced after a few weeks (Blumgart et al. 1971). The architecture of the regenerated liver is indistinguishable from the normal.

It now appears that the complex process involved in regulating proliferation of hepatocytes is largely controlled by humoral factors. The hormones of the pituitary, adrenal, parathyroid, and thyroid have been shown to influence the regenerative response (Hays 1974; Shulte-Herman 1974). The release by the liver itself of humoral substances that inhibit or stimulate liver regeneration has been implicated (Labrecque and Pesch 1975; Makowka et al. 1983). There is solid evidence that the splanchnic organs, especially the pancreas, are involved in releasing factors that preserve the integrity of the hepatocyte (Starzl and Terblanche 1979) and promote its proliferation (Duguay and Orloff 1976; Bucher and Swaffield 1975). The exact nature of the regulatory system remains to be elucidated. For wider background information the reader is referred to reviews by Becker (1973), Hays (1974), Starzl and Terblanche (1979), Karran and Eagles (1979), and Bucher and McGowan (1979). Criteria of liver regeneration involve a wide range of morphological, histological, chemical, and isotopic tracer techniques. Early methods were based on determinations of weight and volume of the organ, coupled with measurements of

Alcohol Related Diseases in Gastroenterology
Edited by H. K. Seitz and B. Kommerell

protein and DNA content. Mitotic index enjoyed some popularity but was plagued with sampling errors and subjectivity in the counting; these methods have been largely replaced by more accurate isotopic techniques based on the rate of incorporation of ^{3}H-thymidine into DNA (Bucher 1963). This point requires emphasis in view of certian conflicting observations to be discussed later.

Liver regeneration is also a matter of clinical relevance, since it is the common mechanism by which a patient will recover from a liver injury, be it infectious, surgical, toxic, or traumatic. In the Western world, alcohol is the most important etiological factor associated with liver disease. In North America, 80% of all liver cirrhosis is thought to be related to abusive alcohol consumption (Garceau 1963). In large urban areas, cirrhosis of the liver is the third major cause of death in patients between 35 and 54 years old (Galambos 1979). Liver injury is explained by a combination of a number of factors deleterious to the liver: intracellular accumulation of acetaldehyde, microsomal activation of hepatotoxins, alterations in the redox state, and enhancement of lymphocyte cytotoxicity (Lieber 1978). Ethanol by itself, however, appears to be a rather benign hepatotoxin: a man of average size who drinks 170 g ethanol a day stands only a 50% chance of developing liver cirrhosis (Pequignot 1963). In the strictly controlled environment of the laboratory, it has been very difficult to induce cirrhosis in rats by alcohol consumption only (Lieber et al. 1963). If ethanol by itself does not constantly produce irreversible liver damage, it appears to increase the vulnerability of liver cells to additional trauma, be it malnutrition (Lieber and Rubin 1969), infection, or chemical toxins (Joly and Hétu 1978; Hétu et al. 1983). These observations have led to speculations that ethanol might interfere with the capacity of the liver to regenerate. The available information on the subject is reviewed.

Effect of Ethanol on Liver Regeneration

Studies on the effects of ethanol on liver regeneration in the rat are summarized in Tables 1–3. Protocols differ in the dose of ethanol administered, the time of ethanol administration, and the times at which incorporation of DNA precursors were assessed following partial hepatectomy. Of 13 studies, 11 reported an inhibition of tritiated thymidine incorporation into hepatic DNA, while two found no such inhibition. The measurement of parameters such as liver mass, total liver protein, and DNA content in three studies did not lead to the conclusion that ethanol inhibits liver regeneration following hepatic resection; total liver mass and DNA content were the same as in controls when measured 4–15 days after surgery (Frank et al. 1979; Pösö et al. 1980b; Orrego et al. 1981).

The studies reviewed are grouped on the basis of the timing and frequency of ethanol administration in relation to partial hepatectomy. Data on the effect of acute ethanol administration within 24 h of a 67% hepatic resection are summarized in Table 1. In Table 2 are summarized results of studies on the effects of short-term (up to 6 days) ethanol administration following partial hepatectomy. Results of the effect of chronic (10 days to 8 months) ethanol administration prior to partial hepatic resection and summarized in Table 3.

Table 1. Acute ethanol administration

Reference	Ethanol (g/kg)	No. of doses	Time of ethanol relative to hepatectomy (h)	Time of killing following post hepatectomy (h)	Findings[a]
Craig (1975)	~4	1	+0.5, +16, +22	20–24	No inhibition of incorporation
Weesner et al.	7	1	–	–	60% suppression of α-fetoprotein blood levels 18 h after ethanol
Wands et al. (1979)	6	1	−4 to +20	24	No inhibition if ethanol given between −4 h and +8 h and at +20 h relative to hepatectomy. Inhibition of incorporation when ethanol given at +12 h and +16 h after hepatectomy
	6	2	0, +4	24	35% inhibition of incorporation
	6	2	+4, +8	24	58% inhibition of incorporation
	6	3	0, +4, +8	24	80% inhibition of incorporation
Frank et al. (1979)	8	1	0, +16	24	~ 70% and ~ 50% inhibition of incorporation
	8	1	+22	24	No inhibition of incorporation
	2.67	3	0, +8, +16	–	> 85% inhibition of incorporation
Dreosti et al. (1981)	2	1	+14	23	27% inhibition of incorporation
	5	1	+14	23	45% inhibition of incorporation
	5	1	–	–	Ethanol administered to dams at 16 days of gestation. Livers of fetuses studied 2 h after ethanol. 44% inhibition of incorporation found
	5	2	–	–	Same as above. Interval between the two doses: 7 h. Liver of fetuses studied 2 h after last dose of ethanol. 67% inhibition of incorporation
Makowka et al. (1981)	2	2	+1, +3	12–42	Inhibition of incorporation between 18 h and 42 h after hepatectomy. Maximal (89%) inhibition at 24 h. Inhibition partially reversed by regenerating liver cytosol, which increases three to fourfold thymidine incorporation

[a]Incorporation refers to ^{3}H-thymidine incorporation into hepatic DNA

One can summarize the existing data by stating that ethanol appears to exert a dual effect. On the one hand, ethanol inhibits the early incorporation of DNA precursors into hepatic DNA; on the other hand, it apparently fails to inhibit hepatic regeneration, since total liver mass and DNA content of ethanol-fed animals do not differ from those of controls 14 days following partial hepatectomy.

Effect of Acute Ethanol Administration

In six studies (Table 1), ethanol was administered in one to three doses at various time periods in relation to surgery. Ethanol was given 4 h prior to partial hepatic resection, at the time of surgery, or 0.5 h to 22 h after partial hepatectomy. In all studies, the major criterion of liver regeneration was the incorporation of ^{3}H-thymidine into hepatic DNA. In all but one study, incorporation of labeled thymidine into hepatic DNA was assessed between 20 and 24 h after surgery. In one report, incorporation studies were carried out up to 42 h after hepatic resection (Makowka et al. 1981). In one study, the effect of ethanol was assessed by its effect on blood α-fetoprotein (α-FP) levels in nonhepatectomized animals. Maximal suppression of α-FP was observed 18 h after administration of a large single dose of ethanol (Weesner et al. 1978). In another study, the effect of ethanol administration to pregnant mothers on the incorporation of ^{3}H-thymidine into hepatic DNA of fetuses was studied 2 h following the administration of a single or the latter of two doses of ethanol. Maternal administration of ethanol was found to inhibit incorporation of thymidine into fetal hepatic DNA (Dreosti 1981).

Craig (1975) could not observe any inhibition of ^{3}H-thymidine incorporation into hepatic DNA after administering a single dose of approximately 4 g ethanol per kilogram 0.5 h, 16 h, or 22 h after partial hepatectomy. The lack of inhibition by ethanol when administered 0.5 h or 22 h after surgery is in agreement with the results of other studies. Wands et al. (1979) found no inhibition of incorporation when a single dose (6 g/kg) of ethanol was administered prior to surgery or up to 8 h after partial hepatic resection; no inhibition of incorporation of thymidine was observed if ethanol was given 20 h following surgery.

Frank et al. (1979) reported that the administration of a single sublethal dose (8 g/kg) of ethanol 22 h following hepatic resection did not inhibit ^{3}H-thymidine incorporation into hepatic DNA. When multiple doses of ethanol (8 g/kg) were administered 0 and 4 h, 4 and 8 h, and 0, 4, and 8 h after surgery, incorporation of the precursor into hepatic DNA was inhibited 35%, 58%, and 80% respectively (Frank et al. 1979). The administration of two low doses of ethanol (2 g/kg) as early as 3 h following hepatic resection led to inhibition of thymidine incorporation into hepatic DNA which was maximal 18 h after surgery (Makowka et al. 1981). The lack of inhibition of liver DNA precursor incorporation 16 h after ethanol administration cannot be explained from the results of other studies except by the fact that a single dose of 4 g/kg might not have sufficiently elevated blood ethanol concentrations. However, Dreosti et al. (1981) observed a 27% inhibition of incorporation by administering a single dose of ethanol (2 g/kg) 14 h after partial hepatic resection. This group also reported a 45% inhibition of thymidine incorporation into hepatic DNA following a single dose of 5 mg ethanol per kilogram administered 14 h after surgery. After the administration of a single dose of ethanol, no inhibition of incorporation could be observed if ethanol was administered earlier than 8 h and at or later than 20 h following partial hepatectomy. When two or three doses of ethanol are administered, the inhibitory effect of ethanol has been observed even when given at the time of surgery with doses of 6 g/kg or given 1 h after partial hepatectomy with doses as low as 2 g/kg. One can most probably conclude that acute ethanol administration inhibits the incorporation of thymidine into hepatic DNA following a 67% hepatic resection.

Effect of Short-Term Ethanol Administration After Partial Hepatectomy

In two studies (Table 2), ethanol was administered daily following a 67% hepatic resection for a period of 4–6 days up to the time the animals were killed. In both studies, ^{3}H-thymidine incorporation studies were carried out and total liver mass and DNA content were measured. Frank et al. (1979) observed that a daily dose of 8 g/kg administered at 8-h intervals inhibited thymidine incorporation into hepatic DNA between 18 and 48 h after partial hepatectomy. Pösö et al. (1980b) administered ethanol by gastric intubation of liquid diets four times per day. The daily dose of ethanol was estimated from blood ethanol concentrations to maintain continuous intoxication; doses ranged from 3.7 to 11.4 g/kg per day. Ethanol depressed incorporation of thymidine into DNA of regenerating livers on days 2 and 3 following partial hepatic resection. The inhibition was most pronounced (60%) on day 2 following surgery and this was associated with a significant decrease in mitotic activity. In the same studies, ethanol was found to inhibit the incorporation of ^{14}C-leucine into liver proteins on the 2nd day of regeneration.

Table 2. Short-term ethanol administration after partial hepatectomy

Reference	Ethanol (g/kg/d)	Time of ethanol administration after hepatectomy	Findings[a]
Frank et al. (1979)	8	Up to 4 days	Ethanol inhibits incorporation between 18 h and 48 h after hepatectomy when given daily from the time of surgery on. Maximal (90%) inhibition 24 h after hepatectomy. 4 days after surgery, liver weight, protein, and DNA content reconstituted in ethanol-fed rats. Conclusion: no inhibition by ethanol of liver regeneration
Pösö et al. (1980b)	3.7–11.4	Up to 6 days	Ethanol depresses incorporation and mitotic index on days 2 (60%) and 3 after hepatectomy. Ethanol depresses leucine incorporation into liver proteins 48% on day 2 after surgery. 5 days after hepatectomy liver weight, protein, and DNA content not decreased in ethanol rats. Conclusion: no inhibition by ethanol of liver regeneration – inhibition by ethanol of the induction of DNA replication

[a] Incorporation refers to ^{3}H-thymidine incorporation into hepatic DNA

While these two sets of experiments observed an inhibition of thymidine incorporation into hepatic DNA, neither study revealed a decrease in total liver weight, protein, and DNA content after 4 or 6 days of regeneration in ethanol-fed animals. These experiments led to the conclusion that while ethanol may inhibit the early phase of DNA replication, it does not inhibit liver regeneration (Frank et al. 1979; Pösö et al. 1980b).

Effect of Chronic Ethanol Administration

In five studies (Table 3), ethanol was administered for periods ranging from 10 days to 8 months. In four studies, ethanol administration in daily doses as low as 5 g/kg was associated with the inhibition of thymidine incorporation into DNA following partial hepatic resection. In all studies, ethanol was also administered following surgery until the animals were killed. Weesner et al. (1978) reported that daily administration of ethanol for 8 months depressed α-FP blood levels by 57% in nonhepatectomized rats. Wands et al. (1979) found that after 30 days of administration, ethanol inhibited thymidine incorporation into hepatic DNA when partial hepatectomy was performed mid-morning and the incorporation studies were carried out 24 h later. No such effect could be observed if surgery was performed at night and the incorporation experiments carried out the following night. Since continuous presence of ethanol in blood was suggested to be needed for the observation of an ethanol inhibition of thymidine incorporation (Pösö et al. 1980b), it is possible that significant ethanol blood concentrations were only present following a period of nocturnal feeding. However, Duguay et al. (1981) observed that a 3-week period of ethanol feeding was associated with a depression and a delay of thymidine incorporation into hepatic DNA and of ^{3}H-thymidine labeling of liver cell nuclei. When the ethanol diet was withdrawn and replaced by the control diet 1 or 3 days prior to partial hepatectomy, and when ethanol was not administered following surgery, the number of ^{3}H-labeled nuclei of hepatocytes was still decreased by more than 50% 24 h after hepatic resection. After a 7-day period of abstinence from alcohol, the ethanol inhibition of labeled thymidine incorporation into liver cell nuclei could no longer be observed. In another set of experiments, the same authors reported that when phenobarbital was incorporated in the liquid diets to replace ethanol, a 3-week administration of phenobarbital did not inhibit thymidine incorporation into hepatic DNA. Orrego et al. (1981) administered ethanol in liquid diets for 10 days and found no inhibition of thymidine incorporation into hepatic DNA and no decrease in cell labeling in spite of the continued administration of ethanol postoperatively. They further observed that total liver mass, protein, and DNA content were not decreased when compared to controls 7 and 14 days following partial hepatectomy. They confirmed the previous observation by Frank and Houck (1980), who failed to observe a difference in total liver weight and DNA content between ethanol-fed and control animals 24 h after partial liver resection in rats fed liquid diets for 2 months. The latter authors, however, reported a decreased incorporation of thymidine incorporation into hepatic DNA in regenerating livers of ethanol-fed animals. The same group reported similar findings after short-term ethanol administration following partial hepatectomy (Frank et al. 1979).

Table 3. Chronic ethanol administration

Reference	Ethanol (g/kg/d)	Time of ethanol administration prior to hepatectomy	Time of ethanol administration after hepatectomy	Findings[a]
Weesner et al. (1978)	5	8 months	–	α-Fetoprotein blood levels depressed 57%
	5	8 months	Up to 5 days	86% inhibition of incorporation
Wands et al. (1979)	–	30 days	24 h	55% inhibition of incorporation if surgery performed between 8.00 and 10.00 a. m. No inhibition if performed 12 h later
Frank and Houck (1980)	Not stated: probably 10–12	2 months	24 h	Ethanol inhibition of thymidine (94%) and cytidine (20%) incorporation. Liver weight and total DNA greater in ethanol animals. Conclusion: no inhibition by ethanol of liver regeneration
Duguay et al. (1981)	10–15	3 weeks	1, 2, and 3 days	Inhibition of incorporation and 24 h delay in regeneration activity. Number of labeled nuclei depressed more than 50% when ethanol withdrawn 3 days prior to hepatectomy and no ethanol given after surgery. This latter effect not further observed when ethanol withdrawn 7 days prior to surgery
Orrego et al. (1981)	Not stated: probably 10–12	10 days	1, 7, and 14 days	No difference in incorporation and percentage of labeled cells between ethanol-fed and control rats at any time. No decrease in liver weight, protein, and DNA content in ethanol-fed rats. Increase in the surface area of hepatocytes at days 7 and 14 after ethanol

[a] Incorporation refers to ^{3}H-thymidine incorporation into hepatic DNA

Here again, one could conclude, on the basis of radiotracer methods, that ethanol may inhibit and delay the early phase of liver regeneration. Ethanol would not inhibit the whole process of liver regeneration, since total liver mass and DNA content appear to be completely restored in ethanol-fed animals.

Discussion

Time Course of Liver Regeneration

Except for one report (Craig 1975), all studies show that acute ethanol administration inhibits the thymidine incorporation into DNA in regenerating liver following partial hepatectomy. In normal rats, the course of regeneration manifested by the rate of incorporation of labeled thymidine into DNA follows a well-known pattern. A 14–16 h delay after partial hepatectomy is followed by a sharp peak of activity at around 24 h. The activity then drops abruptly and continues at a lower level (Bucher and Swaffield 1975). In the rat, the normal liver mass and volume is restored within 7–10 days. The use of radiolabeled precursors has revealed that parenchymal cells, accounting for 60%–65% of the cell population but for 95% of the liver volume, start incorporating the label 12 h after liver resection, show a peak of activity at around 24 h, and show little activity after 54–60 h. The regenerative activity recorded in the ductal and the littoral cells shows a 12–24 h delay (Fabrikant 1968). DNA synthesis precedes mitotic activity. The peak uptake of DNA precursors, at 18 h after hepatectomy, and the peak mitotic activity, at 30 h, are separated by a lag phase of 12 h. After partial hepatectomy the G_1 phase has a duration of 12–15 h in the first cycle and of 5 h in subsequent cycles. The *S* phase, preceding DNA synthesis, lasts 8 h and the *M* phase, 3.5 h. The total cell cycle of normally 21.5 h is shortened to 16.5 h after the first cell cycle following hepatic resection (Karran and Eagles 1979). Ethanol has been shown to inhibit thymidine incorporation into DNA after the administration of a single dose 14–20 h following liver resection. In the first 16 h following partial hepatectomy in the rat, a number of biochemical events take place before a spurt of hepatic DNA labeling is recorded (Bucher 1967a, b; Whitfield et al. 1979). This preparation phase is dependent on protein and RNA synthesis. Ethanol interferes with the preparatory phase of DNA snythesis.

Ethanol Blood Levels and Inhibition of Liver Regeneration

A relationship between blood ethanol concentrations and this inhibitory effect of ethanol has been suggested. Wands et al. (1979) could find no inhibition of thymidine incorporation with ethanol blood levels between 37 ± 12 and 105 ± 28 mg/gl; an inhibitory effect was, however, found with blood ethanol levels greater than 144 ± 4 mg/gl and up to 476 ± 16 mg/dl. Greater degrees of inhibition tended to be recorded with higher blood ethanol levels. Pösö et al. (1980b) reported inhibition of thymidine incorporation into hepatic DNA with blood levels of 216 ± 74 mg/dl. These authors selected doses of ethanol to maintain continuous levels of ethanol in the blood of the animals throughout the experiments. In two sets of experiments Dreosti et al. (1981) reported 27% and 45% inhibition of thymidine incorporation into hepatic DNA with blood concentrations of 168 ± 32 mg/dl and 324 ± 71 mg/dl respectively. These authors, however, do not state the time at which blood ethanol levels were measured in relation to the partial liver resection. The same authors administered ethanol to pregnant mothers and studied the incorporation of thymidine into hepatic DNA of the fetuses. Maternal ethanol blood levels of 144 ± 27 mg/dl and 373 ± 92 mg/dl were associated with inhibition of 44% and 67% respectively of thymidine incorporation into fetal hepatic DNA. In vitro, ethanol inhibits the incorporation of thymidine into the DNA of liver cells in culture 25% at 500 mg/dl, 63% at 1,500 mg/dl, and 62% at 2,000 mg/dl (Dreosti et al. 1981). Alcoholic intoxication or related metabolic alterations interfere with the incorporation of DNA precursor in regenerating livers or even in liver cell cultures. However, after 3 weeks of ethanol administration, the number of ^{3}H-labeled nuclei of hepatocytes was found to be decreased by more than 50% in regenerating livers after 1 day or 3 days of abstinence from alcohol prior to partial hepatectomy. No blood alcohol was detectable 1 day after ethanol withdrawal (Duguay et al. 1981). These latter observations may, however, be explained by a difference between the acute and chronic models of alcohol intoxication.

Ethanol does not need to be metabolized to exert its inhibitory effect on DNA precursor incorporation in regenerating livers. This ethanol effect is not suppressed by pyrazole (300 mg/kg). A single dose of 2.67 mg ethanol per kilogram administered with pyrazole inhibits thymidine incorporation into DNA 99% (Rayyes and Frank 1976). Frank et al. (1979) also reported that when hepatic ethanol metabolism was suppressed with pyrazole, ^{3}H-DNA labeling was still inhibited while pyrazole by itself showed no inhibitory effect.

Acetaldehyde, however, when administered to pregnant mothers, interferes with thymidine incorporation into fetal hepatic DNA. The doses needed to achieve an effect similar to that of ethanol are much smaller than those of ethanol. Acetaldehyde depresses thymidine incorporation by 60% and 80% at doses of 50 and 100 mg/kg respectively. Similar observations were made using a liver cell culture model (Dreosti et al. 1981). The results of the experiments measuring the incorporation of ^{3}H-thymidine into DNA in regenerating livers suggest that ethanol greatly impairs hepatic regeneration. However, when total liver weight and DNA content were measured in such experiments, the results did not appear to show that ethanol impaired liver regeneration. The rate of liver growth found in control and experimental animals was comparable to that found in normal rats by other investigators (Frank et al. 1979), leading to the conclusion that acute (Frank et al. 1979) or chronic (Orrego et al. 1981) ethanol administration does not impair liver regeneration. What are the factors that could explain such conflicting results?

Isotopic Tracer Techniques

When injected in vivo, ^{3}H-thymidine is incorporated as a unit into DNA of regenerating liver; but thymidine is degraded by biological systems and can label amino acids, tricarboxylic acid cycle intermediates, glycogen, RNA, proteins, and lipids. Only a small fraction is found in nuclear DNA; the bulk of the label is found in the proteins and lipids of the microsomes and of mitochondria. In the cytosol, ^{3}H-thymidine can label triglycerides (Schneider and Greco 1971), but nonspecific binding of ^{3}H-thymidine to lipids and proteins could not explain the differences measured between control and experimental animals. Eight percent of the total hepatic ^3H label was associated with extracted lipids in the livers of control animals, compared to 5% in those of ethanol-fed rats. Similar amounts of ^3H label were measured in the total protein precipitate in both groups of animals (Frank et al. 1979). When studied, metabolism of ^3H-thymidine was not different in control and ethanol-fed animals. In one series of experiments, the supernatant fraction from hepatic homogenates following TCA precipitation was counted and subjected to chromatography. In both supernatants, ^3H-thymidine and its metabolites were recovered in the same proportion in ethanol-fed and control animals (Frank et al. 1979). In the studies where it was sought, ethanol administration was not found to alter either the intracellular nonspecific labeling by ^3H-thymidine or the metabolism of the precursor (Frank et al. 1979). Ethanol does not change the optimal time for ^3H-thymidine administration. Whether administered at 1, 2, or 3 h prior to killing in rats partially hepatectomized 24 h earlier, ^3H-thymidine incorporation reached a plateau between 1 and 3 h both in the control and in the ethanol-fed animals (Frank et al. 1979). The route of administration of the labeled precursors was not a source of variability. Whether they were administered intravenously or intraperitoneally, the same specific activity of DNA was obtained and the same inhibition of hepatic DNA labeling was observed after ethanol (Frank et al. 1979). Pösö et al. (1980b) raised the point that ethanol might accelerate the peak of thymidine incorporation. When rats are subjected to stressful procedures at the time of hepatectomy or during the early period of regeneration, the onset of the regenerative process tends to be rather delayed (Bucher and Swaffield 1973). Frank et al. (1979) observed that acute ethanol administration seemed to delay the peak of thymidine incorporation from 24 to 48 h. A similar delay was also observed following chronic ethanol administration (Duguay et al. 1981). In two studies, the observed inhibition of ^3H-thymidine incorporation into hepatic DNA in one set of conditions was later shown to either decrease after addition of regenerating rat liver cytosol (Makowa et al. 1981) or disappear following 7 days of abstinence from ethanol (Duguay et al. 1981). It is highly unlikely that methodological errors related to the use of ^3H-thymidine may explain the fact that the inhibition of its incorporation into hepatic DNA has been confirmed by many laboratories under as many experimental conditions. It appears safe to conclude that acute and chronic ethanol administration interfere at least with the preparatory phase of liver regeneration.

Liver Weight and Chemical Methods

In a review of the methods for the evaluation of the regenerative response, restoration of the liver mass has been described as the simplest but probably the least reliable index of regeneration. Liver weight may change with deposition of lipids, glycogen, or other materials not directly related to the proliferative response. Precise weighing is difficult because, since the liver is a "blood reservoir", the variable amounts of blood contained may equal over 50% of the weight of the drained organ (Bucher 1963). The experimental models of ethanol administration are particularly vulnerable to such criticisms.

Acute and chronic ethanol administration in normal rats cause fatty infiltration of the liver and increase the liver mass. This increase can also be explained by an accumulation of the export proteins of the hepatocytes, in turn responsible for an increase in cell volume (Baraona et al. 1975, 1977). Total liver DNA does not change. The osmotic effect of the increased protein content of the liver cells attracts water, which accounts for 60% of the increase in liver weight. The increase in liver weight cannot be accounted for by the summation of the effects of the increased proteins, lipids, ions, or DNA. The existence of a yet unknown low-molecular-weight substance has been postulated (Israel et al. 1979). Liver regeneration by itself following partial hepatic resection is also associated with lipid infiltration. By 18 h, triglycerides increase fivefold. Liver regeneration enhances polyploidization (Bucher 1963), which is itself associated with an increase in the protein content (Gaub et al. 1981; Sweeney et al. 1979).

Acute ethanol administration following partial hepatectomy appears to add to the triglyceride accumulation which normally occurs during hepatic regeneration. The total triglyceride content of reconstituted livers of ethanol-fed rats was found to be fourfold, fivefold, and sixfold greater than in controls at 24, 72, and 96 h respectively following partial hepatic resection (Frank et al. 1979). Ethanol thus seems to exaggerate total lipid accumulation following partial liver resection (Pösö et al. 1980b). Chronic ethanol administration also strikingly enhances total lipid accumulation after partial hepatectomy. In controls, total lipids increase following hepatectomy but return to normal values 14 days after surgery. In ethanol-fed animals a striking increase in liver lipid accumulation can be observed 24 h and 7 days after partial hepatectomy. It continues to rise for 14 days, when it reaches a value fivefold that of controls. This lipid accumulation is more than double the fat accumulation observed in the sham-operated ethanol-fed animals. In controls, light microscopy shows fat droplets around the hepatic venules with a maximum at 7 days after partial hepatectomy. In ethanol-fed rats, large confluent fatty cysts are distributed through the hepatic lobule (Orrego et al. 1981). Continued administration of ethanol following partial hepatic resection thus appears to enhance the toxicity associated with chronic ethanol administration in animals without liver resection.

Liver protein accumulates more rapidly in regenerating livers of partially hepatectomized animals administered ethanol. Eighty percent of the original hepatic protein content is restored 96 h after liver resection, compared to 57% in controls. The effect of ethanol and of hepatic regeneration of total liver protein content would appear to be additive (Frank et al. 1979). Acute ethanol administration has a dual effect on liver proteins. On the 2nd day after partial hepatectomy, ethanol decreases the incorporation of ^{14}C-leucine into hepatic proteins. This early inhibition is short-lived, since 5 days after hepatectomy the liver protein content is greater in the ethanol-fed animal. Through water retention, protein accumulation contributes to increase the size and weight of the liver (Pösö et al. 1980b). The early effect of ethanol on protein synthesis after partial hepatectomy has been documented with particular reference to two enzymes with short half-lives, namely ornithine decarboxylase (ODC) and tyrosine aminotransferase (TAT). The early inhibitory effect of ethanol on the stimulation of ODC and TAT by liver regeneration probably takes place at the transcriptional level (Pösö et al. 1980a; Pösö and Pösö 1981). Ethanol blocks both the synthesis and the intracellular degradation of these enzymes (Pösö and Pösö 1980). Ethanol also prevents the decrease in activity of alanine aminotransferase in regenerating livers (Pösö 1980). The overall effect of ethanol is an accumulation of protein in the regenerating liver. It has been suggested that delayed synthesis, decreased degradation, and secretion of intracellular liver proteins could be a general phenomenon that could markedly contribute to the pathological accumulation of protein in the liver after chronic alcohol consumption (Pösö 1980).

Chronic ethanol administration prior to and following partial hepatectomy is also associated with a significant increase in total protein liver content which persisted at 24 h and was even increased at 14 days after partial hepatectomy in comparison to values in control animals (Orrego et al. 1981). The surface area of the hepatocytes is closely related to their volumes, and thus to the water retention caused by protein accumulation. After partial hepatic resection, the increase in hepatocyte size seen in controls rapidly returned to normal within 7–14 days, while the hepatocytes remained enlarged in the ethanol-fed rats. In the livers of these animals the cell volume did not increase, presumably because hepatocyte expansion in the alcohol-fed animal has a limit (Orrego et al. 1981).

The liver mass following partial hepatectomy and acute ethanol administration is greater in the ethanol-fed rats than in control animals 72 and 96 h after surgery (Frank et al. 1979). Liver weight gain was lower in the ethanol-fed animals in the first 3 days following partial hepatic resection and greater thereafter. The size of the hepatocytes 5 days after partial hepatectomy was, however, greater in ethanol-fed rats than in controls. The increase in liver mass 5 days after hepatectomy in ethanol-fed animals cannot be explained only by the increase in total liver lipids (Pösö et al. 1980b). The rate of liver growth found in control animals following partial hepatectomy, 58% at 96 h, is similar to observations by previous investigators. With ethanol, 80% of the initial liver weight was reconstituted 96 h after hepatic resection (Frank et al. 1979). Following partial hepatectomy, liver weight increases faster also in animals chronically fed ethanol (Orrego et al. 1981). If the increase in liver cell mass is the least reliable index of regeneration in normal rats, one may legitimately question the reliability of the same index of liver regeneration in animals fed ethanol in view of the fact that markers of ethanol toxicity, such as lipid accumulation and protein retention, appear to be enhanced following partial hepatic resection. Any misinterpretation of the liver weight gain will be reflected in any other parameter calculated with the use of the total liver weight values. When using the total liver weight, Frank et al. (1979) reported that the reconstitution of total liver weight and protein content is greater in ethanol-fed animals at 96 h (respectively 80% vs 57% and 80% vs 58%). The implicit percentage error in calculations using total liver weight possibly overshadows, or at least strongly minimizes, any detectable difference between control and ethanol-fed animals. Moreover, in all the studies reviewed, alcohol was shown to partially suppress (but not totally inhibit) liver regeneration. Any parameter measured late (e.g., 14 days) after liver resection may fail to reveal a significant effect. Compensatory phenomena may eventually mask any significant changes found at earlier stages following resection.

The regenerative response induced by partial hepatectomy significantly increases hepatic DNA concentration. Total hepatic DNA reconstituted at 4 days was 98% in ethanol-treated animals and 87% in controls (Frank et al. 1979, 1980). Previous investigators found that 80% of original DNA content was restored 96 h after surgery in untreated rats (Hopkins et al. 1973). Whether such rapid reconstitution of total liver DNA in ethanol-fed rats, as early as 4 days after partial resection, is accurate cannot be assessed. In the rat, the normal liver mass and volume are restored within 7–10 days (Karran and Eagles 1979). It is, however, surprising that, in spite of the demonstration of decreased ^{3}H-thymidine incorporation into hepatic DNA in almost all of the studies reviewed, the total liver DNA content of ethanol-fed rats is completely reconstituted earlier than that of their pair-fed controls or of untreated rats.

Partial hepatectomy in animals chronically fed ethanol showed a significantly lower DNA content per gram of liver at 7 and 14 days when compared to controls, but total DNA of the regenerated livers of ethanol-fed animals was not different from those of controls at any time (Orrego et al. 1981). In the same study, following partial hepatic resection, no inhibition and no delay of thymidine incorporation into DNA, as determined chemically and by autoradiography, or of the percentage of cells undergoing mitosis could be observed. The data on ^{3}H-thymidine incorporation into DNA reported in these studies are difficult to compare with those of others, since no absolute values were given; results were expressed in % dpm in the liver homogenate incorporated into the nuclear fraction to correct for differences in total ^{3}H-thymidine uptake by the liver. The authors underline the fact that a reduction in labeling of DNA in ethanol-fed animals should have resulted in a marked reduction in total DNA. This was clearly not evident and therefore shows the possibility of error in using exclusively the incorporation of labeled thymidine into DNA without correcting for specific activities in the cell (Orrego et al. 1981). When results of ^{3}H-thymidine incorporation into hepatic DNA are expressed as a ratio of dpm nuclei/dpm homogenate, and when subsequently the ratios obtained in controls and experimental animals are compared, a decreased incorporation into hepatic DNA in the experimental animal, to give the same ratio as that of the controls, must be associated with a proportional decrease in nonspecific labeling. Frank et al. (1979) have reported no significant differences in nonspecific labeling in ethanol-fed rats. We know of no other studies using such a complex model as the partially hepatectomized rat chronically fed with ethanol, in which labeling of cellular constituents following ^{3}H-thymidine administration is reported. It is therefore impossible to draw conclusions in the light of such discrepancies between studies.

In the review of the methods for the evaluation of the regenerative response previously cited (Bucher 1963), the mitotic index is considered rather insensitive at an early stage of regeneration but the rate of increase of the cell population is rated as a valid measurement of regeneration. By contrast, net increments in various components (dry weight of tissue, total protein, RNA, DNA, enzymes, etc.) determined by chemical methods cannot be detected at low levels of regenerative activity. Chemical methods are considered 50 times less sensitive an index of growth than the mitotic count. Therefore, these methods have been largely replaced by isotopic tracer techniques. Thymidine is specific; it significantly labels only DNA of liver. In vivo, its use is considered more suitable for surveying regenerative stimuli over a broad range of experimental conditions. However, no distinction can be made between parenchymal and nonparenchymal cells. Nevertheless, incorporation of

^{3}H-thymidine into DNA fraction still appears to be a valid specific measure of DNA synthesis (Goldspink and Goldberg 1973). It is clear that each means employed to determine regenerative activity has drawbacks, and a combination of methods may be desirable.

Conclusion

A number of confirmed observations lead to the conclusion that acute ethanol administration to partially hepatectomized rats is associated with an inhibition and possibly a delay in ^{3}H-thymidine incorporation into hepatic DNA. This is associated with a decrease in protein synthesis early after partial hepatic resection. Not only ethanol by itself but also acetaldehyde may explain such an effect. The presence of significant concentrations of ethanol in blood at the time of DNA labeling appears to be important. Ethanol exerts its effect in the preparatory phase of DNA synthesis. The metabolic signs of ethanol liver toxicity are enhanced when ethanol is administered to rats during liver regeneration as shown by the striking increase in liver mass, protein, and lipid content. Some questions, however, remain unanswered. The measurement of total liver mass and DNA content in the days that follow partial hepatic resection does not confirm the inhibitory effect of ethanol on liver regeneration. On the contrary, these data show that partially hepatectomized animals fed with ethanol show a more rapid liver restitution than their pair-fed controls. The use of such parameters may render difficult the measurement of small but significant differences between controls and ethanol-fed animals during liver regeneration. The use of relatively crude indices of liver regeneration should be validated in the experimental model of alcohol administration before any firm conclusion is drawn. While observations in the model of chronic ethanol administration to partially hepatectomized rats generally agree with those made in the acute model, the persistence of inhibition of thymidine incorporation into hepatic DNA in the absence of ethanol in the blood suggests significant differences between the acute and chronic models. With chronic ethanol administration, metabolic indicators of ethanol liver toxicity appear also to be more profoundly altered after partial hepatectomy. Whether such findings preclude the use of certain parameters of liver regeneration to assess significant differences between ethanol-fed and control animals is still a matter of debate and surely deserves to be elucidated.

References

Baraona E, Leo MA, Borowsky SA, Lieber CS (1975) Alcoholic hepatomegaly: accumulation of protein in the liver. Science 190: 794–795

Baraona E, Leo MA, Borowsky SA, Lieber CS (1977) Pathogenesis of alcohol-induced accumulation of protein in the liver. J Clin Invest 60: 546–554

Becker FF (1973) Humoral aspects of liver regeneration. In: Lobue J, Gordon ES (eds) Humoral control of growth and differentiation, Vol 1. Academic Press, New York, pp 249–256

Blumgart LH, Leach KG, Karran SJ (1971) Observation on liver regeneration after right hepatic lobectomy. Gut 12: 922–928

Bucher NLR (1963) Regeneration of mammalian liver. Int Rev Cytol 15:245–301
Bucher NLR (1967a) Experimental aspects of hepatic regeneration I. N Engl J Med 277:686–696
Bucher NLR (1967b) Experimental aspects of hepatic regeneration II. N Engl J Med 277:738–746
Bucher NLR, Malt RA (1971) Regeneration of the liver and kidney. Little, Brown, Boston
Bucher NLR, McGowam JA (1979) Regeneration, part 2. Regulatory mechanisms. In: Wright R, Alberti KGMM, Karran S, Millward-Sadler GH (eds) Liver and biliary disease. Saunders, London, pp 210–227
Bucher NLR, Swaffield MN (1973) Regeneration of liver in rats in the absence of portal splanchnic organs and a portal blood supply. Cancer Res 33:3189–3194
Bucher NLR, Swaffield MN (1975) Synergistic action of glucagon and insulin in regulation of hepatic regeneration. In: Weber G (ed) Advances in enzyme regulation, Vol 13. Pergamon, New York, pp 281–293
Craig J (1975) Effects of ethanol and ethionine on DNA synthesis during experimental liver regeneration. J Stud Alcohol 36:148–157
Dreosti IE, Ballard J, Belling GB, Record IR, Manuel SJ, Hetzel BS (1981) The effect of ethanol and acetaldehyde on DNA synthesis in growing cells and on fetal development in the rat. Alcoholism (NY) 5:357–362
Duguay LR, Orloff MJ (1976) Regulation of liver regeneration by the pancreas in dogs. Surg Forum 27:355–356
Duguay L, Coutu D, Hétu C, Joly JG (1981) Inhibition of liver regeneration by chronic alcohol administration. Gut 23:8–13
Fabrikant JI (1968) The kinetics of cellular proliferation in regenerating liver. J Cell Biol 36:551–565
Frank WO, Houck C (1980) Does chronic ethanol consumption impair hepatic regeneration. Gastroenterology 78:1167
Frank WO, Rayyes AN, Washington A, Holt PR (1979) Effect of acute ethanol administration upon hepatic regeneration. J Lab Clin Med 402–413
Galambos JT (1979) Cirrhosis. Saunders, Philadelphia (Major problems in internal medicine, vol. XVII)
Garceau AJ (1963) The natural history of cirrhosis. I. Survival with oesophageal varices. N Engl J Med 268:469–473
Gaub J, Fauerholdt L, Keiding S, Kondrup J, Petersen P, Wantzin GL (1981) Cytophotometry of liver cells from ethanol-fed rats: ethanol causes increased polyploidization and protein accumulation. Eur J Clin Invest 11:235–237
Goldspink DR, Goldberg AL (1973) Problems in the use of (Me-^{3}H)thymidine for the measurement of DNA synthesis. Biochim Biophys Acta 299:521–532
Hays DM (1974) Surgical research aspects of hepatic regeneration. Surg Gyn Obst 139:609–619
Hétu C, Dumont A, Joly JG (1983) Effect of chronic ethanol administration on bromobenzene liver toxicity in the rat. Toxicol Appl Pharmacol 67:166–177
Hopkins HA, Campbell HA, Barbiroli B, Potter VR (1973) Thymidine kinase and deoxyribonucleic acid metabolism in growing and regenerating livers from rats on controlled feeding schedules. Biochem J 136:955–966
Israel Y, Khanna JM, Orrego H, Rachamin G, Wahid S, Britton R, MacDonald A, Kalant H (1979) Studies on metabolic tolerance to alcohol, hepatomegaly and alcoholic liver disease. Drug Alcohol Depend 4:109–118
Joly JG, Hétu C (1978) Mechanisms of faster appearance of acetaminophen toxicity after chronic ethanol feeding. Gastroenterology 74:1158
Karran S, Eagles C (1979) Regeneration, part 1. Physical aspects. In: Wright R, Alberti KGMM, Karran S, Millward-Sadler GH (eds) Liver and biliary disease. Saunders, London, pp 197–210
Karran SJ, Leach KG, Blumgart LH (1974) Assessment of liver regeneration in rat using gamma camera. J Nucl Med 15:10–16
Labrecque DR, Pesch LA (1975) Preparation and partial characterization of hepatic regenerative stimulator substances (SS) from rat liver. J Physiol 248:273–284

Lieber CS, Rubin E (1969) Alcoholic fatty liver. N Engl J Med 280: 705–708
Lieber CS (1978) Pathogenesis and early diagnosis of alcoholic liver injury. N Eng J Med 298: 888–893
Lieber CS, Jones DP et al. (1963) Fatty liver hyperlipemia and hyperuricemia produced by prolonged alcohol consumption and adequate dietary intake. Trans Assoc Am Physicians 76: 289
Makowka L, Falk RE, Hardy M, Falk JA, Langer B, Blendis L, Phillipps MJ (1981) Reversal of ethanol-induced inhibition of hepatic regeneration (REG) by regenerating rat liver cytosol (RLC). Pharmacologist 23: 221
Makowka L, Falk RE, Teodorczyk-Injeyan J, Venturi D, Rotstein LE, Falk W, Langer B, Blendis LM, Philipps MJ (1983) The effect of liver cytosol on hepatic regeneration and tumor growth. Cancer 51: 2181–2190
Morley CGD, Kingdon HS (1972) Use of ^{3}H-thymidine for measurement of DNA synthesis in rat liver – a warning. Anal Biochem 45: 298–305
Orrego H, Crossley IR, Saldivia V, Medline A, Varghese G, Israel Y (1981) Long-term ethanol administration and short- and long-term liver regeneration after partial hepatectomy. J Lab Clin Med 221–230
Pequignot G (1963) Les enquêtes par interrogatoires permettent-elles de déterminer la fréquence de l'étiologie alcoolique des cirrhoses du foie. Bull Acad Nat Med 147: 90–97
Pösö H (1980) Inhibition of protein degradation in regenerating rat liver by ethanol treatment. Pharmacol Biochem Behav 13 suppl. 1: 83–88
Pösö H, Pösö AR (1980) Stabilization of tyrosine aminotransferase and ornithine decarboxylase in regenerating rat liver by ethanol treatment. FEBS Lett 113: 211–214
Pösö H, Pösö AR (1981) Inhibition of RNA and protein synthesis by ethanol in regenerating rat liver: evidence for transcriptional inhibition of protein synthesis. Acta Pharmacol Toxicol (Copenh) 49: 125–129
Pösö R, Pösö H, Väänänen H, Salaspuro M (1980a) Inhibition of the synthesis of macromolecules by ethanol in regenerating rat liver. Adv Exp Med Biol 132: 551–560
Pösö H, Väänänen H, Salaspuro MP, Pösö AR (1980b) Effects of ethanol on liver regeneration after partial hepatectomy in rats. Med Biol 58: 329–336
Rayyes AN, Frank WO (1976) Suppression of hepatic regeneration by ethanol in the rat. Gastroenterology 70: 991
Schneider WC, Greco AE (1971) Incorporation of pyrimidine deoxyribonucleosides into liver lipids and other components. Biochim Biophys Acta 228: 610–626
Shulte-Hermann R (1974) Induction of liver growth by xenobiotic compounds and other stimuli. Crit Rev Toxicol 3: 79–158
Starzl T, Terblanche J (1979) Hepatotrophic substances. In: Popper H. Schaffner F (eds) Progress in liver disease, vol 5, Grune and Stratton, New York, pp 135–151
Sweeney GD, Cole FM, Freeman KB, Patel HV (1979) Heterogeneity of rat liver parenchymal cells. J Lab Clin Med 94: 718–725
Wands JR, Carter EA, Bucher NLR, Isselbacher KJ (1979) Inhibition of hepatic regeneration in rats by acute and chronic ethanol intoxication. Gastroenterology 77: 528–531
Weesner R, Mendenhall C, Morgan D, Kessler V, Kromme C (1978) Suppression of liver (L) regeneration in ethanol (E) treated rats. Gastroenterology 75: 993
Whitfield JR, Boynton AL, MacManus JP, Sikorska M, Tsang BK (1979) The regulation of cell proliferation by calcium and cyclic AMP. Mol Cell Biochem 27: 155–179

15 Ethanol and the Immune System

F. Paronetto

Immunopathology Laboratory, Veterans Administration Medical Center,
130 West Kingsbridge Road, Bronx, NY 10468, USA

Introduction

Although the metabolic pathways of ethanol are well understood, the pathogenesis of tissue damage and the development of cirrhosis are far from clear. While alcohol is considered by some investigators to be a hepatotoxic substance that produces a dose- and time-related hepatotoxic damage (Peguignot et al. 1974; Lelbach 1976; Rankin 1977), several observations militate against this hypothesis: the variation of hepatotoxic damage in patients with similar alcohol intake is considerable and only a small percentage of alcoholic patients develop cirrhosis and/or alcoholic hepatitis (Galambos 1974). In the search for alternative explanations, alterations of the immune response have been assuming a growing importance and a number of altered immune reactions have been observed in patients with prolonged alcohol abuse.

Alcohol has been considered an unlikely immunogen because of its small molecule; nevertheless, alcohol or its metabolites may act like haptenes and modify normal proteins, rendering them antigenic. The advent of monoclonal antibodies, with the possibility of defining the phenotypic expression of mononuclear cells, allows a more precise dissection and recognition of the type of cells involved in liver damage and has begun to attract more investigators to the field. The major challenge will be to separate primary pathogenetic events from secondary inconsequential phenomena.

Immune Alterations in Alcoholic Subjects Without Alcoholic Liver Disease

It has been an old observation that alcoholic patients are more prone to a variety of infections (Adams and Jordan 1984), but it was not determined whether this finding is related to concomitant malnutrition, environmental factors, liver diseases, or a direct effect of alcohol on the immune system. Well-controlled studies are lacking, but scattered information indicates a decrease in sensitization to dinitrochlorobenzene (DNCB) (Straus and Berenyi 1973; Lundy et al. 1975) and a decrease in

Alcohol Related Diseases in Gastroenterology
Edited by H. K. Seitz and B. Kommerell

antibody formation after challenge with the antigen keyhole limpet, hemocyanin (Gluckman et al. 1977). These data, if confirmed, may explain the increased susceptibility of alcoholic patients to a variety of infections. Experimental work has suggested that prolonged administration of alcohol produces atrophy of the thymus, a decreased antibody response to antigenic stimulation, a decrease in T cells and an increase in B cells (Tennenbaum et al. 1969; Grossman et al. 1979; Yamamoto 1983). Arthus and delayed skin reactions are reduced by chronic ethanol administration. In contrast, allergic encephalomyelitis is enhanced, suggesting a dual opposite effect on immune response (Jancović et al. 1984). Alcohol is said to have a depressing effect on the reticuloendothelial system (RES), inducing a variety of alterations related to a direct activity on the macrophages (Guarnieri and Laurenzi 1968; Laurenzi and Guarnieri 1966). Pulmonary macrophages exhibit a decreased clearance of injected microorganisms (Stillman 1924). In vitro studies have indicated a deleterious effect of alcohol on tissue macrophages and on the ability of lymphocytes to react to the mitogen phytohemagglutinin (Tisman and Herbert 1972). It is not clear whether the alterations of the immune system, if they exist, are mediated by ethanol itself or by acetaldehyde, which is a primary metabolic product of ethanol with high reactivity with membranes and proteins. It has been suggested that acetaldehyde may combine with hepatic membranes without inducing functional alterations of hepatocytes but activating complement (Barry 1983) or inducing antigenic changes in the membranes. This hypothesis is particularly appealing because it may explain why lesions morphologically indistinguishable from alcoholic hepatitis may be observed in conditions other than excessive alcohol intake, e.g., after jejunoileal bypass operations for obesity. It has been suggested that bacteria in the gastrointestinal tract may produce acetaldehyde that may reach the liver and may initiate hepatocellular damage.

Genetic Factors in Alcoholic Liver Disease

The observation that severe ALD affects a minority of alcoholic patients has prompted a number of laboratories throughout the world to investigate whether an association may exist between histocompatibility antigens (HLA) and severe liver diseases. These analyses are particularly difficult because of the importance of selecting the proper controls in order to avoid false results related to population stratification. The observations so far seem to differ in various countries and a common increase in frequency of a HLA phenotype was not identified. In Norway Bw40 was reported to be increased in patients with cirrhosis (Bell and Nordhagen 1978), while in Chile B13 was frequent in the same class of patients (Melendez et al. 1979). In England and Switzerland the association between B8 and alcoholic hepatitis and cirrhosis was repeatedly reported (Bailey et al. 1976; Morgan et al. 1980; Bron et al. 1982) and B8 was frequently observed with a more accelerated form of liver disease (Saunders et al. 1982). These findings are particularly important because B8 is also associated with a variety of autoimmune diseases and increased cellular and humoral immune reactivity. Unfortunately, these observations were not registered in other countries (Gluud et al. 1980; Rada et al. 1981; Faizallah et al. 1982), and therefore it is possible that those studies may have uncovered isolated subgroups of patients with heightened reactivity to alcohol and abnormal clinical manifestations. It is of interest that when one combines data from various investigators the frequencies of Aw32, B8, B13, and B27 are higher in the alcoholic cirrhotics than in controls, suggesting a weak association between these antigens and alcoholic cirrhosis and explaining the variations in different studies (Eddleston and Davis 1982). Data on the HLA-DR typing are few, and one study indicated a marked increase in HLA-DR2 (Tait and MacKay 1982), suggesting additional studies on the frequency of HLA-DR terminants. Variations of aldehyde dehydrogenase isoenzymes have been reported (Cavulli et al. 1971; Agarwal et al. 1981) and may have a genetic basis; thus if genetic alterations are found in alcoholic patients these may be outside the HLA system. Association of HLA and other ethanol-related diseases, such as alcoholic pancreatitis and alcoholic cardiomyopathy, have been reported, however (Fauchet et al. 1979; Kachru et al. 1980), suggesting that further investigations on the immunopathology of these diseases may be rewarding.

Serologic Alterations

Autoantibodies

Antibodies to a variety of autoantigens have been described in patients with ALD. Antinuclear and anti-smooth-muscle antibodies, the characteristic antibodies of autoimmune chronic active hepatitis, have been observed in patients with ALD, especially those bearing the HLA B8 and DR3 haplotypes. Antibodies to liver-specific protein (LSP), which have been observed in many patients with chronic hepatitis, may also be frequently detected in ALD patients (Manns et al. 1980; Perperas et al. 1981). These antibodies were said to correlate with chronic portal and periportal inflammation. LSP antibodies may not be species specific because they were observed also to react with the kidney antigen that is similar to LSP (Behrens and Paronetto 1979). Antibodies to liver membranes of the IgG and IgA class have been described, using as substrate isolated rabbit hepatocytes (Burt et al. 1982). Of particular interest are the antibodies reported to react against liver membrane from alcohol-treated animals (MacSween et al. 1981; Anthony et al. 1983). These antibodies seem to recognize a new antigen on the surface of hepatocytes. These observations, however, could not be confirmed in another laboratory (Krogsgaard et al. 1982). Still controversial is the presence of antibodies directed against alcohol hyalin. These have been observed in one laboratory (Kanagasundaram et al. 1977; Kanagasundaram and Leevy 1979) but negated in another (Kehl et al. 1981).

Immune Complexes

The frequent observation of autoantibodies in patients with ALD has suggested the possibility that these antibodies are circulating as part of an immune complex. Indeed, immune complexes have been observed frequently in alcoholic patients (Thomas et al. 1978; Penner et al. 1978; Abrass et al. 1980; Brown et al. 1983), and more recent work is being aimed at the identification of the antigens involved in these complexes. Alcoholic hyaline has been shown to be a component of circulating immune complexes (Gorindasajan et al. 1982) and has been localized, often being associated with IgG and complement (Burns et al. 1983) in the renal glomeruli of alcoholic patients. A recent report indicates that high molecular immunoglobulin may circulate in an aggregate form (Stoltenberg and Soltis 1984). These IgG aggregated IgG may be responsible for a variety of abnormalities (Theofilopoulos deposited in the vascular system. The presence of immune complexes and/or aggregated IgG may be responsible for a variety of abnormalities (Theofilopoulos and Dixon 1980), including an altered modulation of lymphocytes, activation of monocytes and the complement system, and damaging immune reactions in many tissues.

Antibodies to Hepatitis B Virus

Besides antibodies to autoantigens, antibodies to hepatitis B virus have been described with increased frequence in patients with ALD (Mills et al. 1979; Naito et al. 1977; Vargas et al. 1979; Beaugrand et al. 1980; Chaudhuri 1980; Goudeau et al. 1981; Hislop et al. 1981; Mills et al. 1981; Gluud et al. 1982; Orillac et al. 1983; Villa et al. 1982; Boron et al. 1983; Saunders et al. 1983). There is disagreement whether markers of hepatitis B virus are found more frequently in alcoholic cirrhosis than in fatty liver and alcoholic hepatitis. Most of the reports indicate that there seems to be no difference between the various groups of alcoholics.

It has been suggested that hepatitis B virus in patients with ALD may aggravate liver lesions, but statistical analysis indicates that hepatitis B virus does not increase the influence of ethanol on the production of cirrhosis (Chevilotte et al. 1983). The reason for the high frequency of antibodies is not clear but it may be related to environmental factors or lifestyle.

Immunosuppressive Serum Factors

Besides humoral antibodies and immune complexes, patients with ALD may develop humoral factors that inhibit transformation of lymphocytes by mitogens, a rosette-inhibitory factor (RIF), and a serum-inhibitory factor (SIF) (Grauer et al. 1984; Behrens et al. 1982): all these factors may indirectly contribute to the alteration of cell-mediated immunity.

Alteration of Peripheral Blood Mononuclear Cells

Sensitization to Autoantigens and/or Alcohol

Early studies had indicated that specific reactivities of mononuclear cells to LSP, hyalin, and alcohol and acetaldehyde are enhanced in patients with alcoholic hepatitis but not in alcoholic patients with steatosis (Sorrell and Leevy 1972; Newberry et al. 1973; Meyer zum Büschenfelde et al. 1975; Zetterman et al. 1976; Actis et al. 1978; Kawanishi et al. 1979; Cooksley et al. 1980). These abnormal reactions were not specific for ALD but shared with other types of chronic hepatitis.

T Cell Subpopulations

Investigations with monoclonal antibodies have confirmed that the T cells are decreased in ALD. Extensive work on the distribution of the various phenotypically different lymphocytes has not appeared as yet, but preliminary data seem to indicate that the helper/suppressor cell ratio is not altered in ALD (Routhier et al. 1980; Thomas 1981). The function of the helper or suppressor cells has not been extensively investigated, and so far the results are contradictory (Kawanishi et al. 1981; Wands et al. 1981).

Cytotoxicity of Mononuclear Cells

More attention has been drawn by the cytotoxic activity of mononuclear cells against autologous liver cells. This phenomenon has been confirmed in a number of laboratories, not only in patients with ALD (Kakumu and Leevy 1977; Actis et al. 1983; Izumi et al. 1983; Hütteroth et al. 1983) but also in experimental animals after prolonged administration of ethanol (Lue et al. 1981). Mononuclear cells are the effector cells and, at least in experimental animals, liver cells from alcoholic animals seem to be particularly vulnerable to the cytotoxic activity of mononuclear cells.

Cytotoxicity can be mediated by three types of cells that induce tissue damage in different ways (Paronetto 1982). The first type of effector cells are T-lymphocytes, which produce damage by interaction with the antigen and a histocompatibility antigen. A second type of cytotoxicity is mediated by K cells, which recognize an antibody on the surface of liver cells by means of receptors for the Fc portion of immunoglobulins. The third type of cytotoxicity is mediated by natural killer cells (NK cells), which recognize unusual determinants on cells.

In ALD, K cells have been demonstrated to be involved in cytotoxicity against autologous hepatocytes (Actis et al. 1983), but more recent information indicates that T cells may also participate, although to a lesser extent (Izumi et al. 1983; Hütteroth et al. 1983). Thus in ALD one may postulate that sensitized T-lymphocytes may recognize membrane antigens modified by ethanol, or that K cells may express cytotoxicity by interacting with antibodies located on the liver cell membrane.

Immunopathology of Liver Lesions

It has been recognized that in vitro phenomena may not reflect what happens in the liver itself. Therefore, attention has been recently directed toward the hepatic inflammatory cells and the hepatocytes of alcoholic patients.

Steatosis

The early lesions in alcoholic patients are characterized by steatosis without inflammatory reaction, and it is unlikely that they are mediated by immune reactions. The more severe and progressive lesions of alcoholic hepatitis have more recently attracted the attention of immunopathologists, who have found some similarity of immune alterations between alcoholic hepatitis and chronic active hepatitis (Brunt et al. 1974; Goldberg et al. 1977; Paronetto 1981).

Mononuclear Cells in Alcoholic Hepatitis

The typical lesion of alcoholic hepatitis is characterized by the presence of ballooned hepatocytes and Mallory bodies surrounded by neutrophils. It has recently been recognized that lymphocytes, even if not so numerous as neutrophils, are present in discreet numbers (French et al. 1979), and in vitro experiments have indicated that mononuclear cells, in the presence of Mallory bodies, may induce the formation of a factor chemotactic for neutrophils and mononuclear cells (Peters et al. 1983).

T-lymphocytes are localized in the liver parenchyma (Burbige et al. 1976) (Fig. 1), but it is uncertain whether the lymphocytes are only contributing to the neutrophil accumulation or are also involved in tissue-damaging reactions. Indeed, a recent report has indicated that lymphocytes with the T8 (cytotoxic, suppressor) phenotype are present in alcoholic hepatitis (Si et al. 1983). Activated lymphocytes may also produce factors with fibrogenetic activity (Johnson and Ziff 1976; Neilson et al. 1980), thus contributing to fibrosis and eventually to the cirrhotic process. We do not have information as to whether NK or K cells are also present in alcoholic hepatitis.

Mononuclear Cells in Alcoholic Cirrhosis

The more advanced lesions of ALD are characterized by septum formation and cirrhosis, frequently active, with unsharp borders between portal tracts, septa, and parenchyma. In these active lesions, in areas of piecemeal necrosis, an abundant number of lymphocytes have been described with characteristics of the T8 (cytotoxic, suppressor) phenotype (Si et al. 1984), while cells bearing the phenotype characteristics for NK and K cells were scarce. Thus these observations suggest that, at least in the active cirrhotic lesions, T cells are the predominant cells and T-cell-mediated cytotoxic reactions may be of importance. Certainly, more work is

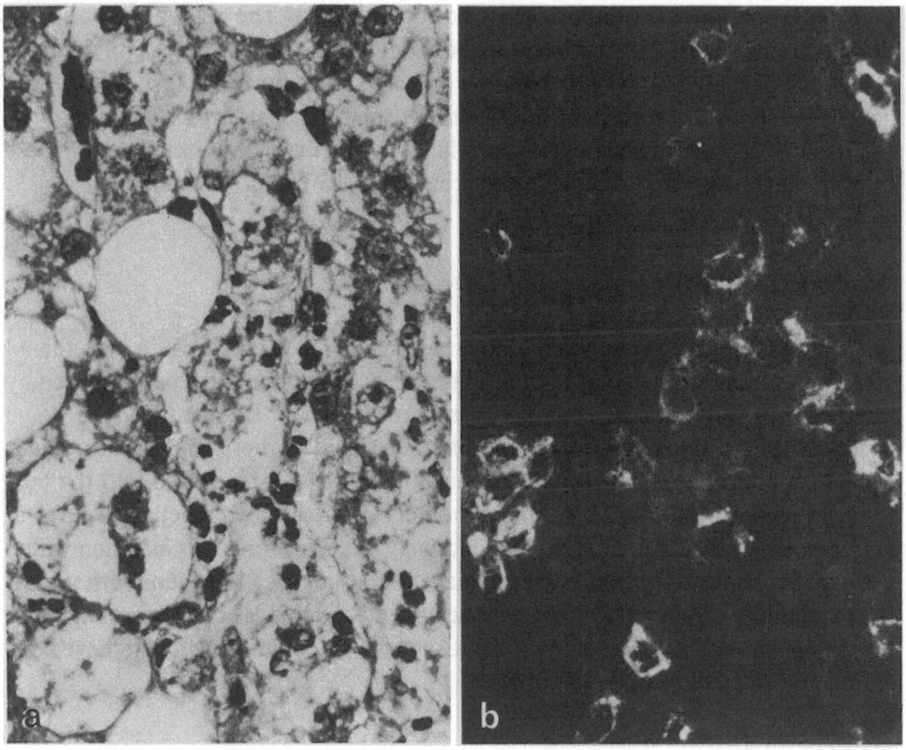

Fig. 1a, b. Liver biopsy of a patient with alcoholic hepatitis. **a** Hepatocytes with Mallory bodies are surrounded by neutrophils and rare mononuclear cells. Hematoxylin-eosin, × 250. **b** A cryostat section treated with antibodies to T cell shows several T cells in the area of hepatocellular damage (anti-T-cells [Leu 1] antibodies, followed by biotinylated antimouse antibodies and flouresceinated avidin, × 560)

needed to define better the function of the inflammatory infiltrate in alcoholic cirrhosis.

Class I Histocompatible Antigens and Immunoglobulin Deposition in the Liver

Besides mononuclear cells, deposition of immunoglobulins or formation of neoantigen in the liver may contribute to the initiation of liver damage. Therefore, the demonstration of class I antigens in hepatocytes is of particular interest. These antigens are not seen in normal hepatocytes, but they appear in liver cells in chronic hepatitis and in alcoholic hepatitis (Barbatis et al. 1981), and it is possible that they may trigger cytotoxic lymphocytes that are activated by the simultaneous presence of antigens and class I HLA.

A more specific finding in liver tissue seems to be the deposition of IgA in a continuous pattern (Kater et al. 1979; Swerdlow et al. 1982). These immunoglobulins are of the IgA_2 subclass and they have probably been derived from the gastrointestinal tract (Swerdlow and Chowdhury 1983). The significance of these findings is not clear. It has been suggested that IgA may be part of immune complexes or may saturate Kupffer cells and alter their function. IgG and IgA are also detected on the surface of mechanically isolated hepatocytes (Trevisan et al. 1983) and IgG-positive hepatocytes correlated with activity of transaminases. These findings have suggested that IgG may contribute to liver cell damage.

Alcoholic Liver Disease and Viral Infections

It should also be kept in mind that lesions in alcoholics may not be caused by ethanol directly but by some metabolites of alcohol or by concomitant infections induced by viruses, especially non-A, non-B viruses or hepatitis B virus. Indeed some observations in French alcoholics who had developed cirrhosis and hepatocellular carcinoma indicate that the genome of hepatitis B virus is integrated into the host DNA (Brechot et al. 1982). In addition, the previously mentioned high level of antibodies to hepatitis B virus in alcoholics may indicate a previous infection with a virus that is known to produce liver damage. Superimposed infections with viruses has to be ruled out when livers of alcoholic patients develop lesions resembling those seen in patients with chronic active hepatitis, where autoimmune reactions have been considered of pathogenetic importance.

Alcoholic Chronic Active Hepatitis

A cluster of papers have attempted to delineate the entity of alcoholic chronic active hepatitis, which is characterized by a morphological picture resembling that seen in alcoholic fibrosis and autoimmune chronic active hepatitis, but which is related to ethanol because it improves upon discontinuation of alcohol intake (Goldberg et al. 1977; Nei et al. 1983; Crapper et al. 1983). Certainly, more detailed immunopathological studies of this entity are warranted.

Conclusions

A large body of observations has indicated that ethanol or acetaldehyde may trigger a variety of immune reactions in the serum, leukocytes, and liver of alcoholic patients. While there is scarce information that alcohol per se has a direct damaging effect on the cells and organs involved in the immune responses in man, there is better evidence that cytotoxic reactions mediated by mononuclear cells may take place in some patients following the direct toxic effect of alcohol or its metabolites.

The presence of autoantibodies and immune complexes may explain some generalized manifestations and the alterations of immune reactivity, while the recently demonstrated neoantigens may be the source of an immune reaction directed against the liver. In addition, ALD may be complicated by viral infections, especially hepatitis B virus. The advent of monoclonal antibodies that will be able to define more precisely the type of cell involved in tissue-damaging immune reactions coupled with a better functional study of the cells present in the liver may help us better to understand the mechanism of tissue damage in ALD.

References

Abrass CK, Border WA, Hepner G (1980) Non-specificity of circulating immune complexes in patients with acute and chronic liver disease. Clin Exp Immunol 40: 292–298

Actis GC, Ponzetto A, Rizzetto M, Verme G (1978) Cell-mediated immunity to acetaldehyde in alcoholic liver disease demonstrated by leukocyte migration test. Dig Dis 23: 883–886

Actis G, Mieli-Vergani G, Portmann B, Eddleston ALWF, Davis M, Williams R (1983) Lymphocyte cytotoxicity to autologous hepatocytes in alcoholic liver disease. Liver 3: 8–12

Adams HG, Jordan C (1984) Infections in the alcoholic. Med Clin North Am 68: 179–200

Agarwal DP, Harada S, Goedde HW (1981) Racial differences in biological sensitivity to ethanol: the role of alcohol dehydrogenase and aldehyde dehydrogenase isoenzymes. Alcoholism 5: 12–16

Anthony RS, Farquharson M, MacSween RNM (1983) Liver membrane antibodies in alcoholic liver disease II. Antibodies to ethanol-altered hepatocytes. J Clin Pathol 36: 1302–1308

Bailey RJ, Krasner N, Eddleston ALWF, Williams R, Tee DEH, Doniach D, Kennedy LA, Batchelor JR (1976) Histocompatibility antigens, autoantibodies, and immunoglobulins in alcoholic liver diseases with cirrhosis. Br Med J 2: 727–729

Barbatis C, Woods J, Morton JA, Fleming KA, McMichael A, McGee JO'D (1981) Immunohistochemical analysis of HLA (A,B,C) antigens in liver disease using a monoclonal antibody. Gut 22: 985–991

Barry RE (1983) The pathogenesis of hepatitis in alcohol abuse and jejunoileal bypass. Lancet 2: 489–490

Beaugrand M, Dupont T, Ferrier JP, Courouce AM (1980) Étude des marqueurs sériques du virus B chez 77 malades atteints de cirrhose alcoolique comparés à des temoins. Gastroenterol Clin Biol 4: 89A

Behrens UJ, Paronetto F (1979) Studies on "liver-specific" antigens. I. Evaluation of the liver specificity of "LSP" and "LP-2". Gastroenterology 77: 1045–1052

Behrens U, Friedrich I, Vernace S, Schaffner F, Paronetto F (1982) Lymphocyte responsiveness per unit volume of blood in patients with chronic nonalcoholic and alcoholic liver disease. Plasma inhibitory factors and functional defects of responder cells. J Clin Lab Immunol 8: 143–152

Bell H, Nordhagen R (1978) Association between HLA-BW40 and alcoholic liver diseases with cirrhosis. Br Med J 1: 882

Boron VP, Boron-Kaczmarska A, Sokolewicz-Bobrowska E, Borzuchowska A, Borowski T, Bernadska A, Sidun Z, Kiwerski H, Prokopowicz D, Szpakowicz T, Szadkowska D, Nadowska K, Lotocka K, Grochowska A, Kozlowska H (1983) Chronischer Alkoholismus und die Infektion mit dem Hepatitis-B-Virus. Dtsch Z Verdau Stoffwechselkr 43: 215–218

Brechot C, Nalpas B, Courouce AM, Doramel G, Callard P, Carnot F, Tiollais P, Berthelot P (1982) Evidence that hepatitis B virus has a role in liver-cell carcinoma in alcoholic liver disease. N Engl J Med 306: 1384–1387

Bron B, Kubski D, Widmann JJ, Fliedner VV, Jeannet M (1982) Increased frequency of DR3 antigen in alcoholic hepatitis and cirrhosis. Hepatogastroenterology 29: 183–186

Brown SE, Steward MW, Viola L, Howard CR, Murray-Lyon IM (1983) Chronic liver disease: the detection and characterization of circulating immune complexes. Immunology 49: 673–683

Brunt PW, Kew MC, Scheuer PJ, Sherlock S (1974) Studies in alcoholic liver disease in Britain. Gut 15: 52–58

Burbige EJ, Tarder G, Denton T, Harkin C, Luehrs D, Rourke E, French SW (1976) Predominantly T-cell infiltrate in the liver in alcoholic hepatitis. Gastroenterology 71: 900

Burns J, D'Ardenne AJ, Morton JA, McGee JOD (1983) Immune complex nephritis in alcoholic cirrhosis: detection of Mallory body antigen in complexes by means of monoclonal antibodies to Mallory bodies. J Clin Pathol 36: 751–755

Burt AD, Anthony RS, Hislop WS, Bouchier IAD, MacSween RNM (1982) Liver membrane antibodies in alcoholic liver disease: 1. Prevalence and immunoglobulin class. Gut 23: 221–225

Cavulli N, Manenti F, Gallo M, Salvioli GG (1971) Alcohol-drug interaction in man: alcohol and tolbutamide. Eur J Clin Invest 1: 421–424

Chaudhuri AKR (1980) Evidence for a role of hepatitis virus B and other viral antibodies in alcoholic cirrhosis. Gastroenterology 79: 1115

Chevillotte G, Durbec JP, Gerolami A, Berthezene P, Bidart JM, Camatte R (1983) Interaction between hepatitis B virus and alcohol consumption in liver cirrhosis. Gastroenterology 85: 141–145

Cooksley WGE, Halliday JW, Halliday WJ, Burnett W, Kerr JF, Row H, Roberts I, Powell LM (1980) A prospective study of CMI to liver antigens as assessed by leukocyte adherence inhibition (LAI) in acute and chronic liver disease. Gastroenterology 79: 1102

Crapper RM, Bhathaland PS, MacKay IR (1983) Chronic active hepatitis in alcoholic patients. Liver 3: 327–337

Eddleston ALWF, Davis M (1982) Histocompatibility antigens in alcoholic liver disease. Br Med Bull 38: 13–16

Faizallah R, Woodrow JC, Krassner NK, Walker RJ, Morris AI (1982) Are HLA antigens important in the development of alcohol-induced liver disease? Br Med J 285: 533–534

Fauchet R, Genetet B, Gosselin M, Gastard J (1979) HLA antigens in chronic active pancreatitis. Tissue Antigens 13: 163–169

French SW, Burbige EJ, Tarder G, Bourke E, Harkin CG (1979) Lymphocyte sequestration by the liver in alcoholic hepatitis. Arch Pathol Lab Med 103: 146–152

Galambos JT (1974) Alcoholic hepatitis. In: Schaffner F, Sherlock S, Leevy C (eds) The liver and its disease. Thieme, Stuttgart, pp 225–265

Gluckman SJ, Dvorak VC, MacGregor RR (1977) Host defenses during prolonged alcohol consumption in a controlled environment. Arch Intern Med 137: 1539–1545

Gluud C, Aldershvile J, Dietrichson O, Hardt F, Iversen K, Juhl E, Nielsen JO, Ryder LP, Skinhoj P, Svejgaard A (1980) A human leucocyte antigens in patients with alcoholic liver cirrhosis. Scand J Gastroenterol 15: 337–341

Gluud C, Aldershvile J, Henriksen J, Kryger P, Mathiesen L (1982) Hepaitis B and A virus antibodies in alcoholic steatosis and cirrhosis. J Clin Pathol 35: 693–697

Goldberg SJ, Mendenhall CL, Connell AM, Chedid A (1977) "Nonalcoholic" chronic active hepatitis in the alcoholic. Gastroenterology 72: 598–604

Goudeau A, Maupas P, Dubois F, Corsager P, Bougnoux P (1981) Hepatitis B infection in alcoholic liver disease and primary hepatocellular carcinoma in France. Prog Med Virol 27: 26–34

Govindarajan S, Tingberg H, Radvan G (1982) Circulating immune complexes in alcoholic hepatitis. Clin Research 30: 995A

Grauer W, Brattig NW, Schomerus H, Frösner G, Berg PA (1984) Immunosuppressive serum factors in viral hepatitis. III. Prognostic relevance of rosette inhibitory factor and serum inhibition factor in acute and chronic hepatitis. Hepatology 4: 15–19

Grossman CJ, Roselle G, Sholiton LJ, Mendenhall CL (1979) The effect of chronic ethanol (E) on rat thymus: a possible mechanism for the altered immune response of the alcoholic. Gastroenterology 77: A14

Guarneri JJ, Laurenzi GA (1968) Effect of alcohol on the mobilization of alveolar macrophages. J Lab Clin Med 72: 40–51

Hislop WS, Follett EAC, Bouchier IAD, MacSween RNM (1981) Serologic markers of hepatitis B in patients with alcoholic liver disease: multicenter survey. J Clin Pathol 34: 1017–1019

Hütteroth TH, Poralla T, Meyer zum Büschenfelde KH (1983) Cellular cytotoxicity against autologous hepatocytes in alcoholic liver disease. Hepatology 3: 842

Izumi N, Hasumura Y, Takeuchi J (1983) Lymphocyte cytotoxicity for autologous human hepatocytes in alcoholic liver disease. Clin Exp Immunol 53: 219–224

Jancović BD, Marcović BM, Nešić K (1984) Dual effect of ethanol intoxication on the immune responses in the rat. Fed Proc 43: 756

Johnson RL, Ziff M (1976) Lymphokine stimulation of collagen accumulation. J Clin Invest 58: 240–252
Kachru RB, Proskey AJ, Telischi M (1980) Histocompatibility antigens and alcoholic cardiomyopathy. Tissue Antigens 15: 398–399
Kakumu S, Leevy CM (1977) Lymphocyte cytotoxicity in alcoholic hepatitis. Gastroenterology 72: 594–597
Kanagasundaram N, Leevy CM (1979) The antigenic moiety in alcoholic hyalin (AH): its identification and detection of AHAb by radioimmunoassay. Gastroenterology 77: A20
Kanagasundaram N, Kakumu S, Chen T, Leevy CM (1977) Alcoholic hyalin antigen (AHAg) and antibody (AHAb) in alcoholic hepatitis. Gastroenterology 73: 1368–1373
Kater L, Jöbsis AC, Baart de la Faille-Kuyper EH, Vogten AJM, Grijm R (1979) Alcoholic hepatic disease: specificity of IgA deposits in liver. Am J Clin Pathol 71: 51–257
Kawanishi H, Ibrahim M, Wong KM, MacDermott RP, Sheagren JN (1979) Stimulation of lymphotoxin production in vitro by liver specific protein in patients with active alcoholic and chronic active liver diseases. Gastroenterology 76: 1166
Kawanishi H, Tavassolie H, MacDemott RP, Sheagren JN (1981) Impaired concanavalin A-inducible suppressor T-cell activity in active alcoholic liver disease. Gastroenterology 80: 510–517
Kehl A, Schober A, Junge U, Winckber K (1981) Solid-phase radioimmunoassay for detection of alcoholic hyalin antigen (AHAH) and antibody (anti-AH). Clin Exp Immunol 43: 215–221
Krogsgaard K, Tage-Jensen U, Gluud C (1982) Liver membrane antibodies in alcoholic liver disease. Lancet 1: 1365–1366
Laurenzi GA, Guarneri JJ (1966) A study of the mechanisms of pulmonary resistance to infection: the relationship of bacterial clearance to ciliary and alveolar macrophage function. Am Rev Respir Dis 93: 134–145
Lelbach WK (1976) Epidemiology of alcoholic liver disease. In: Popper H, Schaffner F (eds) Progress in liver diseases, vol V. Grune and Stratton, New York, pp 494–515
Lue SL, Paronetto F, Lieber CS (1981) Cytotoxicity of mononuclear cells and vulnerability of hepatocytes in alcoholic fatty liver of baboons. Liver 1: 264–267
Lundy J, Raaf JH, Deakins S, Wanebo HJ, Jacobs DA, Lee T, Jacobowitz D (1975) The acute and chronic effects of alcohol on the human immune system. Surg Gynecol Obstet 147: 212–214
MacSween RNM, Anthony RS, Farquharson M (1981) Antibodies to alcohol-altered hepatocytes in patients with alcohol liver disease. Lancet 2: 803–804
Manns M, Meyer zum Büschenfelde KH, Hess G (1980) Autoantibodies against liver specific membrane lipoprotein in acute and chronic liver diseases: studies on organ-, species-, and disease-specificity. Gut 21: 955–961
Meléndez M, Vargas-Tank L, Fuentes C, Armas-Merino R, Castillo D, Wolff C, Wegmann ME, Soto J (1979) Distribution of HLA histocompatibility antigens, AB0 blood groups and Rh antigens in alcoholic liver disease. Gut 20: 288–290
Meyer zum Büschenfelde KH, Ablerti A, Arnold W, Freudenberg J (1975) Organ-specificity and diagnostic value of cell-mediated immunity against a liver-specific membrane protein: studies in hepatic and non-hepatic diseases. Klin Wochenschr 53: 1061–1067
Mills PR, Rennington TH, Kay P, MacSween RNM, Watkinson G (1979) Hepatitis B_s antibody in alcoholic cirrhosis. J Clin Pathol 32: 778–782
Mills PR, Follet EAC, Urquhart GED, Clements G, Watkinson G, MacSween RN (1981) Evidence for previous hepatitis B virus infection in alcoholic cirrhosis. Br Med J 8: 282–437
Morgan MY, Ross MGR, Ng CM, Adams DM, Thomas HC, Sherlock S (1980) HLA-B8, immunoglobulins, and antibody responses in alcohol-related liver disease. J Clin Pathol 33: 488–492
Naito K, Tsuji T, Nozaki H, Nagashima H (1977) High prevalence of hepatitis B virus infection among heavy drinkers in Japan. Microbiol Immunol 21: 735–738
Nei J, Matsuda Y, Takada A (1983) Chronic hepatitis induced by alcohol. Dig Dis 28: 207–215
Neilson EG, Jimenez SA, Phillips MS (1980) Cell-mediated immunity in interstitial nephritis. III T lymphocyte-mediated fibroblast proliferation and collagen synthesis: an immune mechanism for renal fibrogenesis. J Immunol 125: 1708–1714

Newberry WM, Shorey JW, Sanford JP, Combes B (1973) Depression of lymphocyte reactivity to phytohemagglutinin by serum of patients with liver disease. Cell Immunol 6: 87–97

Orillac J, Torraza E, Ballesteros A, Uscanga L, Galván E, Kershenobich D (1983) Hepatitis B virus infection among alcoholics. Gastroenterology 84: 1388

Paronetto F (1981) Immunologic factors in alcoholic liver disease. Seminars in liver disease. 1: 232–243

Paronetto F (1982) The immunoregulatory network in liver diseases: an overview. Liver 2: 255–259

Penner E, Albini B, Milgrom F (1978) Detection of circulating immune complexes in alcoholic liver disease. Clin Exp Immunol 34: 28–31

Pequignot G, Chabert C, Eydoux H, Courcol MA (1974) Increased risk of liver cirrhosis with intake of alcohol. Rev Alcohol 20: 191–202

Perperas A, Santoulas T, Portmann B, Eddleston ALWF, Williams R (1981) Autoimmunity to a liver membrane lipoprotein and liver damage in alcoholic liver disease. Gut 22: 149–152

Peters M, Liebman HA, Tong MJ, Tinberg HM (1983) Alcoholic hepatitis: granulocyte chemotatic factor from Mallory body-stimulated human peripheral blood mononuclear cells. Clin Immunol Immunopathol 28: 418–430

Rada RT, Knodell RG, Troup GM, Kellner R, Hermanson SM, Richards M (1981) HLA antigen frequencies in cirrhotic and noncirrhotic male alcoholics: a controlled study. Alcoholism 5: 188–191

Rankin JK (1977) The natural history and management of the patient wich alcoholic liver disease. In: Fisher MM, Rankin JK (eds) Alcohol and the liver. Plenum, New York, pp 365–381

Routhier G, Goldstein G, Tidman N, Janossi G, Thomas HC, Sherlock S (1980) T-lymphocyte subpopulations (identified by monoclonal antibodies) in chronic liver disease. Gastroenterology 79: 1122

Saunders JB, Wodak AD, Haines A, Powell-Jackson PR, Portmann B, Davis M, Williams R (1982) Accelerated development of alcoholic cirrhosis in patients with HLA-B8. Lancet 1: 1381–1384

Saunders JB, Wodak AD, Morgan-Capner P, White YS, Portmann B, Davis M, Williams R (1983) Importance of markers of hepatitis B virus in alcoholic liver disease. Br Med J 286: 1851–1854

Si L, Whiteside TL, Schade RR, Van Thiel DH (1983) Lymphocyte subsets studied with monoclonal antibodies in liver tissues of patients with alcoholic liver disease. Alcoholism 7: 431–435

Si L, Whiteside TL, Van Thiel DH, Rabin BS (1984) Lymphocyte subpopulations at the site "piecemeal" necrosis in end stage chronic liver diseases and rejecting liver allografts in cyclosporine-treated patients. Lab Invest 50: 341–347

Sorrell MF, Leevy CM (1972) Lymphocyte transformation and alcoholic liver injury. Gastroenterology 63: 1020–1025

Stillman EG (1924) Persistence of inspired bacteria in the lungs of alcoholized mice. J Exp Med 40: 353–366

Stoltenberg PH, Soltis RD (1984) The nature of IgG complexes in alcoholic liver disease. Hepatology 4: 101–106

Straus B, Berenyi MR (1973) Infection and immunity in alcoholic cirrhosis. Mt Sinai J Med 40: 631–640

Swerdlow MA, Chowdhury LN (1983) IgA subclasses in liver tissues in alcoholic liver disease. Am J Clin Pathol 80: 283–289

Swerdlow MA, Chowdhury LN, Horn I (1982) Patterns of IgA deposits in liver tissue in alcoholic liver disease. Am J Clin Pathol 77: 259–266

Tait BD, MacKay IR (1982) HLA and alcoholic cirrhosis. Tissue Antigens 19: 6–10

Tennenbaum JI, Ruppert RD, St. Pierre RL, Greenberger NJ (1969) The effect of chronic alcohol administration on the immune responsiveness of rats. J Allergy 44: 272–281

Theofilopoulos AN, Dixon FJ (1980) Immune complexes in human diseases. Am J Pathol 100: 531–594

Thomas HC (1981) T cell subsets in patients with acute and chronic HBV infection, primary biliary cirrhosis and alcohol induced liver disease. Int J Immunopharmacol 3: 301–305

Thomas HC, DeVilliers D, Potter B, Hodgson H, Jain S, Jewell DP, Sherlock S (1978) Immune complexes in acute and chronic liver disease. Clin Exp Immunol 31: 150–157

Tisman G, Herbert V (1972) In vitro myelosuppression and immunosuppression by ethanol. J Clin Invest 52: 1410–1414
Trevisan A, Cavigli R, Meliconi R, Stefanini GF, Zotti S, Rugge M, Noventa F, Betterle C, Realdi G (1983) Detection of immunoglobulins G and A on the cell membrane of hepatocytes from patients with alcoholic liver disease. J Clin Pathol 36: 530–534
Vargas V, Pedreira JD, Villaseca J, Ruiz J, Esteban R (1979) Hepatocellular carcinoma and hepatitis B virus markers in Europe. Lancet 1: 721–722
Villa E, Rubbiani L, Barchi T, Ferretti I, Grisendi A, DePalma M, Bellentani S, Manenti F (1982) Susceptibility of chronic symptomless HBsAg carriers to ethanol-induced hepatic damage. Lancet 2: 1243–1244
Wands JR, Dienstag JL, Weake JR, Koff RS (1981) In vitro studies of enhanced Ig synthesis in severe alcoholic liver disease. Clin Exp Immunol 44: 396–404
Yamamoto H (1983) Chronological changes in the immune response and liver tissue of mice by long-term alcohol administration. Jpn J Gastroenterol 80: 2375–2383
Zetterman RK, Luisada-Opper A, Leevy CM (1976) Alcoholic hepatitis: cell-mediated immunological response to alcoholic hyalin. Gastroenterology 70: 382–284

16 Pathology of Alcoholic Liver Disease with Special Emphasis on Alcoholic Hepatitis

H. Ishii, I. Okazaki, and M. Tsuchiya

Department of Internal Medicine Division of Gastroenterology and Department of Preventive Medicine and Public Health, School of Medicine, Keio University, Tokyo 160, Japan

Introduction

It has been widely recognized that alcoholism is one of the major sociomedical problems of modern society. It is also known that excessive alcohol intake is a factor in a significant percentage of all deaths, roughly 10% in the United States. It is associated with half of all traffic deaths, many involving teenagers. Cirrhosis of the liver, which ranks among the ten leading causes of death, is largely attributable to excessive alcohol consumption (Lieber 1978; Eckardt et al. 1981).

The social and economic burdens associated with excessive alcohol consumption are enormous. Alcohol abuse and alcoholism are estimated by the Alcohol, Drug

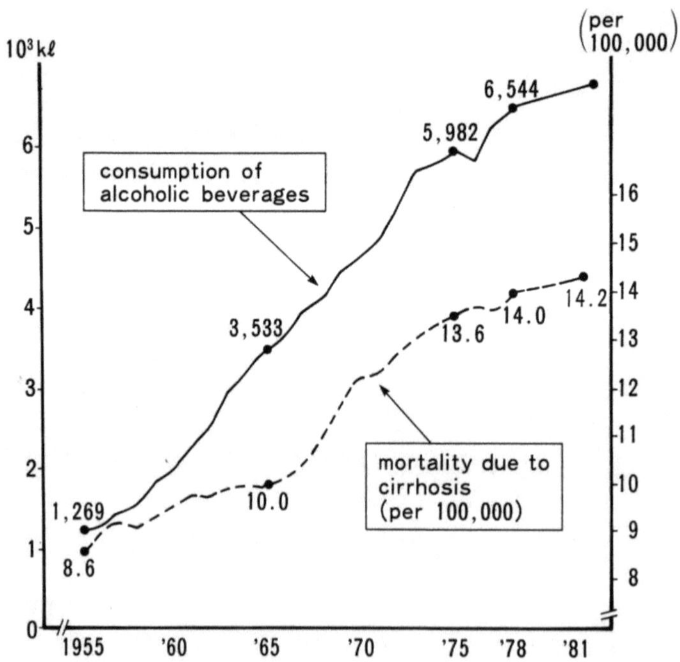

Fig. 1. Changes in alcohol consumption and mortality due to hepatic cirrhosis in Japan (1955–1981)

Alcohol Related Diseases in Gastroenterology
Edited by H. K. Seitz and B. Kommerell

Abuse and Mental Health Administration (USA) to have cost the nation nearly 43 billion dollars in 1975, including medical costs, lost production, motor vehicle accidents, violent crimes, fire loses, and so forth (Berry and Boland 1979).

In Japan, it is now generally recognized that alcoholism or excessive alcohol consumption is a major growing medical problem confronting our society. Indeed, there has been a tremendous increase in alcohol consumption accompanied in parallel by an increased number of heavy drinkers or alcoholics over the last 20 years. Actually, alcohol consumption increased from 1.2*l* per capita in 1955 to approximately 6.5 *l* per capita in 1981, according to the official statistics. At present it is roughly estimated that there are approximately 2 million alcoholics or problem drinkers in Japan. It is also clear that in Japan there has been a steady increase in the mortality from hepatic cirrhosis in the past 20 years, concomitant with the increase in alcohol consumption (Fig. 1).

In the present chapter, we will describe the histopathology of alcoholic liver diseases, alcoholic hepatitis, alcoholic fatty liver, and alcoholic cirrhosis and the relation between these lesions. We will further discuss the characteristics of alcoholic liver diseases in Japan with special reference to the histopathological observations and immunological abnormalities.

Histopathology of Alcoholic Liver Disease

Fatty Liver

The initial lesion of alcoholic liver disease is a fatty liver. Fatty liver, the most common hepatic lesion associated with alcoholism, is produced in 90%–100% of all patients taking excessive amounts of alcohol (Leevy 1962). The normal liver weighs about 1.2–1.5 kg, whereas the alcoholic fatty liver weighs 2.0–2.5 kg, and in rare cases massive accumulation of fat may occur, resulting in liver weighs of 4.0–6.0 kg (Lieber 1982). Gross appearance of fatty liver is yellowish with enlarged dull edge.

The fatty change ranges from vacuolization of a few liver cells, often centrilobular (Fig. 2) or midzonal, to severe involvement of all parts of the lobules. The liver cells usually contain a single droplet of fat which displaces the nucleus to one side of the cells, and when the cell membranes between adjacent hepatocytes rupture, fatty cysts are formed. In livers with fatty change, focal necrosis, liver cell lysis, acidophilic bodies, and lipogranulomas may be seen (Christoffersen et al. 1971). Sometimes, hepatocytes are surrounded by mononuclear cells, which indicate a mild inflammatory response. These formations are called "lipogranulomas". Increased hepatic lipid content may be demonstrated by biochemical measurements before it becomes histologically apparent (Lieber 1982). Moreover, the collagen content of fatty liver increases slightly even when fibrosis cannot be observed microscopically (Feinman and Lieber 1972).

Cholestasis is a relatively infrequent and inconspicuous finding in alcoholic liver disease, except in severe alcoholic hepatitis and in the terminal stages of cirrhosis (Christoffersen and Poulsen 1979). Characteristic ultrastructural changes include

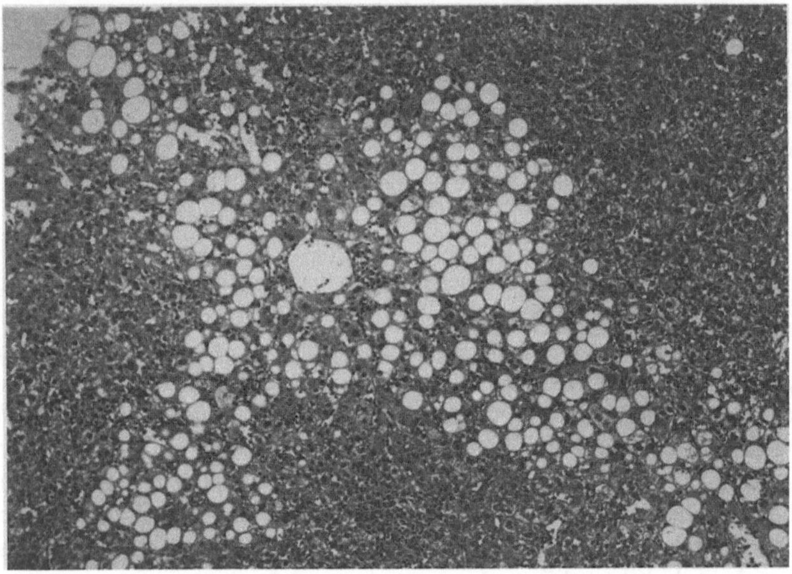

Fig. 2. Fatty liver in an alcoholic. Large lipid vacuoles are seen around the central vein H & E, × 80

enlarged and distorted mitochondria with shortened cristae containing crystalline inclusions (Svoboda and Manning 1964). The endoplasmic reticulum shows vacuolar dilatation and proliferation. Despite adequate nutrition, alcohol can induce these alterations in baboons (Rubin and Lieber 1974) and in both alcoholic and nonalcoholic human volunteers (Lane and Lieber 1966; Rubin and Lieber 1968).

The lesion is reversible within a few weeks to months, provided the patient consumes an adequate diet and avoids alcoholic beverages.

Alcoholic Hepatitis

Alcoholic hepatitis is characterized by the appearance of necrosis with an inflammatory reaction, including polymorphonuclear cells. The liver shows fat, focal or larger areas of necrosis, Mallory bodies, and aggregates of neutrophils, especially around the Mallory bodies in the centrilobular regions of Rappaport zone 3 (Gerber and Popper 1972) (Fig. 3).

The Mallory body, alcoholic hyalin, was first described by Mallory (1911) as follows: an irregular, coarse, hyaline meshwork which stains deeply with eosin. They are classically ropy or serpiginous in contour and partially encircle, in horseshoe fashion, the hepatocyte's nucleus. They are PAS negative and Luxol-fast blue positive. Cells containing these bodies may be surrounded by numerous polymorphonuclear leukocytes (Fig. 3). Mallory bodies can be regarded as a diagnostic hallmark (Hall and Ophulus 1925; Popper et al. 1955; Edmundson et al. 1963;

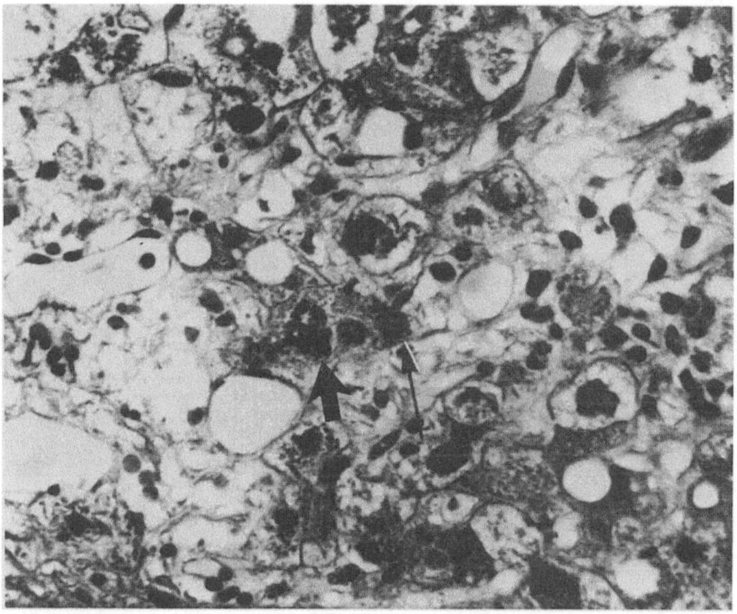

Fig. 3. Mallory bodies and fatty change in acute alcoholic hepatitis. Lytic necrosis of hepatocytes and Mallory bodies are prominent in the ballooned cells *(short arrows)* associated with polymorphonuclear leucocytic infiltration. Bile thrombus is also seen in a pseudoductule in the periportal area *(long arrow)*. H & E, × 500

Lishner et al. 1971; Harinasuta and Zimmerman 1971), but are not always present (Birschbach et al. 1974).

The Mallory bodies gradually become less obvious, and finally disappear within weeks or months following an acute episode of alcoholic hepatitis (Zakim et al. 1982). Although the real origin of Mallory bodies has not yet been resolved, it is interesting that the antitubulin agents colchicine and griseofulvin also induce Mallory body formation in mice (Denk and Eckerstorfer 1977; Denk et al. 1979).

Ultrastructural changes in alcoholic hepatitis are similar but more severe than those seen in fatty liver. Collagenous material may be found in the space of Disse and may be associated with loss of microvilli that normally project from the hepatocytes (Edmondson et al. 1967; Klion and Schaffner 1968). Electron microscopically the Mallory body is fibrillar, is not surrounded by a limiting membrane, and is readily distinguished from giant mitochondria (Biava 1964; Flax and Tisdale 1964; Iseri and Gottlieb 1971; Yokoo et al. 1972; French and Davis 1975; Petersen 1977). Centrally, granular deposits of amorphous dense material may obscure the filamentous architecture. Yokoo et al. (1972) subclassified Mallory bodies into three types. Type I is characterized by filaments which run parallel to each other (Fig. 4). Their thickness is variable, averaging 14.1 nm in diameter. Type II filaments consist of randomly oriented fibrils averaging 15.2 mm thick, which have a tubular appearance. Type III is composed of granular or homogeneous electron-dense materials. Figure 4 shows type I Mallory body filaments with type II filaments partially.

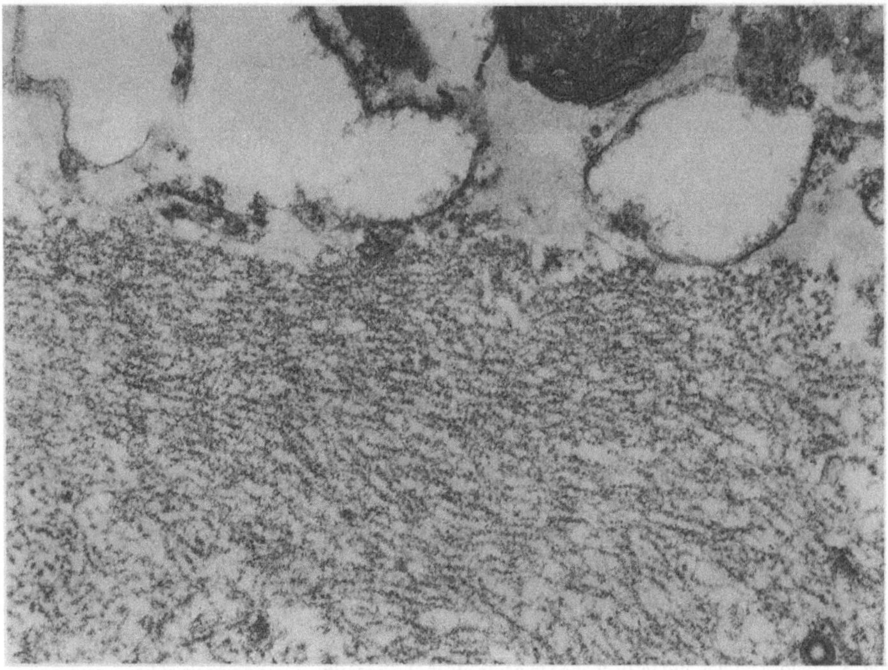

Fig. 4. Electron micrograph of Mallory body. Type I Mallory body filaments run parallel and interconnect. They show the beaded appearance of the filaments, and merge with type II filaments at left bottom corner. × 59,500 (Courtesy of Dr. French, Ottawa)

Mallory bodies were isolated from three autopsied cases of alcoholic hepatitis (French and Davies 1975). The filamentous character of the Mallory body became more obvious. The average diameter of the in situ Mallory body filaments was 8 nm. The average diameter of the isolated filaments was also 8 nm. Their branching and tubularity were also quite similar. It is likely, therefore, that isolated Mallory bodies are in fact derived from in situ Mallory bodies. Two types of filaments were noted near the plasma membranes in glycerinated cells. The first filaments were termed "microfilaments", and formed a lattice network of fine interconnecting filaments near the plasma membrane, averaging 5.2 nm. The second filaments were intermediate filaments averaging 6 nm in thickness and were rigid, straight, and tubular. The latter resembled the type II Mallory body and the intermediate filaments did not bind heavy meromyosin (HMM) in either liver or parotid cells (French and Davies 1975).

According to French and Davies (1975), Mallory body filaments appear to be more likely to represent pathological accumulations of intermediate filaments than microfilaments or thick filaments for the following reasons: (a) they have similar morphological characteristics such as thickness, branching, and tubularity; (b) they occur in the same location within the cell; (c) they do not bind HMM; (d) they are insoluble in solutions of varying ionic strength and pH; (e) they are soluble in 2 M guanidine HCl; and (f) they appear to contain high molecular weight proteins

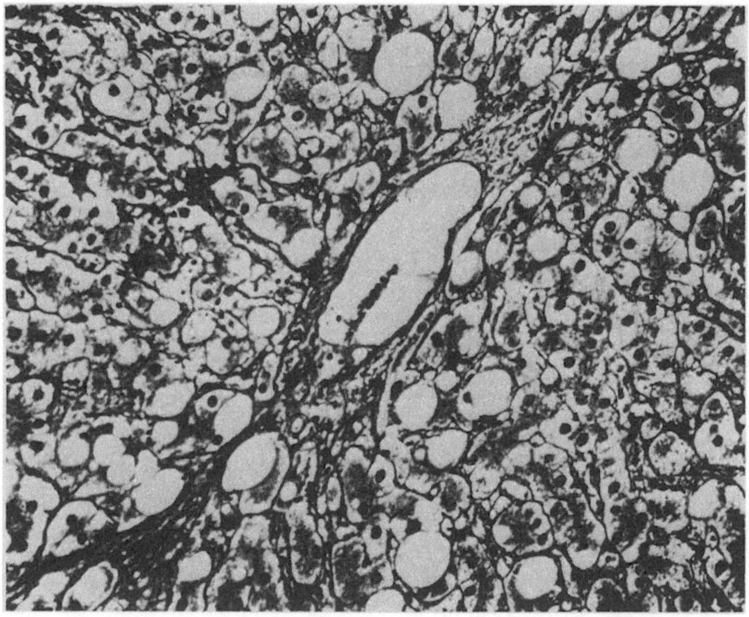

Fig. 5. Thickening of the central vein wall, so-called central hyaline sclerosis, in alcoholic hepatitis. Silver impregnation, × 250

(French and Davies 1975). Those filaments appear to connect to plasma membranes, vesicles, and liver cell nuclei as do normal intermediate filaments with various cell organelles (French 1981). Moreover, recent biochemical studies have revealed a set of polypeptides resembling in their molecular weights prekeratin components of epidermal origin (Franke et al. 1979). Subsequent immunofluorescence microscopical studies using antibodies to cytokeratin polypeptides from hepatocytes revealed that Mallory body formation reflects in alteration of the cytokeratin filament skeleton (Denk et al. 1981). On the other hand, other immunohistochemical observation indicates that Mallory bodies contain unique antigenic determinants not present in prekeratin and that Mallory bodies are not simply composed of intermediate filament proteins (Morton et al. 1980).

Varying degrees of fibrosis are also seen in most of the patients with alcoholic hepatitis. When collagen deposition is increased and becomes visible morphologically, it usually appears first around the central venules, resulting in so-called pericentral or perivenular fibrosis (Edmondson et al. 1963; Popper 1977; Van Waes and Lieber 1977a), as shown in Fig. 5. Then the fibrosis extends into the lobules and this is often called "wiremesh fibrosis" or "creeping fibrosis". In some patients, there may be a striking centrilobular fibrosis, so-called sclerosing hyaline necrosis (Edmondson et al. 1963; Van Waes and Lieber 1977a). These observations suggest a causal relation between alcoholic hepatitis and the development of cirrhosis (Nakano et al. 1982; Lieber 1983). In Fig. 6, portal-central and intralobular fibrosis are seen in acute alcoholic hepatitis.

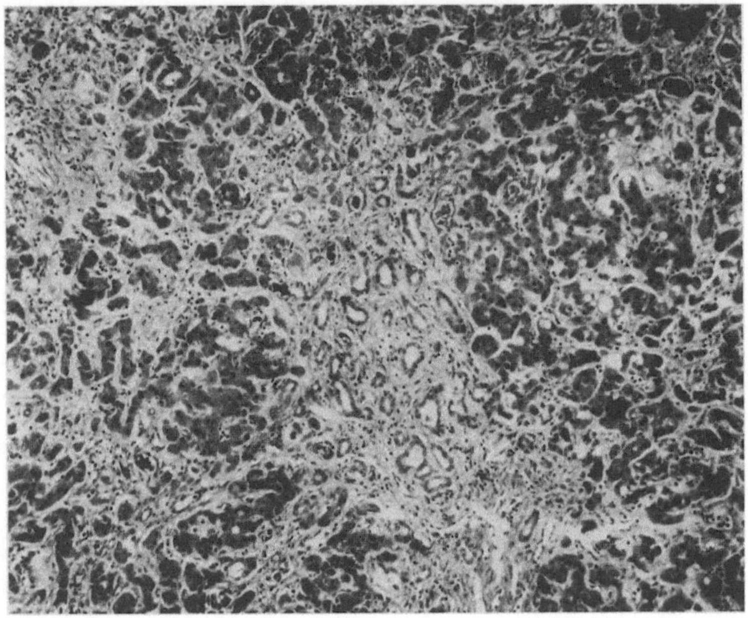

Fig. 6. Sclerosing hyaline necrosis with massive portal-central and intralobular fibrosis in acute alcoholic hepatitis. Diffuse inflammatory cell infiltration, fatty metamorphosis, and bile ductular proliferation in the fibrous septa are also seen. H & E, × 100

Alcoholic Liver Cirrhosis

The final stage of the rather long course of alcoholic liver disease is hepatic cirrhosis. Microscopically, "micronodular or Laennec's cirrhosis" and "macronodular cirrhosis" are known as the end-stage lesions of alcoholic liver disease. The former is the early cirrhotic liver with fine uniform nodules (1–5 mm) on the surface, and the latter is the late cirrhotic liver with the various sizes of nodules ranging from 5 to 50 mm and deep scars between the nodules.

Microscopically, the architecture of liver is remodeled by forming bands of connective tissue between the nascent portal tracts and central zones. The cirrhosis is characteristically accompanied by fatty change. This varies greatly in severity and often diminishes or disappears in the terminal stages of the disease (Porta et al. 1967). Liver cell necrosis in established cirrhosis of the alcoholic takes various forms. The lesion of alcoholic hepatitis with liver cell swelling, Mallory bodies, neutrophil leukocytes, and pericellular fibrosis often continues into the stage of cirrhosis if the patient continues to drink (Christoffersen and Poulsen 1979).

Clinical Picture of Alcoholic Hepatitis and Its Linkage with Alcoholic Cirrhosis

The clinical picture of alcoholic hepatitis ranges from mild to severe. The most frequent signs and symptoms of the milder cases include anorexia, fatigue, intermittent fever, right epigastric pain, and hepatomegaly. Those of fatal cases are fever, anorexia, nausea, abdominal pain, jaundice, tender hepatomegaly, ascites, edema of lower extremities, and hepatic coma. It develops in patients who have been continuously consuming unusually large amounts of alcoholic beverages. On admission to hospital, the patient is usually febrile and dehydrated and has an alcoholic fetor. The early hospital course is often complicated by delirium tremens. Intense jaundice with pruritus in febrile patients may simulate extrahepatic biliary tract obstruction (Afshami et al. 1978).

Characteristic laboratory findings include elevation of erythrocyte sedimentation rate, leukocytosis (polymorphonuclear dominance), serum enzymes such as gamma-glutamyltranspeptidase (GGTP), glutamic-oxaloacetic transaminase (GOT), glutamic-pyruvic transaminase (GPT) in all cases, and alkaline phosphatase (ALP) in 80%. GOT to GPT ratios or GGTP to ALP ratios are increased. The increased serum activity of GGTP may be partly due to the induction of the enzyme (Ishii et al. 1976, 1978, 1980). A better correlation between the degree of liver necrosis and inflammation and the serum level of glutamic acid dehydrogenase (GDH) has been observed (Van Waes and Lieber 1977b). GDH is one of the enzymes originating from the mitochondria of hepatocytes. A high value of serum GDH suggests a liver injury more severe than simple fatty liver (Lieber 1982). The depression of serum albumin and prolongation of the prothrombin time and the elevation of serum immunoglobulin IgA have been shown to correlate with the severity of the histologic lesion (Helman et al. 1971; Iturriaga et al. 1977).

During the course of the disease, the lesion of alcoholic hepatitis develops as a complication of fatty liver. Great attention has been paid to this lesion in the past few years. This form of hepatic injury is also known as "alcoholic steatonecrosis". At present, alcoholic cirrhosis is thought to develop in response to inflammation and necrosis of the liver, namely alcoholic hepatitis. Data from Brunt et al. (1974) suggest that alcoholic hepatitis is precirrhotic, whereas fatty change is not. However, in contrast to the evidence gathered from clinical data, laboratory studies on baboons fed ethanol for prolonged periods have suggested that alcoholic hepatitis is not necessarily a prerequisite in the development of alcoholic cirrhosis. Popper and Lieber (1980) studied sequential liver biopsies from 18 baboons exposed to ethanol with nutritionally adequate liquid diet for up to 6 years. All baboons fed ethanol first developed fatty liver, then fibrosis including perivenular sclerosis, but there was no florid picture of alcoholic hepatitis; yet in one-third of the animals cirrhotic changes developed.

Therefore, it has been suggested that, independent of necrosis or inflammation, ethanol itself may have a direct effect on the metabolism of collagen. As has already been stated, at the earlier stages of alcoholic liver diseases, collagen is detectable only by chemical analysis (Feinman and Lieber 1972). The role of increased collagen synthesis was also suggested by measurement of increases in activity of hepatic peptidylproline hydroxylase in rats and baboons (Feinman and Lieber 1972). Moreover, Mann et al. (1979) found in man that hepatic peptidylproline hydroxylase activity was significantly increased in patients with alcoholic hepatitis or in all stages of alcoholic liver diseases.

The accumulation of hepatic collagen during the development of cirrhosis could theoretically be associated with increased synthesis, decreased degradation, or both. The rate of hepatic fibrous tissue degradation has not been directly measured in human alcoholic liver diseases or in any of the animal models of alcoholic liver disease. Therefore, its exact role in pathogenesis of hepatic fibrosis remains to be determined. Although the mechanisms of collagen degradation in the liver are complex, Okazaki et al. (1977, 1983) and Maruyama et al. (1981) have recently developed a quantitative system for neutral collagenase activity in liver tissues of baboon and human. With this system, baboon collagenase activity was found to be enhanced already at the fatty liver stage (Maruyama et al. 1982).

Another possible mechanism whereby alcohol consumption may be linked to collagen formation is the increase in tissue lactate level secondary to alcohol metabolism (Lieber 1981). The oxidation of ethanol by alcohol dehydrogenase results in the transfer of hydrogen to nicotinamide-adenine dinucleotide (NAD). The resulting enhanced NADH/NAD ratio, in turn, produces an increased lactate/pyruvate ratio resulting in hyperlactacidemia because of both decreased utilization and enhanced production of lactate by the liver. Lindy et al. (1971) studied the relationship of lactate, lactate dehydrogenase, and peptidylproline hydroxylase activity and found that increased concentrations of lactate are associated with increased peptidylproline hydroxylase activity in vivo.

Kowaloff et al. (1977) found that proline oxidase, the first enzyme of the proline degradative pathway, is inhibited by lactate, which in combination with enhanced peptidylproline hydroxylase activity may contribute to hepatic fibrogenesis in alcoholic patients. A more detailed well-reviewed article on hepatic fibrogenesis will be found elsewhere (Lieber 1983).

More recently, Rohde et al. (1979) have developed an assay for type III procollagen peptide which is cleaved off by specific proteases during the conversion of procollagen to collagen (Fessler and Fessler 1978). It has been suggested that the level of the aminoterminal propeptide of type III procollagen in serum reflects connective tissue formation activity in the liver (Rohde et al. 1979). Their results revealed that peptide levels in alcoholic liver disease correlated well with inflammation and necrosis observed in liver biopsies, but not with the grades of hepatic fibrosis. Since then, Frei et al. (1982) have reported the close correlation between peptide levels in chronic liver diseases and the grades of hepatic fibrosis. Niemelä et al. (1983) stressed the usefulness of this assay for the detection of alcoholic hepatitis with cirrhosis by showing a markedly increased serum level of this peptide.

We measured this peptide in the sera of 25 patients with alcoholic liver diseases and concomitantly measured collagenase activity in the homogenates of liver biopsy specimens in order to clarify collagen metabolism from both aspects of collagen synthesis and degradation. Collagenase activity and type III procollagen peptide levels in various alcoholic liver diseases are shown in Table 1. Close correlation between collagenase activity and type III procollagen peptide was observed in

Table 1. Hepatic collagenase activity and serum type III procollagen peptide content in alcoholic liver disease

Alcoholic liver diseases	*n.*	Collagenase activity (units)[a]	Serum type III procollagen peptide (ng/ml)
Alcoholic fatty liver	3	4.77 ± 1.42[b]	8.1 ± 2.91[b]
Alcoholic CPH[c]	9	4.92 ± 0.90	11.8 ± 3.72 } **
Alcoholic CAH[d]	10	5.81 ± 1.51 } **	20.0 ± 10.9
Alcoholic liver cirrhosis	3	4.48 ± 0.34 } **	32.8 ± 4.05 } *

Significant difference: *$p < 0.01$, **$p < 0.05$
[a]1 unit: 1μg collagen degraded/h/mg protein
[b]mean ± SD
[c]CPH (chronic persistent hepatitis)
[d]CAH (chronic active hepatitis)

alcoholic CAH. Increased collagen synthesis and decreased degradation were concomitantly observed in alcoholic liver cirrhosis, while increased collagen synthesis with increased degradation was observed in alcoholic liver fibrosis. Thus the combined determination of collagenase activity and type III procollagen peptide is useful when evaluating a pathophysiological state of collagen metabolism in chronic liver diseases.

Alcoholic Liver Diseases in Japan

Epidemiology

In Japan, alcohol consumption and the numbers of heavy drinkers are increasing gradually and alcoholism has become a major and growing medical problem. On a nationwide basis, cirrhosis now ranks as the fourth largest cause of death in the active age group of 40 to 60 years old, and the incidence of alcoholic cirrhosis has continued to increase and constitutes approximately 30% of all hepatic cirrhosis.

According to the annual report of the Internal Revenue Service of Japan, alcohol consumption increased from 1.2 *l* per capita in 1955 to approximately 6.5 *l* per capita in 1981, as shown in Fig. 1. Another statistic from the Department of Health and Welfare clearly shows that an increase in alcohol consumption is accompanied by an increased number of problem drinkers (approximately 1.75 million problem drinkers in 1981).

In 1977 the consumption of alcohol was studied in 1,182 residents in the Tokyo area, who ranged from 15 to 69 years old. Thirty percent of male residents took some alcoholic beverage every day, whereas 3.6% of females did. Among them, 11% of males and 5% of females took more than 490 g alcohol per week. Heavy drinking was seen in 3.4% of the residents.

In 1980 we studied 2,439 cases who visited the Department of Internal Medicine, University of Occupational and Environmental Health. Four percent of total cases were heavy drinkers (6% of males and 2% of females). Based on these data there may be approximately 2 million alcoholics or problem drinkers in Japan.

Table 2. Histological diagnosis of liver disease in 178 alcoholics

Histological diagnosis	Patients	
	(*n*)	(%)
Alcoholic hepatitis	20	(11.2)
"Chronic active hepatitis"	48	(27.0)
"Chronic persistent hepatitis"	43	(24.2)
Liver cirrhosis	20	(11.2)
Fatty liver	19	(10.7)
Hepatic fibrosis	17	(9.5)
Nonspecific reactive hepatitis	11	(6.2)
Total	178	(100.0)

In Japan, patients with alcoholic hepatitis, strictly diagnosed on the basis of the criteria of the Fogarty International Nomenclature (Leevy et al. 1976), are not so commonly encountered as in European countries and the United States. Karasawa et al. (1979) studied 130 Japanese alcoholic patients compared with 238 American alcoholics and found 12 typical cases of acute alcoholic hepatitis with Mallory bodies in Japan, whereas there were 94 typical cases in the United States. In contrast to the remarkably small numbers of acute alcoholic hepatitis, 46% of Japanese alcoholic patients had chronic hepatitis. More recently, we have performed histopathological studies of the liver in Japanese alcoholics (Ishii et al. 1982).

In the National Institute on Alcoholism (National Kurihama Hospital), 178 chronic alcoholics with abnormal liver function tests were examined histologically by needle liver biopsy, as shown in Table 2. The number of patients with alcoholic hepatitis was 20, constituting 11.2% of the alcoholics. Therefore, typical cases with alcoholic hepatitis are relatively uncommon in Japan. Although the lesion of alcoholic hepatitis develops to cirrhosis, another route leading to cirrhosis without typical alcoholic hepatitis seems to exist.

Histopathological Characteristics

As stated previously, prevalence of acute alcoholic hepatitis is relatively low in Japan. According to the report of Karasawa et al. (1979), acute alcoholic hepatitis was seen in 22 (17%) of 130 Japanese alcoholic liver diseases, whereas it was seen in 155 of 238 (65%) American cases. In our study on 178 Japanese alcoholics who were negative for serum HBsAg and had no past history of blood transfusion, typical alcoholic hepatitis with centrilobular necrosis and neutrophilic infiltration was observed in 20 cases (11%).

The incidence of Mallory body among alcoholics appears to be lower in Japan than in the United States. Mallory bodies were observed in five (25%) of 20 cases with alcoholic hepatitis in our series, whereas in the United States and European countries Mallory bodies are found in at least 30% of patients with alcoholic hepatitis (Birschback et al. 1974; Christoffersen and Poulsen 1979). Karasawa reported that

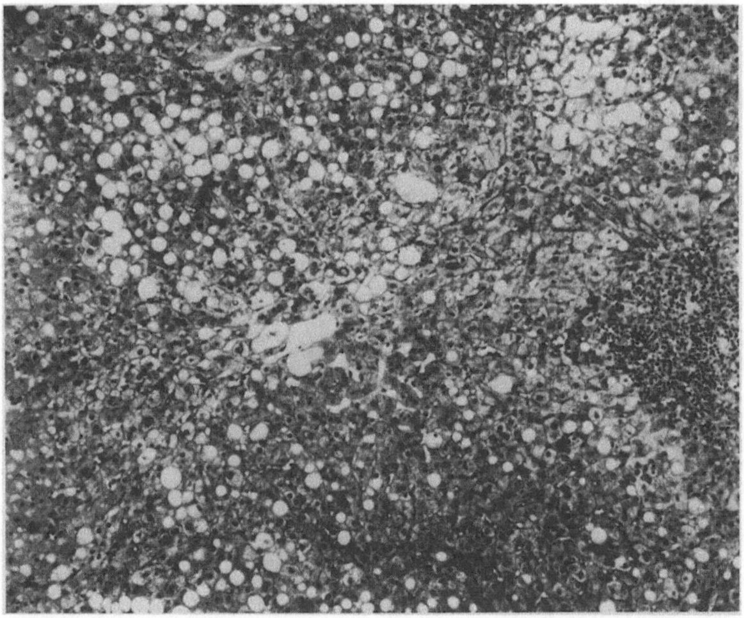

Fig. 7. Alcoholic hepatitis with portal inflammatory cell infiltration. There are moderate to marked mononuclear inflammatory cell infiltrate and fibrosis in the portal tract (seen at the right end of the figure), coexisting with diffusely distributed large fat droplets, spotty necrosis, and ballooning degeneration in the lobule. H & E, × 100

Mallory bodies were seen in 9.2% of 130 Japanese heavy drinkers who showed severe liver function derangement, whereas 39.5% of 238 heavy drinkers in the United States had Mallory bodies. This low incidence of Mallory bodies appears to be another characteristic feature of alcoholic liver diseases in Japan.

Moreover, we have noticed frequently in the liver of alcoholics the infiltrations of small round cells in the periportal area, some of these being accompanied by so-called piecemeal necrosis (Figs. 7 and 8). In our series of 178 alcoholics, 91 cases (51%) were found to have histologic features of chronic hepatitis characterized by mononuclear cell infiltration in the portal tract with periportal fibrosis (Fig. 7). Moreover, piecemeal necrosis was found in 48 of these 91 patients. However, these cases also frequently revealed characteristic features of alcoholic liver disease, such as pericellular and perivenular fibrosis, fatty degeneration, and ballooning of parenchymal cells. In view of these findings, it is of interest to note that Galambos (1972) and French et al. (1977) also observed similar findings rather frequently in patients with alcoholic hepatitis. French et al. (1977) stated that Mallory bodies were found in all cases of alcoholic hepatitis with cirrhosis and in 62% without cirrhosis for an overall frequency of 77%. Moreover, French noted that focal lymphocytic infiltration and piecemeal necrosis were frequently encountered, sometimes making it difficult to distinguish alcoholic hepatitis from chronic aggressive hepatitis.

Thus the presence of hepatic fibrosis without inflammatory process as well as a significant number of cases with "chronic active hepatitis in alcoholics" and "chronic

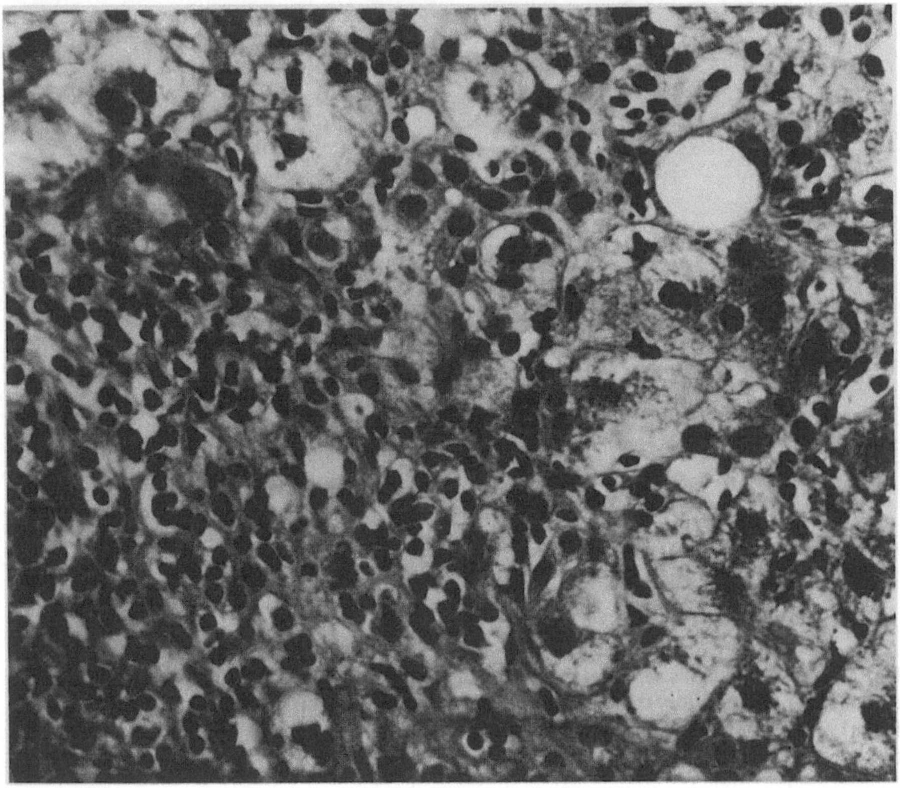

Fig. 8. Piecemeal necrosis seen in the liver of alcoholic hepatitis. A high-power view of the periportal area shows the destruction of limiting plate by a number of lymphocytes and plasma cells. H & E, × 500

persistent hepatitis in alcoholics" appears to be another feature of alcoholic liver diseases in Japan.

In another series of studies, Nei et al. (1983) studied chronic hepatitis induced by alcohol and they mentioned that hepatic fibrosis characterized by pericellular fibrosis without apparent hepatic necrosis and inflammation is the most common histologic feature of alcoholic liver disease in Japan.

Several possible mechanisms have been considered to explain the association of alcoholism with chronic hepatitis (Goldberg et al. 1977). Since the manifestation of chronic hepatitis was most frequently encountered in viral hepatitis, it is conceivable that alcoholism may increase the susceptibility of the patients to viral hepatitis, resulting in the initiation of chronic liver disease. Although we excluded the alcoholic patients with positive HBsAg from our series, approximately 20%–25% of our patients were positive either for anti-HBs or anti-HBc. Moreover, concomitant viral infection due to non-A, non-B hepatitis virus should also be considered.

Furthermore, since histological manifestation characterized by mononuclear cell infiltration and piecemeal necrosis resembles the lesion frequently observed in the autoimmune-type liver disease, possible involvement of immunological abnormalities has been discussed (Paronetto 1981) as one of the pathogenetic factors in alcoholic liver disease.

Possible Involvement of Hepatitis B Virus Infection in Alcoholic Liver Disease

The association of hepatitis B virus with the development of chronic hepatitis and cirrhosis has been discussed since Pettigrew et al. (1972) observed that stimulation of lymphocytes by HBsAg caused significant blastoid transformation in the group with chronic alcoholic liver disease, suggesting cell-mediated reactivity to the virus. Several studies (Mills et al. 1979, 1981; Goudeau et al. 1981; Basile et al. 1981; Villa et al. 1982; Buffet et al. 1982) showed the higher prevalence of hepatitis B virus infection among alcoholic patients with liver disease. Chalmers et al. (1981), however, failed to demonstrate an increase in hepatitis B markers in alcoholic cirrhosis in London and West Yorkshire. This discrepancy might be due to the difference in population and geographical area. Most of these authors suggested that alcoholics have a greater risk of hepatitis B virus infection, probably as a consequence of their social and hygienic circumstances. The question is whether this factor is related to the development of chronic alcoholic liver disease.

In Japan, viral hepatitis has been thought to be a major cause of liver cirrhosis because of the higher prevalence (2.5%) of HBsAg (Red Cross Report 1975) among Japanese than in European countries and the United States.

Inoue (1979) examined 55 patients with alcoholic liver injury in relation to hepatitis B virus infection and revealed three cases with persistent HBs antigenemia, whereas six cases (16%) of 38 studied were positive for anti-HBs and five cases (21%) of 24 studied were positive for anti-HBc. Only three cases with persistent HBs antigenemia were occasionally associated with severe hepatic changes. Inoue suggested that HBs infection might not have continuous influence on the progression of hepatic changes, although the exposure to HBs might have modified the clinicopathological features of these cases with positive anti-HBs and/or anti-HBc.

We have studied the prevalence of anti-HBs in 172 alcoholics with various liver diseases. As shown in Table 3, average prevalence of anti-HBs in alcoholic liver disease was 20.9% and there was no significant difference in the incidence of anti-HBs between various histological findings; 25% in alcoholic hepatitis, 27.1% in chronic active hepatitis, 20.9% in chronic persistent hepatitis, 25% in cirrhosis, and 26.3% in fatty liver.

Furthermore, we have also investigated the prevalence of anti-HBc among these alcoholics (Table 3). The average incidence of anti-HBc was 23.8% in these patients. However, all of these anti-HBc titers were lower than 2×10^9 when assayed by the immune adherence hemagglutination (LAHA) method (except for a case of fatty liver which showed anti-HBc titers of 2×10^{11}). It should be pointed out that these incidences of hepatitis B virus markers in our study do not differ from that found among the general population in Japan.

Therefore, these data suggest that hepatitis B virus infection is not contributory to the histological findings of the livers of these alcoholics, although possible involvement of non-A, non-B hepatitis virus was not ruled out.

Table 3. Hepatitis-B-virus-associated antibodies in alcoholics[a]

Histological diagnosis	Anti-HBs		Anti-HBc	
	No. examined	% positive	No. examined	% positive[b]
Alcoholic hepatitis	20	25.0	12	8.3
Alcoholic CAH	48	27.1	34	29.4
Alcoholic CPH	43	20.9	34	23.5
Liver cirrhosis	20	25.0	13	23.1
Fatty liver	19	26.3	15	26.7
Hepatic fibrosis	17	23.5	9	11.1
Nonspecific reactive hepatitis	5	0	5	40.0
Total	172	20.9	122	23.8

CAH, chronic active hepatitis; *CPH*, chronic persistent hepatitis

[a] Alcoholics in this study are all HBsAg negative. Anti-HBs and Anti-HBc were determined by radioimmunoassay and IAHA method respectively

[b] Anti-HBc titers in positive cases were all less than 2×10^9 except for one case of fatty liver which showed a titer of 2×10^{11}, suggesting that these weakly positive anti-HBc titers (less than 2×10^7) indicate only past infection with hepatitis B virus

Immunological Aspects of Alcoholic Liver Diseases

A variety of humoral and cellular immunological alterations have been reported in alcoholic liver diseases.

Humoral Immunological Alterations

Although the elevation of immunoglobulin (Ig) levels are observed in alcoholic patients with cirrhosis, early IgA elevation has been considered a characteristic of alcoholic liver diseases (Paronetto 1981). Takagi (1979) measured serum IgA level in 94 patients with chronic alcoholism and revealed that IgA was significantly increased in 40% of chronic alcoholics without cirrhosis.

Secretory IgA is also elevated in the serum (Delacroix et al. 1982), because secretory IgA is not produced by lymphoid cells, but by hepatocytes and other epithelial cells (Paronetto 1981). Bjøneboe and Prytz (1976) suggested that increased stimulation of Ig-producing cells due to increaesd absorption of antigen from the intestinal mucosa and decreased clearance of antigen by the reticuloendothelial system makes levels of Ig increase in alcoholic liver diseases. The mechanism of increased Ig levels in alcoholic liver diseases has been discussed from the aspect of production by B cells (Cherrick et al. 1958; Havens 1959; Bjøneboe et al. 1972; Wands et al. 1978; Mutchnick et al. 1981; Smith et al. 1980), but remains unknown.

Various autoantibodies, such as anti-smooth muscle antibody, antifibroblast antibody, anti-liver membrane antibody, anti-alcoholic hyaline antibody, have been detected frequently in the sera of patients with alcoholic liver diseases. The high frequency of antibodies to a variety of antigens may result in an increased incidence of circulating immune complexes. However, the antigen composition of these immune complexes and their pathogenetic significance are not fully understood or clarified. The deposition of immune complex may occur in various organs (Theofilopoulos and Dixon 1980).

Cellular Immunological Alterations

A cellular immunological alteration has been reported in patients with alcoholic liver diseases. We examined a delayed hypersensitivity reaction to purified protein derivative (0.05 μg/ml) or tuberculin and/or DNCB (2,4-dinitrochlorobenzene) in 47 alcoholic patients and revealed a positive reaction in 29% and 76% of the patients respectively (Takagi 1979). Skin sensitization to DNCB was reported to be impaired in patients with alcoholic liver disease but not in alcoholics without liver involvement (Lang et al. 1980). However, we observed that DNCB test was abnormal in 77% of the chronic alcoholics without liver injury, therefore suggesting that abnormal cellular immunity appears to exist in chronic alcoholics irrespective of the presence of liver disease. These results were consistent with those of Lundy et al. (1975).

Lymphocyte response to the mitogen phytohemagglutinin (PHA) has been reported to be decreased (Sorrell and Leevy 1972; Takagi 1979). The number and percentage of T-lymphocytes are decreased in alcoholic liver diseases, especially in patients with alcoholic hepatitis and active cirrhosis (Bernstein et al. 1974; Lundy et al. 1975; Thomas et al. 1976). In our study the decreased population of T-lymphocytes was observed in 34% of chronic alcoholics regardless of the presence of liver disease.

Purified isolates of Mallory bodies were added to lymphocytes from patients with alcoholic liver disease and it was found that patients with alcoholic hepatitis have a significant increase in the percentage of migration inhibition compared to healthy subjects (Zetterman and Leevy 1975). Mallory bodies may stimulate some immune responses in the progression of alcoholic liver disease (Leevy et al. 1979), but this is still the subject of controversy. Further studies are needed to elucidate the pathogenic role of alcoholic hyalin.

Recently, evidence has been accumulated that antibody directed against a liver membrane lipoprotein in cooperation with cytotoxic lymphocytes may have an important role in the pathogenesis of chronic active hepatitis (Jensen et al. 1978). Jensen et al. (1978) detected antibody to liver-specific membrane protein in 29 of 30 patients with chronic active hepatitis, and the titer of antibody to the lipoprotein showed a significant correlation with the activity of the disease. Moreover, Cochrane et al. (1977) examined whether an autoimmune reaction to liver-specific proteins occurs in alcoholic liver disease.

They demonstrated the cytotoxic effect of lymphocytes on isolated hepatocytes in 15 of 17 patients with alcoholic hepatitis and they were also able to block the cytotoxic reaction by using a liver-specific membrane protein, thus suggesting that sensitization to this protein is responsible for the cytotoxicity. Moreover, the demonstration of cytotoxicity using non-T-cell fraction indirectly suggested an

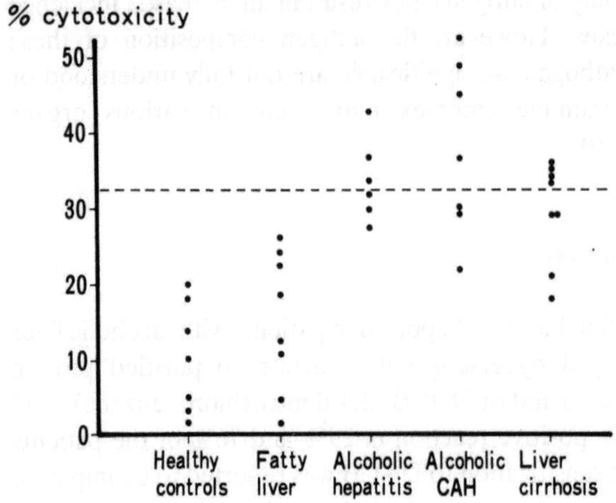

Fig. 9. Antibody-dependent cell-mediated cytotoxicity against Chang liver cell in the sera of 29 patients with various forms of alcoholic liver disease. The *dotted line* indicates the upper limit of the normal range (mean + 3SD). *CAH,* chronic active hepatitis

involvement of an antibody-dependent cell-mediated cytotoxicity (ADCC). On the basis of these observations, we have recently examined the prevalence of anti-hepatocyte (Chang liver cell) antibody by I^{125} protein A assay in sera from 46 patients with various alcoholic liver diseases (Ishii et al. 1982; Tsuchimoto 1982). Anti-Chang liver cell membrane antibodies were found in 78% of alcoholic hepatitis and in 90% of chronic active hepatitis in the alcoholic, whereas the antibody was found in only 10% of fatty liver. The antibody titers were well correlated with the severity of periportal cell infiltration. We have further examined whether this antibody could mediate ADCC against Chang liver cells. It was indeed shown (Fig. 9) that ADCC was significantly enhanced in patients with alcoholic hepatitis as well as in alcoholic patients with chronic active hepatitis, as compared to control subjects. Thus these results suggest that ADCC against hepatocytes induced by anti-liver-cell-membrane antibody are involved at least partly in alcoholic liver disease.

Genetic Factors in Alcoholic Liver Disease

Although various degrees of liver damage result from prolonged and excessive consumption of alcohol and a number of statistical and experimental studies have clarified the direct hepatotoxicity of alcohol and its metabolites, it is unequivocal that only a relatively small portion (10%–30%) of alcoholics develop cirrhosis. These facts lead us to the assumption that among other mechanisms, including malnutrition, there may also be an immunogenetic mechanism in the etiology of alcoholism or alcoholic liver disease. As already briefly reviewed (in the previous section of this chapter), a variety of immunologic abnormalities have been described in chronic alcoholics.

Although histocompatibility antigen (HLA antigen) as an immunogenetic marker is very important for the clarification of individual susceptibility to disease, the association between HLA antigens and alcoholic liver disease is somewhat controversial. In England HLA-B8 was found to be associated with alcoholic hepatitis and cirrhosis (Bailey et al. 1976; Morgan et al. 1980). The association of Bw40, B13, and cirrhosis has been reported (Bell and Nordhagen 1978; Melendez et al. 1979). On the contrary, Scott (1977) and Gluud et al. (1980) found no significant association between HLA antigens and alcoholic cirrhosis.

We have studied 77 chronic alcoholics with liver diseases to evaluate the HLA antigen association (Miyamoto et al. 1983). There were no significant differences in HLA antigen phenotype frequencies between the patients and control subjects regarding A and C loci. Prevalence of HLA-B40 complex (B40. 48. 13) and DRW9 tended to increase among chronic alcoholics. Furthermore, when alcoholics were divided according to whether they had liver cirrhosis or not, the cirrhosis group (42 cases) revealed a significantly higher frequency of HLA-DRW9 as compared to controls. There was also a tendency of B40 complex to increase in frequency in the cirrhosis group. Haplotype frequency and linkage disequilibrium parameters of HLA-B40 48-DRW9 were significantly higher than those of controls. These data suggest that HLA-DRW9 and/or HLA-B40-DRW9 might be closely associated with susceptibility to developing alcoholic cirrhosis and that autoimmune mechanisms might be partly involved in the etiology.

Conclusion

Although "classic" alcoholic hepatitis is clinically important lesion probably linking alcoholic fatty liver with cirrhosis, the incidence of the disease is relatively uncommon especially in Japan.

Other forms of alcoholic liver disease including chronic hepatitis and hepatic fibrosis should also be considered as possible precursor lesions of alcoholic cirrhosis.

Histopathology, fibrogenesis, and immunological abnormalities of alcoholic liver diseases were discussed.

References

Afshami P, Littenberg GD, Wollman J, Kaplowitz N (1978) Significance of microscopic cholangitis in alcoholic liver disease. Gastroenterology 75: 1045–1050

Bailey RJ, Krasner N, Eddleston ALWF, Williams R, Tee DEH, Doniach D, Kennedy LA, Bachelor JR (1976) Histocompatibility antigens, autoantibodies, and immunoglobulins in alcoholic liver disease. Br Med J 2: 727–729

Basile A, Vitale G, Macor C, Scudeller G (1981) Hepatitis B virus infection in alcoholic cirrhosis. Br Med J 282: 1705–1706

Bell H, Nordhagen R (1978) Association between HLA-BW40 and alcoholic liver diseases with cirrhosis. Br Med J 1: 822

Bernstein IM, Webster KH, Williams RC Jr, Strickland RG (1974) Reduction in circulating T lymphocytes in alcoholic liver disease. Lancet 2: 488–490

Berry RE Jr, Boland JP (1979) The economic cost of alcohol abuse. Free Press, New York

Biava C (1964) Mallory alcoholic hyaline: a heretofore unique lesion of hepatocellular ergastoplasm. Lab Invest 13: 301–320

Birschbach HR, Harinasuta W, Zimmerman HJ (1974) Alcoholic steatonecrosis. II. Prospective study of prevalence of Mallory bodies in biopsy specimens and comparison of severity of hepatic disease in patients with and without this histological feature. Gastroenterology 66: 1195–1202

Bjøneboe M, Prytz H (1976) The mononuclear phagocyte functions of the liver. In: Ferguson A, MacSween RNM (eds) Immunologic Aspects of the liver and intestinal tract. Garden City Press, Lancaster, pp 251–289

Bjøneboe M, Prytz H, Orskov F (1972) Antibodies to intestinal microbes in serum of patients with cirrhosis of the liver. Lancet 1: 58–60

Brunt PW, Kew MC, Shemer PJ, Sherlock S (1974) Studies in alcoholic liver disease in Britain. I. Clinical and pathological patterns related to natural history. Gut 15: 52–58

Buffet C, Attali P, Papoz L, Chaput JC, Etienne JP (1982) Evidence for previous hepatitis B virus infection in alcoholic cirrhosis. Digest Dis Sci 27: 473–474

Chalmers DM, Bullen AW, Levi AJ (1981) Evidence for previous hepatitis B virus infection in alcoholic cirrhosis. Br Med J 282: 819

Cherrick GR, Potheir L, Dufour JJ, Sherlock S (1958) Immunologic response to tetanus toxoid inoculation in patients with hepatic cirrhosis. N Engl J Med 261: 340–342

Christoffersen P, Poulsen H (1979) Alcoholic liver disease. In: MacSween RNM, Anthony PP, Scheuer PJ (eds) Pathology of the liver. Churchill Livingstone, Edinburgh, pp 232–247

Christoffersen P, Braendstrup O, Juhl E, Poulsen H (1971) Lipogranulomas in human liver biopsies with fatty change. Acta Pathol Microbiol Scand [A] 79: 150–158

Cochrane AMG, Moussouros A, Portmann B, McFarlane IG, Thomson AD, Eddleston ALWF, Williams R (1977) Lymphocyte cytotoxicity for isolated hepatocytes in alcoholic liver disease. Gastroenterology 72: 918–923

Delacroix DL, Reynaert M, Pauwels S, Geubel AP, Vaerman JP (1982) High serum levels of secretory IgA in liver disease. Dig Dis Sci 27: 333–340

Denk H, Eckerstorfer R (1977) Colchicine-induced Mallory body formation in the mouse. Lab Invest 36: 563–565

Denk H, Franke WW, Eckerstorfer R (1979) Formation and involution of Mallory bodies ("alcoholic hyalin") in murine and human liver revealed by immunofluorescence microscopy with antibodies to prekeratin. Proc Natl Acad Sci USA 76: 4112–4116

Denk H, Franke WW, Dragosics B, Zeiler I (1981) Pathology of cytoskeleton of liver cells: demonstration of Mallory bodies (alcoholic hyalin) in murine and human hepatocytes by immunofluorescence microscopy using antibodies to cytokeratin polypeptides from hepatocytes. Hepatology 1: 9–20

Eckardt MJ, Harford TC, Kaelber CT, Parker ES, Rosenthal LS, Ryback RS, Salmoiraghi GC, Vanderveen EV, Warren K (1981) Health hazards associated with alcohol consumption. JAMA 246: 648–666

Edmondson HA, Peters RL, Reynolds TB, Kuzma DT (1963) Sclerosing hyaline necrosis of the liver in the chronic alcoholic. A recognizable clinical syndrome. Ann Intern Med 59: 646–673

Edmondson HA, Peters RL, Frankel HH, Borowsky S (1967) The early stage of liver injury in the alcoholic. Medicine 46: 119–129

Feinman L, Lieber CS (1972) Hepatic collagen metabolism: effect of alcohol consumption in rats and baboons. Science 176: 795

Fessler JH, Fessler LI (1978) Biosynthesis of procollagen. Annu Rev Biochem 47: 129–162

Flax MH, Tisdale WA (1964) An electron microscopic study of alcoholic hyalin. Am J Pathol 44: 441–453

Franke WW, Denk H, Schmid E, Osborn M, Weber K (1979) Ultrastructural biochemical and immunologic characterization of Mallory bodies in griseofulvin-treated mice. Fimbriated rods of filaments containing prekeratin-like polypeptides. Lab Invest 40: 207–220

Frei A, Zimmerman A, Weigand K (1982) Serum procollagen-III-aminopeptide (pp) as index of liver fibrosis – new morphometric evidence. Hepatology 2: 733

French SW (1981) The Mallory body: structure, composition and pathogenesis. Hepatology 1: 76–83

French SW, Davies PL (1975) The Mallory body in the pathogenesis of alcoholic liver disease. In: Khanna JM, Israel Y, Kalent H (eds) Alcoholic liver pathology. Addiction Research Foundation, Toronto, pp 113–114

French SW, Sim JS, Franks KE (1977) Alcoholic hepatitis. In: Fisher MM, Rankin JG (eds) Alcohol and the liver. Plenum, New York, pp 261–288

Galambos JT (1972) Natural history of alcoholic hepatitis. III. Histological changes. Gastroenterology 63: 1026–1035

Gerber MA, Popper H (1972) Relation between central canals and portal tracts in alcoholic hepatitis. A contribution to the pathogenesis of cirrhosis in alcoholics. Human Pathol 3: 199–207

Gluud C, Aldershvile J, Dietrichson O, Hardt F, Iversen K, Juhl E, Nielsen JO, Ryder LP, Skinhoj P, Srejgaard A (1980) A human leucozyte antigen in patients with alcoholic liver cirrhosis. Scand J Gastroenterology 15: 337–341

Goldberg SJ, Mendenhall CL, Connell AM, Chedid A (1977) "Nonalcoholic" chronic hepatitis in the alcoholic. Gastroenterology 72: 598–604

Goudeau A, Maupas P, Dubois F, Coursaget P, Bougnoux P (1981) Hepatitis B infection in alcoholic liver disease and primary hepatocellular carcinoma in France. Prog Med Virol 27: 26–34

Hall EM, Ophulus W (1925) Progressive alcoholic cirrhosis. Report of four cases. Am J Pathol 1: 477–493

Harinasuta U, Zimmerman HJ (1971) Alcoholic steatonecrosis. I. Relationship between severity of hepatic disease and presence of Mallory bodies in the liver. Gastroenterology 60: 1036–1046

Havens WP Jr (1959) Liver disease and antibody formation. Int Arch Allergy Appl Immunol 14: 75–83

Helman RA, Temko MH, Nye SW, Fallon HJ (1971) Alcoholic hepatitis: natural history and evaluation of prednisolone therapy. Ann Intern Med 74: 311–321

Inoue K (1979) The clinicopathological features of the alcoholic liver injury in Japan and its etiological relationship to hepatitis B virus. Gastroenterol Jpn 12: 230–240

Iseri OA, Gottlieb LS (1971) Alcoholic hyaline and megamitochondria as separate and distinct entities in liver disease associated with alcoholism. Gastroenterology 60: 1027–1035

Ishii H, Kanno T, Shigeta Y, Yasuraoka S, Kano S, Takeshita E, Tsuchiya M (1976) Hepatic gamma glutamyl transpeptidase activity: its activation by chronic ethanol administration. Gastroenterology 71: A-20/913

Ishii H, Yasuraoka S, Shigeta Y, Kamiya T, Okuno F, Miyamoto K, Tsuchiya M (1978) Hepatic and intestinal gamma glutamyl transpeptidase activity: its activation by chronic ethanol administration. Life Sci 23: 1393–1398

Ishii H, Okuno F, Shigeta Y, Tsuchiya M (1980) Significance of serum gamma glutamyl transpeptidase as a marker of alcoholism. Pharmacol Biochem Behav 13: 95–99

Ishii H, Inagaki Y, Tsuchimoto K, Yamauchi H, Takagi S, Shigeta Y, Tsuchiya M (1982) Increased incidence of "non alcoholic" chronic hepatitis in alcoholics: evidence for autoantibodies to hepatocyte and antibody dependent cell-mediated cytotoxicity (ADCC). Hepatology 2: 704

Iturriaga H, Pereda T, Estevez A, Ugarte G (1977) Serum immunoglobulin changes in alcoholic patients, Ann Clin Res 9: 39–43

Jensen DMBS, McFarlane IG, Portmann BS, Eddleston ALWF, Williams R (1978) Detection of antibodies detected against a liver-specific membrane lipoprotein in patients with acute and chronic active hepatitis. N Engl J Med 299: 1–7

Karasawa T, Shikata T, Kushida T, Kaneda H, Ohknbo H (1979) Pathological study on alcoholic liver diseases in high and low incidence countries: comparison between Cincinnati, USA and Tokyo, Japan. Kanzo 20: 115–123

Klion FM, Schaffner F (1968) Ultrastructural studies in alcoholic liver disease. Digestion 1: 2–14

Kowaloff FM, Phang JM, Granger AS, Downing SJ (1977) Regulation of proline oxidase activity by lactate. Pro Natl Acad Sci USA 74: 5368–5371

Lane BP, Lieber CS (1966) Ultrastructural alterations in human hepatocytes following ingestion of ethanol with adequate diets. Am J Pathol 49: 593–603

Lang JM, Ruscher H, Hasselmann JP, Grandjean P, Bigel P, Mayer S (1980) Decreased autologous rosette-forming T lymphocytes in alcoholic cirrhosis. Absence of correlation with other T cell markers and with delayed cutaneous hypersensitivity. Int Arch Allergy Appl Immunol 61: 337–343

Leevy CM (1962) Fatty liver: a study of 270 patients with biopsy proven fatty liver and a review of the literature. Medicine 41: 249–276

Leevy CM, Popper H, Sherlock S (1976) Diseases of the liver and biliary tract. Standardization of

nomenclature, diagnostic criteria and diagnostic methodology. DHEW Publication No. (NIH) 76-725
Leevy CM, Kanagasundaram N, Matsumoto K, Chen T (1979) Alcoholic hyalin and immunologic reactivity. In: Eddleston ALWF, Weber JCP, Williams R (eds) Immune reactions in liver disease. Pitman Medical, Tunbridge Wells, pp 195-207
Lieber CS (1978) A public health strategy against alcoholism and its complications. Am J Med 65: 722-725
Lieber CS (1981) Metabolism and metabolic effects of ethanol. Semin Liver Dis 1: 189-202
Lieber CS (1982) Medical disorders of alcoholism: pathogenesis and treatment. Saunders, Philadelphia, p 263
Lieber CS (1983) Precursor lesions of cirrhosis. Alcoholism 18: 5-20
Lindy S, Pedersen FB, Turto H, Uitoo J (1971) Lactate, lactate dehydrogenase and protocollagen proline hydroxylase in rat skin autograft. Hoppe Seylers Z Physiol Chem 352: 113-118
Lishner MW, Alexander JF, Galambos JT (1971) Natural history of alcoholic hepatitis. I. The acute disease. Am J Dig Dis 16: 481-494
Lundy J, Raaf JH, Deakins S, Wanebo HJ, Jacobs DA, Lee T-d, Jacobowitz D, Spear C, Oettgen HF (1975) The acute and chronic effects of alcohol on the human immune system. Surg Gynecol Obstet 141: 212-219
Mallory FB (1911) Cirrhosis of the liver. Five different types of lesions from which it may arise. Bull Johns Hopkins Hospital 22: 69-75
Mann SW, Fuller GC, Rodil JV, Vidins EI (1979) Hepatic prolyl hydroxylase and collagen synthesis in patients with alcoholic liver disease. Gut 20: 825-832
Maruyama K, Feinman L, Okazaki I, Lieber CS (1981) Direct measurement of neutral collagenase activity in liver homogenates from baboons and humans. Biochim Biophys Acta 658: 124-131
Maruyama K, Feinman L, Fainsilber Z, Nakano M, Okazaki I, Lieber CS (1982) Mammalian collagenase increases in early alcoholic liver disease and decrease with cirrhosis. Life Sci 30: 1379-1384
Melendez M, Vargas-Tank L, Fuentes C, Armas-Merino R, Castillo D, Wolff C, Wegmann ME, Soto J (1979) Distribution of HLA histocompatibility antigens, AB0 blood groups and RH antigens in alcoholic liver disease. Gut 20: 288-290
Mills PR, Pennington TH, Kay P, McSween RNM, Watkinson G (1979) Hepatitis Bs antibody in alcoholic cirrhosis. J Clin Pathol 32: 778-782
Mills PR, Follett EAC, Urguhart GeD, Clements G, Watkinson G, MacSween RNM (1981) Evidence for previous hepatitis B virus infection in alcoholic cirrhosis. Br Med J 282: 437-438
Miyamoto K, Ishii H, Takata H, Takagi S, Shigeta Y, Sekiguchi S, Suyama K, Kohno H, Tsuchiya M (1983) Association of HLA-B40 and DRW 9 with Japanese alcoholic liver cirrhosis. Pharmacol Biochem Behav [Suppl 1] 18: 467-471
Morgan MY, Ross MGR, Ng CM, Adams DM, Thomas HC, Sherlock S (1980) HLA-B8, immunoglobulins, and antibody responses in alcohol-related liver disease. J Clin Pathol 33: 488-492
Morton JA, Fleming KA, Trowell JM, McGee JO'D (1980) Mallory bodies – immunohistochemical detection by antisera to unique non-prekeratin components. Gut 21: 727-733
Mutchnick MG, Lederman HM, Missirian A, Hohnson AG (1981) In vitro synthesis of IgG by peripheral blood lymphocytes in chronic liver disease. Clin Exp Immunol 43: 370-375
Nakano M, Worner TM, Lieber CS (1982) Perivenular fibrosis in alcoholic liver injury: ultrastructure and histologic progression. Gastroenterology 83: 777-785
Nei J, Matsuda Y, Takada A (1983) Chronic hepatitis induced by alcohol. Dig Dis Sci 28: 207-215
Niemelä O, Risteli L, Sotaniemi EA (1983) Aminoterminal propeptide of type III procollagen in serum in alcoholic liver disease. Gastroenterology 85: 254-259
Okazaki I, Feinman L, Lieber CS (1977) Hepatic mammalian collagenase: development of an assay and demonstration of reversal activity after ethanol consumption. Gastroenterology 73: 1236
Okazaki I, Feinman L, Maruyama K, Fainsilber Z, Nakano M, Lieber CS (1983) Development of an assay for mammalian hepatic collagenase and study of the effect of ethanol on enzyme activity. In: Lieber CS (ed) Biological approach to alcoholism. U.S. Department of Health and Human Services (ADM 83-1261), Washington, D.C., pp 392-398

Paronetto F (1981) Immunologic factors in alcoholic liver disease. Semin Liver Dis 1: 232–243
Petersen PU (1977) Alcoholic hyalin, microfilaments and microtubules in alcoholic hepatitis. Acta Pathol Microbiol Scand 85: 384–394
Pettigrew NM, Goudie RB, Russell RI (1972) Evidence for a role of hepatitis virus B in chronic alcoholic liver disease. Lancet 2: 724
Popper H (1977) The pathogenesis of alcoholic cirrhosis. In: Fisher MM, Rankin JG (eds) Alcohol and the liver. Plenum, New York, pp 289–305
Popper H, Lieber CS (1980) Histogenesis of alcoholic fibrosis and cirrhosis in the baboon. Am J Path 98: 695–716
Popper H, Szanto PB, Parathasarathy M (1955) Florid cirrhosis – a review of 35 cases. Am J Clin Pathol 25: 889–901
Porta EA, Hartroft WS, de la Iglesia FA (1967) Structural and ultrastructural hepatic lesions associated with acute and chronic alcoholism in man and experimental animals. In: Maickel RP (ed) Biochemical factors in alcoholism. Pergamon, Oxford, pp 201–238
Rohde H, Vargas L, Hahn E, Kalbfleisch H, Bruguera M, Timpl R (1979) Radioimmunoassay for type III procollagen peptide and its application to human liver disease. Eur J Clin Invest 9: 451–459
Rubin E, Lieber CS (1968) Alcohol-induced hepatic injury in nonalcoholic volunteers. N Engl J Med 278: 869–876
Rubin E, Lieber CS (1974) Fatty liver, alcoholic hepatitis and cirrhosis produced by alcohol in primates. N Engl J Med 290: 128–135
Scott BB, Rajah SM, Losowsky MS (1977) Histocompatibility antigens in chronic liver disease. Gastroenterology 72: 122–125
Smith WI Jr, Van Thiel DH, Whiteside T, Janoson B, Magovern I, Puet T, Rabin BS (1980) Altered immunity in male patients with alcoholic liver disease: evidence for defective immune regulation. Alcoholism Clin Exp Res 4: 199–206
Sorrell MF, Leevy CM (1972) Lymphocyte transformation and alcoholic liver injury. Gastroenterology 63: 1020–1025
Svoboda DJ, Manning RT (1964) Chronic alcoholism with fatty metamorphosis of the liver. Am J Pathol 44: 645–662
Takagi S (1979) Altered humoral and cell-mediated immunity in chronic alcoholics. Its relation to the development of liver injury. J. Keio Med. Soc. 76: 184–195
Theofilopoulos AN, Dixon FI (1980) Immune complexes in human diseases. Am J Pathol 100: 531–594
Thomas HC, Freni M, Sanchez-Tapias J, DeVilliers D, Jain S, Sherlock S (1976) Peripheral blood lymphocyte populations in chronic liver disease. Clin Exp Immunol 26: 222–227
Tsuchimoto K (1982) Detection of antibody and antibody dependent cell-mediated cytotoxicity (ADCC) against Chang liver cell in alcohol liver disease. Jpn J Gastroenterol 79: 1581–1588
Van Waes L, Lieber CS (1977a) Early perivenular sclerosis in alcoholic fatty liver: an index of progressive liver injury. Gastroenterology 73: 646–650
Van Waes L, Lieber CS (1977b) Glutamate dehydrogenase, a reliable marker of liver cell necrosis in the alcoholic. Br Med J 2: 1508–1510
Villa E, Rubbiani L, Barchi T, Ferretti I, Grisendi A, De Palma M, Bellentani S, Manerti F (1982) Susceptibility of chronic symptomless HBsAg carriers to ethanol-induced hepatic damage. Lancet 2: 1243–1244
Wands JR, Dienstag JL, Weake JR, Koff RS (1978) Proliferative and secretory B cell activity in severe alcoholic liver disease. Gastroenterology 75: 992
Yokoo H, Minick OT, Batti F, Kent G (1972) Morphologic variants of alcoholic hyalin. Am J Pathol 69: 25–40
Żakim D, Boyer TD, Montgomery C, Kanas N (1982) Alcoholic liver disease. In: Zakim D, Boyer T (eds) A texbook of liver disease. Saunders, Philadelphia, pp 739–789
Zetterman RK, Leevy CM (1975) Immunologic reactivity and alcoholic liver disease. Bull N Y Acad Med 51: 533–544

17 Clinical and Therapeutic Aspects of Alcoholic Liver Disease

C. L. Mendenhall and the VA Cooperative Study on Alcoholic Hepatitis*

Veterans Administration Medical Center, 3200 Vine Street, Cincinnati, OH 45220, USA

Introduction

The socioeconomic health impact of alcoholism is enormous. In the United States in 1975, $ 12.74 billion was spent for health and medical expenses associated with alcoholism (Berry et al. 1977). The total number of deaths directly related to alcoholism in 1977 ranged between 18,066 and 34,724 (National Institute on Alcohol Abuse and Alcoholism 1978), 85% of these being associated with alcoholic liver disease. Indeed, death from cirrhosis ranks fourth among causes of death in the United States (US Bureau of Census 1975).

Fortunately, not all alcoholics develop severe liver disease. Many persons imbibe heavily for many years or even a lifetime without developing significant liver disease. Others succumb after a much lesser exposure. It is estimated that only 10%–15% of chronic alcoholics develop cirrhosis (Leevy 1967; Tashiro and Lipscomb 1963; Gorwitz et al. 1970; Pell and D'Alonzo 1968). The true incidence is unknown, since many have only minimal laboratory and clinical manifestations in spite of moderate to severe pathology (Orrego et al. 1981).

Alcoholic liver disease encompasses a spectrum of clinical and pathologic changes. Based on histologic features, three major types of injury are observed: (a) alcoholic fatty liver, (b) alcoholic hepatitis, and (c) alcoholic cirrhosis. Most commonly, fatty liver and alcoholic hepatitis are regarded as reversible forms of this process, while established cirrhosis is regarded as irreversible. These changes are not mutually exclusive, and most typically all three may be present in the same individual. I will not elaborate further on either the pathologic features or the pathophysiology and biochemistry of these changes, since entire chapters elsewhere in this volume are devoted to these descriptions.

Not all liver disease developing in alcoholics is classic for alcoholism. Levin et al. (1979) reported a 43% (62/145) incidence of nonalcoholic liver disease in heavy drinkers. In our Veterans' Administration (VA) Cooperative study, 2,074 alcoholics with liver disease were screened. Of these, only 21% had uncomplicated alcoholic hepatitis. However, the true incidence is certainly higher, since many were excluded (35%) because their conditions were mild and biopsy confirmation was not available. In a retrospective study of 202 consecutive biopsy specimens at the Martinez VA

* The list of participants is given in the Acknowledgements at the end of this chapter

Alcohol Related Diseases in Gastroenterology
Edited by H. K. Seitz and B. Kommerell

Hospital, French et al. (1977) reported an 18.7% incidence of classic alcoholic liver disease (alcoholic hepatitis and alcoholic fatty liver). Cirrhosis of undetermined etiology was seen in 8%; 14% had chronic active hepatitis and 41% had miscellaneous nonalcoholic pathology. This would suggest that alcohol predisposes the liver to other forms of pathology.

Demographic Characteristics and Alcohol History

A cooperative study involving six VA Medical Centers has recently been completed in which natural history, risk factors, and treatment of alcoholic hepatitis were evaluated in 363 patients (Mendenhall et al. 1983). The diagnosis of alcoholic hepatitis was based on clinical and laboratory changes characteristic of this disease, using the criteria recommended by the International Association for the Study of the Liver (Nomenclature, Diagnostic Criteria and Diagnostic Methodology for Diseases of the Liver and Biliary Tract 1976).

The stratification of patients into those with mild, moderate, and severe disease was based on predictor variables previously reported as reliable indicators of severity (Mendenhall and Goldberg 1977; Maddrey et al. 1978); these comprised bilirubin and prothrombin time. Those with mild disease consisted of patients with compensated liver disease, with bilirubin less than or equal to 5 mg/dl. Those with moderate disease consisted of patients with liver decompensation, bilirubin more than 5 mg/dl, but with no or only mild prothrombin time abnormality (less than 4 s prolonged). In patients with severe disease, the bilirubin was greater than 5 mg/dl and the prothrombin time was prolonged by 4 s or more. Because the diagnosis of alcoholic hepatitis may be more difficult in milder cases, all patients with compensated liver disease (bilirubin less than 5 mg/dl) were required to have histological confirmation of their disease process before admission into the study. However, so as not to exclude those patients with more severe disease, the absolute requirement for histological confirmation was not required for the initial inclusion of the more severely ill patients. Some of our observations generated from this study are reported here. Mean age at the time of admission into our study was 50.18 years; mean alcohol consumption was 227.91 g daily with a mean gram-years of 5476. No significant difference could be detected in the daily alcohol consumption among severity groups; however, gram-years of drinking did approach significance when mild disease was compared to severe illness. Racial distribution reflected that of the general patient population admitted to the hospital, suggesting that no racial predisposition exists.

The relationship of disease to grams of alcohol consumed varies considerably between individuals. Rydberg and Skerfuing (1977) observed the toxic range of ethanol consumption to vary from 35 to 200 g/day. For most, toxicity is not observed until more than 80 g/day is consumed (Lelbach 1976). This represents drinking in excess of eight 12-oz 6% beers, a liter of 12% wine, or half a pint of 80 proof whisky per day. Duration of drinking also represents an important factor in evaluating the risk for liver disease. On the basis of existing reports on the association of alcohol consumption, the occurrence of liver disease (Lelbach 1975), and survival (Powell and Klatskin 1968), Rankin (1976) has calculated that the consumption of 80 g ethanol daily in a 70-kg man would begin to produce severe liver disease in 6 years and cirrhosis in 10 years. By increasing ethanol consumption to 160 g daily, the process would be accelerated so that severe liver disease would be anticipated in 3 years and cirrhosis in 6 years. A more detailed consideration of the epidemiologic aspects of alcoholism and alcoholic disease is contained elsewhere in this text.

Since our patient population was obtained from six VA Medical Centers, no female alcoholics were available to study. However, our observations at the Cincinnati University Hospital suggest that approximately a 6 : 4 ratio of males to females exists. Similar observations have been reported by others both in the United States (Maddrey et al. 1978; Levin et al. 1979; Helman et al. 1971) and in Europe (Theodossi et al. 1982; Copenhagen Study Group for Liver Disease 1969).

Sex appears to represent a serious risk factor for females. This increased susceptibility was first reported by Spain (1945) and supported by numerous subsequent observations (Schaffner and Popper 1970; Mikkelsen et al. 1968; Kramer et al. 1968). The increased susceptibility is manifest by both a younger age of onset (Lischner et al. 1971) and a higher mortality (Mikkelsen et al. 1968; Phillips and Davidson 1954) in females than in males. These observations appear to be especially true for black females (Kramer et al. 1968; Lischner et al. 1971; Popper et al. 1969).

Clinical Features

The milder forms of alcoholic fatty liver, alcoholic hepatitis, and inactive alcoholic cirrhosis may all be present without symptoms and cannot be diagnosed without a liver biopsy. In a study of 270 patients with fatty liver, Leevy (1962) reported that hepatomegaly was the most common clinical abnormality observed in 75% of cases. However, jaundice and ascites were present in more than 10% of cases.

Table 1. Clinical features of alcoholic hepatitis with and without cirrhosis

	Without cirrhosis (n = 90)	With cirrhosis (n = 65)
Age (years)	49.2 ± 9.1	49.9 ± 9.0
Race		
White	64	36
Black	68	32
Hispanic	30	70
Ascites	48	65
Encephalopathy	38	55
Anorexia	61	48

All values are expressed as percentage of total except for age, which is given as the mean ± standard deviation

In our patients with alcoholic hepatitis, 90.6% had hepatomegaly; ascites was seen in 66.9% and some degree of encephalopathy was seen in 48.9%. Shown in Table 1 are data on 65 patients with biopsy-proven alcoholic hepatitis plus cirrhosis and 90 patients with alcoholic hepatitis without cirrhosis. Note that the clinical findings are quite similar, so that they are poor diagnostic criteria for cirrhosis.

Figure 1 lists the presenting symptomatology with respect to the clinical severity of the disease. As with fatty liver, these are nonspecific signs and symptoms. This is especially true of patients with mild disease, where anorexia, weight loss, and symptoms of alcohol withdrawal predominate. In the more severe forms of the disease, findings of liver decompensation and the various complications of the liver disease predominate, i.e., jaundice, ascites, encephalopathy, and portal hypertension.

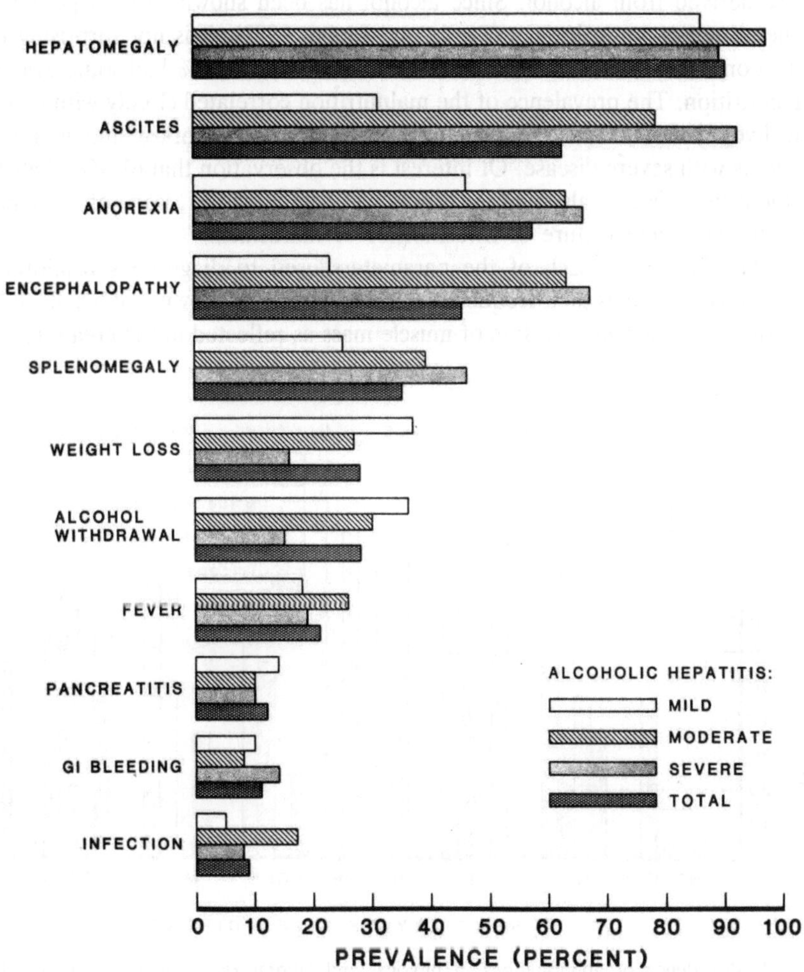

Fig. 1. Clinical features associated with varying severity of alcoholic hepatitis. *GI*, gastrointestinal

Nutritional and Immunologic Status

Malnutrition has long been associated with chronic alcoholism (Korsten and Lieber 1979; Patek 1979; Leevy et al. 1965; Mezey and Faillace 1971; Spies and DeWolf 1933; Strauss 1953; Leevy et al. 1970; Jacobs and Sorrell 1981; Patek et al. 1975; Halsted and Tamura 1979). The role of these nutritional deficiencies in the initiation and progression of the alcoholic liver injury is still controversial. Using established criteria to diagnose and classify protein calorie malnutrition (Blackburn et al. 1977), 284 alcoholic patients with liver disease were carefully evaluated for their nutritional status. An additional 21 alcoholics without clinical liver disease were matched for age, sex, and alcohol consumption and studied nutritionally as a comparison group (Mendenhall et al. 1983).

By recall history, most alcoholics consumed more calories than their estimated energy requirements. However, in most instances more than 50% of their calories were derived from alcohol. Since alcohol has been shown to be a poor nutrient generating "empty calories" (Pirola and Lieber 1972), it is not surprising to find nutritional deficits. In our patients with liver disease, 100% had some evidence of malnutrition. The prevalence of the malnutrition correlated closely with severity of the liver disease, with a 72% incidence of both kwashiorkor and marasmus in patients with severe disease. Of interest is the observation that 61.9% of alcoholics without liver disease also had nutritional deficits. However, none of these patients had the complete picture of kwashiorkor or marasmus.

The changes in each of the parameters used to diagnose kwashiorkor and marasmus, as well as their frequency of occurrence, are shown graphically in Fig. 2. Fat storage abnormalities, loss of muscle mass as reflected in the creatinine height

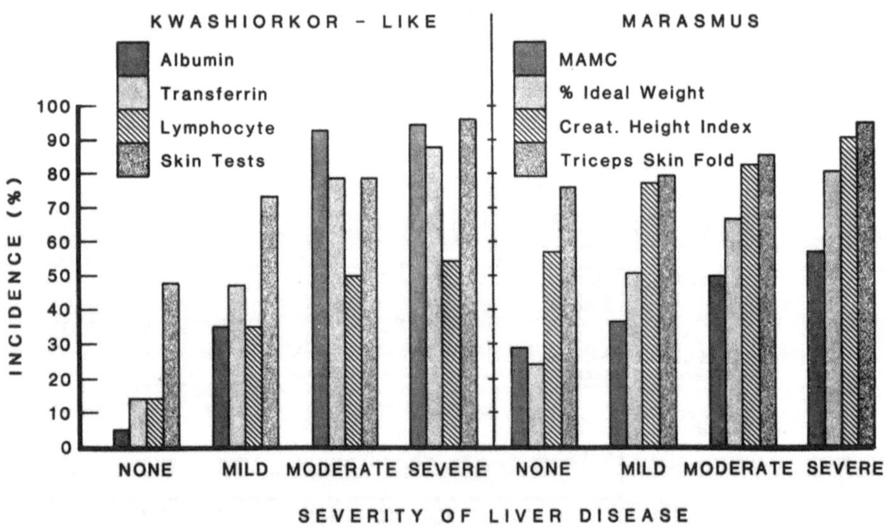

Fig. 2. Prevalence of abnormalities in physical and laboratory parameters used to diagnose marasmus and kwashiorkor-like nutritional disease observed in patients with varying severity of alcoholic hepatitis. *MAMC*, midarm muscle circumference

index, and immunologic dysfunction as reflected in the skin tests were the most frequently observed abnormalities. These abnormalities were also observed in a significant percentage of patients without liver disease, suggesting that the malnutrition preceded the liver injury and was not merely an epiphenomenon.

Associated with malnutrition is a suppression of immune function manifeste by decreased lymphocyte counts and altered skin test responsiveness. This may be of clinical significance in predisposing the alcoholic to infections (Busch 1971; Cherubin et al. 1981; Conn and Fessel 1971; Nolan 1965; Feingold 1976; Rhodes et al. 1969). In our series, 9% of the patients had diagnosed bacterial infections.

Of special interest is the immune responsiveness to hepatitis B infections and to vaccination against hepatitis B. Thirty percent of our patients had circulating antibodies to hepatitis B surface antigen (HBs) and an additional 10% converted during the follow-up period. This suggested increased susceptibility and/or exposure to this infection. When vaccination against hepatitis B was attempted, response to vaccine was suppressed in all of the patients studied. Mean antibody titers after three innoculations were 13 radioimmunoassay (RIA) units in the alcoholic patients with liver disease compared to 19,453 RIA units in age- and sex-matched controls (Lybecker et al. 1983). Alcoholics without liver disease gave a mixed picture, with an adequate response being observed in only 50%. A more detailed description of immunologic changes is described elsewhere in this text.

Vitamins

Vitamin deficiencies are frequent in alcoholics and especially in the alcoholic with liver disease (Youmans 1951; Leevy et al. 1960; Baker et al. 1964; Gabuzda and Davidson 1962). Leevy et al. (1965) reported folate, pyridoxine, and thiamine to be the most frequent vitamin deficiencies among 172 alcoholics. The prevalence of the deficiency varied with severity of the liver disease, with multiple deficiencies (two or more) being observed in up to 49% of the patients.

Serum levels for folate, pyridoxine, thiamine, vitamin B_{12}, vitamin A, and beta carotene were measured in our VA patients with alcoholic hepatitis. Many patients had normal or high values as a result of prior outpatient vitamin therapy. This was especially true of thiamine. Only 9% of our patients (33/370) had low levels at the time of examination. The most frequent abnormality was seen with pyridoxine, for which low levels were observed in 58% of patients. These changes did not correlate well with the severity of liver injury. Vitamin A gave the best correlation with the severity of the liver injury, with mean values of 44 μg/dl, 33 μg/dl, and 27 μg/dl observed in patients with mild, moderate, and severe liver disease respectively ($p < 0.05$). However, low values were observed in only about half (52%) of patients even with severe disease. The pathophysiology and biochemistry for these vitamin abnormalities are more complex than simply poor dietary intake. Animal and human studies have indicated that alcoholism impairs the intestinal absorption of folate (Halsted et al. 1967; Halsted et al. 1971; Romero et al. 1981), pyridoxine (Baker et al. 1975), thiamine (Tomasulo et al. 1968; Thomson et al. 1970), and vitamin A

Table 2. Serum vitamins in alcoholic hepatitis

	Overall ($n = 370$)	Mild ($n = 151$)	Moderate ($n = 116$)	Severe ($n = 103$)
		Severity of disease[a]		
Pyridoxine	29 ± 2 (58%)	36 ± 1 (32%)	26 ± 2 (75%)	23 ± 1 (77%)
Folate	8.1 ± 0.6 (34%)	8.1 ± 0.6 (38%)	7.3 ± 0.5 (36%)	8.9 ± 0.7 (27%)
Vitamin A	36.3 ± 2 (27%)	44 ± 2 (15%)	33 ± 2 (37%)	27 ± 1 (52%)
Betacarotene	72 ± 4 (19%)	85 ± 4 (14%)	66 ± 3 (20%)	62 ± 4 (38%)
Thiamine	71 ± 6 (9%)	60 ± 3 (10%)	69 ± 5 (4%)	89 ± 10 (13%)
Vitamin B_{12}	930 ± 86 (37%)	417 ± 22 (7%)	1,014 ± 73 (51%)	1,507 ± 162 (64%)

[a] Values represent the observed serum concentration mean ± SEM; (%) indicates the prevalence of abnormal observations. Normal values are: 30–80 ng/ml for pyridoxine; 5–24 ng/ml, folate; 25–70 µg/dl, vitamin A; 80–300 µg/dl, betacarotene; 25–75 ng/ml, thiamine; 100–800 pg/ml, vitamin B_{12}

(Russell 1980); decreases the hepatic uptake of folate (Horne et al. 1979) and vitamin A (Russell 1980); decreases the hepatic storage of vitamin A (Russell 1980), folate (Cherrick et al. 1965), and thiamine (Sorrell et al. 1974), especially when other vitamin and mineral deficiencies are present (Smith et al. 1973; Nishino and Itokawa 1977; Howard et al. 1974); and alters the metabolism of pyridoxine (Hines 1975; Veitch et al. 1975; Parker et al. 1979; Rossouw et al. 1977; Mitchell et al. 1976), folate (Brown et al. 1973), and vitamin A (Russell 1980; Raskin et al. 1976).

Vitamin B_{12} represented a special case in which values were elevated, probably representing release from liver stores as a result of cellular injury (Nelson and Doctor 1960). Indeed, a direct correlation was observed with the severity of the liver injury and serum B_{12} levels ($p < 0.01$). Table 2 lists the vitamin changes and their frequency of occurrence.

Laboratory Changes

Since long-term alcoholism produces a spectrum of injury ranging from none to severe debilitating disease involving multisystems, it is not surprising that the laboratory changes are varied and typically nonspecific. In Leevy's series (1962) with fatty liver, the most frequent abnormality was hypoalbumenia, which may have been more a reflection of malnutrition than the liver disease. Anemia was present in 22%, hyperglycemia in 11%, and hypoglycemia in 4%. Of the biochemical tests commonly performed to evaluate liver injury, aspartic transaminase (SGOT) was the most frequent abnormality (63%), followed by bilirubin (32%) and alkaline phosphatase (28%).

In our VA patients with biopsy-proven alcoholic hepatitis, the serum SGOT, cholyl-glycine, and IgA were all abnormal in more than 85% of the cases. Mean values ± SEM were 98 ± 6.0 mUnits for SGOT, 1,572 ± 129 µg/dl for cholyl-glycine, and 800 ± 45 mg/dl for IgA. The serum alanine transaminase (SGPT) was abnormal in 61% of cases, but the degree of abnormality was less than that seen with SGOT. The reason for this difference in response by these two similar enzymes in patients with alcoholic liver disease is unclear. However, some (Cohen and Kaplan 1979) have used the SGOT/SGPT ratio as a diagnostic test to differentiate mild alcoholic hepatitis from viral liver disease. A SGOT/SGPT ratio greater than 2.0 is considered highly indicative of alcoholic hepatitis. In our series this was unreliable, with the frequency of an SGOT/SGPT ratio greater than 2.0 depending on the severity of the liver injury. With biopsy-proven mild disease, this occurred in only 36% of the patients, while in moderate or severe disease it occurred in 71% of the cases. In patients treated with corticosteroids, reversal of the ratio was common. In this instance, selective enzyme induction (Rosen et al. 1959) produced a reversal of the ratios in 30%–40% of such patients within 2–4 weeks.

All of the other common chemistries for liver dysfunction were altered to greater or lesser degrees. Tests of cholestasis, bilirubin, and alkaline phosphatase were elevated in 65% and 81% of patients respectively. Prothrombin time was prolonged beyond the control even after vitamin K therapy in 84%. Other laboratory tests not commonly considered to be parameters of liver injury were also frequently abnormal. Amont the hematological tests, anemia with low hematocrits and hemoglobins was seen in 85%; white cell abnormalities were seen in 47%, 8% with leukopenia and 39% with leukocytosis. Mean corpuscular volume was increased in 79% but did not correlate with serum folate levels; mean $\pm$ SEM = $102 \pm 1.4\ mm^3$. On admission, blood urea nitrogen (BUN) and creatinine were elevated in 16% and 14% respectively and tended to correlate with the clinical severity of the liver disease. Abnormalities in carbohydrate metabolism were present in 13%; 1% had hypoglycemia and 12% had hyperglycemia. Creatinine phosphokinase (CPK) was used as a screen for alcoholic muscle injury. Although mean values were within normal limits, 64 ± 7 IU/l, 15% of cases did exhibit a transient elevation greater than 100 IU/l. Amylase was elevated in 20% of cases with alcoholic hepatitis. To evaluate the incidence of pancreatic dysfunction, a small group, 43 alcoholics admitted for detoxification without clinical signs and symptoms of either alcoholic liver disease or acute pancreatitis, were tested for amylase isoenzymes, elastase, and trypsin (Weesner et al. 1983). Forty-seven percent had laboratory evidence of some pancreatic injury. This was usually an elastase elevation. Only 2% had abnormal pancreatic amylase while 7% had salivary amylase elevations.

Clinical Course

During the early course of their disease, it is not uncommon for patients with alcoholic liver disease to show clinical and laboratory deterioration (Hardison and Lee 1966; Lischner et al. 1971; Helman et al. 1971; Sabesin et al. 1978). In many instances, this is the result of complications associated with the underlying alcoholism and/or liver disease such as pancreatitis, gastrointestinal bleeding, infections, or hepatorenal syndrome. Marshall et al. (1983) recently reviewed 103 patients with biopsy-proven alcoholic liver disease without associated complications capable of causing deterioration. When their histology was examined, this phenomenon was found to be almost exclusively associated with alcoholic hepatitis rather than alcoholic fatty liver or alcoholic cirrhosis, but did not correlate histologically with the presence or absence of Mallory bodies. In this series, 38% had continued deterioration after hospitalization. The greatest deterioration generally occurred within 2 weeks of hospitalization, with the peak at day 9. It was manifeste primarily by laboratory changes rather than clinical findings such as developing or worsening ascites and encephalopathy. In our VA study, which involved only patients with alcoholic hepatitis, of those with moderate to severe disease 75% showed such deterioration. It was manifest primarily as enzyme increases in serum transaminases and alkaline phosphatase, which increased in 47% and 40% of the patients respectively. Bilirubin increased in only 19%. The most frequently observed change was in BUN, which increased in 68% of the cases.

The natural progression of the disease is one in which alcoholism without liver disease evolves into fatty liver and/or alcoholic hepatitis and finally to cirrhosis and death. These are histologic diagnoses and frequently do not correlate with the clinical symptomatology and severity of the disease. This failure to correlate clinical severity with histologic severity has been previously noted (Orrego et al. 1981; Orrego and Israel 1979; Rankin et al. 1978; Mendenhall and Cooperative Study Group on Alcohol Hepatitis 1981) and remains one of the unexplained facets of the disease (Phillips 1983). Of our VA patients with mild disease (Mendenhall et al. 1983), all had some increased fat and hepatic fibrosis; 48% had cirrhosis. Moderate and severe disease groups both had a higher incidence of cirrhosis (74% and 81% respectively), but the increased incidence was not useful in predicting either clinical symptomatology or prognosis.

Few studies have been performed in which serial biopsies have been obtained to show histologic resolution time and/or progression of the disease. Our laboratory reported (Mendenhall 1968) the disappearance of severe fatty liver in 2–11 weeks (mean 4.2 weeks) after hospitalization, depending on therapy used. Leevy (1962) reported long-term sequential histologic observations on 49 patients who initially had only alcoholic fatty liver. In those who continued to drink, he observed a 64% (15/25) incidence of fatty liver progressing to cirrhosis in 1.5–8 years, with an additional 16% (4/25) developing hepatic fibrosis. All of those who progressed continued their alcoholism and had malnutrition. An additional 24 patients stopped drinking and improved their nutritional status. All of these (24/24) mobilized their liver fat and returned to normal in 2–6 weeks.

In the case of alcoholic hepatitis, the picture is somewhat different, reflecting the more serious nature of the injury (Galambos 1972). When patients stopped drinking, the time required for healing ranged from 4 months to 3 years. In a significant percentage healing did not occur, and the lesion progressed to cirrhosis in spite of their sobriety. Only 29% returned to normal histologically.

Data on the histologic progression are not available from our VA study, since sequential biopsies were not obtained from most of the patients. However, sequential changes in the clinical course were closely monitored. Initially, we had anticipated a high incidence of progression from mild alcoholic hepatitis to moderate and severe disease. However, this was not observed. Only seven of 166 (4.2%) of the mild cases progressed to more severe clinical disease over a $3^1/_2$-year period. This

Table 3. Complications contributing to death in alcoholic hepatitis

Hepatic coma	55.8%
Gastrointestinal bleeding	30.8%
Hepatorenal syndrome	27.8%
Infection	14.9%
Miscellaneous	13.7%

Values are expressed as percent of total (214 mortalities) taken from VA Cooperative Study on Alcoholic Hepatitis pooled with data previously reported (Harinasuta et al. 1967; Brunt et al. 1974; Mendenhall and Goldberg 1977). Totals exceed 100% since more than one complication may terminally be present in the same patient

unusually low incidence may be the result of enrollment into the study, with the monthly follow-up visits altering natural history of the disease. Indeed, historically mean ethanol consumption decreased from 256 g/day initially to 44 g/day at 12 months, with a 28% incidence of patients abstaining.

Mortality varied with the clinical severity of the disease. In mild disease with bilirubin less than 5 mg/dl, less than 1% mortality was observed during the initial 30 days of enrollment; with moderate disease, initial mortality increased to 12%; and with severe disease, a further increase to 34% was observed. The overall mortality among these patients was 13%. This variation in mortality based on disease severity probably accounts for the marked variations reported by various investigators, ranging from 0 (Beckett et al. 1962; Green 1965) to 80% (Edmondson et al. 1963). Seventy-five percent of our acute mortality occurred in the initial 3 weeks of hospitalization; medium survival time was 13 days. Complications leading to death are usually multiple. In our patients, hepatorenal syndrome was the single most common precipitating complication. Table 3 lists our data in conjunction with those reported by others (Mendenhall and Cooperative Study Group on Alcohol Hepatitis 1981; Harinasuta et al. 1967; Brunt et al. 1974; Mendenhall and Goldberg 1977) so as to total 214 deaths.

Prognostic risk factors for predicting survival were of only limited help. Ascites, hepatic encephalopathy, esophageal varices, and azotemia were frequent accompanying features. However, their frequency was of sufficient magnitude in the surviving patients to make them less helpful in predicting acute mortality. Of 40 laboratory and clinical features evaluated by discriminant analysis to predict 30-day mortality, five correlated significantly: age, disease severity (based on bilirubin and prothrombin time, findings of renal dysfunction (BUN and oliguria), serum vitamin B_{12}, and prior exposure to hepatitis B as manifest by antibodies to HBs. The long-term annual mortality was even more striking. With mild disease, 9% died within the initial year; with moderate disease, 30% died; and with severe disease, 58% died. This will be discussed in greater detail when therapy is considered.

Treatment

The injurious effects of excess ethanol consumption and its essential role in the pathogenesis of the resulting liver injury are undeniable (Lieber et al. 1963;

Schaffner and Popper 1970; Brunt et al. 1974; Popper 1971; Mendenhall and the Cooperative Study Group on Alcohol Hepatitis 1981). Hence the mainstay in any treatment program must be abstinence. Although there is no evidence that occasional drinking of small amounts is harmful, the psychodynamics of the chronic alcoholic makes complete abstinence the more desirable goal. Unfortunately, this is usually difficult to achieve.

Diet therapy represents the second most important mainstay for treatment. As indicated earlier in this chapter, nutritional deficits are very frequent and develop even in alcoholics without liver disease. Experimental evidence suggests that nutritional repletion is essential for improvement of the liver disease (Phillips 1983; Phillips et al. 1952; Phillips and Davidson 1954). When alcohol was given up, malnourished alcoholics did not show improvement of their liver disease until an adequate diet was also provided. Conversely, alcohol administration did not prevent improvement of the liver disease so long as nutrition was adequate (Patek and Post 1941; Summerskill et al. 1957; Reynolds et al. 1965; Volewiler et al. 1948; Erenoglu et al. 1964). Experimental use of a variety of toxins has repeatedly shown malnutrition to render the liver more susceptible to toxicity (Davis and Whipple 1919). Furthermore, nutritional supplementation protects against these toxins (Von Glahn and Flinn 1939; Kensler et al. 1941; Goldschmidt et al. 1939; Miller and Wipple 1940; Smith 1939). Animal studies in which ethanol served as the toxin produced similar results (Stanko et al. 1978; Porta et al. 1968; Koch et al. 1969; Patek et al. 1976). Koch and associates (1969), by means of a high-protein "super diet", offered protection against hepatic injury from alcohol even when ethanol constituted 30%–45% of caloric intake.

In our study, patients were hospitalized for 30 days of treatment, thus effecting abstinence from ethanol, provided with a well-balanced 2,500 kcal diet, and given multivitamins. However, calorie counts during the initial weeks of hospitalization indicated that because of the anorexia associated with the disease, only 63.1% ± 3.9% of energy requirements were consumed by the moderately ill patients (range 27%–101%, median 61%), and 53.4% ± 4.5% of the energy need for the severely ill (range 33%–94%, median 56%).

In view of this information, it is not surprising that not every patient showed nutritional improvement. Indeed, 19% showed some deterioration. Older studies have reported the therapeutic use of diets rich in protein and vitamin B complex with encouraging results. Patek and associates (1941, 1948) treated 115 patients with a high-protein diet (114 g) plus vitamins. With this regimen, he observed a significant improvement in survival over 5 years of follow-up. Unfortunately, his controls were collected retrospectively and from different hospitals than the treated patients, making the comparison of controls to treated less meaningful. Most recently, Galambos and associates (1979) and others (Nasrallah and Galambos 1980) have reported the use of amino acid therapy and parenteral hyperalimentation in the treatment of acute alcoholic hepatitis. They observed a significant decrease in acute mortality and improvement in several laboratory parameters of liver injury. However, the number of patients studied was small and the initial nutritional status unreported. These two studies do, however, strongly suggest the important and beneficial role of nutritional therapy in this disease.

To evaluate the role of hyperalimentation as therapy, we have recently studied

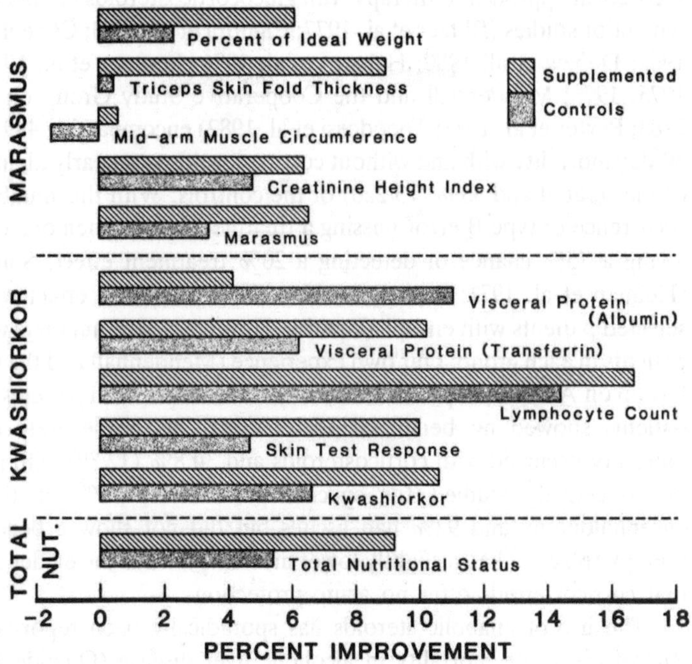

Fig. 3. Changes during 30 days of treatment in the physical and laboratory parameters used to diagnose marasmus and kwashiorkor-like nutritional disease. Values indicate the difference between initial values and those after 30 days of treatment expressed as percent of lower limits of normal

14 patients in which caloric intake was actively encouraged and nutritional therapy given orally as liquid diet supplements (Hepatic-Aid, American McGaw), so that caloric consumption exceeded 3,000 kcal/day and daily energy consumption exceeded estimated energy requirements. On this regimen, the degree of nutritional improvement exceeded that of controls by 70.4%; the most marked improvement was seen in those parameters associated with marasmus, which exceeded control values by 98%. Figure 3 graphically depicts these changes. Recovery time from the liver injury appeared to be improved by nutritional therapy. Only one of 11 patients who received supplement did not recover by the 5th week. Of the controls, 32% (21/64) had not yet recovered by the 5th week. Acute mortality in this small group was not significantly improved during the initial week of hospitalization. However, no deaths occurred in this group after the initial week of treatment, compared to 13 among the controls (17%). Thus it appears that nutritional therapy must be given for at least 1 week to produce an effect. These observations suggest that nutritional support to cover caloric and protein needs should be provided in amounts sufficient to meet the energy needs of the patient. Since vitamin deficiencies in this population are very frequent and typically multiple (Leevy et al. 1965) and since tissue levels are difficult and expensive to obtain, therapy is usually given empirically. In our study, decavitamins (Lannett Co., Philadelphia) were used.

Assessment of the efficacy of specific therapy is considerably more complicated. Since severe alcoholic liver disease is characterized by inflammation and necrosis, a theoretical approach to therapy with glucocorticosteroids has been proposed. Of the ten major studies (Blitzer et al. 1977; Campra et al. 1973; Copenhagen Study Group 1969; Depew et al. 1980; Helman et al. 1971; Lesener et al. 1978; Maddrey et al. 1975, 1978; Mendenhall and the Cooperative Study Group on Alcohol Hepatitis 1983; Porter et al. 1971; Theodossi et al. 1982) encompassing 449 patients, the acute 30-day mortality with and without corticosteroids was nearly identical: 38% (86/229) of the treated and 33% (73/220) of the controls. With this number of patients, the occurrence of type B error missing a treatment effect when one exists) is quite low, giving a 75% chance of detecting a 20% treatment effect. Some have suggested (Helman et al. 1971; Lesener et al. 1978) a beneficial effect with this therapy in selected patients with encephalopathy. However, their numbers were very small: 13 patients in each group. Our own experience (Mendenhall and the Cooperative Study Group on Alcohol Hepatitis 1983) with 82 moderately ill patients and 84 severely ill patients showed no beneficial acute effect. In the severely ill, 33.3% (18/45) mortality occurred with corticosteroids and 30.8% (12/39) with placebo. Although not selectively studied for specific parameters, 72% of these patients had encephalopathy and 92% had ascites but did not show a beneficial response to therapy (placebo had a slightly lower mortality). Thus the evidence is overwhelming that corticosteroids offer no acute protection.

The use of anabolic steroids has sporadically been reported to be beneficial against the acute mortality in alcoholic liver disease (Quadde 1969; Leevy 1968, 1970; Mendenhall 1968), the rationale being an increase in protein anabolism in a condition characterized by accelerated tissue catabolism. Indeed, such agents have been shown to facilitate improvement in defective formation of coagulation factors (Tamburro and Leevy 1967), proliferation of hepatic endoplasmic reticulum (Jabbari and Leevy 1967), and accelerated removal of excess hepatic fat (Mendenhall 1968, 1974; Leevy 1962). Indeed, when treatment is indicated for severe fatty liver, anabolic steroids represent specific effective therapy for this condition. In our initial pilot study with 50 alcoholic hepatitis patients (Mendenhall and Goldberg 1977), such treatment did not alter acute mortality but did diminish the duration of hospitalization, suggesting an accelerated rate of recovery.

In the more extensive VA Cooperative Study, oxandrolone (Anavar, GD Searle) 80 mg/day was used for 30 days on patients with moderate and severe disease. Again, acute 30-day survival was not improved over that of controls: 25% (21/84) vs 20% (16/81). Thus it appears that neither anabolic steroids nor glucocorticosteroids significantly reduces acute mortality.

Long-term mortality, however, did appear to be beneficially affected. Early, in the initial 6 months following treatment, both anabolic and glucocorticosteroids improved survival (Mendenhall and the Cooperative Study Group on Alcoholic Hepatitis 1983). When this population was studied by disease severity, it was noted that no improvement was present in the severely ill patients. However, in moderately ill patients, it was quite evident that by the 4th month after treatment both prednisolone and oxandrolone improved survival (see Fig. 4). For prednisolone this improvement was of short duration, being most marked at 6 months. The beneficial effects of oxandrolone were more enduring, with a 15% difference in survival still present at 36 months. The mechanism responsible for this beneficial effect has not been established. Nor has optimal dosing been established. Both agents improved

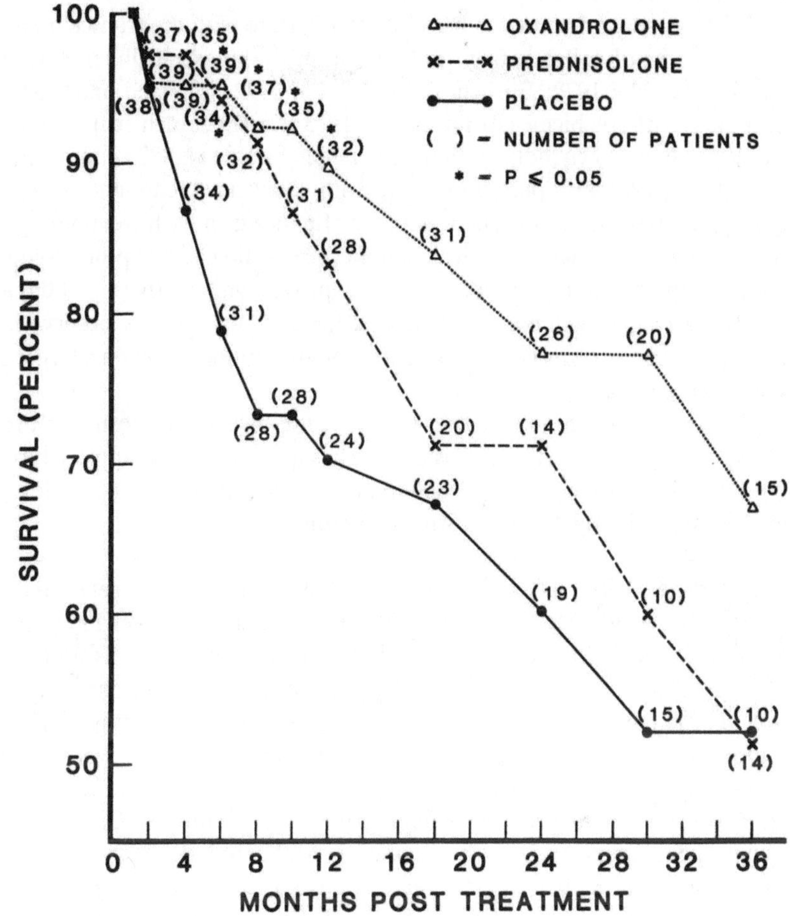

Fig. 4. Life table analysis for moderately ill patients with alcoholic hepatitis treated for 30 days with either oxandrolone, prednisolone, or placebo

nutrition beyond that seen with placebo. Many unanswered questions remain.

From in vitro and in vivo animal studies, chronic alcoholism has been shown to induce a hypermetabolic state in the liver (Videla et al. 1973; Berstein et al. 1973) which mimics the effects of thyroxine (Israel et al. 1973). This hypermetabolic state can be depressed by treatment with propylthiouracil (Israel et al. 1975). On this basis, propylthiouracil has been used to treat acute alcoholic hepatitis (Orrego et al. 1977, 1979). The effectiveness of therapy was evaluated by improvement in a composite clinical and laboratory severity index. Although propylthiouracil therapy did lead to a significantly more rapid rate of normalization, no difference in mortality was observed. The efficacy of such therapy remains to be proven.

Since cirrhosis and hepatic fibrosis are frequent early pathologic changes associated with alcoholic liver injury, agents that inhibit collagen synthesis may be useful in therapy. D-Penicillamine, a drug that retards the cross-linking of collagen molecules, has been used in the treatment of acute alcoholic liver disease (Resnick et al. 1974). A second approach has been to interfere with the transcellular movement and transport of collagen from the cytoplasm to the extracellular space (Digelman and Peterkofsky 1972; Ehrlich and Bornstein 1972; Olmsted and Borisy 1973) by treatment with colchicine (Rojkind et al. 1973; Kershenobich et al. 1976). In a 4-year study involving 43 cirrhotic patients (Kershenobich et al. 1979), mortality improved from 40% (8/20) with placebo to 17% with colchicine treatment (1 mg/day, 5 days per week). Because of the small number of patients in each treatment group, these results did not reach statistical significance. Clinical symptomatology (ascites, encephalopathy, and splenomegaly) did improve significantly ($p < 0.05$) and in 15% (3/20) histologic evidence of decreased hepatic fibrosis was observed. Although these preliminary observations are encouraging, confirmation on a larger population sample is needed.

In conclusion, treatment for the acute, life-threatening forms of the disease is disappointing at best. Therapy is primarily supportive. New approaches to therapy for long-term survival are encouraging, suggesting the need for additional well-controlled studies to confirm their efficacy.

Acknowledgements. Data for much of this chapter were generated from VA Cooperative Study No. 119, funded by the Cooperative Studies Program of the VA Medical Research Service. The following were participants in this study: Chairman's Office (VAMC, Cincinnati, Ohio): C. L. Mendenhall (Chairman), K. L. Richardson. Participating VA Medical Centers (Investigators, Study Nurses, Central Laboratory Support): Cincinnati, Ohio: S. J. Goldberg, R. E. Weesner, N. J. Jaeger, C. Kromme; E. Orange, New Jersey: T. Kiernan, C. M. Leevy, K. Ciasulli, H. Baker, B. DeAngelis; Louisville, Kentucky: C. Tamburro, B. Miller, H. P. Fortwengler, S. Summers; Omaha, Nebraska: M. Sorrell, R. Zetterman, D. Burnett, J. Sneckenburg, K. Knott; San Juan, Puerto Rico: P. Garcia-Pont, E. Mercado, R. Lluevers; Washington, DC: L. Seeff, B. Cuccherini, Z. Bales. Pathology Evaluation: L. Rabin (Washington, DC), T. Chen (E. Orange, New Jersey), A. Chedid (Chicago, Illinois). Central Administration Cooperative Studies Program (Washington, DC): J. A. Hagans (Chief), P. Huang. Biostatistical Support (Cooperative Studies Program Coordinating Center, VAMC, Hines, Illinois): W. G. Henderson (Chief), S. Anderson, M. E. Vitek, J. Foregger. Cooperative Studies Program Clinical Research Pharmacy Coordinating Center (VAMC, Albuquerque, New Mexico): M. R. Sather (Chief), J. H. Weber. Executive Committee: C. L. Mendenhall (Chairman, Cincinnati, Ohio), S. Anderson (Hines, Illinois), H. Baker (E. Orange, New Jersey), L. Rabin (Washington, DC), L. Seeff (Washington, DC), C. Tamburro (Louisville, Kentucky). Ad Hoc Members: J. Van Eeckhout (Albuquerque, New Mexico), S. Goldberg (Cincinnati, Ohio), C. M. Leevy (Newark, New Jersey), R. Zetterman (Omaha, Nebraska). Operations Committee: F. Schaffner (Chairman, New York, New York), C. Davidson (Cambridge, Massachusetts), S. French (Martinez, California), P. Feigl (Seattle, Washington). Human Rights Committee: J. Umbel McKoy (Chairperson, Hines, Illinois), E. Perez (Chicago Heights, Illinois), W. Dix (Chicago, Illinois), D. Hall (Villa Park, Illinois), N. Cahill (Chicago, Illinois).

References

Baker H, Frank O, Zetterman RK et al. (1975) Inability of chronic alcoholics with liver disease to use food as a source of folates, thiamin and vitamin B_6. Am J Clin Nutr 28: 1377–1380
Baker H, Frank O, Zieffer H et al. (1964) Effect of hepatic disease on liver B-complex vitamin titers. Am J Clin Nutr 14: 1–6
Beckett AG, Livingstone AV, Hill KR (1962) Acute alcoholic hepatitis without jaundice. Br Med J 2: 580–582
Bernstein J, Videla L, Israel Y (1973) Metabolic alterations produced in the liver by chronic ethanol administration. Changes related to energetic parameters of the cell. Biochem J 134: 515–521
Berry RE, Boland JP, Smart CN et al. (1977) The economic costs of alcohol abuse and alcoholism – 1975. National Institute on Alcohol Abuse and Alcoholism, Bethesda
Blackburn GL, Bistrian BR, Maini BS et al. (1977) Nutritional and metabolic assessment of the hospitalized patient. J Parent Enter Nutr 1: 11–22
Blizter BL, Mutchnick MG, Joshi PH et al. (1977) Adrenocorticosteroid therapy in alcoholic hepatitis: a prospective double-blind randomized study. Dig Dis Sci 22: 477–484
Brown JP, Davidson GE, Scott JM et al. (1973) Effect of diphenylhydantoin and ethanol feeding on the synthesis of rat liver folates from exogenous pteroylglutamate (^{3}H). Biochem Pharmacol 22: 3287–3289
Brunt PW, Kew MC, Scheuer PJ et al. (1974) Studies in alcoholic liver disease in Britain. Gut 15: 52–58
Busch LA (1971) Human listeriosis in the United States. J Infec Dis 123: 328–331
Campra JL, Hamlin EM, Kirshbaum RJ et al. (1973) Prednisone therapy of acute alcoholic hepatitis: report of a controlled trial. Ann Intern Med 79: 625–631
Cherrick GR, Baker H, Frank O et al. (1965) Observations on hepatic avidity for folate in Laennec's cirrhosis. J Lab Clin Med 66: 446–451
Cherubin CE, Marr JS, Sierra MF et al. (1981) Listeria and gram-negative bacillary meningitis in New York City, 1972–1979. Am J Med 71: 199–209
Cohen JA, Kaplan MM (1979) The SGOT/SGPT ratio: an indicator of alcoholic liver disease. Dig Dis Sci 24: 835–838
Conn H, Fessel J (1971) Spontaneous bacterial peritonitis and cirrhosis: variations on a theme. Medicine 50: 161–197
Copenhagen Study Group for Liver Diseases (1969) Effect of prednisone on the survival of patients with cirrhosis of the liver. Lancet I: 119–122
Davis NC, Whipple GH (1919) The influence of fasting and various diets on the liver injury effected by chloroform anesthesia. Arch Intern Med 23: 612–635
Depew W, Boyer T, Omata M et al. (1980) Double-blind controlled trial of prednisolone therapy in patients with severe acute alcoholic hepatitis and spontaneous encephalopathy. Gastroenterology 78: 524–529
Digelman RF, Peterkofsky B (1972) Inhibition of collagen secretion from bone and cultured fibroblasts by microtubular disruptive drugs. Proc Natl Acad Sci USA 69: 892–896
Edmondson HA, Peters RL, Reynolds TB et al. (1963) Sclerosing hyaline necrosis of the liver in the chronic alcoholic: a recognizable clinical syndrome. Ann Intern Med 59: 646–673
Ehrlich HR, Bornstein P (1972) Microtubules in transcellular movement of procollagen. Nature 238: 257–260
Erenoglu E, Edreira JG, Patek AJ (1964) Observations on patients with Laennec's cirrhosis receiving alcohol while on controlled diets. Ann Intern Med 60: 814–823
Feingold AO (1976) Association of tuberculosis and alcoholism. South Med J 69: 1336–1337
French SW, Sim JS, Franks KE et al. (1977) Alcoholic hepatitis. In: Fisher MM, Rankin JG (eds) Alcohol and the liver. Plenum, New York, pp 261–288 (Hepatology research and clinical issues, vol 3)
Gabuzda G, Davidson CS (1962) Tryptophan and nicotinic acid metabolism in patients with cirrhosis of the liver. Am J Clin Nutr 11: 502–508
Galambos JT (1972) Alcoholic hepatitis: its therapy and prognosis. In: Popper H, Schaffner F (eds) Progress in liver disease. Grune and Stratton, New York, pp 567–588
Galambos JT, Hersh T, Fulenwider JT et al. (1979) Hyperalimentation in alcoholic hepatitis. Am J Gastroenterol 72: 535–541

Goldschmidt S, Vars HM, Ravdin IS (1939) The influence of foodstuffs upon the susceptibility of the liver to injury by chloroform and the probable mechanism of their action. J Clin Invest 18: 277–289
Gorwitz K, Bahn A, Warthen FJ et al. (1970) Some epidemiological data on alcoholism in Maryland. Q J Stud Alcohol 31: 423–443
Green JR (1965) Subclinical acute liver disease of the alcoholic. Aust Ann Med 14: 111–124
Halsted CH, Tamura T (1979) Folate deficiency in liver disease. In: Davidson CS (ed) Problems in liver diseases. Stratton, New York, pp 91–100
Halsted CH, Griggs RC, Harris JW (1967) The effect of alcohol on the absorption of folic acid (H^3-PGA) evaluation by plasma levels and urine excretion. J Lab Clin Med 69: 116–131
Halsted CH, Robles EA, Mezey E (1971) Decreased jejunal uptake of labeled folic acid (^3H-PGA) in alcoholic patients: roles of alcohol and nutrition. N Engl J Med 285: 701–706
Hardison WG, Lee FI (1966) Prognosis in acute liver disease of the alcoholic patient. N Engl J Med 275: 61–66
Harinasuta U, Chomet B, Ishak K et al. (1967) Steatonecrosis – Mallory body type. Medicine 46: 141–162
Helman RA, Temko MH, Nye SW et al. (1971) Alcoholic hepatitis: natural history and evaluation of prednisolone therapy. Ann Intern Med 74: 311–321
Hines JD (1975) Hematologic abnormalities involving vitamin B_6 and folate metabolism in alcoholic subjects. Ann NY Acad Sci 252: 316–327
Horne DW, Briggs WT, Wagner C (1979) Studies on the transport mechanism of 5-methyltetrahydrofolic acid in freshly isolated hepatocytes: effect of ethanol. Arch Biochem Biophys 196: 557–565
Howard LC, Wagner C, Schenker S (1974) Malabsorption of thiamin in folate deficient rats. J Nutr 104: 1024–1032
Israel Y, Videla L, MacDonald A et al. (1973) Metabolic alterations produced in the liver by chronic ethanol administration. Comparison between the effects produced by ethanol and by thyroid hormones. Biochem J 134: 523–529
Israel Y, Videla L, Bernstein J (1975) Liver hypermetabolic state after chronic ethanol consumption: hormonal interrelationships and pathogenic implications. Fed Proc 34: 2052–2059
Jabbari M, Leevy CM (1967) Protein anabolism and fatty liver of the alcoholic. Medicine 46: 131–139
Jacobs RM, Sorrell MF (1981) The role of nutrition in the pathogenesis of alcoholic liver disease. Semin Liver Dis 1: 224–253
Kensler CJ, Sugiura K, Young NF et al. (1941) Partial protection of rats by riboflavin with casein against liver cancer caused by dimethylaminoazobenzene. Science 93: 308–310
Kershenobich D, Uribe M, Suarez GI et al. (1976) Treatment of cirrhosis with colchicine: a randomized trial. Gastroenterology 70: 986
Kershenobich D, Uribe M, Suarez GI et al. (1979) Treatment of cirrhosis with colchicine: a double-blind randomized trial. Gastroenterology 77: 532–536
Koch OR, Porta EA, Hartroft WS (1969) A new experimental approach in the study of chronic alcoholism. V. Super diet. Lab Invest 21: 298–303
Korsten MA, Lieber CS (1979) Nutrition in the alcoholic. Med Clin North Am 63: 963–972
Kramer K, Kuller L, Fisher R (1968) The increasing mortality attributed to cirrhosis and fatty liver in Baltimore (1957–1966). Ann Intern Med 69: 273–282
Leevy CM (1962) Fatty liver: a study of 270 patients with biopsy proven fatty liver and a review of the literature. Medicine 41: 249–278
Leevy CM (1967) Clinical diagnosis, evaluation and treatment of liver disease in alcoholics. Fed Proc 26: 1474–1481
Leevy CM (1968) Cirrhosis in alcoholics. Med Clin North Am 52: 1445–1455
Leevy CM (1970) Hard liquor and cirrhosis. In: Engle E, Larsson J (eds) Alcoholic cirrhosis and other toxic hepatopathies. Nordiska, Stockholm, pp 283–285 (Skandia international symposium)
Leevy CM, George WS, Ziffer H et al. (1960) Pantothenic acid, fatty liver and alcoholism. J Clin Invest 39: 1005
Leevy CM, Baker H, TenHove W et al. (1965) B-complex vitamins in liver disease of the alcoholic. Am J Clin Nutr 16: 339–346

Leevy CM, Thompson A, Baker H (1970) Vitamins and liver injury. Am J Clin Nutr 23: 493–498

Lelbach WK (1975) Cirrhosis in the alcoholic and its relation to the volume of alcohol abuse. Ann NY Acad Sci 252: 85–105

Lelbach WK (1976) Epidemiology of alcoholic liver disease. In: Popper H, Schaffner F (eds) Progress in liver disease, vol 5. Grune and Stratton, New York, pp 494–515

Lesener HR, Bozymski EM, Fallon HF (1978) Treatment of alcoholic hepatitis with encephalopathy: comparison of prednisolone with caloric supplements. Gastroenterology 74: 169–173

Levin DM, Baker AL, Riddell RH et al. (1979) Nonalcoholic liver disease: overlooked causes of liver injury in patients with heavy alcohol consumption. Am J Med 66: 429–434

Lieber CS, Jones DP, Mendelson J, et al. (1963) Fatty liver, hyperlipemia and hyperuricemia produced by prolonged alcohol consumption despite adequate dietary intake. Trans Assoc Am Physicians 76: 289–301

Lischner MW, Alexander JF, Galambos JT (1971) Natural history of alcohol. I. The acute disease. Am J Dig Dis 16: 481–494

Lybecker LA, Mendenhall CL, Marshall LE et al. (1983) Response to hepatitis B vaccine (HBVac). Hepatology 3: 807

Maddrey WC, Boitnott JK, Weber FL et al. (1975) Corticosteroid treatment of active alcoholic liver disease – a controlled trial. Gastroenterology 69: 843

Maddrey WC, Boitnott JK, Bedine MS et al. (1978) Corticosteroid therapy of alcoholic hepatitis. Gastroenterology 75: 193–199

Marshall JB, Burnett DA, Zetterman RK et al. (1983) Clinical and biochemical course of alcoholic liver disease following sudden discontinuation of alcoholic consumption. Alcoholism 7: 312–315

Mendenhall CL (1968) Anabolic steroid therapy as an adjunct to diet in alcoholic hepatic steatosis. Am J Dig Dis 13: 783–791

Mendenhall CL (1974) Augmented release of hepatic triglycerides with anabolic steroids in patients with fatty liver. Am J Dig Dis 19: 122–126

Mendenhall CL, Cooperative Study Group on Alcoholic Hepatitis (1981) Alcoholic hepatitis. Clin Gastroenterol 10: 417–441

Mendenhall CL, Cooperative Study Group on Alcoholic Hepatitis (1983) Survival after steroid treatment for alcoholic hepatitis. Hepatology 3: 850

Mendenhall CL, Goldberg SJ (1977) Risk factors and therapy in alcoholic hepatitis (AH). Gastroenterology 72: 1100

Mendenhall CL, Anderson S, Weesner RE et al. (1983) A clinical study of protein calorie malnutrition associated with alcoholic hepatitis. Am J Med

Mezey E, Faillace LA (1971) Metabolic impairment and recovery time in acute ethanol intoxication. J Nerv Ment Dis 153: 445–452

Mikkelsen WP, Turrill FL, Kern WH (1968) Acute hyalin necrosis of the liver: a surgical trap. Am J Surg 116: 266–272

Miller LL, Whipple GH (1940) Chloroform liver injury increases as protein stores decrease. Am J Med Sci 199: 204–216

Mitchell D, Wagner C, Stone WJ et al. (1976) Abnormal regulation of plasma pyridoxal 5′-phosphate in patients with liver disease. Gastroenterology 71: 1043–1049

Nasrallah SM, Galambos JT (1980) Aminoacid therapy of alcoholic hepatitis. Lancet II: 1276–1277

National Institute on Alcohol Abuse and Alcoholism (1978) Third special report to the US Congress on Alcohol and Health from the Secretary of Health, Education and Welfare. DHEW Pub. No. 79-832. Supt of Docs, US Govt Printing Office, Washington, DC

Nelson RS, Doctor VM (1960) Hepatic and serum vitamin B_{12} content in liver disease. Gastroenterology 38: 188–193

Nishino K, Itokawa Y (1977) Thiamin metabolism in vitamin B_6 or vitamin B_{12} deficient rats. J Nutr 107: 775–782

Nolan J (1965) Alcohol as a factor in the illness of university service patients. Am J Med Sci 249: 135–142

Nomenclature, diagnostic criteria and diagnostic methodology for diseases of the liver and biliary tract (1976) Proceedings 22. Fogarty International Center. US Government Printing Office. Washington DC, pp 6–7

Olmsted JB, Borisy GG (1973) Microtubules. Annu Rev Biochem 42: 507–540
Orrego H, Israel Y (1979) Biochemical, morphological and clinical correlates of alcoholic liver disease. In: Thurman RG (ed) Alcohol and aldehyde metabolizing systems – IV. Plenum, New York, pp 497–508
Orrego H, Kalant H, Israel Y et al. (1977) Effect of propylthiouracil in treatment of alcoholic liver disease. Gastroenterology 73: A39/1237
Orrego H, Kalant H, Israel Y et al. (1979) Effect of short term therapy with propylthiouracil in patients with alcoholic liver disease. Gastroenterology 76: 105–115
Orrego H, Israel Y, Blendis LM (1981) Alcoholic liver disease: information in search of knowledge? Hepatology 1: 267–283
Parker TH, Marshall JP, Roberts RK et al. (1979) Effect of acute alcohol ingestion on plasma pyridoxal 5'-phosphate. Am J Clin Nutr 32: 1246–1252
Patek AJ (1979) Alcohol, malnutrition and alcoholic cirrhosis. Am J Clin Nutr 32: 1304–1312
Patek AJ, Post J (1941) Treatment of cirrhosis of the liver by a nutritious diet and supplements rich in vitamin B complex. J Clin Invest 20: 481–505
Patek AJ, Post J, Ratnoff OD et al. (1948) Dietary treatment of cirrhosis of the liver. Results in 124 patients observed during a 10 year period. JAMA 138: 543–549
Patek AJ, Toth MG, Saunders GA et al. (1975) Alcohol and dietary factors in cirrhosis: an epidemiological study of 304 alcoholic patients. Arch Intern Med 135: 1053–1057
Patek AJ, Bowry SC, Sabesin SM (1976) Minimal hepatic changes in rats fed alcohol and high casein diet. Arch Pathol Lab Med 100: 19–24
Pell S, D'Alonzo CA (1968) The prevalence of chronic disease among problem drinkers. Arch Environ Health 16: 679–684
Phillips GB (1983) Acute hepatic insufficiency of the chronic alcoholic – revisited. Am J Med 75: 1–3
Phillips GB, Davidson CS (1954) Acute hepatic insufficiency of the chronic alcoholic: clinical and pathology study. Arch Intern Med 94: 585–604
Phillips GH, Gabuzda GJ, Davidson CS (1952) Comparative effects of a purified and an adequate diet on the course of fatty cirrhosis in the alcoholic. J Clin Invest 31: 351–356
Pirola RC, Lieber CS (1972) The energy cost of the metabolism of drugs including ethanol. Pharmacology 7: 185–196
Popper H (1971) The problem of hepatitis. Am J Gastroenterol 55: 335–346
Popper H, Davidson CS, Leevy CM et al. (1969) The social impact of liver disease. N Engl J Med 281: 1455–1458
Porta EA, Koch OR, Gomez-Dumm CLA et al. (1968) Effects of diet protein on the liver of rats in experimental alcoholism. J Nutr 94: 437–446
Porter HR, Simon FR, Pope CF et al. (1971) Corticosteroid therapy in severe alcoholic hepatitis: a double-blind drug trial. N Engl J Med 284: 1350–1355
Powell WJ, Klatskin G (1968) Duration of survival in patients with Laennec's cirrhosis. Influence of alcohol withdrawal and possible effects of recent changes in general management of the disease. Am J Med 44: 406–420
Quadde F (1969) Glucocorticoid in liver diseases. Acta Med Scand [Suppl 500] 186: 49–54
Rankin JG (1976) The natural history and management of the patient with alcoholic liver disease. In: Fisher MM, Rankin JG (eds) Alcohol and the liver. Plenum, New York, pp 365–381
Rankin JGD, Orrego-Matte H, Deschenes J et al. (1978) Alcoholic liver disease: the problem of diagnosis. Alcoholism 4: 327–338
Raskin NH, Sligar KP, Steinberg RH (1976) A pathophysiologic role for alcohol dehydrogenase: is retinol its "natural" substrate? Ann NY Acad Sci 273: 317–327
Resnick RH, Boitnott J, Iber FT et al. (1974) Preliminary observations of δ-penicillamine therapy in acute alcoholic liver disease. Digestion 11: 257–265
Reynolds TB, Redeker AG, Kuzma OT (1965) Role of alcohol in pathogenesis of alcoholic cirrhosis. In: McIntyre N, Sherlock S (eds) Therapeutic agents and the liver. Blackwell, Oxford, pp 131–142
Rhodes RJ, Hames GH, Campbell MD (1969) The problem of alcoholism among hospitalized tuberculous patients. Am Rev Respir Dis 99: 440–442
Rojkind M, Uribe M, Kershenobich D (1973) Colchicine and the treatment of liver cirrhosis (letter). Lancet I: 38–39

Romero JJ, Tamura T, Halsted CH (1981) Intestinal absorption of (^{3}H) folic acid in the chronic alcoholic monkey. Gastroenterology 80: 99–102

Rosen F, Roberts NR, Nichol CA (1959) Glucocorticosteroids and transaminase activity. I. Increased activity of glutamic-pyruvic transaminase in four conditions associated with gluconeogenesis. J Biol Chem 243: 476–480

Rossouw JE, Labadarios D, McConnell JB et al. (1977) Plasma pyridoxal phosphate levels in fluminant hepatic failure and the effects of parenteral supplementation. Scand J Gastroenterol 12: 123–127

Russell RM (1980) Vitamin A and zinc metabolism in alcoholism. Am J Clin Nutr 33: 2741–2749

Rydberg U, Skerfuing S (1977) Toxicity of ethanol. A tentative risk evaluation. In: Gross EM (ed) Alcohol intoxication and withdrawal. Plenum, New York, pp 403–418

Sabesin SM, Hawkins HL, Bertram PD et al. (1978) Alcoholic hepatitis. Gastroenterology 74: 276–286

Schaffner F, Popper H (1970) Alcoholic hepatitis in the spectrum of ethanol-induced liver injury. Scand J Gastroenterol 5: 69–78

Smith JC, McDaniel EM, Fan FF et al. (1973) Zinc: a trace element essential in vitamin A metabolism. Science 181: 954–955

Smith MI (1939) The influence of diet on the chronic toxicity of selenium. US Public Health Rep 54: 1441–1453

Sorrell MF, Baker H, Barak AJ (1974) Release by ethanol of vitamins into rat liver perfusates. Am J Clin Nutr 27: 743–745

Spain DM (1945) Portal cirrhosis of the liver: review of 250 necropsies with reference to sex differences. Am J Clin Pathol 15: 215–218

Spies TD, Dewolf HF (1933) Observations on the etiological relationship of severe alcoholism to pellagra. Am J Med Sci 186: 521–525

Stanko RT, Mendenlow H, Shinozuka H et al. (1978) Prevention of alcohol-induced fatty liver by natural metabolites and riboflavin. J Lab Clin Med 91: 228–235

Strauss MB (1953) The etiology of "alcoholic" polyneuritis. Am J Med Sci 189: 378–382

Summerskill WHJ, Wolfe SJ, Davidson CS (1957) Response to alcohol in chronic alcoholics with liver disease. Clinical, pathological and metabolic changes. Lancet I: 335–340

Tamburro C, Leevy CM (1967) Protein clotting factor synthesis in liver disease. Gastroenterology 52: 325

Tashiro M, Lipscomb WR (1963) Mortality experience of alcoholics. Q J Stud Alcohol 24: 203–212

Theodossi A, Eddleston ALWF, Williams R (1982) Controlled trial of methylprednisolone therapy in severe acute alcoholic hepatitis. Gut 23: 75–79

Thomson AD, Baker H, Leevy CM (1970) Patterns of ^{35}S-thiamine hydrochloride absorption in the malnourished alcoholic patient. J Lab Clin Med 76: 34–45

Tomasulo PA, Kater MH, Iber FL (1968) Impairment of thiamine absorption in alcoholism. Am J Clin Nutr 21: 1341–1344

US Bureau of the Census, Dept of Commerce (1975) Statistical abstract of the United States – 1975. US Govt Printing Office, Washington, p 65

Veitch RL, Lumeng L, Li T (1975) Vitamin B_6 metabolism in chronic alcohol abuse: the effect of ethanol oxidation on hepatic pyridoxal 5′-phosphate metabolism. J Clin Invest 55: 1026–1032

Videla L, Bernstein J, Israel Y (1973) Metabolic alterations produced in the liver by chronic ethanol administration. Increased oxidative capacity. Biochem J 134: 507–514

Volwiler W, Jones CM, Mallory TB (1948) Criteria for the measurement of results of treatment in fatty cirrhosis. Gastroenterology 11: 164–182

VonGlahn WC, Flinn FB (1939) The effect of yeast on the incidence of cirrhosis produced by lead arsenate. Am J Pathol 15: 771–789

Weesner RE, Mendenhall CL, Giannella RA et al. (1983) A comparison of serum elastase I (E), trypsin (T) and pancreatic isoamylase (PI) in patients with acute pancreatitis (AP) and chronic alcoholism. Dig Dis Sci 28: 952

Youmans JB (1951) Deficiencies of the water-soluble vitamins. In: Handbook of nutrition. Blakiston, New York, pp 533–576

18 Ethanol and the Endocrine System*

D. H. Van Thiel and J. S. Gavaler

Department of Gastroenterology, University of Pittsburgh School of Medicine, 1000 J Scaife Hall, Pittsburgh, PA 15261, USA

Introduction

Hypoandrogenization is commonly seen in chronic alcoholic men: 70%–80% of such men experience reduced libido and/or impotence (Van Thiel 1983a; Van Thiel et al. 1982). Reproductive as well as Leydig cell failure is also common in such men, with 70%–80% of them demonstrating gross testicular atrophy as well as infertility after years of alcohol abuse (Van Thiel 1982, 1983a; Van Thiel and Lester 1980, 1976a; Van Thiel et al. 1982). Histological studies of the testicular tissue obtained from chronically alcoholic men usually demonstrate marked seminiferous tubular atrophy and loss of mature germ cells (Van Thiel et al. 1974a). Many of the residual, less mature germ cells have an abnormal morphology.

Evidence for hyperestrogenization is also present in such men but it occurs less often than does the hypoandrogenization. Thus a female escutcheon and palmar erythema are seen in 50%, spider angiomata in 40%, and gynecomastia in 20% (Van Thiel 1979; Van Thiel and Lester 1979; Van Thiel et al. 1974a; Lester et al. 1979). These "estrogenic" signs of chronic alcoholism, unlike the transient impotence experienced with an acute alcoholic bout, persist in the absence of intoxication and are due, in large measure, to alcohol-induced permanent tissue injury.

Until recently, liver disease was considered to be of primary importance in the pathogenesis of these evidences of sexual dysfunction in alcoholic men (Lloyd and Williams 1948; Weichsilbaum 1910). However, during the past 10 years, this concept has been severely challenged and a diametrically opposite point of view of the pathogenesis of the sexual dysfunction occurring in alcoholic men has gained currency. This change in thinking has occurred as a result of the demonstration that sexual dysfunction can be present in alcoholic men with a wide spectrum of morphological alcohol-associated hepatic injuries, varying from essentially normal liver to that of severe alcoholic hepatitis and/or cirrhosis (Van Thiel et al. 1974a). Moreover, testosterone concentrations can be shown to fall in normal male volunteers within hours of their ingesting sufficient amounts of alcohol to produce a hangover (Van Thiel 1983a; Van Thiel et al. 1978a). In addition, many of the

* This work was supported in part by grants from the National Institutes of Alcohol Abuse and Alcoholism #AA04425 and the Gastroenterology Medical Foundation of Southwestern Pennsylvania

Alcohol Related Diseases in Gastroenterology
Edited by H. K. Seitz and B. Kommerell

features of the syndrome of alcohol-induced sexual dysfunction can be produced in experimental animals and appear in such animals at a time when hepatic biochemical function and morphological appearance are altered only minimally. Thus the concept that the sexual changes observed in chronically alcoholic men are the result of alcohol abuse per se, rather than the indirect consequence of alcohol-induced liver disease, has gained considerable credence (Van Thiel et al. 1975a, 1979a; Gavaler et al. 1980).

The Alcoholic Male

As noted earlier, confirmatory evidence for this concept has been developed in experimental animals. A dose-dependent decline in plasma testosterone has been reported in mice receiving graded amounts of alcohol over a 5-day period (Badr and Barthe 1974). Alcohol-induced chronic testicular injury characterized by atrophy, loss of germ cells, and reduced testosterone levels has been reported in alcohol-fed rats after only 6 weeks of alcohol administration (Van Thiel et al. 1975a; Gavaler et al. 1980). Both adult and weanling rats, fed a diet in which ethanol accounted for 36% of their caloric intake but which contained sufficient vitamins, trace minerals, and proteins for normal growth and development, become hypogonadal (Van Thiel et al. 1975a, 1979a; Gavaler et al. 1980). The hypogonadism seen in these animals was characterized by atrophy of androgen-dependent target organs and testes and destruction of the germinal epithelium.

The most direct evidence that alcohol, and possibly acetaldehyde, may disturb testicular function has been developed in studies utilizing the isolated perfused rat testes (Cobb et al. 1978, 1980, 1981c; Van Thiel et al. 1981). In these studies, testes were perfused with a defined tissue culture medium which contained chorionic gonadotropin. When alcohol or acetaldehyde was added to the perfusion medium, at concentrations comparable to those found in the blood of chronic alcoholic and acutely intoxicated individuals, testosterone production and secretion were markedly reduced.

Specific mechanisms by which alcohol adversely affects testicular function are being unraveled slowly. Alcohol may interfere with testicular vitamin A activation, which is essential for normal spermatogenesis (Van Thiel and Lester 1974; Van Thiel et al. 1974b; Van Thiel and Lester 1976b). In addition, alcohol metabolism may shift the testicular balance between nicotinamide-adenine dinucleotide (NAD) and reduced NAD (NADH) as it does in the liver, thereby secondarily inhibiting testosterone biosynthesis (Van Thiel et al. 1978a, 1981; Ellingboe and van Anelli 1979; Gordon et al. 1980; Chiao et al. 1981). Similarly, acetaldehyde, either produced directly in the testes as a result of testicular metabolism of ethanol or entering the testes from the plasma as a result of ethanol metabolism by the liver, may have a deleterious effect upon testicular mitochondria, organelles which are critical for steroidogenesis (Lester et al. 1979; Cobb et al. 1981c). Thus it has been reported that the conversion of cholesterol to pregnenolone, a reaction that occurs in mitochondria, is inhibited as a result of exposure of mitochondria to either ethanol or acetaldehyde. In addition, several groups have demonstrated reduced activity of several microsomal enzymes which are important for testosteronogenesis, particularly 17α-hydroxylase, 3β-hydroxysteroid dehydrogenase/isomerase, and desmolase (Gordon et al. 1980; Chiao et al. 1981; Johnston et al. 1981). Not only is testicular testosterone production inhibited as a result of alcohol exposure, but recent studies have demonstrated that ethanol interferes with gonadotropin [luteinizing hormone (LH)] binding to testicular tissue and that chronic alcohol exposure is associated with a hypothalamic-pituitary defect in gonadotropin secretion (Bhalla et al. 1976, 1979; Gavaler et al., to be published). Thus alcoholic individuals not only have inappropriately low plasma gonadotropin concentrations for the degree of their gonadal failure but also demonstrate inadequate responses to exogenous stimuli which normally provoke gonadotropin release, such as clomiphene and luteinizing-hormone-releasing factor (GNRH) (Van Thiel 1983a; Van Thiel et al. 1974a, 1978a). Similarly, inadequate gonadotropin responses can be demonstrated in chronically alcohol-fed rats and in normal rats following alcohol administration (Van Thiel et al. 1979a, 1980). The foregoing suggests that chronic alcohol ingestion induces gonadal injury through its direct effects on the gonad as well as through indirect effects exerted at the level of the hypothalamus and pituitary (Gavaler et al., to be published).

As noted earlier, in addition to being hypogonadal, chronically alcoholic men also are often grossly feminized (Van Thiel et al. 1982; Van Thiel 1982, 1983a; Van Thiel and Lester 1976a, 1980). Thus palmar erythema, spider angiomata, a female escutcheon, and gynecomastia are common physical findings in such men. Biochemical evidence of hyperestrogenization in such men is documented by reported increases in estrogen-responsive proteins, such as sex-steroid-binding globulin and estrogen-responsive neurophysin (Van Thiel and Lester 1977; Van Thiel et al. 1975b). It is probable also that the observed increase in prolactin levels seen in cirrhosis are related to increased estrogen levels or estrogen responsiveness (Van Thiel and Lester 1976, 1978; Van Thiel et al. 1975b, 1978b, 1979b). Moreover, because testicular atrophy can be produced by estrogen administration, the testicular damage in alcoholic men with cirrhosis can, at least in part, be ascribed to a hyperestrogenemic state. However, when plasma estradiol levels are actually measured in chronic alcoholic men, they are found to be either normal or only slightly increased (Van Thiel et al. 1982; Van Thiel 1982, 1983a; Van Thiel and Lester 1976a, 1980). In contrast, plasma estrone levels are moderately increased (Van Thiel and Lester 1976a; Van Thiel et al. 1975b; Van Thiel and Loriaux 1979).

This finding of normal to near normal estrogen levels in the presence of androgen deficiency is paradoxical in that estrogens can only be produced by conversion from preformed androgens. Thus the mechanism of normal or moderately increased plasma estrogen levels in the presence of markedly reduced plasma androgen levels requires further explanation. Preliminary results would suggest that the hyperestrogenemic state is a result of both the direct effects of alcohol and the indirect effects of alcohol mediated through the development of liver disease (Van Thiel 1979; Van Thiel and Lester 1979). Contrary to what was initially expected, the metabolic clearance rate for estradiol in men with Laennec's cirrhosis is normal, not reduced.

Evidence has accumulated to suggest that adrenal overproduction of weak androgens and estrogen precursors regularly occurs in chronically alcoholic men (Gordon et al. 1975; Oliva et al. 1975). Moreover, signs and symptoms of adrenocorticoid hyperresponsiveness resembling Cushing's syndrome have been described in such men (Smals and Kloppenborg 1977; Rees et al. 1977; Smals et al. 1977). Thus these patients not infrequently develop loss of peripheral muscle mass, truncal obesity, hypertension, facial erythema, increased plasma cortisol and androstenedione levels, loss of the normal diurnal variation of plasma cortisol, and failure to show dexamethasone suppression. The mechanism responsible for this overproduction of adrenocortical precursors of estrone is as yet uncertain. Recent studies would suggest, however, that ethanol and acetaldehyde directly stimulate the adrenal cortex by activating adenyl cyclase (Cobb et al. 1979, 1981a,b; Cobb and Van Thiel 1982; Van Thiel 1983b). Thus, in an isolated perfused rat adrenal system, concentrations of ethanol and acetaldehyde observed in the plasma of intoxicated males have been shown to increase corticosterone secretion. Moreover, in clinical studies weakly androgenic steroids such as androstenedione and dehydroepiandrosterone sulfate have been shown to undergo aromatization to estrogens in various tissues, including skin, fat, muscle, bone, and brain (Whercat et al. 1969; Grennet 1970; Cron et al. 1977; Horn and Manthei 1965; Pekkanen et al. 1978; Gordon et al. 1979; Van Thiel et al. 1980; Eagon et al. 1980, 1981a,b; Eagan and Van Thiel 1982; Jung and Russfield). Such peripheral aromatization has been shown to be enhanced in men with Laennec's cirrhosis (Gordon et al. 1975; Oliva et al. 1975). In addition, aromatase activity in the liver and presumably other tissues is enhanced in experimental animals as a result of ethanol administration (Gordon et al. 1979a). Thus a normal metabolic clearance rate and an increased production rate of androstenedione and estrone account for the observed increases in the plasma concentrations of these two steroids seen in alcoholic individuals (Van Thiel and Loriaux 1979; Gordon et al. 1975; Olivo et al. 1975).

Compounding this overproduction of adrenal androgens which can be converted to estrogens, portosystemic shunting, which occurs as a consequence of alcoholic liver disease, has been shown to allow steroidal estrogen precursors such as androstenedione and dehydroepiandrosterone, which are secreted into the systemic circulation, to escape the confines of the enterohepatic circulation. As a result, these steroids are converted to estrogens at peripheral sites where aromatase activity is increased (Van Thiel 1979; Lester et al. 1979; Van Thiel et al. 1980). The slight increase in plasma estrogen levels observed in alcoholics therefore reflects the levels that reflux back into the plasma from these peripheral sites after aromatization has occurred (Van Thiel 1979; Lester et al. 1979; Van Thiel et al. 1980).

Compounding the effect of this increased peripheral aromatization of androgens to estrogens is the observation that, at least in the liver, cytosolic estrogen receptor activity is enhanced in chronic alcohol-fed animals and presumably also in man (Eagon et al. 1980, 1981a,b; Eagon and Thiel 1982). Normal male rat liver contains one-third the amount of cytoplasmic estrogen receptor present in normal female liver. After castration of the male, cytoplasmic estrogen receptor activity increases toward female values. Treatment of the castrated rat with dehydrotestosterone prevents the castration-induced change toward the female pattern. Moreover, chronic alcohol feeding of otherwise normal male rats is associated with a decline in the hepatic cytosolic content of a male-specific estrogen binding protein and an increase in the classic estrogen receptor, thus in effect converting the male alcohol-fed rat liver to that of a female in terms of its cytosolic estrogen-binding characteristics. Such increased estrogen receptor activity and reduced levels of the nonreceptor male-specific estrogen-binding protein allows the male liver obtained from alcohol-fed animals to hyperrespond to the normal or only moderately increased plasma estrogen levels present.

The Alcoholic Female

In contrast to the male, the alcoholic female is not superfeminized but instead shows severe gonadal failure commonly manifested as oligomenorrheal amenorrhea, loss of secondary sex characteristics (such as breast and pelvic fat accumulation), and infertility (Van Thiel 1982; Van Thiel and Lester 1980). Histologic studies of the ovaries obtained at autopsy from chronically alcoholic women who have died of cirrhosis while still in their reproductive years (20–40 years of age) have shown a paucity of developing follicles and few or no corpora lutea, thus documenting reproductive failure (Jung and Russfield 1972). Moreover, these findings have been reproduced recently in an animal model (Gavaler et al. 1980; Van Thiel et al. 1977). Endocrine failure of the ovaries of alcoholic women is manifested by reduced plasma levels of estradiol and progesterone, loss of secondary sex characteristics, and ovulatory failure (Van Thiel 1982, 1983a; Van Thiel and Lester 1980). The biochemical mechanisms for such endocrine failure are probably the same as those occurring within the testes of the male, as the pathways for steroidogenesis are the same in the gonads of the two sexes and an alcohol dehydrogenase has been reported to be present within the ovary as well as the testes.

Hypothalamic Pituitary Injury

In addition to demonstrating evidence of primary gonadal failure, chronic alcoholics, whether male or female, demonstrate evidence of a central hypothalamic-pituitary defect in gonadotropin secretion. Thus in both sexes, despite severe gonadal injury, follicle-stimulating hormone (FSH) levels, although increased, are well below levels expected for the degree of reproductive failure present (Van Thiel 1982; Van Thiel and Lester 1980). Further, in both sexes, despite the marked reduction in sex steroid levels, LH concentrations range from normal to only moderately increased (Van Thiel 1982, 1983a; Van Thiel and Lester 1976a, 1980; Van Thiel et al. 1982).

Interestingly, acute ethanol administration to normal women has not resulted in changes in plasma levels of FSH, LH, estradiol, or progesterone when given during the follicular or luteal phase of the menstrual cycle (Mendelson et al. 1981, to be published). The effect of ethanol upon the midcycle gonadotropin surge remains to be studied, however.

Effects of Ethanol on the Hypothalamic-Pituitary-Adrenal Axis

The earliest animal studies examining the effect of alcohol on the hypothalamic-pituitary-adrenal axis were based upon indirect evidence of adrenal activation and suggested but did not prove that ethanol altered adrenal cortical function. Hion-Jon was the first to demonstrate histologic changes in rabbit adrenal gland after acute and chronic alcohol administration (Hion-Jon 1928). Subsequently, others demonstrated that alcohol administration either by gavage or intraperitoneal injection altered adrenocortical activity as determined by reductions in adrenal ascorbic acid and cholesterol content (Smith 1951; Farber and Duncan 1951; Czaja and Kalant 1961; Ellis 1966). Moreover, such studies also demonstrated that these changes could be prevented by prior hypophysectomy. This latter observation suggested that the primary effect of ethanol on the hypothalamic-pituitary-adrenal axis was at the level of the hypothalamus and pituitary rather than at the level of the adrenal gland (Ellis 1966). Additional evidence for a primary hypothalamic-pituitary effect of ethanol was the observation that pretreatment with dexamethasone blocked the expected adrenal response to ethanol administration (Kakihana et al. 1968). Moreover, it could be shown that alcohol administration resulted in an acute decrement in pituitary adrenocorticotropic hormone (ACTH) content (Noble et al. 1971).

Ellis (1966) was the first actually to measure corticosterone levels in response to ethanol treatment. Using dogs and rats, he was able to demonstrate increases in corticosterone levels and no change in the metabolic clearance rate of corticosterone following ethanol administration. Alcohol given chronically to mice has been shown to disturb the normal diurnal variation in plasma corticosterone levels (Kakihana and Moore 1976). Similarly, in normal human subjects, ethanol at intoxicating doses produces an immediate increase in plasma cortisol (Fazekas 1966; Mendelson and Stein 1966; Jenkins and Connoly 1968; Merry and Marks 1969). The cortisol response, moreover, appears to parallel the blood ethanol level. This response is probably due to a hypothalamic stress reaction and pituitary release of ACTH. Such stress responses can be blocked in experimental animals by prior pituitary stock transection, morphine administration, or dexamethasone. In contrast, ethanol intoxication and withdrawal have been shown to be associated with an enhanced adrenergic activity in the brain, a factor known to inhibit hypothalamic release of corticotropin-releasing factor and thereby impairment of the normal ACTH stress response (Tabakoff and Hoffman 1980; Littleton 1978).

A fascinating area of recent investigation is the study of the effect of endogenous opiate agonists (enkephalins and endorphins) on the control of pituitary releasing factors. Activation of such opiate agonists appears to inhibit corticotropin-releasing factor (CRF) and thereby ACTH release and might contribute, at least in part, to the reduced stress-induced ACTH release seen during alcohol intoxication and withdrawal (Morley et al. 1980). Moreover, the observation that the specific opiate antagonist naloxone improves the psychomotor impairment and depressed level of consciousness induced by ethanol in humans strengthens such a hypothesis (Jeffcoate et al. 1979; Jefferys et al. 1980).

Because acute exposure to ethanol causes an increase in plasma cortisol levels in normal persons, the effect of sustained ethanol levels on adrenal function is of interest. Plasma cortisol levels remain elevated in mice chronically fed alcohol, and the presence of ethanol disturbs the normal diurnal variation of corticosterone levels in such animals (Kakihana and Moore 1976; Kakihana et al. 1971; Tabakoff et al. 1978). Studies using human subjects given ethanol for 4–29 days have similarly demonstrated persistent elevations of plasma cortisol in response to alcohol administration. Moreover, a pseudo-Cushing's syndrome has been recognized recently in chronically alcoholic individuals (Smals and Kloppenborg 1977; Rees et al. 1977; Smols et al. 1977). Patients manifesting this syndrome have many of the classic features of true Cushing's syndrome and exhibit increased plasma cortisol levels and abnormal responses to dexamethasone. Despite such evidence for direct ethanol-induced adrenocortical hypersecretion, most investigators have found no pathologic changes at autopsy in the adrenal glands of patients dying of Laennec's cirrhosis.

In the 1960s, Peterson reported decreased urinary excretion of glucocorticoids in cirrhotic individuals (Peterson 1960). Zumoff and co-workers subsequently demonstrated that there was a distinctly abnormal pattern of urinary cortisol metabolites in cirrhotic patients with a decreased glucuronidate fraction (34% vs 54% for normals) as well as a marked increase in the fraction excreted as cortolones (Zumoff et al. 1967). Moreover, recent studies by Gordon et al. have demonstrated that the A-ring reductase activity of the liver is reduced in cirrhotic individuals (Gordon et al. 1979b). Despite these various findings, the plasma levels of cortisol observed in cirrhosis have generally been reported to be normal. However, the normal diurnal pattern of glucocorticoid plasma levels is frequently absent in such patients. In addition, the metabolic clearance rate of cortisol is decreased in cirrhosis, while the volume of distribution for cortisol is increased slightly, suggesting that the production rate of cortisol is reduced somewhat (Peterson 1960).

The urine of patients with Laennec's cirrhosis, especially those with edema, contains increased amounts of aldosterone. Consistently, an increased secretion rate and increased plasma levels of aldosterone are reported in such individuals (Bongiovanni and Eisenmenger 1951; Chant and Shipley 1953; Leutscher et al. 1954; Vlick et al. 1958; Coppage et al. 1972; Laragh et al. 1962; Wolff et al. 1962; Lommer et al. 1968; Wright et al. 1975). It is well known that chronic liver disease is associated with a significant decrease in hepatic blood flow. As a result, the metabolic clearance rate of aldosterone is reduced and the half-life of the hormone is prolonged (Largh et al. 1962; Mendelson et al. 1971). It should be noted, however, that the resultant elevated plasma levels of aldosterone in cirrhotic individuals reflect not only a decreased metabolic clearance but also an increased secretion rate corresponding, at least in part, to an increased level of activity of the renin-angiotensin system.

The state of the hypothalamic-pituitary-adrenal axis during alcohol withdrawal has been studied extensively. Cessation of chronic alcohol administration to mice is followed by an abstinence syndrome characterized by hyperactivity and seizures and increased plasma corticosterone levels (Kakihana et al. 1971; Tabakoff et al. 1978; Peterson 1960; Zumoff et al. 1967; Gordon et al. 1979b; Bongiovanni and Eisenmenger 1951; Chant and Shipley 1953; Leutscher et al. 1954; Vlick et al. 1958; Coppage et al. 1972; Laragh et al. 1962; Wolff et al. 1962; Lammer et al. 1968; Wright et al. 1975; Goldstein 1972). Chronic alcoholics also experience hypercortisolism during ethanol withdrawal (Mendelson and Stein 1966; Mendelson et al. 1971; Stokes 1973). Interestingly, the pituitary response of ACTH in individuals experiencing alcohol withdrawal is frequently abnormal (Pohorecky 1974). In contrast to the apparent lack of ACTH reserve seen with withdrawal, adrenal reserve of cortisol, when stimulated by exogenous ACTH, is normal (Marks and Wright 1977). Alcoholics who abstain from drinking often show persistent abnormalities of the hypothalamic pituitary adrenal axis when tested during abstinence (Oxenkrug 1978).

In addition to altering adrenocortical function, alcohol also enhances the adrenomedullary secretion of catecholamines and induces an increase in adrenal medullary content of phenolethanolamine-*N*-methyltransferase activity (Oxenkrug 1978; Pohorecky 1974; Matunaga 1942; Klingman and Goodal 1957; Perman 1958; Pohorecky et al. 1974; Walsh and Truitt 1970). Studies in rats have suggested that acetaldehyde is principally responsible for these changes in catecholamine secretion and synthesis occurring in response to alcohol administration (Walsch and Truitt 1970).

Effects of Alcohol on the Hypothalamic-Pituitary-Thyroidal Axis

The most consistent effect of alcohol on the function of the thyroid is moderately decreased serum thyroxine (T4) levels and markedly decreased serum triiodothyronine (T3) levels (Israel et al. 1979b; Chopra et al. 1974; Orrego et al. 1979a; Israel et al. 1973; Nomura et al. 1975). It appears that the low T3 levels seen in alcoholics and normals after acute alcohol administration primarily reflect hepatic injury and reduced hepatic deiodination of T4 to T3, as T4 levels and thyroid-stimulating hormone (TSH) levels are only moderately reduced or are normal (Namura et al. 1975; Green et al. 1977; Yadav et al. 1978; Wright 1978; Van Thiel et al. 1979c). Despite the considerable data suggesting that alcohol adversely effects the hypothalamic-pituitary-thyroidal axis indirectly through hepatic injury, some data exist to show that ethanol also increases the uptake of iodine by the thyroid (Yadav et al. 1978). Moreover, Wright and Van Thiel et al. have reported a diminished TSH response to thyrotropin-releasing factor (TRH) in chronic alcoholics (Wright 1978; Van Thiel et al. 1979c).

Recent studies indicate that the administration of TRH to alcohol-naive male rats may antagonize the acute hypnotic and hypothermic effects of ethanol. In addition, Israel et al. have suggested that alterations in thyroid hormone metabolism may play a role in the clinical management of alcoholic hepatitis (Israel et al. 1975, 1979a; Orrego et al. 1979b). They have reported that propylthiouracil reverses the adverse effects of alcohol on the liver. Moreover, the therapeutic value of propylthiouracil was found to be greatest in those patients with the most severe liver disease.

The Effects of Alcohol on Growth Hormone and Prolactin

Considerable evidence has accrued to suggest that ethanol ingestion blocks stimulated growth hormone release both in normal individuals and chronic alcoholics. Based on histologic examinations of laboratory animals, early studies suggested that alcohol feeding produced degenerative changes in the pituitary (Schanez-Calvo 1941). More recently, the acute administration of large amounts of ethanol to laboratory rats has been shown to abolish spontaneous growth hormone secretion (Redmond 1980). In contrast, chronic alcoholics with cirrhosis have elevated basal levels of growth hormone and frequently show abnormal growth hormone responses to stimuli such as TRH (Van Thiel et al. 1978c). Torro and co-workers have reported no effect of a single dose of ethanol on prolactin levels (Torro et al. 1973). Similarly, Earll and co-workers have found no change in prolactin levels following ethanol ingestion (Earll et al. 1976). In contrast, Loosen and Prange reported reduced basal prolactin levels and minimally impaired prolactin responses to TRH during alcohol withdrawal by alcohol abusers (Loosen and Prange 1977).

It is of some interest that Jung and Russfield have reported an increased prevalence of prolactin-secreting cells in the pituitaries examined at autopsy of male and female patients who died of alcoholic liver disease (Jung and Russfield). Consistent with such a finding, Ylikahri et al. noted increased prolactin responses to TRH in alcoholics, a finding which has been confirmed by Van Thiel et al. 1978b; Ylikahri et al. 1976). Van Thiel and his co-workers have reported that individuals with cirrhosis have elevated basal prolactin levels and reduced prolactin responses to TRH. In contrast, individuals with fatty liver demonstrate reduced basal prolactin levels and exaggerated prolactin responses to TRH. In contrast to the human studies cited above, preliminary studies in the rat have suggested that ethanol administration increases plasma prolactin levels.

Effects of Alcohol on Vasopressin and Oxytocin

Acute alcohol administration to man and animals produces an immediate diuresis which seems to occur as a result of inhibition of vasopressin secretion (Kleeman 1972). Thus several groups have reported reduced vasopressin levels in normal volunteers and animals after acute ethanol administration (Linkola et al. 1977, 1978; Helderman et al. 1978). Little or no tolerance to the antidiuretic effects of ethyl alcohol or its inhibition of vasopressin secretion has been observed in chronic alcoholics (Marquis et al. 1975; Sereny et al. 1966). The site of alcohol suppression of vasopressin release appears to be at the level of the hypothalamus. Thus stimuli applied directly to the superoptic nucleus can override the diuretic effects of alcohol. Moreover, alcohol inhibits the electrically evoked discharges of the superoptic nucleus and prevents the degranulation produced by large doses of sodium chloride (Raiba 1960). More recently, Marguis and co-workers in rats and Linhold and co-workers in human subjects were able, using radioimmunoassay, to demonstrate transient suppression of antidiuretic hormone during the phase of increasing blood levels of alcohol followed by a decrease in vasopressin levels during the period of maximal alcohol levels (Marquis et al. 1975).

On the basis if bioassay data, ethanol appears to inhibit the release of oxytocin (Wagner and Fuchs 1968). No data, however, are available as yet on the actual blood levels of oxytocin during ethanol administration and withdrawal.

Effects of Alcohol on Calcium Metabolism

A dose-related acute sustained hypocalcemic effect of alcohol, which is unrelated to the alterations produced in serum phosphate, has been reported recently in both dogs and rats (Peng et al. 1972). Parathormone administration does not reverse this hypocalcemic effect (Peng and Gitelman 1974). Tracer studies utilizing radioactive calcium have suggested that the fall in calcium is related to a shift of the ion out of the serum which is unrelated to any change in pH, magnesium, or binding protein levels (Peng et al. 1972). Studies in normal volunteers and in alcoholic subjects have suggested that alcohol produces an increased urinary secretion of both calcium and magnesium (Kalbfleisch et al. 1963; Jones et al. 1969). Both in vitro and in vivo, ethanol has been shown to reduce duodenal transport of calcium (Krawitt 1973, 1974, 1975; Krawitt et al. 1975).

The defect in calcium homeostasis found in chronic alcoholics with hypocalcemia, however, is most likely due to an ethanol-induced depression of serum magnesium levels (Martin et al. 1959; Fankushen et al. 1964). Magnesium is well known to be important in the regulation of parathormone secretion and the mediation of the hormone's peripheral action (Rasmussen 1974; Estep et al. 1969). It must be said, however, that both malabsorption of calcium and a primary dietary deficiency of vitamin D are probably responsible, at least in part, for the hypocalcemia seen in individuals with advanced alcoholic cirrhosis. Finally, Williams and co-workers have reported recently that the oral administration of alcohol to normal volunteers results in an increased calcitonin level which may play a role, at least in part, in the pathogenesis of the hypocalcemia seen with alcohol use (Williams et al. 1981).

It is of some interest that a decreased density of tibial bone has been reported in rats fed ethanol chronically when compared to pair-fed control animals (Saville and Lieber 1965). In this regard, it is important to note that alcoholic patients have a decreased bone mass and that they have an increased incidence of fractures (Nilsson 1970). Finally, it should be said that the hypophosphatemia which is seen commonly in alcoholics may contribute, at least in part, to the pathogenesis of the metabolic bone disease seen in such individuals (Stein et al. 1966; Knockel 1977).

Effects of Alcohol on the Endocrine Pancreas

Singh and Patel and Shah et al. have reported that alcohol administration increases insulin secretion in response to orally or parenterally administered glucose loads (Singh and Patel 1976; Shah et al. 1977). Mikkila and Taskinen also observed that alcohol pretreatment augments the insulin release in normal volunteers to an administered glucose load (Nikkila and Taskinen 1975). This observation has been confirmed further by Metz and co-workers, who have reported that pretreatment with alcohol potentiates the plasma insulin response to a glucose load (Metz et al. 1969). Similar studies and results have been reported by Friedenberg et al. (1971), Phillips and Safrit (1971), and McMonagle and Thalig (1975) using normal volunteers and diabetics. Interestingly, alcohol pretreatment also increases the insulin response to other secretagogues such as tolbutamide (Kuhl and Anderson 1974; Adreani et al. 1974).

In contrast to these acute effects of alcohol on the endocrine pancreas, glucose intolerance is common in individuals with established alcoholic liver disease, particularly those with portal hypertension (Phillips and Safrit 1971; Lochner et al. 1967; Kreisberg et al. 1970; Lefebvre et al. 1970; Samols and Holdworth 1968). Both insulin and glucagon levels are increased in such individuals, with glucagon levels being increased to a greater extent than are the insulin levels (McMonagle and Felig 1975; Marco et al. 1973; Collins and Croffard 1969; Slamaan et al. 1969; Sherwin et al. 1974). It is thought that the glucose intolerance in such individuals is due primarily to the hyperglucagonemia. It is important to remember, however, that abnormal glucose tolerance can occur in chronic alcoholic individuals as a result of chronic pancreatitis as well as of advanced liver disease.

In patients with well-compensated cirrhosis, the levels of branched chain amino acids in muscles and plasma are depressed and the plasma clearance of branched chain amino acids is increased. Sherwin and co-workers observed that the degree of depression of branched chain amino acids present in the plasma of such patients might be related, at least in part, to the degree of portal systemic shunting observed in these patients (Sherwin et al. 1974). As noted earlier, portosystemic shunting is accompanied by hyperglucagonemia and hyperinsulinemia. Sherwin postulates that the hyperinsulinemia in such patients explains the depression of plasma branched chain amino acids and the increased catabolism of these amino acids in peripheral tissues, especially in skeletal muscle. In contrast, Soeters and Fischer propose that the role of hyperglucagonemia in this regard might be more important than is the hyperinsulinemia, and that it contributes pathogenetically to branched chain amino acid uptake by fat rather than by muscle (Soeters and Fischer 1976).

The consequences of portosystemic shunting on plasma insulin levels as well as branched chain amino acids have also been studied by Iwasaki and co-workers (Iwasaki et al. 1980). They noted that the ratio of insulin C peptide to insulin was significantly lower in patients with cirrhosis than controls but not in patients with idiopathic portal hypertension. Similarly, when compared to control values, plasma immunoreactive insulin was significantly increased in the patients with cirrhosis but not in the patients with idiopathic portal hypertension, suggesting that a decreased degradation of insulin occurs in patients with cirrhosis.

Finally, the syndrome of fasting alcoholic hypoglycemia is known to develop within 6–36 h after ingestion of alcohol by individuals who are previously malnourished or fasting. No age group appears to be exempt from this alcohol-induced reaction, but children seem to be particularly susceptible to its occurrence. The most important clues to the correct diagnosis of such cases are an accurate history of ethanol use, a blood alcohol level, blood glucose levels, and a fasting insulin determination made on blood collected at the time of the hypoglycemic episode. Frienkel and co-workers, Field and co-workers, and Mark and Medd have reported that alcohol is capable of producing severe and conceivably fatal hypoglycemia (Freinkel et al. 1963; Field et al. 1963; Arky and Frienkel 1966; Marks and Medel 1964; Arky et al. 1968). In addition to children, individuals who utilize sulfonylurea drugs and atheletes are at particular risk for the development of severe hypoglycemia following the combination of alcohol ingestion and fasting or strenuous exercise.

As must be obvious from the above, ethanol and possibly acetaldehyde, the first metabolic product of ethanol metabolism, have numerous effects upon the endocrine system. For some organs it acts to enhance function, but for most it inhibits function. Moreover, although much is known about these effects, much more yet remains to be learned.

References

Adreani D, Tamburrano G, Javicoli M (1974) Alcohol hypoglycemia: hormonal changes. In: Adreani D, Lefebvre P, Mark V (eds) Hypoglycemia: proceedings of the European symposium, Rome. Thieme, Stuttgart, p 99

Arky RA, Frienkel N (1966) Alcohol hypoglycemia V. N Engl J Med 274: 426–431

Arky RA, Veverbrants E, Abramson EA (1968) Irreversible hypoglycemia: a complication of alcohol and insulin. JAMA 206: 554–556

Badr FM, Bartke A (1974) Effect of ethyl alcohol on plasma testosterone levels in mice. Steroids 23: 921–928

Bhalla VK, Haskell J, Grier H, Mahesh VB (1976) Gonadotropin binding factors. J Biol Chem 4947–4957

Bhalla VK, Chen CJH, Gramnaprakasam MS (1979) Effects of in vivo administration of human chorionic gonadotropin and ethanol on the process of testicular receptor depletion and replenishment. Life Sci 24: 1315–1324

Bongiovanni AM, Eisenmenger WJ (1951) Adrenocortical metabolism in chronic liver disease. J Clin Endocrinol 11: 152–172

Chant JJ, Shipley EG (1953) The mechanism of sodium retention in cirrhosis of the liver. J Clin Invest 32: 560–567

Chiao Y-B, Johnston AB, Gavaler JS, Van Thiel DH (1981) Effect of chronic ethanol feeding on testicular content of enzymes required for testosteronogenesis. Alcoholism 5: 230–235

Chopra IJ, Solomon DH, Chopra U, Young RT, Chunateco GN (1974) Alterations in circulating thyroid hormones and thyrotropin in hepatic cirrhosis. Evidence for erythroidism despite subnormal serum triiodothyronine. J Clin Endocrinol Metab 39: 501–507

Cobb CF, Van Thiel DH (1982) Mechanism of ethanol-induced adrenal stimulation. Alcoholism 6,2: 202–206

Cobb CF, Ennis MF, Van Thiel DH, Gavaler JS, Lester R (1978) Acetaldehyde and ethanol are direct testicular toxins. Clin Toxicol 29: 641–644

Cobb CF, Van Thiel DH, Ennis MF, Gavaler JS, Lester R (1979) Is acetaldehyde an adrenal stimulant? Curr Surg 36: 431–434

Cobb CF, Ennis MF, Van Thiel DH, Gavaler JS, Lester R (1980) Isolated testes perfusion: a method using a cell- and protein-free perfusate useful for the evaluation of potential drug and/or metabolic injury. Metabolism 29: 71–79

Cobb CF, Van Thiel DH, Gavaler JS, Lester R (1981a) Effects of ethanol and acetaldehyde on the rat adrenal. Metabolism 30: 537–543

Cobb CF, Van Thiel DH, Gavaler JS (1981b) Isolated rat adrenal perfusion: a new method to study adrenal function. Metabolism 31: 347–353

Cobb CF, Gavaler JS, Van Thiel DH (1981c) Is ethanol a testicular toxin? Clin Toxicol 18: 149–154

Collins JR, Croffard OB (1969) Glucose intolerance and insulin resistance in patients with liver disease. Arch Intern Med 124: 142–146

Coppage WS, Island DP, Cooner AE, Liddle GW (1972) The metabolism of aldosterone in normal subjects and in patients with hepatic cirrhosis. J Clin Invest 41: 1672–1679

Crou CE, Cornell NW, Vecch RL (1977) The rate of ethanol metabolism in isolated rat hepatocytes. Alcoholism 1: 43–45

Czaja C, Kalant H (1961) The effect of acute alcohol intoxication or adrenal ascorbic acid and cholesterol in the rat. Can J Biochem Physiol 39: 327–332

Eagon PK, Van Thiel DH (1982) E2T/ratio: an update. In: Langer M, Chiandussi L, Chopra IJ, Martini L (eds) The endocrines and the liver. Academic, New York, p 153

Eagon PK, Fisher SE, Imhoff AF, Porter LE, Stewart RR, Van Thiel DH, Lester R (1980) Estrogen binding proteins of male rat liver: influences of hormonal changes. Arch Biochem Biophys 201: 486–499

Eagon PK, Porter LE, Gavaler JS, Egler KM, Van Thiel DH (1981a) Effect of ethanol feeding upon levels of a male-specific hepatic estrogen binding protein: a possible mechanism for feminization. Alcoholism 5: 183–187

Eagon PK, Zdunek JR, Van Thiel DH, Singletary BK, Egler KM, Gavaler JS, Porter LE (1981b) Alcohol-induced changes in hepatic estrogen binding proteins: a mechanism explaining feminization in alcoholics. Arch Biochem Biophys 211: 48–54

Earll JM, Gaunt K, Earll LA, Diuh YY (1976) Effect of ethyl alcohol on urine calcium and prolactin in man. Aviat Space Environ Med 47: 808–812

Ellingboe J, Van Anelli CC (1979) Ethanol inhibits testosterone biosynthesis by direct action on Leydig cells. Res Commun Chem Pathol Pharmacol 30: 87–102

Ellis FW (1966) Effect of ethanol on plasma corticosterone levels. J Pharmacol Exp Therap 153: 121–128

Estep H, Shaw WA, Watlington C et al. (1969) Hypocalcemia due to hypomagnesemia and reversible parathyroid hormone unresponsiveness. J Clin Endocrinol 29: 842–848

Fankushen D, Roskin D, Dimich A, Wallack S (1964) The significance of hypomagnesemia in alcoholic patients. Am J Med 37: 802–806

Farber JC, Duncan GM (1951) The effect of acute alcohol intoxication in the adrenal glands of rats and guinea pigs. Q J Stud Alcohol 12: 355–360

Fazekas JG (1966) Hydrocortisone content of human blood and alcohol content of blood and urine after wine consumption. Q J Stud Alcohol 27: 439–446

Field JB, Williams HE, Mortimore GE (1963) Studies on the mechanism of ethanol induced hypoglycemia. J Clin Invest 42: 497–503

Freinkel N, Singer DL, Arky RA (1963) Alcohol hypoglycemia I. J Clin Invest 42: 1112–1133

Fridenberg R, Metz R, Mako M, Surmaczynska V (1971) Differential plasma insulin response to glucose and glucagon stimulation following ethanol priming. Diabetes 20: 397–403

Gavaler JS, Van Thiel DH, Lester R (1980) Ethanol, a gonadal toxin in the mature rat of both sexes: similarities and differences. Alcoholism 4: 271–276

Gavaler JS, Gay V, Egler KM, Van Thiel DH (to be published) Evaluation of the differential in vivo toxic effects of ethanol and acetaldehyde on the hypothalamic-pituitary-gonadal axis using 4-methylpyrazole. Alcoholism

Goldstein DB (1972) An animal model for testing effects of drugs on alcohol withdrawal reactions. J Pharmacol Exp Ther 183: 14–22

Gordon GG, Olivo J, Rafii F, Southren AL (1975) Conversion of androgens to estrogens in cirrhosis of the liver. J Clin Endocrinol Metab 40: 1018–1026

Gordon GG, Southren AL, Vittek J, Lieber CS (1979a) The effect of alcohol ingestion on hepatic aromatase activity and plasma steroid hormones in the rat. Metabolism 28:20–24

Gordon GG, Vittek J, Ho R (1979b) Effect of chronic alcohol use on hepatic testosterone 5 alpha A-ring reductase in the baboon and in the human being. Gastroenterology 77:110–114

Gordon GG, Vittek J, Southren AL (1980) Effects of chronic alcohol ingestion on the biosynthesis of steroids in rat testicular homogenate in vitro. Endocrinology 106:1880–1885

Green JRB, Stritcher EJ, Mowat AG (1977) Thyroid function and thyroid regulation in erythroid men with chronic liver disease. Evidence of multiple abnormalities. Clin Endocrinol 7:453–458

Grennet N (1970) Oxidation of extra mitochondrial NADH by rat liver mitochondria: possible role of Acyl-SCoA elongation enzymes. Biochem Biophys Res Commun 41:909–917

Helderman J, Vestal R, Rowe J, Tobin J, Andres R, Robertson G (1978) The response of arginine vasopressin to intravenous ethanol and hypertoxic saline in man. J Gerontol 33:39–47

Hion-Jon V (1928) The influence of alcohol on the endocrine glands. Fol Neuropathol Eston 3:288–301

Horn RS, Manthei RW (1965) Ethanol metabolism in chronic protein deficiency. J Pharmacol Exp Ther 147:385–390

Israel Y, Videla L, MacDonald A, Bernstein J (1973) Metabolic alterations produced in the liver by chronic ethanol administration. Comparison between effects produced by ethanol and by thyroid hormones. Biochem J 134:523–529

Israel Y, Videla L, Fernandez-Videla Y, Bernstein J (1975) Effects of chronic ethanol treatment and thyroxine administration on ethanol metabolism and liver oxidative capacity. Exp Therap 192:565–571

Israel Y, Walfish PG, Orrego H (1979a) Thyroid hormones in alcoholic liver disease. Effect of treatment with 6*N* propylthiouracil. Gastroenterology 76:116–122

Israel Y, Walfish PG, Orrego H (1979b) Thyroid hormones in alcoholic liver disease. Effect of treatment with 6-*N*-propylthiouracil. Gastroenterology 76:116–121

Iwaski Y, Sato H, Ohkubo A (1980) Effect of spontaneous portal systemic shunting on plasma insulin and amino acid concentrations. Gastroenterology 78:677–683

Jeffcoate WJ, Cullen MH, Herbert M, Hasting AG, Walder CP (1979) Prevention of effects of alcohol intoxication by naloxone. Lancet 2:1157–1159

Jefferys DB, Flanagan RJ, Volans GN (1980) Reversal of ethanol-induced coma with naloxone. Lancet 1:308–309

Jenkins JS, Connolly J (1968) Adrenocortical response to ethanol in man. Br Med J 2:804–806

Johnston DE, Chiao Y-B, Gavaler JS, Van Thiel DH (1981) Inhibition of testosterone synthesis by ethanol and acetaldehyde. Biochem Pharmacol 30:1827–1831

Jones JE, Shane SR, Jacob WH, Flink EB (1969) Magnesium balance studies in chronic alcoholism. Ann NY Acad Sci 162:934–939

Jung Y, Russfield AB (1972) Prolactin cells in the hypophysis of cirrhotic patients. Arch Pathol 94:265–270

Kalbfleisch JM, Lindenman RD, Ginn HE, Smith NO (1963) Effects of ethanol administration of urinary excretion of magnesium and other electrolytes in alcoholic and normal subjects. J Clin Invest 42:1471–1475

Kakihana R, Moore JA (1976) Circadian rhythm of corticosterone in mice: the effect of chronic consumption of alcohol. Psychopharmacology (Berlin) 46:301–305

Kakihana R, Noble EP, Butte JC (1968) Corticosterone response to ethanol in inbred strains of mice. Nature 218: 360–366

Kakihana R, Butte JC, Hathaway A, Noble EP (1971) Adrenocortical response to ethanol in mice: modification by chronic ethanol consumption. Acta Endocrinol 67:753–664

Kleeman CR (1972) Water metabolism. In: Maxwell H, Kleeman CR (eds) Clinical disorders of fluid and electrolyte metabolism. McGraw Hill, New York, pp 215–295

Klingman GI, Goodal MC (1957) Urinary epinephrine and levarterenol excretion during acute sublethal alcohol intoxication in dogs. J Pharmacol Exp Ther 121:313–318

Knockel JP (1977) The pathophysiology and clinical characteristics of severe hypophosphatemia. Arch Intern Med 137:203–220

Krawitt EL (1973) Ethanol inhibits intestinal calcium transport in rats. Nature 243:88–89

Krawitt EL (1974) Effect of acute ethanol administration on duodenal calcium transport. Proc Soc Exp Biol Med 146: 406–408

Krawitt EL (1975) Effect of ethanol ingestion on duodenal calcium transport. J Lab Clin Med 85: 665–671

Krawitt EL, Sampson HW, Katagiri CA (1975) Effect of 1,25-dihydroxycholecalciferol on ethanol mediated suppression of calcium absorption. Calcif Tissue Res 18: 119–124

Kreisberg RA, Owen WC, Siegal AM (1970) Hyperlacticalcidemia in man: ethanol-phenformin synergism. J Clin Endocrinol 34: 29–35

Kuhl C, Anderson O (1974) Glucose- and tolbutamine-mediated insulin response after preinfusion with ethanol. Diabetes 23: 821–826

Laragh JH, Cannon PJ Ames RP (1962) Angiotensin II and renal sodium transport; natriuresis and diuresis in patients with cirrhosis and ascites. J Clin Invest 41: 1375–1380

Lefebvre A, Adler H, Lieber CS (1970) Effect of ethanol on ketone metabolism. J Clin Invest 49: 1775–1782

Lester R, Eagon PK, Van Thiel DH (1979) Feminization of the alcoholic: the estrogen/testosterone ratio (E/T). Gastroenterology 76: 415–417

Leutscher JA, Johnson BB, Dowdy AA (1954) Observations on the sodium retaining corticoid in the urine of children and adults in relation to sodium balance and oedema. J Clin Invest 33: 1441–1445

Linkola J, Fyhrquist F, Forsander O (1977) Effects of ethanol in urinary arginine vasopressin excretion in two rat strains selected for their different ethanol preferences. Acta Physiol Scand 101: 126–128

Linkola J, Ylikahri R, Fyhrquist F, Wallenius M (1978) Plasma vasopressin in ethanol intoxication and hangover. Acta Physiol Scand 104: 180–187

Littleton J (1978) Alcohol and neurotransmitters. Clin Endocrinol Metab 7: 369–384

Lloyd CW, Williams RH (1948) Endocrine changes associated with Laennec's cirrhosis of the liver. Am J Med 4: 315–330

Lochner A, Wulff J, Madison CC (1967) Ethanol induced hyperglycemia. Metabolism 16: 1–10

Lommer D, Dusterdiech G, Jahnecke J (1968) Sekretion, Plasmakonzentration, Verteilung, Stoffwechsel und Ausscheidung von Aldosteron beim Gesunden und Kranken. Klin Wochenschr 46: 741–746

Loosen PT, Prange AJ (1977) Alcohol and anterior pituitary secretion. Lancet 2: 985

Marco J, Diego J, Villaneuva ML (1973) Elevated plasma glucagon levels in cirrhosis. N Engl J Med 289: 1107–1111

Marks V, Medel WE (1964) Alcohol induced hypoglycemia. Br J Psychol 110: 228–233

Marks V, Wright JW (1977) Endocrinologic and metabolic effects of alcohol. Proc R Soc Med 70: 337–344

Marquis C, Marchetti J, Burlet C, Bowlange M (1975) Sécrétion urinaire et hormone antidiurétique chez des rats soumis à une administration répétée d' éthanol. C R Soc Biol (Paris) 54: 555–562

Martin HE, McCuskey C, Tupikoua N (1959) Electrolyte disturbance in acute alcoholism. Am J Clin Nutr 7: 191–194

Matunaga H (1942) Experimentelle Untersuchungen über den Einfluß des Alkohols und den Kohlenhydratstoffwechsel: über die Wirkung des Alkohols auf den Blutzuckerspiegel und den Glykogengehalt der Leber mit besonderer. J Exp Med 44: 130–135

McMonagle J, Felig P (1975) Effects of ethanol ingestion on glucose tolerance and insulin secretion in normal and diabetic subjects. Metabolism 24: 625–632

Mendelson JH, Stein S (1966) Serum cortisol levels in alcoholic and nonalcoholic subjects after experientially induced alcohol intoxication. Psychosom Med 28: 616–621

Mendelson JH, Ogata M, Mello NK (1971) Adrenal function and alcoholism: serum cortisol. Psychosom Med 33: 145–149

Mendelson JH, Mello NK, Ellingboe J (1981) Acute alcohol intake and pituitary gonadal hormones in normal females. J Pharmacol Exp Ther 218: 23–26

Mendelson JH, Mello NK, Ellingboe J, Baoli S (to be published) Alcohol effects on plasma luteinizing hormone levels in postmenopausal women. J Pharmacol Exp Ther

Merry J, Marks V (1969) Plasma hydrocortisone response to ethanol in chronic alcoholics. Lancet 1: 921

Metz R, Berger S, Mako M (1969) Potentiation of the plasma insulin response to glucose by prior administration of alcohol. Diabetes 18: 517–522
Morley JE, Baranetsky NG, Wingert TD (1980) Endocrine effects of naloxone-induced opiate receptor blockade. J Clin Endocrinol Metab 50: 251–257
Nikkila EA, Taskinen MR (1975) Ethanol-induced alterations of glucose tolerance, postglucose hypoglycemia, and insulin secretion in normal, obese and diabetic subjects. Diabetes 24: 933–943
Nilsson BE (1970) Conditions contributing to fracture of the femoral neck. Acta Clin Scand 136: 383–384
Noble EP, Kakihana R, Butte JC (1971) Corticosterone metabolism in alcohol adapted mice. In: Roach MK, McIssac MW, Creven PJ (eds) Biological aspects of alcohol. University of Texas Press, Dallas, p 389
Nomura S, Pittman CS, Chamber JB (1975) Reduced peripheral conversion of thyroxine to triodothyronine in patients with hepatic cirrhosis. J Clin Invest 56: 643–649
Olivo J, Gordon GG, Rafii F, Southren AL (1975) Estrogen metabolism in hyperthyroidism and in cirrhosis of the liver. Steroids 26: 47–56
Orrego H, Kalant H, Israel Y (1979a) Effect of short term therapy with propylthiouracil in patients with alcoholic liver disease. Gastroenterology 76: 105–110
Orrego H, Kalant H, Israel Y (1979b) Effect of short term therapy with propylthiouracil in patients with alcoholic liver disease. Gastroenterology 76: 105–109
Oxenkrug GF (1978) Dexamethasone test in alcoholics. Lancet 2: 795
Pekkanen L, Eriksson K, Sihvonen ML (1978) Dietary induced changes in voluntary ethanol consumption and ethanol metabolism in the rat. Br J Nutr 40: 103–107
Peng TC, Gitelman HJL (1974) Ethanol-induced hypermagnesemia and inhibition of the serum calcium-raising effect of parathyroid hormone in rats. Endocrinology 94: 608–612
Peng T, Cooper CW, Munson PL (1972) The hypocalcemia effect of alcohol in rats and dogs. Endocrinology 91: 586–591
Perman ES (1958) The effect of ethyl alcohol on the secretion from the adrenal medulla in man. Acta Physiol Scand 44: 241–247
Perman ES (1960) The effect of ethyl alcohol on the secretion from the adrenal medulla of the cat. Acta Physiol Scand 48: 323–328
Peterson RE (1960) Adrenocortical steroid metabolism and adrenal cortical function in liver disease. J Clin Invest 39: 320–331
Phillips GB, Safrit HF (1971) Alcoholic diabetes: induction of glucose intolerance with alcohol. JAMA 217: 1513–1519
Pohorecky LA (1974) Effects of ethanol on central and peripheral noradrenergic neurons. J Pharmacol Exp Ther 189: 380–391
Pohorecky LA, Jaffe LS, Berkeley HA (1974) Effects of ethanol on the adrenal medulla of the rat. Pharmacology 12: 340–346
Raiba N (1960) Effect of ethanol on cytological changes induced by salt load in nucleus supraopticus of rat. Proc Soc Exp Biol Med 103: 387–391
Redmond CP (1980) Effect of ethanol on endogenous rhythms of growth hormone secretion. Alcoholism 4: 50–56
Rees LH, Besser GM, Jeffcoate WJ (1977) Alcohol induced pseudo Cushing's syndrome. Lancet 1: 726–728
Rosmussen H (1974) Parathyroid hormone calcitonin and the calciferols. In: Williams RH (ed) Textbook of endocrinology, 5th edn. Saunders, Philadelphia, p 686
Samaan NA, Stone DB, Eckhard RD (1969) Serum glucose, insulin and growth hormone in chronic hepatic cirrhosis. Arch Intern Med 124: 149–154
Samols E, Holdsworth D (1968) Disturbances in carbohydrate metabolism: liver disease. In: Dickens F, Randle PJ, Whelan JW (eds) Carbohydrate metabolism and its disorders, vol 12. Academic, New York, p 289
Saville PD, Lieber CS (1965) Effect of alcohol on growth, bone density and muscle magnesium in the rat. J Nutr 87: 477–484
Schanez-Calvo R (1941) Die Alkoholvergiftung und ihre Folgen für das Ernährungssystem. Virchows Arch 308: 14–21
Sereny G, Rapaport A, Hudson A (1966) The effect of alcohol withdrawal on electrolyte and acid-base balance. Metabolism 15: 896–899

Shah JH, Wongsurawat N, Aran PO (1977) Effect of ethanol on stimulus-induced insulin secretion and glucose tolerance. A study of mechanisms. Diabetes 26: 271–277

Sherwin R, Joshi P, Hendler R (1974) Hyperglucagonemia in Laennec's cirrhosis. N Engl J Med 290: 239–242

Singh SP, Patel DG (1976) Effects of ethanol on carbohydrate metabolism. I. Influence on oral glucose tolerance test. Metabolism 25: 239–243

Smals AG, Kloppenborg PW (1977) Alcohol induced Cushingoid syndrome. Lancet I (8026): 1369 (letter)

Smals AG, Njo KT, Knoben JM, Ruland CM, Kloppenborg PW (1977) Alcohol induced Cushingoid syndrome. J R Coll Physicians Lond 12: 36–41

Smith JJ (1951) The effect of alcohol on the adrenal ascorbic acid and cholesterol content of the rat. J Clin Endocrinol 11: 792–795

Soeters PB, Fischer JE (1976) Insulin, glucagon, amino imbalance and hepatic encephalopathy. Lancet 2: 880–882

Stein JZ, Smith WO, Ginn HE (1966) Hypophosphatemia in acute alcoholism. Am J Med Sci 252: 78–83

Stokes PE (1973) Adrenocortical activity in alcoholics during chronic drinking. Ann NY Acad Sci 215: 77–82

Tabakoff B, Hoffman PL (1980) Alcohol and neurotransmitters. In: Rigter H, Crabbe J (eds) Alcohol tolerance and dependence. Elsevier, Amsterdam, pp 201–226

Tabakoff B, Jaffe RC, Ritzmann RF (1978) Corticosterone concentrations in mice during ethanol drinking and withdrawal. J Pharm Pharmacol 30: 371–374

Torro G, Kolodny RC, Jacobs LS (1973) Failure of alcohol to alter pituitary and target organ hormone levels. Clin Res 21: 505

Ulick S, Larangh GJH, Lieberman S, Loeb RR (1958) The isolation of a urinary metabolite of aldosterone and its use to measure the rate of secretion of aldosterone by the adrenal cortex of man. Trans Assoc Am Physicians 71: 225–230

Van Thiel DH (1979) Feminization of chronic alcoholic men: a formulation. Yal J Biol Med 52: 219–225

Van Thiel DH (1982) Disorders of the hypothalamic-pituitary-gonadal axis in patients with liver disease. In: Zakim D, Boyer TD (eds) Hepatology, a textbook of liver diseases. Saunders, Philadelphia, pp 516–528

Van Thiel DH (1983a) Ethanol: its adverse effects upon the hypothalamic-pituitary-gonadal axis. J Lab Clin Med 101: 21–33

Van Thiel DH (1983b) Adrenal response to ethanol: a stress response? In: Pohorecky LA, Brick J (eds) Stress and alcohol use. Elsevier, New York, pp 23–27

Van Thiel DH, Lester R (1974) Sex and alcohol. N Engl J Med 291: 251–253

Van Thiel DH, Lester R (1976a) Alcoholism: its effect on hypothalamic pituitary gonadal function. Gastroenterology 71: 318–327

Van Thiel DH, Lester R (1976b) Sex and alcohol: a second peek. N Engl J Med 295: 835–836

Van Thiel DH, Lester R (1977) Hypothalamic-pituitary-gonadal dysfunction in patients with alcoholic liver disease. In: Davidson CS (ed) Problems in liver disease. Stratton, New York, pp 289–298

Van Thiel DH, Lester R (1978) Further evidence for hypothalamic-pituitary dysfunction in alcoholic men. Alcoholism 2: 265–270

Van Thiel DH, Lester R (1979) The effect of chronic alcohol abuse on sexual function. Clin Endocrinol Metab 8: 499–510

Van Thiel DH, Lester R (1980) Hypothalamic-pituitary-gonadal function in liver disease. Viewpoints Dig Dis 12: 13–16

Van Thiel DH, Loriaux DL (1979) Evidence for an adrenal origin of plasma estrogens in alcoholic men. Metabolism 28: 536–541

Van Thiel DH, Lester R, Sherins RF (1974a) Hypogonadism in alcoholic liver disease: evidence for a double defect. Gastroenterology 67: 1188–1199

Van Thiel DH, Gavaler JS, Lesrer R (1974b) Ethanol inhibition of vitamin A metabolism in the testes: possible mechanism for sterility in alcoholics. Science 186: 941–942

Van Thiel DH, Gavaler JS, Lester R, Goodman MD (1975a) Alcohol-induced testicular atrophy: an experimental model for hypogonadism occurring in chronic alcoholic men. Gastroenterology 69: 326–332

Van Thiel DH, Gavaler JS, Lester R, Loriaux DL, Braunstein GD (1975b) Plasma estrone, prolactin, neurophysin and sex steroid binding globulin in chronic alcoholic men. Metabolism 24: 1015–1019

Van Thiel DH, Gavaler JS, Sherins RJ, Lester R (1977) Alcohol-induced ovarian failure in the rat. J Clin Invest 61: 624–632

Van Thiel DH, Lester R, Vaitukaitis J (1978a) Evidence for a defect in pituitary secretion of luteinizing hormone in chronic alcoholic men. J Clin Endocrinol Metab 47: 499–507

Van Thiel DH, McClain CJ, Elson MK, McMillin MJ (1978b) Hyperprolactinemia and thyrotropin releasing factor (TRH) responses in men with alcoholic liver disease. Alcoholism 2: 344–348

Van Thiel DH, Gavaler JS, Wight WI, Abuid (1978c) Thyrotropin releasing hormone (TRH) induced growth hormone (hGH) responses in cirrhotic men. Gastroenterology 75: 66–70

Van Thiel DH, Gavaler JS, Cobb CF, Sherins RJ, Lester R (1979a) Alcohol-induced testicular atrophy in the adult male rat. Endocrinology 105: 888–895

Van Thiel DH, Smith WI, McClain CJ, Lester R (1979b) Abnormal prolactin and growth hormone responses to thyrotropin releasing hormone in chronic alcoholic men. Curr Alcohol 5: 71–79

Van Thiel DH, Smith WI, Wight C, Abuid J (1979a) Elevated basal and abnormal thyrotropin releasing hormone induced thyroid stimulating hormone secretion in chronic alcoholic men with liver disease. Alcoholism 3: 302–308

Van Thiel DH, Gavaler JS, Slone FL, Cobb CF, Smith WI Jr., Bron KM, Lester R (1980) Is feminization in alcoholic men due in part to portal hypertension: a rat model. Gastroenterology 78: 81–91

Van Thiel DH, Cobb CF, Herman GB, Perez HA, Estes L, Gavaler JS (1981) An examination of various mechanisms for ethanol-induced testicular injury studies utilizing the isolated perfused rat testes. Endocrinology 109: 6, 2009–2015

Van Thiel DH, Gavaler JS, Cobb CF, Chiao Y-B (1982) Effects of ethanol upon the hypothalamic-pituitary gonadal axis. In: Langer M, Chiandussi L, Chopra IJ, Martini L (eds) The endocrines and the liver. Academic, New York, pp 117–134

Wagner G, Fuchs AR (1968) Effects of ethanol on uterine activity during suckling in post partum women. Acta Endocrinol 58: 133–141

Walsh MJ, Truitt EG (1970) Acetaldehyde mediation in the mechanism of ethanol-induced changes in neuroepinephrine metabolism. Mol Pharmacol 6: 416–424

Weichsilbaum A (1910) Über Veränderungen der Hoden bei chronischem Alkoholismus. Verh Dtsch Ges Pathol 14: 234–239

Whercat AF, Orishimo MW, Nelson J (1969) The location of different synthetic systems for fatty acid in inner and outer mitochondrial membranes from rabbit heart. J Biol Chem 244: 6498–6506

Williams GA, Bowser EN, Hargis GK (1981) Effect of ethanol on parathyroid hormone and calcitonin secretion in man. Proc Soc Exp Biol Med 159: 187–191

Wolff HP, Lommer D, Torbica M (1962) Studies in plasma aldosterone metabolism in some heart, liver and kidney diseases. Schweiz Med Wochenschr 95: 387–392

Wright JW (1978) Endocrine effects of alcohol. Clin Endocrinol Metab 7: 351–356

Wright J, Merry J, Fry D, Marks V (1975) Pituitary function in chronic alcoholism. Adv Exp Med Biol 59: 253–255

Yadav HS, Chandhurij BN, Mukherjee SK (1978) Effect of ethyl alcohol on thyroidal iodine trapping and renal clearance of 131-I label in rats. Indian J Med Res 58: 1421–1427

Ylikahri RH, Huttunen M, Harkonen M (1976) Effect of alcohol on anterior-pituitary secretion of trophic hormones. Lancet 1: 1535

Zumoff B, Bradlow HG, Gallagher TF, Hellman L (1967) Cortisol metabolism in cirrhosis. J Clin Invest 46: 1735–1741

19 Effect of Ethanol on Intestinal Morphology, Metabolism, and Function

E. Mezey

Department of Medicine of The John Hopkins Hospital, The John Hopkins University, School of Medicine, 600 N Wolfe Street; Baltimore, MD 21205, USA

Chronic alcoholism is a frequent cause of diarrhea, malabsorption, malnutrition, and weight loss. Although the principal cause of malnutrition and weight loss is decreased dietary intake, malabsorption of nutrients may also play a significant contributory role. Both the acute and chronic ingestion of ethanol have varied effects on the morphology, metabolism, and function of the intestine. Furthermore, the effects of ethanol on the intestine often vary depending on the nutritional status.

Absorption and Metabolism of Ethanol by the Intestine

The absorption of ethanol starts in the stomach and continues in the upper small intestine. Immediately after the oral ingestion of ethanol, high concentrations of ethanol, similar to those of alcoholic beverages, are reached in the stomach and jejunum (Fig. 1); thereafter, as ethanol is absorbed, its concentration in the upper gastrointestinal tract decreases rapidly, reaching levels that are in equilibrium with the vascular space. Ethanol is detected in the blood as early as 5 min after ingestion and maximum concentrations are reached in 30–90 min. The concentrations of ethanol reached in the ileum following alcohol ingestion are parallel to levels in the vascular space, suggesting that ethanol enters the ileum from the vascular space rather than traveling the length of the intestine (Halsted et al. 1973a).

The rate of ethanol absorption is decreased by delayed gastric emptying and by the presence of food in the stomach, and is increased after gastric surgery with gastroenterostomies. Caffeine (Siegers et al. 1972) and disulfiram (Lamboeuf et al. 1975) decrease ethanol absorption in the rat by their effect of depressing the rate of gastric emptying. Carbohydrates (Broitman et al. 1976), amino acids, and dipeptides (Hajjar et al. 1981b) enhance ethanol absorption in the perfused rat jejunum in vivo. The increased ethanol absorption results from the increased water flow and solvent drag that accompany the active absorption of nutrients. No increases in ethanol absorption occurred after the addition to the jejunal perfusates of carbohydrates such as fructose of 3-*O*-methylglucose whose absorption does not require energy. Among the disaccharides tested, sucrose and maltose but not lactose increased ethanol absorption. It was postulated that the failure of lactose to stimulate ethanol

Alcohol Related Diseases in Gastroenterology
Edited by H. K. Seitz and B. Kommerell

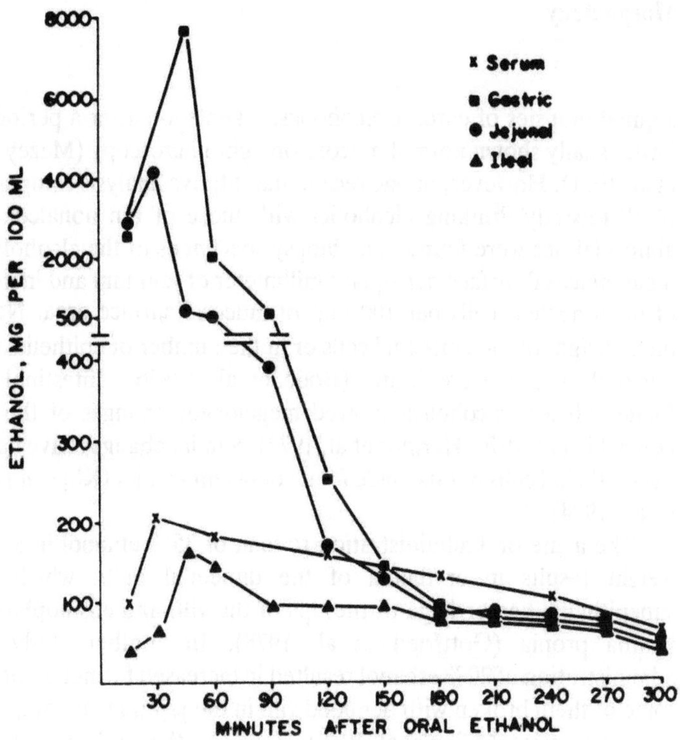

Fig. 1. Concentrations of ethanol in the blood, stomach, jejunum, and ileum after the oral administration of 20% w/v ethanol in a total dose of 0.8 g/kg body weight to one patient. (Halsted et al. 1973a)

absorption may be related to its relatively slow hydrolysis compared to that of the other disaccharides.

Ethanol has been shown to be metabolized by the rat stomach and small intestine (Carter and Isselbacher 1974; Lamboeuf et al. 1981). Increases in acetaldehyde and acetate levels were detected in afferent vessels after instillation of ethanol into the ligated rat stomach or duodenum (Lamboeuf et al. 1981). Alcohol dehydrogenase, the principal enzyme which catalyzes the oxidation of ethanol, is present in the mucosa of the stomach, jejunum, and ileum in both the rat (Mistilis and Garske 1969) and man (Spencer et al. 1964). In addition, bacteria present in the gastrointestinal tract contain alcohol dehydrogenase and produce endogenous ethanol by fermentation (Krebs and Perkins 1970). Microsomal ethanol-oxidizing activity has also been demonstrated in the intestinal mucosa (Seitz et al. 1979). The amount of ethanol metabolized by the mucosa of the gastrointestinal tract is probably minimal because the quantity of alcohol dehydrogenase enzyme present in the gastrointestinal mucosa is small and its affinity for ethanol lower (K_m for ethanol higher) than is the case for the enzyme in the liver (Mezey and Halsted 1979). However, some of the effects of ethanol on the intestine, such as its effect in enhancing triglyceride synthesis, are linked to its metabolism in the intestinal mucosa.

Morphology

Jejunal biopsies of chronic alcoholics, taken soon after a period of binge drinking, have usually shown normal mucosa on light microscopy (Mezey et al. 1970; Halsted et al. 1971). However, in one recent quantitative analysis comparing jejunal biopsies of 18 recently drinking alcoholics with those of ten nonalcoholic controls, some abnormalities were found. The biopsy specimens of the alcoholics had decreases in mean mucosal surface per square millimeter of jejunum and increases in the number of mononuclear cells per 100 μm^2 of mucosal surface area. No differences in the mean height of the epithelial cells or in the number of epithelial cells per 100 μm^2 of mucosal surface were found (Bode et al. 1982b). Intestinal biopsies of severe folate-deficient alcoholics showed megalocytic changes of the surface epithelium (Bianchi et al. 1970; Hermos et al. 1972). Similar changes have been demonstrated in the epithelial cells of rats made folate deficient by diet (Klipstein et al. 1973; Howard et al. 1974).

The acute oral administration to man of 35% ethanol in a dose of 1 g/kg body weight results in erythema of the duodenal bulb, which on biopsy reveals subepithelial hemorrhage of the tips of the villi and eosinophilic infiltration of the lamina propia (Gottfried et al. 1978). In another study, acute intragastric administration of 20% ethanol resulted in increased formation of subepithelial blebs; some of them broken with denuded villi in the jejunum as compared with findings in control biopsies (Millan et al. 1980). However, the epithelial cells remained intact. It was suggested that these changes, which were also observed to a lesser extent in control biopsies, indicate an increased susceptibility of the mucosa to damage produced by the suction biopsy. The observed changes decreased toward a lesser damage found in controls in 2 h after peak ethanol concentration. In rats, the intragastric administration of ethanol in a dose of 3 g/kg results in acute hemorrhagic erosions of the jejunal villi, the severity of the histologic response correlating with the concentration of ethanol, which varied from 5 to 50 g/100 ml (Baraona et al. 1974). In another study, perfusion of the rabbit jejunum with 3% ethanol resulted in the formation of subepithelial blebs, exfoliation of enterocytes, and dilated lacteals and occasional localized subepithelial hemorrhage; an increase in the number of subepithelial blebs, with rupture of some of them and an increase in hemorrhagic lesions, occurred when the jejunum was perfused with 6% ethanol (Buell and Beck 1983).

Chronic administration of ethanol to man with an adequate diet did not result in any changes in the intestinal mucosa by light microscopy (Halsted et al. 1971). In the rat, however, chronic administration of ethanol with an adequate diet results in shortening of the villi and decreases in the number of epithelial cells lining the villi (Baraona et al. 1974), as well as decreases in enterocyte height and increased mitotic activity in the crypt cells (Zucoloto and Rossi 1979). In one study, the changes were more pronounced in the ileum than the jejunum, suggesting that factors in addition to direct topical contact of the intestine to high concentrations of ethanol present in the upper intestine, such as blood-borne ethanol or its metabolites or nutritional alterations, contributed to the abnormalities (Zucoloto and Rossi 1979). Chronic alcohol feeding to rats and to human volunteers for 2–3 months results in ultrastructural changes in the intestinal epithelial cells similar to those produced in hepatocytes. These changes consist of mitochondrial enlargement with distortion of the matrix and cristae, dilatation and proliferation of the smooth endoplasmic reticulum, and decreased rough endoplasmic reticulum with fewer attached ribosomes (Rubin et al. 1972; Rossi and Zucoloto 1977).

Permeability

Intoxicated alcoholic patients with a mean blood ethanol level of 241 mg% had increased gastrointestinal permeability, as evidenced by increased urinary excretion of an oral load of polyethyleneglycol (PEG)-400, which returned to normal 2 days later when the patients were sober (Robinson et al. 1981). Ethanol in vitro in a concentration of 0,45 *M* or 2.1% did not affect permeability to PEG by everted jejunal sacs (Fox et al. 1978). Chronic administration of 20% ethanol by gavage to rats in a daily dose of 6 g for 4–8 weeks resulted in disintegration of apical junctional complexes between adjacent epithelial cells and increased intestinal permeability to horseradish peroxidase (Worthington et al. 1978).

Blood Flow and Metabolism

Ethanol in concentrations ranging from 3% to 6% was shown to increase jejunal blood flow in segments of dog (Siregar and Chou 1982) and rabbit (Buell and Beck 1983) jejunum. Furthermore, it was demonstrated that the increase in blood flow is restricted to the luminal layer consisting of mucosa and submucosa, with no change in blood flow to the muscularis plus serosa (Buell and Beck 1983). The increased jejunal blood flow did not alter oxygen consumption (Siregar and Chou 1982). In other studies, acute administration of ethanol resulting in hemorrhagic erosions of the villi in rats reduced oxygen consumption by jejunal slices, while administration of the same acute ethanol dose to animals after chronic ethanol feeding resulted in increases in oxygen consumption (Baraona et al. 1974).

The acute administration of ethanol increased mitosis in the crypt cells of the jejunum and ileum, but decreased thymidine kinase activity and did not affect the incorporation of tritiated thymidine into intestinal DNA (Baraona et al. 1974). Chronic ethanol feeding increased mitosis in the crypt cells, thymidine kinase activity, and incorporation of tritiated thymidine into DNA (Baraona et al. 1974). Increased incorporation of tritiated thymidine into DNA after chronic ethanol feeding was also demonstrated in isolated intestinal cells (Seitz et al. 1982). The acute intragastric administration of ethanol inhibited the intestinal activity of ornithine decarboxylase (Pösö 1980); this enzyme is rate controlling in the synthesis of polyamincs, which are closely associated with regenerative processes in mammalian tissues, Neither the acute or chronic administration of ethanol affects the protein and DNA concentrations of the intestine (Baraona et al. 1974).

Intestinal Metabolism

Glucose

Chronic alcoholics who ingested beer were found to have increases in the jejunal activity of the glycolytic enzyme glucokinase but decreases in the gluconeogenic enzyme fructose-1,6-diphosphatase. The activities of hexokinase, ketohexokinase, 6-phosphofructokinase, and glucose-6-phosphate dehydrogenase were normal (Bode et al. 1982a). It was suggested that the enzyme changes observed may result from an increase in carbohydrate intake, since glycolytic enzymes show an adaptive increase while fructose-1,6-diphosphatase falls in response to carbohydrate (Rosensweig et al. 1971). Administration of 45 g ethanol a day to normal volunteers for 7 days resulted in 35%–45% decreases in the glycolytic enzymes hexokinase, fructose-1-phosphate aldolase, and fructose-1,6-diphosphate aldolase, and a 58% decrease in the activity of the gluconeogenetic enzyme fructose-1,6-diphosphatase. By contrast, ethanol resulted in an increase in the activity of pyruvate kinase. Folate given with the ethanol reversed the inhibitory effect of ethanol on all the glycolytic enzymes and further increased the activity of pyruvate kinase, but had no effect on fructose-1,6-diphosphatase (Greene et al. 1975).

Lipids

The acute administration of ethanol has been shown to result in increases in triglyceride (Carter et al. 1971) and cholesterol synthesis by intestinal slices (Middleton et al. 1971), triglyceride content of the small-intestinal mucosa, and lymphatic output of triglycerides, cholesterol, and phospholipids (Mistilis and Ockner 1972). The increased triglyceride synthesis could be partially suppressed by pyrazole, while the increased cholesterol synthesis (Carter et al. 1971) was only demonstrable when ethanol remained in the intestinal lumen (Middleton et al. 1971), suggesting that the effect of ethanol may be mediated by its metabolism in the intestinal mucosa. Intraduodenal infusion of ethanol for 5 h increased the activity of the lipid reesterifying enzymes acyl-CoA: monoglyceride acyltransferase and acyl-CoA synthetase in the rat jejunum (Rodgers and O'Brien 1975). In one study, however, ethanol in vitro inhibited palmitate and acetate oxidation, CO_2 production, and triacylglycerol synthesis by intestinal slices, but enhanced esterification of fatty acids with ethanol (Baraona et al. 1975). The acute administration of ethanol in a dose of 3 g/kg body weight also inhibited fatty acid oxidation and triglyceride synthesis in the jejunum but not in the ileum. By contrast, chronic ethanol feeding increased fatty acid oxidation and triglyceride synthesis in both the jejunum and ileum, and those increases were associated with an increase in palmitoyl-CoA synthetase activity, suggesting fatty acid activation. It has also been suggested that the observed increases in intestinal lipid synthesis and lipid output by the lymph may contribute to the hyperlipemia and fatty infiltration of the liver induced by ethanol (Mistilis and Ockner 1972). Chronic ethanol feeding, however, inhibited the increase in lymph flow and lipid output of intragastrically administered labeled tripalmitin that occurs after a dietary lipid challenge or the acute administration of ethanol (Baraona and Lieber 1975).

Intestinal Enzymes

Disaccharidases

Alcoholism results in a marked enhancement of lactose intolerance in black but not in white individuals. Decreases in intestinal lactase activity were found in all 11 black alcoholic patients following heavy alcohol ingestion, but in only five of the ten black nonalcoholic controls (Perlow et al. 1977). Seventy percent of the black alcoholic patients who were lactase deficient had abnormal lactose tolerance tests, most of them associated with symptoms of abdominal colic and diarrhea. No significant differences in lactase activity or lactose tolerance tests were found between white alcoholics and nonalcoholic subjects. Sucrase activity was decreased in 33% of alcoholics of both races. The abnormal activities of both disaccharidases rose after 2 weeks of abstinence from alcohol in five alcoholics of both races who had repeat intestinal biopsies. In another study of ten white beer drinkers, jejunal sucrase and maltase activities were decreased, while lactase activity was normal (Madzarova-ra-Nohejlova 1971). Chronic ethanol administration to rats as 15% of the drinking water together with a normal diet resulted in increases in the jejunal activities of sucrase, maltase, and lactase. Similar increases in the activities of these enzymes occurred after feeding a high carbohydrate-low protein diet, but the timing of the increase differed. After ingestion of the diet, maximal increases in sucrase and maltase occurred at 2 weeks and of lactase at 4 weeks, while after ethanol all maximal increases occurred at 4 weeks (Raul et al. 1982). In another study, the administration of ethanol in varying concentrations in the drinking water resulted in a decrease in the intestinal activities of sucrase and alkaline phospholase (Eloy et al. 1979). These decreases became maximal at 30 days after initiation of the ethanol administration and did not recover to normal after 8 days of withdrawal from ethanol. In vitro incubation of brush border membranes of hamster jejunum with ethanol concentration ranging from 0.8% to 6.4% resulted in decreases in the activities of sucrase, maltase, and lactase. These decreases correlated with the ethanol concentration and length of exposure of the brush border membranes to ethanol. In addition, ethanol resulted in decreased affinity for substrate (higher K_m) of sucrase and lactase but not of maltase (Dinda et al. 1979).

Enzymes Involved in Intestinal Transport

Effects of ethanol have been demonstrated in vitro in rat jejunum on enzymes that regulate water and electrolyte absorption and secretion. Ethanol in concentrations ranging from 0.5% to 4.6% inhibits Na^+-K^+-ATPase activity, the sodium pump necessary for absorption of sodium and water, in basolateral preparations of the rat jejunum (Hoyumpa et al. 1977). Ethanol in vitro increased the fluidity, measured by electron magnetic resonance using a *N*-oxyl 4′-4-dimethyloxazolidine derivative of 5-ketostearic acid, of both basolateral and brush border membranes (Gray et al. 1980). Chronic ethanol feeding had no effect on membrane fluidity, but made the membranes resistant to change on exposure to ethanol in vitro. Ethanol in concentrations of 11%–13%, which are higher than concentrations found in vivo in the intestine after the ingestion of alcoholic beverages, stimulates adenyl cyclase (Greene et al. 1971). This enzyme catalyzes the conversion of ATP to cyclic AMP, a compound that mediates intestinal secretion. The ATP content of the intestine has been found to be decreased after the acute and chronic administration of ethanol in the rat (Carter and Isselbacher 1973). The effect of ethanol in decreasing ATP concentration was also demonstrated in vitro with guinea pig small intestine, although it was not found after chronic ethanol consumption in this species (Krasner et al. 1976a).

Microsomal Enzymes

Although the liver is the principal site of drug metabolism, the mixed function oxidases present in the intestine probably contribute significantly to the metabolism of drugs and xenobiotics. The chronic administration of ethanol in the rat resulted in proliferation of the smooth endoplasmic reticulum and increases in cytochrome *P*-450, reduced nicotinamide-adenine dinucleotide phosphate (NADPH)-cytochrome *c* reductase, and the microsomal ethanol-oxidizing system (Seitz et al. 1979). Also, increases in the microsomal activities of aniline hydroxylase, benzphetamine demethylase, 7-ethoxycoumarine deethylase (Joly and Hetu 1980), benzo(a)pyrene hydroxylase (Seitz et al. 1980), and gamma-glutamyltranspeptidase (Ishii et al. 1978) have been demonstrated. Intestinal microsomes after chronic ethanol feeding had increased content of phospholipids and oleic acid but decreased linoleic acid; in addition an increased amount and affinity of binding to the fluorescent probe 1-anilinonaphthalene-8-sulfonic acid suggested that structural changes of the microsomal membranes had occurred (Koivusaari et al. 1981). Furthermore, intestinal microsomes from rats after chronic ethanol feeding had an enhanced capacity to activate benzo(a)pyrene, 2-aminofluorene, and tryptophan pyrolyzate to derivatives mutagenic for *Salmonella typhimurium* (Seitz et al. 1980). The greater activation of procarcinogens to carcinogens after alcohol ingestion may play a role in the increased incidence of some visceral cancers in alcoholic patients. As regards the intestine, an association between cancer of the colon and alcohol ingestion has been found in various studies. The consumption of beer is particularly strongly correlated with mortality from carcinoma of the colon and rectum (Breslow and Enstrom 1974; Meyer 1977). No correlation was found between the ingestion of wine and carcinoma of the rectum in one study in France (Meyer 1977).

Bacterial Overgrowth

An increase in small-intestinal flora and an alteration in its composition have been found in patients with cirrhosis and alcoholism. Martini et al. (1956) found increased total coliform count in the duodenum, jejunum, and ileum in one half of the patients studied; in addition, *Streptococcus faecalis* was present in the jejunum of 25%

patients with cirrhosis but in none of the normal subjects examined. Examination of the feces revealed a higher incidence of atypical coliform bacteria such as *Escherichia freundii* in patients with cirrhosis. Lal et al. (1972) reported an increase in urea-splitting bacteria, predominantly *Klebsiella* and *Proteus* strains, in the small and large bowel of 88% of patients with alcoholic cirrhosis. Contamination of the small bowel with these organisms was found in 50% of the patients. However, a control group of hospitalized patients with chronic illness also had a similar increase in the colonization of the intestine with these organisms. In another study, alcoholic patients with and without cirrhosis had a higher incidence of intestinal cultures positive for anaerobes, including more with colony counts greater than 10^5/ml of jejunal juice (Bode et al. 1980). Also, the asporogenic bacilli were found more frequently in the alcoholics. The incidence of positive cultures and the number of bacteria per milliliter of jejunal juice correlated with an increase in the gastric juice pH, which was higher in the alcoholics than in the control group. Intestinal bacterial overgrowth can have a deleterious effect on nitrogen metabolism (Jeejeebhoy 1964) and contribute to catabolism of ingested protein, increased loss of endogenous protein, and diminished absorption of protein. In addition, it can alter the composition of amino acids and nitrogen breakdown products that are available for absorption (Tabaqchali 1970). Also, as mentioned previously, ethanol is formed by bacterial fermentation in the gastrointestinal tract and levels of ethanol are normally detectable in the portal vein (Krebs and Perkins 1970). Furthermore, the incubation of human fecal homogenates with 0.1% ethanol results in the formation of higher alcohols such as *N*-propyl, allyl, *N*-butyl, *t*-butyl, and *N*-amyl alcohols (Levitt et al. 1981). Some of these higher alcohols may play a role in organ damage in alcoholics.

Motility

Ethanol given either orally or intravenously in a dose of 0.8 g/kg body weight was shown to alter the motility of the human small intestine. In the jejunum ethanol inhibited type I waves, which impede forward progress of intestinal contents, while in the ileum there was enhancement of type III waves, which are associated with propulsion of the intestinal contents (Robles et al. 1974). Ethanol was also demonstrated to decrease both the frequency and amplitude of basal motility waves in the rectosigmoid (Berenson and Avner 1981) in man and to disrupt the myoelectric complex in dogs (Angel et al. 1980). The decrease in rectosigmoid motility correlated inversely with blood ethanol concentration, and diarrhea developed in five of 12 subjects within 8 h of completion of the study. Hence, the effects of ethanol on intestinal motility could result in increased transit of intestinal contents and contribute to the diarrhea seen in chronic alcoholics.

Intestinal Absorption

In evaluating the effect of ethanol on absorption, it is important to distinguish acute from chronic effects. In addition, dietary deficiencies often contribute to and in some cases are the principal cause of the abnormalities of absorption observed.

Amino Acids

Studies of the effect of ethanol on amino acid transport are summarized in Table 1. Ethanol in vitro consistently inhibits intestinal transport of amino acids. Ethanol administered intragastrically in a dose of 2.5 g/kg inhibited the absorption of phenylalanine in the rat (Israel et al. 1968). In man, the direct addition of ethanol to intestinal perfusates in a concentration of 2% resulted in a 55% reduction in the uptake of L-methionine (Israel et al. 1969). By contrast, recent experiments in the rat show that ethanol added to the perfused jejunum in concentrations ranging from 2.3% to 4.5% has no effect on the absorption of amino acids (Green et al. 1981). Only perfusions of the jejunum with higher concentrations of ethanol resulting in histologically demonstrable injury to the intestine resulted in decreased amino acid transport. The chronic feeding of ethanol to rats as 36% of the calories of a liquid diet resulted in an increased transport of leucine from the perfused intestine, when expressed per wet weight of intestine or per dry weight of mucosa (Hajjar et al. 1981b). However, the rats fed ethanol chronically had decreases in the dry weight of the mucosa per length of intestine. Chronic ethanol administration to rats (Rodrigo

Table 1. Effect of ethanol on intestinal transport of amino acids

Ethanol	Methodology	Effect	Reference
3%	In vitro, Rat	↓ Transport; MET, PHE, LEU, GLY, ALA, VAL	Chang et al. (1967)
2%	In vitro, Rat	↓ Transport; PHE, MET	Israel et al. (1968)
3%	In vitro, Rabbit	↓ Transport; ALA	Kuo and Shanbour (1978)
2.5 g/kg	In vivo, Rat	↓ Absorption; PHE	Israel et al. (1969)
2%	In vivo perfusion; Man	↓ Transport; MET	Israel et al. (1968)
2.3%–4.6%	In vivo perfusion; Rat	No effect; MET, PHE, LEU, GLY, ALA, LYS	Green et al. (1981)
Chronic (7 weeks)	In vivo perfusion	↑ Transport[a]; leu	Hajjar et al. (1981a)

[a] Only when calculated per wet weight of intestine or per dry weight of mucosa

et al. 1971) or to man (McDonald and Margen 1976) did not change fecal nitrogen excretion. However, increases in fecal nitrogen excretion were found in 52% of 27 chronic alcoholic patients admitted for detoxification to a City Hospital following recent heavy alcohol ingestion (Roggin et al. 1969).

Glucose

Ethanol 450 m*M* (2.0%) inhibits the uptake of 10 m*M* glucose by everted intestinal segments of hamster when placed on the mucosal but not the serosal side, suggesting that the effect is not due to inhibition of basolateral Na^+-K^+-adenosine triphosphatase (ATPase) (Dinda and Beck 1977). The depression of glucose transport obtained when ethanol is added to jejunal perfusates in vivo correlates with the concentration of ethanol added to the perfusate (Fox et al. 1978). Ethanol in concentrations ranging from 1% to 5% also depresses glucose uptake by brush border membrane vesicles of hamster jejunum, and this effect is readily reversible to normal after the membranes are washed free of ethanol (Dinda and Beck 1981). The effect of ethanol 2.5 g/kg given intragastrically to rats on depressing glucose transport by everted jejunal sacs was observed at concentrations of glucose ranging from 1 to 5 m*M*, but was not apparent at higher (20 m*M*) concentrations of glucose (Hoyumpa et al. 1981).

Water and Sodium Transport

A decrease in water and sodium transport by the jejunum has been observed using the triple lumen perfusion technique in alcoholic patients studied a few days after an acute alcohol binge (Halsted et al. 1971; Krasner et al. 1976b). The feeding of ethanol as 36% of the calories for 2 weeks to normal human volunteers resulted in similar decreases in water and sodium transport (Mekhjian and May 1977). These changes, which represent either decreased absorption, increased secretion, or both processes, were more pronounced when the subjects were given ethanol combined with a folate-deficient diet. The acute perfusion of ethanol in concentrations ranging from 2% to 10% in a glucose-free electrolyte solution in man did not result in any changes in water and sodium absorption. This latter observation differs from the findings in acute experiments in animals. Ethanol in a concentration of 2% decreased the net transport of water and glucose and the net mucosal to serosal transport of sodium by everted segments of hamster jejunum (Dinda et al. 1975). In another study using isolated jejunal mucosa from the rabbit in a chamber preparation, ethanol in a concentration of 3% inhibited transport of sodium, 3-*O*-methylglucose, and alanine (Kuo and Shanbour 1978). The effect of ethanol in these animal experiments was only found when it was added to the mucosal but not when added to the serosal side, suggesting that it is due to an inhibition of uptake across the brush border, and not related to the effect of ethanol in depressing Na^+-K^+-ATPase activity in the basolateral membrane. In in vivo studies of perfusion of the hamster jejunum, glucose transport was reduced by ethanol concentrations of 2%, while water transport was only found to be depressed at concentrations of 4.8% (Fox et al.

1978). The difference in results of the acute effects of ethanol between animals and man may be related to the difficulty in maintaining a high intestinal ethanol concentration in the perfusion studies and to the absence of glucose in the perfusion solution in the human studies.

D-Xylose

D-Xylose malabsorption, which is found in 18%–74% of alcoholic patients on admission to hospital, is rapidly reversible to normal after 2 weeks on a normal diet despite continuation of alcohol intake (Mezey 1975). The acute administration of ethanol in a dose of 0.8 g/kg body weight results in decreased D-xylose absorption, while chronic ethanol feeding results in no change (Mezey 1975) or enhances (Lindenbaum and Lieber 1975) D-Xylose absorption. D-Xylose malabsorption has been induced in man by a combination of ethanol and a folate-deficient diet (Halsted et al. 1973b), and in rats by a combination of ethanol and a protein-deficient diet (Broitman et al. 1961).

Thiamine

Immediately following prolonged drinking, chronically malnourished alcoholic patients have decreased thiamine absorption (Tomasulo et al. 1968; Thomson et al. 1970). Following a 6–8 week period of abstinence from alcohol and administration of a high-protein, vitamin-rich diet, thiamine absorption returns to normal (Thomson et al. 1970). The acute or intravenous administration of ethanol in a dose of 1.5 g/kg body weight decreased thiamine absorption in one-third of patients studied (Thomson et al. 1970). Investigations in rats using isolated duodenal loops or everted jejunal sacs demonstrated that ethanol decreases active thiamine absorption, which occurs at low concentrations of thiamine (0.06–2.0 m*M*), but does not affect passive transport at higher concentrations of thiamine (Hoyumpa et al. 1975). Further studies revealed that ethanol did not block uptake of thiamine into the mucosa, but rather blocked the exit of small concentrations of thiamine from the cell to the serosal compartment. This inhibitory effect of ethanol is similar to that obtained with ouabain, a known inhibitor of Na^+-K^+-ATPase (the sodium pump). Recent studies showing that ethanol inhibits the activity of intestinal basolateral membrane Na^+-K^+-ATPase suggest that the inhibitory effect of ethanol on thiamine transport is related to its inhibition of this enzyme (Hoyumpa et al. 1977). Folate deficiency induced in rats resulting in megalocytosis of intestinal epithelial cells was also shown to result in decreased absorption of low, but not of high, concentrations of thiamine (Howard et al. 1974).

Folate

Malabsorption of folic acid has been demonstrated only in malnourished chronic alcoholic patients (Halsted et al. 1967, 1971). Abstinence and a hospital diet for 2 weeks was followed by improvement in the jejunal uptake of folic acid. The acute administration of ethanol in the form of $4^1/_2$ oz whisky over a 1-h period decreased the absorption of orally administered folic acid in only one of five subjects tested (Halsted et al. 1967). The chronic administration of ethanol in a dose of 192 g/day for 2 weeks to abstinent, well-nourished alcoholic patients had no effect on jejunal uptake of folic acid in five patients but decreased it in two (Halsted et al. 1973b). In another study, decreased absorption of folic acid was induced by the administration of a folic-acid-deficient diet and ethanol but not by either alone, and the abnormal absorption was corrected to normal by the administration of folic acid despite the continuation of ethanol (Halsted et al. 1973b). In the monkey, decreased absorption of folic acid has been found after 12 months of ethanol feeding with an adequate diet (Romero et al. 1981).

Vitamin B_{12}

The absorption of vitamin B_{12} measured by the Schilling test was shown to be decreased in six of eight well-nourished alcoholics who received ethanol in daily doses ranging from 158–253 g for 11–38 days (Lindenbaum and Lieber 1975). The two subjects in whom vitamin B_{12} absorption remained unaffected received the lowest doses of ethanol, suggesting that the effect may be dose related. The concomitant administration of pancreatic extract, which has been shown to correct vitamin B_{12} absorption in pancreatic insufficiency, to one patient did not prevent the effect of ethanol in decreasing vitamin B_{12} absorption. Studies in rats also show that chronic ethanol administration decreases intestinal uptake of vitamin B_{12} and that this is associated with decreased binding of the vitamin-B_{12}-intrinsic factor complex by intestinal homogenates (Lindenbaum et al. 1973).

Iron

Increased hepatic iron stores are common in alcoholics. Possible causes for this increased iron deposition are ingestion of alcoholic beverages such as wine that have a high iron content and pancreatic insufficiency or cirrhosis, which are associated with increased iron absorption (Deller 1965; Williams et al. 1967). In a recent study, however, the absorption of nonheme ferric chloride was not found to be greater in nonaemic alcoholic patients with fatty liver, alcoholic hepatitis, and cirrhosis than in control subjects (Chapman et al. 1983). The oral or intravenous administration of ethanol to the subjects in this study resulted in small mean percentage increases in ferric chloride absorption, which, however, were not statistically significant. In other studies, the administration of whisky was found either to increase (Charlton et al. 1964) or decrease (Celada et al. 1978) iron absorption.

Calcium

The acute and chronic administration of ethanol results in a decrease in in vitro calcium transport by the everted rat intestine (Krawitt 1973, 1974). The acute effect which was demonstrated with a high oral dose of ethanol (7.5 g/kg) is associated with necrosis of the intestinal epithelium (Krawitt 1974). Similar changes were demonstrated after direct addition of ethanol to the incubation medium bathing the everted gut sacs. No decrease in calcium absorption was demonstrated after intraperitoneal administration of ethanol (3.0 g/kg), which did not alter intestinal morphology. The effect of chronic ethanol feeding in decreasing calcium could not be reversed by the administration of vitamin D or 25-hydroxyvitamin D, suggesting that ethanol interferes with calcium transport by a mechanism independent of vitamin D (Krawitt 1975). Other studies show that pyrazole partially blocked the inhibitory effect of ethanol on calcium transport, and that on a molar basis acetaldehyde is a more potent inhibitor of intestinal calcium transport than ethanol (Carter and Isselbacher 1979).

Fat

Increased fecal fat excretion was a finding in one-third of 42 alcoholics on admission to the hospital (Mezey et al. 1970). It disappeared in all but one patient when they were placed on an adequate diet, whether or not they were also receiving ethanol. The presence of steatorrhea correlates best with transient pancreatic dysfunction, in particular with decreases in pancreatic lipase output (Roggin et al. 1972; Mezey and Potter 1976). In rats, the acute oral administration of ethanol interfered equally with the absorption of triolein and oleic acids, suggesting that ethanol may also impair fat absorption by a mechanism independent of any effect on lipase activity (Barboriak and Meade 1969). However, in another study, ethanol increased the absorption of high loads of oleic acid from the intestinal perfusate, but had no effect on low loads of oleic acid which were being absorbed at maximal rates in the absence of ethanol (Saunders et al. 1982). Deficiencies in fat-soluble vitamins are common in patients with cirrhosis, whether alcoholic or nonalcoholic. A decreased synthesis of bile acids resulting in diminished bile acid pool and bile acid concentration in the intestine is the principal cause of steatorrhea and decreased absorption of fat soluble vitamins in patients with cirrhosis (Vlahcevic et al. 1971).

Effects of Ethanol on Drug Absorption

The absorption of drugs from the gastrointestinal tract occurs by a passive process. A low degree of ionization and a high lipid/water partition coefficient of the nonionized form favour absorption. The low pH of the stomach favours absorption of weak acids, while in the more alkaline intestine bases are absorbed better.

Ethanol placed in the stomach in concentrations ranging from 1% to 10% or given intravenously to maintain blood levels of 150 mg/dl enhanced the absorption of phenobarbital and pentobarbital, which are weak acids, from the rat stomach. This increased absorption was attributed to an effect of ethanol in enhancing the blood supply to the gastric mucosa, demonstrated indirectly by means of the excretion of neutral red into the stomach after its intravenous administration (Magnussen 1968).

Concentrations of ethanol higher than 10% depressed absorption of the above barbiturates, probably due to the mucosal damage known to be produced. These observations are probably of little pharmacological significance because ethanol had no effect on the intestinal absorption of the barbiturates, which accounts for the major part of drug absorption.

Ethanol increased the intestinal transport of estradiol in isolated sacs of rat duodenum, but decreased the transport of estrone sulfate (Martins and Dada 1982). Also in the rat, ethanol in a concentration of 5% increased theophylline absorption by intestinal loops, while higher concentrations of 20% inhibited theophylline absorption. The increased absorption correlated with increased intestinal net water flux (Koysooko and Leevy 1974). Indeed, increased theophylline absorption could also be demonstrated on exposure of the intestine to compounds such as glycerin and propylene glycol which increase net water flux (Houston and Levy 1975). In man, Schluger et al. (1957) demonstrated a greater absorption of orally administered theophylline when mixed with 20% (w/v) ethanol than when given in the form of choline theophyllinate (Choledyl) or theophylline ethylenediamine (aminophylline). Also in man, administration of powdered diazepam mixed with 30 ml 50% ethanol resulted in higher plasma levels of diazepam than when it was given in water, suggesting that ethanol might enhance the absorption of this drug (Hayes et al. 1977). However, this observation is of doubtful clinical significance because it is the extent of absorption, which was not determined, rather than the rate of absorption which is important. Furthermore, ethanol had no effect on the absorption of diazepam when the latter was given in capsule form, although it appeared to increase the initial rate of chlordiazepoxide absorption (Linnoila et al. 1974). Claims that ethanol enhances the absorption of other drugs are not well documented.

Effects of ethanol on gastric emptying may affect rates of drug absorption. Concentrations of ethanol of 6% or less either accelerate or have no effect on gastric emptying. Concentrations greater than 10% delay gastric emptying (Cooke 1972). The most likely mechanism for the delay in gastric emptying is hyperosmolarity of the ethanol solution ingested; however, the effect of ethanol may also be mediated by hormonal or central nervous system mechanisms. In the rat, ethanol given orally as a 30% solution inhibited isoniazid absorption (Thomas and Solomonraj 1977). In man, ethanol in a concentration of 40% given 30 min before or in conjunction with oxazepam resulted in a delay in the absorption of oxazepam but not in the peak levels of oxazepam reached in the blood and total amount of drug absorbed (Mallach et al. 1975).

References

Angel F, Sava P, Crenner F, Lambert A, Grenier JF (1980) Modifications de la motricité intestinale après injection intraveineuse d'alcool. C R Soc Biol (Paris) 174: 192–198

Baraona E, Lieber CS (1975) Intestinal lymph formation and fat absorption: stimulation by acute ethanol administration and inhibition by chronic ethanol feeding. Gastroenterology 68: 495–502

Baraona E, Pirola RC, Lieber CS (1974) Small intestinal damage and changes in cell population produced by ethanol ingestion in the rat. Gastroenterology 66: 226–234

Baraona E, Pirola RC, Lieber CS (1975) Acute and chronic effects of ethanol on intestinal lipid metabolism. Biochim Biophys Acta 388: 19–28
Barboriak JJ, Meade RC (1969) Impairment of gastrointestinal processing of fat and protein by ethanol in rats. J Nutr 98: 373–378
Berenson MM, Avner DL (1981) Alcohol inhibition of rectosigmoid motility in humans. Digestion 22: 210–215
Bianchi A, Chipman DW, Dreskin A, Rosensweig N (1970) Nutritional folic acid deficiency with megaloblastic change in the small bowel epithelium. N Engl J Med 282: 859–861
Bode JCh, Heidelbach R, Bode Ch, Mannheim W, Dürr HK, Martini GA (1980) Bacterial microflora in the jejunum of chronic alcoholics. In: Stock C, Sarles H (eds) Alcohol and the gastrointestinal tract. INSERM, Paris pp 451–456
Bode JCh, Knüppel H, Schwerk W, Bode Ch (1982a) Activities of cytoplasmic, mitochondrial and brush border enzymes in jejunal mucosa of chronic alcoholics. Z Gastroenterol 4: 228–233
Bode JCh, Knüppel H, Schwerk W, Lorenz-Meyer H, Dürr HK (1982b) Quantitative histomorphometric study of the jejunal mucosa in chronic alcoholics. Digestion 23: 265–270
Breslow NE, Enstrom JE (1974) Geographic correlations between cancer mortality rates and alcohol-tobacco consumption in the United States. JNCI 53: 631–639
Broitman SA, Small MD, Vitale JJ, Zamchek N (1961) Intestinal absorption and urinary excretion of xylose in rats fed reduced protein and thyrosine or alcohol 41: 24–28
Broitman SA, Gottlieb LS, Vitale JJ (1976) Augmentation of ethanol absorption by mono- and disaccharides. Gastroenterology 70: 1101–1107
Buell MG, Beck IT (1983) Effect of ethanol on jejunal regional blood flow in the rabbit. Gastroenterology 84: 81–89
Carter EA, Isselbacher KJ (1973) Effect of ethanol on intestinal adenosine triphosphate (ATP) content. Proc Soc Exp Biol Med 142: 1171–1173
Carter EA, Isselbacher KJ (1974) The metabolism of ethanol to carbon dioxide by stomach and small intestinal slices. Proc Soc Exp Biol Med 138: 817–819
Carter EA, Isselbacher KJ (1979) Acetaldehyde inhibits calcium transport by rat small intestine. Fed Proc 38: 1239
Carter EA, Drummey GD, Isselbacher KJ (1971) Ethanol stimulates triglyceride synthesis in the intestine. Science 174: 1245–1247
Celada A, Rudolf H, Donath A (1978) Effect of a single ingestion of alcohol on iron absorption. Am J Hematol 5: 225–237
Chang T, Lewis J, Glazko AJ (1967) Effect of ethanol and other alcohols on the transport of amino acids and glucose by everted sacs of rat small intestine. Biochim Biophys Acta 135: 1000–1007
Chapman RW, Morgan MY, Boss AM, Sherlock S (1983) Acute and chronic effects of alcohol on iron absorption. Dig Dis Sci 28: 321–327
Charlton RW, Jacobs P, Seftel H, Bothwell TH (1964) Effect of alcohol on iron absorption. Br Med J 2: 1427–1429
Cooke AR (1972) Ethanol and gastric function. Gastroenterology 62: 501–502
Deller DJ (1965) Iron^{59} absorption measurements by whole body counting: studies in alcoholic cirrhosis, hemochromatosis and pancreatitis. Am J Dig Dis 10: 249–258
Dinda PK, Beck IT (1977) On the mechanism of the inhibitory effect of ethanol on intestinal glucose and water absorption. Am J Dig Dis 22: 529–533
Dinda PK, Beck IT (1981) Ethanol-induced inhibition of glucose transport across the isolated brush-border membrane of hamster jejunum. Dig Dis Sci 26: 23–32
Dinda PK, Beck IT, Beck M, McElligott TF (1975) Effect of ethanol on sodium-dependent glucose transport in the small intestine of the hamster. Gastroenterology 68: 1517–1526
Dinda PK, Hurst RO, Beck IT (1979) Effect of ethanol on disaccharidases of hamster jejunal brush border membrane. Am J Physiol 237: E68–E76
Eloy R, Battinger F, Bignon JY, Ananna A, Grenier JF (1979) Intestinal brush border enzymes and chronic alcohol ingestion. Res Exp Med 175: 257–269
Fox JE, Bourdages R, Beck IT (1978) Effect of ethanol on glucose and water absorption in hamster jejunum in vivo. Am J Dig Dis 23: 193–200
Gottfried EB, Korsten MA, Lieber CS (1978) Alcohol-induced gastric and duodenal lesions in man. Am J Gastroenterol 70: 587–592

Gray JP, Hoyumpa AM, Dunn GD, Henderson GI, Wilson FA, Swift LL (1980) Effect of ethanol on fluidity of enterocyte membranes. In: Stock C, Sarles H (eds) Alcohol and the gastrointestinal tract. INSERM, Paris, pp 469–476
Green RS, McDermid RG, Scheig RL, Hajjar JJ (1981) Effect of ethanol on amino acid absorption across in vivo rat intestine. Am J Physiol 241: G176–G181
Greene HL, Herman RH, Kraemer S (1971) Stimulation of jejunal adenyl cyclase by ethanol. J Lab Clin Med 78: 336–342
Greene HL, Stifel FB, Herman RH, Herman YF, Rosensweig NS (1975) Ethanol-induced inhibition of human intestinal enzyme activities; reversal by folic acid. Gastroenterology 67: 434 to 440
Hajjar JJ, Tomicic T, Scheig RL (1981a) Effect of chronic ethanol consumption on leucine absorption in the rat small intestine. Digestion 22: 170–186
Hajjar JJ, Murphy D, Tomicic T, Scheig R (1981b) Effect of amino acids and dipeptides on ethanol absorption across the rat intestine. Pharmacology 23: 177–184
Halsted CH, Griggs RC, Harris JW (1967) The effect of alcoholism on the absorption of folic acid (H^3PGA) evaluated by plasma levels and urine excretion. J Lab Clin Med 69: 116–131
Halsted CH, Robles EA, Mezey E (1971) Decreased jejunal uptake of labelled folic acid (^3HPGA) in alcoholic patients: roles of alcohol and nutrition. N Engl J Med 285: 701–706
Halsted CH, Robles EA, Mezey E (1973a) Distribution of ethanol in the human gastrointestinal tract. Am J Clin Nutr 26: 831–834
Halsted CH, Robles EA, Mezey E (1973b) Intestinal malabsorption in folate deficient alcoholics. Gastroenterology 64: 526–532
Hayes SL, Pablo G, Rodomski T, Palmer RF (1977) Ethanol and oral diazepam absorption. N Engl J Med 296: 186–189
Hermos JA, Adams WH, Liu YK, Trier JS (1972) Mucosa of the small intestine in folate deficient alcoholics. Ann Intern Med 76: 957–965
Houston JB, Levy G (1975) Effect of various alcohols on intestinal net water flux and theophylline absorption in rats. J Pharm Sci 64: 607–610
Howard L, Wagner C, Schenker S (1974) Malabsorption of thiamine in folate deficient rats. J Nutr 104: 1024–1032
Hoyumpa Jr, AM, Breen KJ, Schenker S, Wilson FA (1975) Thiamine transport across the rat intestine. Effect of ethanol. J Lab Clin Med 86: 803–816
Hoyumpa Jr, AM, Nichols SG, Wilson FA, Schenker S (1977) Effect of ethanol administration on intestinal (Na,K) ATPase and intestinal thiamine transport in rats. J Lab Clin Med 90: 1086–1095
Hoyumpa Jr, AM, Patwardham R, Antonson D, Nichols S, Gray JP (1981) Effect of thiamine deficiency and acute ethanol ingestion on jejunal glucose transport in rats. Am J Clin Nutr 34: 14–19
Ishii H, Yasuraoka S, Shigeta Y, Takagi S, Kamiya T, Okuno F, Miyamoto K, Tsuchiya M (1978) Hepatic and intestinal gamma-glutamyltranspeptidase activity: its activation by chronic ethanol administration. Life Sci 23: 1393–1398
Israel Y, Salazar I, Rosenmann E (1968) Inhibitory effects of alcohol on intestinal amino acid transport in vivo and in vitro. J Nutr 96: 499–504
Israel Y, Valenzuela JE, Salazar I, Ugarte G (1969) Alcohol and amino acid transport in the human small intestine. J Nutr 98: 222–224
Jeejeebhoy KN (1964) Endogenous protein loss into the bowel as a cause of hypoalbuminemia. In: Munro HN (ed) The role of the gastrointestinal tract in protein metabolism. Davis, Philadelphia, pp 357–383
Joly JG, Hetu C (1980) Induction of intestinal mucosa mixed function oxidases. In: Stock CH, Sarles H (eds) Alcohol and the gastrointestinal tract. INSERM, Paris, pp 137–142
Klipstein FA, Lipton SD, Schenk EA (1973) Folate deficiency of the intestinal mucosa. Am J Clin Nutr 26: 728–737
Koivusaari U, Norling A, Lang M, Hietanen E (1981) Structural and biotransformational membrane changes in the liver and intestine during chronic ethanol administration. Toxicology 20: 173–183
Koysooko R, Leevy G (1974) Effect of ethanol on intestinal absorption of theophylline. J Pharm Sci 63: 829–834

Krasner N, Carmichel HA, Russell RI, Thompson GG, Cochran KM (1976a) Alcohol and absorption from the small intestins 2. Effect of ethanol on ATP and ATPase activity in guinea-pig jejunum. Gut 17: 249-251

Krasner N, Cochran KM, Russell RI, Carmichel HA, Thompson GG (1976b) Alcohol and absorption from the small intestine. 1. Impairment of absorption from the small intestine in alcoholics. Gut 17: 245-248

Krawitt EL (1973) Ethanol inhibits intestinal calcium transport in rats. Nature 243: 88-89

Krawitt EL (1974) Effect of acute ethanol administration on duodenal calcium transport. Proc Soc Exp Biol Med 146: 406-408

Krawitt EL (1975) Effect of ethanol ingestion on duodenal calcium transport. J Lab Clin Med 85: 665-671

Krebs HA, Perkins JR (1970) The physiological role of liver alcohol dehydrogenase. Biochem J 118: 635-644

Kuo YJ, Shanbour LL (1978) Effects of ethanol on sodium, 3-*O*-methylglucose, and L-alanine transport in the jejunum. Am J Dig Dis 23: 51-56

Lal D, Gorbach SL, Levitan R (1972) Intestinal microflora in patients with alcoholic cirrhosis: urea splitting bacteria and neomycin resistance. Gastroenterology 62: 275-279

Lamboeuf Y, DeSaint Blanquat G, Derache R (1975) Disulfirame et motilité digestive: incidence sur l'absorption de l'ethanol chez le rat. Arch Int Pharmacodyn Ther 213: 163-169

Lamboeuf Y, De Saint Blanquat G, DeRache R (1981) Mucosal alcohol dehydrogenase- and aldehyde dehydrogenase-mediated ethanol oxidation in the digestive tract of the rat. Biochem Pharmacol 30: 542-545

Levitt DD, Doizaki W, Levine AS (1981) Influence of ethanol on colonic bacterial metabolism. Gastroenterology 80: 1339

Lindenbaum J, Lieber CS (1975) Effects of chronic ethanol administration on intestinal absorption in man in the absence of nutritional deficiency. Ann NY Acad Sci 252: 228-234

Lindenbaum J, Saha JR, Shea N, Lieber CS (1973) Mechanism of alcohol-induced malabsorption of vitamin B_{12}. Gastroenterology 64: 702

Linnoila M, Otterström S, Anttila M (1974) Serum chlordiazepoxide, diazepam and thioridazine concentrations after the simultaneous ingestion of alcohol or placebo drink. Ann Clin Res 6: 4-6

Madzarovora-Nohejlova J (1971) Intestinal disaccharidase activity in adults and chronic Pilsen beer drinkers. Biol Gastroenterol 4: 325-332

Magnussen MP (1968) The effects of ethanol on the gastrointestinal absorption of drugs in the rat. Acta Pharmacol Toxicol 26: 130-144

Mallach HJ, Moosmayer A, Gottwald K, Staak M (1975) Pharmakokinetische Untersuchungen über Resorption und Ausscheidung von Oxazepam in Kombination mit Alkohol. Arzneimittelforsch 25: 1840-1845

Martini GA, Phear EA, Ruebner B, Sherlock S (1956) The bacterial content of the small intestine in normal and cirrhotic subjects: relation to methionine toxicity. Clin Sci 16: 35-51

Martins OO, Dada OA (1982) The effect of alcohol on the in vitro absorption of estradiol and estrone sulphate in rat intestine. J Steroid Biochem 16: 265-268

McDonald JT, Margen S (1976) Wine versus ethanol in human nutrition. 1. Nitrogen and calorie balance. Am J Clin Nutr 29: 1093-1103

Mekhjian HS, May ES (1977) Acute and chronic effects of ethanol on fluid transport in the human small intestine. Gastroenterology 72: 1280-1286

Meyer F (1977) Relations alimentation-cancer en France. Gastroenterol Clin Biol 1: 971-982

Mezey E (1975) Intestinal function in chronic alcoholism. Ann NY Acad Sci 252: 215-227

Mezey E, Halsted CH (1979) Effects of alcohol on gastrointestinal and pancreatic function in alcoholics. In: Gastineau CF, Darby W, Turner TB (eds) Fermented food beverages in nutrition. Academic, New York, pp 277-302

Mezey E, Potter JJ (1976) Changes in exocrine pancreatic function produced by altered dietary protein intake in drinking alcoholics. Johns Hopkins Med J 138: 7-12

Mezey E, Jow E, Slavin RE, Tobon F (1970) Pancreatic function and intestinal absorption in chronic alcoholism. Gastroenterology 59: 657-664

Middleton WRJ, Carter EA, Drummey GD, Isselbacher KJ (1971) Effect of oral ethanol administration on intestinal cholesterogenesis in the rat. Gastroenterology 60: 880-887

Millan MS, Morris GP, Beck IT, Henson JT (1980) Villous damage induced by suction biopsy and by acute ethanol intake in normal human small intestine. Dig Dis Sci 25: 513–525
Mistilis SP, Garske A (1969) Induction of alcohol dehydrogenase in liver and gastrointestinal tract. Aust Ann Med 18: 227–231
Mistilis SP, Ockner RK (1972) Effect of ethanol on endogenous lipid and lipoproteins metabolism in small intestine. J Lab Clin Med 80: 34–46
Perlow W, Baraona E, Lieber CS (1977) Symptomatic intestinal disaccharidase deficiency in alcoholics. Gastroenterology 77: 680–684
Pösö H (1980) Inhibition of ornithine decarboxylase in the stomach and small intestine of rat by a single dose of ethanol. Med Biol 58: 117–122
Raul F, Doffoel M, Marescaux J, Bockel R, Grenier JF (1982) Effects of prolonged alcohol administration and a high carbohydrate-low protein diet on the activities of the jejunal brush border enzymes in the rat. Gut 23: 962–967
Robinson GM, Orrego H, Israel Y, Devenyi P, Kapur BM (1981) Low-molecular-weight polyethylene glycol as a probe of gastrointestinal permeability after alcohol ingestion. Dig Dis Sci 26: 971–977
Robles EA, Mezey E, Halsted CH, Schuster MM (1974) Effect of ethanol on motility of the small intestine. Johns Hopkins Med J 135: 17–24
Rodgers JB, O'Brien BA (1975) The effect of acute ethanol treatment on lipid-reesterifying enzymes in the rat small bowel. Am J Dig Dis 20: 354–358
Rodrigo C, Antezana C, Baraona E (1971) Fat and nitrogen balances in rats with alcohol-induced fatty liver. J Nutr 101: 1307–1310
Roggin GM, Iber FL, Kater RMH, Tobon F (1969) Malabsorption in the chronic alcoholic. Johns Hopkins Med J 125: 321–330
Roggin GM, Iber FL, Linscheer WG (1972) Intraluminal fat digestion in the chronic alcoholic. Gut 13: 107–111
Romero JJ, Tamura T, Halsted CH (1981) Intestinal absorption of (^{3}H) folic acid in the chronic alcoholic monkey. Gastroenterology 80: 99–102
Rosensweig NS, Herman RH, Stifel FB (1971) Dietary regulation of small intestinal enzyme activity in man. Am J Clin Nutr 24: 65–69
Rossi MA, Zucoloto S (1977) Effect of chronic ethanol ingestion on the small intestinal ultrastructure in rats. Beitr Pathol 161: 50–61
Rubin E, Ryback BJ, Lindenbaum J, Gerson CD, Walker G, Lieber CS (1972) Ultrastructural changes in the small intestine induced by ethanol. Gastroenterology 63: 801–814
Saunders DR, Sillery J, McDonald GB (1982) Effect of ethanol on transport from rat intestine during high and low rates of oleate absorption. Lipids 17: 356–360
Schluger J, McGinn JT, Hennessy DJ (1957) Comparative theophylline blood levels following oral administration of three different theophylline preparations. Am J Med Sci 233: 296–302
Seitz HK, Korsten MA, Lieber CS (1979) Ethanol oxidation by intestinal microsomes: increased activity after chronic ethanol administration. Life Sci 25: 1443–1448
Seitz HK, Garro AJ, Lieber CS (1980) Enhanced microsomal activation of procarcinogens in the small intestine after chronic ethanol consumption. In: Stock C, Sarles H (eds) Alcohol and the gastrointestinal tract. INSERM, Paris, pp 503–506
Seitz H, Czygan P, Kommerell B (1982) Stimulation of thymidine incorporation in isolated rat intestinal mucosal cells by feeding an ethanol-containing liquid diet. Digestion 23: 65–67
Siegers CP, Strubelt O, Back G (1972) Inhibition by caffeine of ethanol absorption in rats. Eur J Pharmacol 20: 181–187
Siregar H, Chou CC (1982) Relative contribution of fat, protein, carbohydrate, and ethanol to intestinal hyperemia. Am J Physiol 242: G27–G31
Spencer RP, Brody KR, Lutters BM (1964) Some effects of ethanol on the gastrointestinal tract. Am J Dig Dis 9: 599–604
Tabagchali S (1970) The pathophysiological role of small intestinal bacterial flora. Scand J Gastroenterol [Suppl] 6: 139–63
Thomas BH, Salomonraj G (1977) Drug interactions with isoniazid metabolism in rats. J Pharm Sci 66: 1322–1326
Thomson AD, Baker H, Leevy CM (1970) Patterns of ^{35}S-thiamine hydrochloride absorption in the malnourished alcoholic patient. J Lab Clin Med 76: 36–45

Tomasulo PA, Kater RMH, Iber FL (1968) Impairment of thiamine absorption in alcoholism. Am J Clin Nutr 21: 1340–1344
Vlahcevic ZR, Buhac I, Farrar JT Bell C Jr, Swell L (1971) Bile acid metabolism in patients with cirrhosis. 1. Kinetic aspects of cholic acid metabolism. Gastroenterology 60: 491–498
Williams R, Williams HS, Scheuer PJ, Pitcher CS, Loiseau E, Sherlock S (1967) Iron absorption and siderosis in chronic liver disease. Q J Med 36: 151–166
Worthington BS, Meserole L, Syrotuck JA (1978) Effect of daily ethanol ingestion on intestinal permeability to macromolecules. Dig Dis Sci 23: 23–32
Zucoloto S, Rossi MA (1979) Effect of chronic ethanol consumption on mucosal morphology and mitotic index in the rat small intestine. Digestion 19: 277–283

20 Esophageal and Gastric Lesions in the Alcoholic*

M. Wienbeck and W. Berges**

Medizinische Klinik und Poliklinik der Universität Düsseldorf, Moorenstrasse 5,
4000 Düsseldorf 1, Federal Republic of Germany

The oropharynx, esophagus, and stomach are exposed to alcohol immediately after its ingestion, and alcoholic beverages are almost undiluted when they come into contact with the mucosa of these organs. It is therefore not surprising that lesions occur quite frequently in people drinking alcohol. But not all alterations that occur are due to the local action of ethanol. Systemic actions and deficiency states associated with alcoholism may be as important. Because of the different pathomechanisms it is appropriate to distinguish between acute and chronic effects of alcohol in the upper gut, and further between alcohol-associated lesions.

Lesions in the Esophagus

Acute Effects of Alcohol

Mucosal Lesions

Alcohol may damage the esophageal mucosa by an enhanced penetration of cytotoxic agents like H^+ ions and perhaps bile acids into the mucosal cells (Chung et al. 1977). The effect is dose dependent. Increasing doses of ethanol in the diet have been correlated positively with histologic findings of esophagitis (Shirazi and Platz 1978), and in a recent study 40% (v/v) ethanol damaged the mucosal integrity of the rabbit esophagus quite significantly, whereas 20% ethanol was rather harmless (Salo 1983). HCl potentiated this effect. In the clinical situation, however, acute corrosive damage by alcohol in the esophagus appears to be rather rare, probably due to the rapid transit of liquids through the gullet.

Motor Effects

Acute application of ethanol transiently decreases the pressure of the lower esophageal sphincter (Hogan et al. 1972) and its reaction to pentagastrin and protein

* Part of the work reported in the text was supported by a grant of the *Deutsche Forschungsgemeinschaft* (Wi 285)
** The authors gratefully acknowledge the invaluable secretarial help of Ms Beate Herrmann and the skillful technical assistance of Ms Ulrike Altaparmakov

Alcohol Related Diseases in Gastroenterology
Edited by H. K. Seitz and B. Kommerell

stimulation (Mayer et al. 1978). Alcohol also reduces the incidence of primary peristalsis in the distal esophagus (Hogan et al. 1972) and thus impairs esophageal clearance (Kjellen and Tibbling 1978). Since intravenous application of ethanol also disturbs esophageal motility (Hogan et al. 1972), at least part of the alcohol effect is likely to be due to a systemic action.

The lower esophageal sphincter and esophageal peristalsis are thought to be the most important mechanisms protecting the esophagus from damage by gastroesophageal reflux. Therefore, the interference of alcohol with both mechanisms is likely to promote gastroesophageal reflux. Indeed, ingestion of 180 ml 50% ethanol together with a standard meal increased the occurrence of acid reflux episodes in the distal esophagus in 11 out of 12 healthy volunteers (Kaufmann and Kaye 1978).

Saliva Secretion

Within 1 h after acute exposure to alcohol, saliva secretion is reduced by half and its viscosity is increased (Kissin and Kaley 1974). It is conceivable that these changes reduce the buffering capacity and the flushing effect of saliva in the esophagus. Nothing is known, however, about ethanol actions on the function of the mucosal glands in the esophagus itself.

Chronic Effects of Alcohol

Motor Effects

Esophageal motility may be disturbed in chronic alcoholics with alcoholic neuropathy (Winship et al. 1968). In these patients the percentage of peristaltic contractions is decreased, and the strength of the contractions is diminished. This deterioration of esophageal peristalsis may prolong the contact of peptic material with esophageal mucosa and thus contribute to damage by gastroesophageal reflux.

Saliva Secretion

Saliva is an important determinant of esophageal acid clearance (Helm et al. 1983). It is also known that about 12% of chronic alcoholics develop an enlarged parotid gland and that this percentage increases to 50% in patients with alcoholic cirrhosis of the liver (Bode 1980). In a recent study of 24 patients with a history of long-term ethanol consumption of more than 80 g daily, protein output of the parotid gland was reduced, but the total secretory volume was not altered (unpublished results) (Fig. 1). The reduction in bicarbonate output was not significant in comparison to control. The meaning of these observations with to esophageal pathology has yet to be determined.

Esophagitis

Although the previously described disturbances in the esophagus of alcoholics may contribute to the development of esophagitis, there are only a few epidemiological

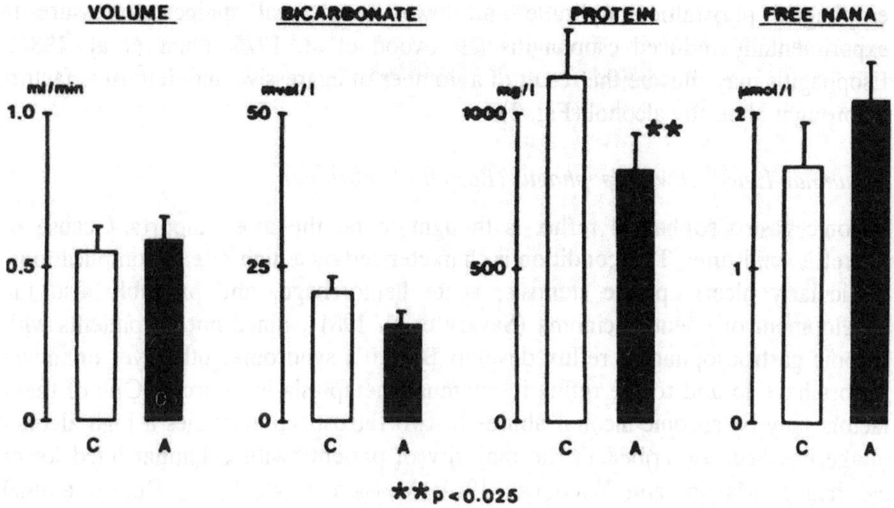

Fig. 1. Output of the parotid gland in 24 alcoholics *(A)* and 14 controls *(C)*. Data are expressed as means + SD. *NANA*, N-acetylneuraminic acid

data available demonstrating a relationship between alcohol consumption and esophagitis. In a follow-up study, medical therapy of esophagitis was more effective in patients abstaining from alcohol than in others (Bucher et al. 1978). It is also common clinical experience that in some alcoholics even severe forms of esophagitis heal within a short time if alcohol is withdrawn. But the evidence that alcohol restriction, in general, is an important factor in the medical treatment of reflux esophagitis is still lacking.

Once established, reflux esophagitis may perpetuate itself (Nier et al. 1983). This hypothesis has been supported by animal studies demonstrating decreases in

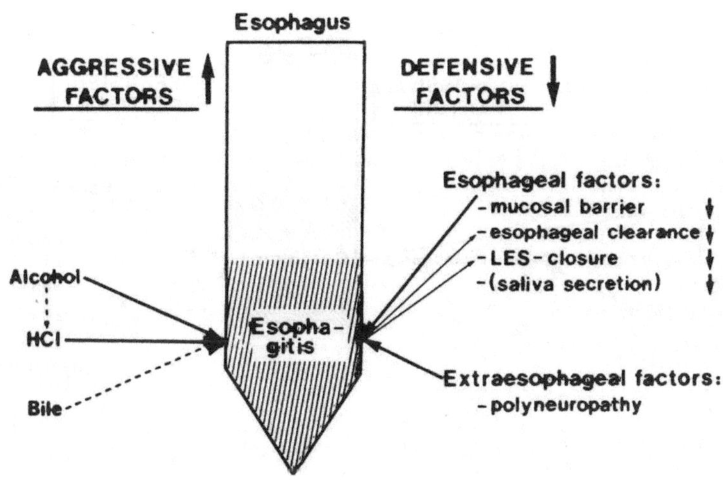

Fig. 2. Possible pathogenetic factors in the development of esophagitis in the alcoholic. *LES*, lower esophageal sphincter

esophageal peristaltic amplitude and lower esophageal sphincter pressure in experimentally induced esophagitis (Eastwood et al. 1975; Sinar et al. 1981). Esophagitis may thus be the result of a number of aggressive and defensive factors all brought about by alcohol (Fig. 2).

Columnar-Lined Lower Esophagus (Barrett's Syndrome)

Chronic gastroesophageal reflux is thought to be the most important cause of Barrett's syndrome. This condition is characterized by a high rate of complications, particularly ulcers, peptic stenosis, acute hemorrhage, and probably also the development of adenocarcinoma (Savary et al. 1981). Since not all patients with chronic gastroesophageal reflux develop Barrett's syndrome, other yet unknown factors have to add to the reflux if columnar metaplasia is to occur. One of these factors may be chronic alcohol abuse. In two retrospective studies a high alcohol intake has been described in the majority of patients with columnar-lined lower esophagus (Martini and Wienbeck 1974; Messian et al. 1978). Recent animal experiments, however, failed to show a major effect of ethanol administration, 2.5 g daily for 6 weeks, on the regeneration of squamous epithelium in the rat forestomach (Bieg et al. 1983).

Carcinoma

Many epidemiological studies support the hypothesis that chronic alcohol consumption markedly contributes to the development of esophageal cancer (Hinds et al. 1980; Tuyns 1979). A look at the geographical distribution of the incidence rates of esophageal cancer in France and the high male to female ratio suggests that esophageal cancer is related to ethanol intake (Audigier et al. 1975). This correlation has been confirmed by prospective studies (Martinez 1969; Tuyns 1979). A contributing role of other geographical factors has largely been ruled out, at least for Europe and North America (Enstrom 1975; Kamionkowski and Fleshler 1965).

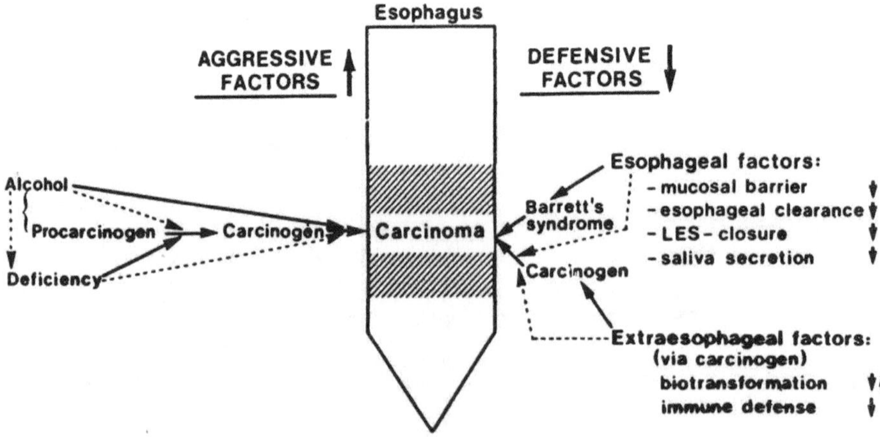

Fig. 3. Possible pathogenetic factors in the development of esophageal cancer in the alcoholic. *LES*, lower esophageal sphincter

Alcohol by itself does not induce cancer, but it may play a permissive role. Multiple actions of ethanol may add to each other and eventually lead to malignancy in the esophagus (Fig. 3). Ethanol has been shown to enhance the action of dimethylbenzanthracene in the production of oral tumors (Lowenfels 1979) and to facilitate the carcinogen-induced alkylation of DNA in the esophagus (Kouros et al. 1983). Alcohol alters the fluidity and composition of cell membranes (Freund 1979), and it increases the permeability of the esophageal mucosa (Salo 1983). It is possible, therefore, that ethanol exerts its carcinogenic effect chiefly by facilitating contact between extrinsic carcinogenic chemicals, such as tar products from smoking, and the contents of the stem cells that are responsible for the integrity of the lining in the human upper digestive tract (Doll and Peto 1981).

Usually, alcoholics are heavy smokers. Alcohol and tobacco appear to act synergistically in the pathogenesis of esophageal cancer (Pottern et al. 1981). It is thus not surprising that the risk of developing carcinoma of the esophagus increases by a factor of 18 in alcoholics drinking more than 80 g ethanol per day and by a factor of 44 if this group, in addition, consumes more than 20 g tobacco daily (Tuyns et al. 1977) (Fig. 4).

In central Africa and other areas of the world, striking differences in the incidence of esophageal cancer have been observed despite a similar level of alcohol consumption (McLashan 1969; Mettlin et al. 1981). This fact has drawn attention to other substances which may contaminate alcoholic beverages, e.g., nitrosamines, benzo(a)pyrene, fusel oil, and other potential cocarcinogens (Gibel et al. 1968;

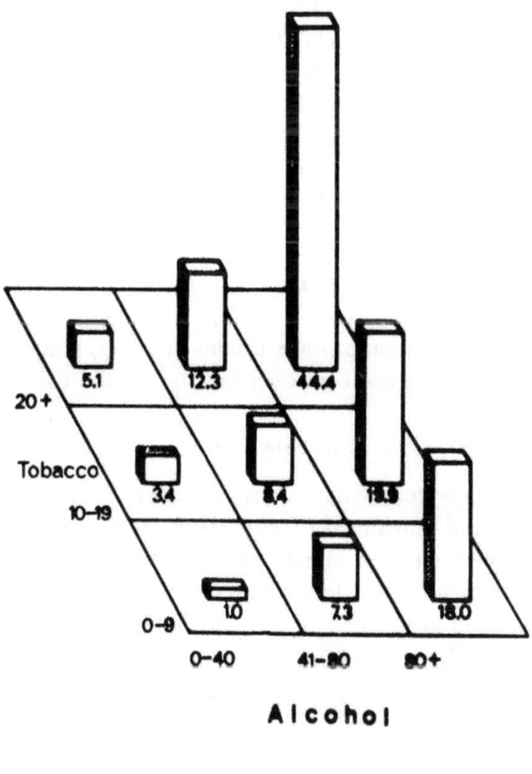

Fig. 4. Risk of developing esophageal cancer as a function of alcohol and tobacco consumption (in grams per day). (Modified from Tuyns et al. 1977)

Lijinski and Epstein 1970; Tuyns 1979). In animal experiments nitrosamines are potent carcinogens, but their role in the induction of carcinoma in man is still controversial.

Alcoholism also may cause a deficiency of substances which are thought to protect against the development of cancer (Martinez 1969; Tuyns 1979). A lack of riboflavin causes neoplasia in mice, probably due to a metabolic activation of carcinogens (Wynder and Chan 1979). Carotene and vitamin A appear to protect against cancer (Willett and McMahon 1984); vitamin-A-deficient hamsters exhibit an enhanced binding of benzopyrene to DNA (Genta et al. 1974). Deficiency states of folic acid, iron, and some minerals, e.g., zinc, have also been discussed with respect to carcinogenesis (Lieber et al. 1979; Mellow et al. 1983). However, most of the hypotheses relating a specific type of deficiency to cancer development are based on animal experiments only. Recent observations of low zinc and vitamin A plasma levels in patients with squamous carcinoma of the esophagus (Willett and McMahon 1984; Mellow et al. 1983), of an inverse relationship between intake of the vitamins A and C and esophageal cancer (Mettlin et al. 1981), and of an inverse association of carcinoma of the esophagus with the level of molybdenum in the soil of the farming land (Yang 1980) brought supportive data from man.

Even other mechanisms of alcohol action have to be considered in esophageal carcinogenesis. Ethanol reduces esophageal motility and may thus prolong the exposure of the esophageal mucosa to (pro)carcinogens. Alcohol promotes gastroesophageal reflux (Kaufmann and Kaye 1978), which eventually may result in Barrett's syndrome; this condition is associated with an increased occurrence of esophageal adenocarcinoma (Savary et al. 1981).

Also, alcoholic liver damage may interfere with the biotransformation of toxic substances in the liver (Lieber et al. 1979). Carcinogens, therefore, could become more active, and they may exert their effects for prolonged periods of time. Further, the immune system of the organism is altered by ethanol (Palmer 1978; Schottenfeld 1979). Alcoholics are prone to infections in spite of their increased serum levels of immunoglobulins, but it is still very speculative to connect alcohol-induced changes in the immune defense mechanisms with carcinogenesis.

Alcohol-Associated Lesions

The association of some lesions in the esophagus with alcohol consumption is very loose and more controversial than in the case of the previous conditions.

Mallory-Weiss Syndrome

Mallory-Weiss syndrome is defined as a mucosal laceration at the gastroesophageal mucosal junction brought about by vomiting, retching and, rarely, coughing. Already in their initial report Mallory and Weiss drew attention to the frequent association of the lesion with alcohol abuse (Mallory and Weiss 1929). Later reports note an incidence of recent alcoholism in 56%–73% of the affected persons (Shay and Johnson 1983). The typical sequence of events is alcohol abuse, nausea, vomiting of gastric contents, and finally hematemesis. But the patient's history is not always that clear.

The mucosal injury is located at the zone of the highest pressure gradient between the abdominal and thoracic cavity during vomiting, i.e., at the gastroesophageal junction. The exact location at, above, or below this junction and the spatial orientation of the tear is very variable. Alcohol-induced mucosal alterations are thought to be a potentially contributing factor to the injury.

Boerhaave's Syndrome and Intramural Hematoma

A rupture of the esophagus (Boerhaave's syndrome) and intramural hematoma are much rarer than the Mallory-Weiss syndrome (Shay and Johnson 1983). The etiopathogenesis of these lesions appears to be very similar to the mucosal tears at the gastroesophageal junction, but less is known about the association of the lesions with alcohol abuse. The lower esophageal sphincter, in general, is normally located. But during vomiting pressure gradients may still cause a distension of the esophagus up to four times its normal diameter (Donner 1976).

Idiopathic Intramural Pseudodiverticulosis

This condition, which was first described by Mendl et al. in 1960, is characterized by multiple tiny intramural outpouchings of the esophageal contour. The degree of accompanying inflammation in the adjacent tissue varies to a great extent. The disease becomes symptomatic as soon as secondary stenosis of the esophageal lumen develops. Although the connection was not specifically looked for in most of the previous studies, we propose that alcohol may play a role in the pathogenesis of the disease. We studied four patients, three of whom were alcoholics (Berges et al. 1980). Our patients also exhibited pronounced disturbances of esophageal motility, but we do not know yet whether these disturbances are the cause or the result of the pseudodiverticulosis. The role of alcohol in these motor disorders has also not yet been clarified.

Lesions in the Stomach

In no other organ have so many effects been falsely associated with alcohol intake than in the stomach. Even nowadays many physicians believe that alcohol, in general, stimulates gastric secretion, delays gastric emptying, and alters the frequency and behavior of peptic ulceration. These beliefs are poorly supported by facts.

The stomach can absorb and metabolize alcohol, but more important, the rate of alcohol absorption is influenced by the speed of gastric emptying. Fat delays gastric emptying and thus retards the rise in blood alcohol levels (Salvesen and Kolberg 1958). Alcohol can also delay the emptying of the stomach (Barboriak and Meade 1970). Therefore, the direct linear correlation found between the concentration of alcohol ingested and the amount absorbed from the stomach (Cooke and Birchall 1969) may be a function of both permeability of the gastric mucosa and delayed gastric emptying. In man a significant amount of extrahepatic alcohol dehydrogenase is found in the stomach (Pestallozzi et al. 1983), which explains the fact that alcohol is partly metabolized already in the stomach.

Acute Effects of Alcohol

Gastric Secretion

Alcohol has been used for a long time to stimulate gastric secretion, but the effect may vary depending on the concentration and composition of the alcohol (Lenz et al. 1983). In general, ethanol concentrations below 10%, and particularly wine and beer, are potent stimulants of acid secretion, whereas concentrations above 20% may even be inhibitory (Davenport 1967). There is still controversy with regard to the mechanisms of the ethanol action. Alcohol by itself does not alter serum gastrin concentrations in man, but beer and wine cause a rise which is apparently due to nonalcoholic ingredients of the beverages (Singer et al. 1983). Ethanol consumption significantly increases gastric adenylate cyclase activity (Seitz et al. 1983a). This increase may be of relevance with respect to the increased acid secretion after alcohol consumption. Histamine release has been proposed as another mechanism of the ethanol action (Dinoso et al. 1976a). Although there is abundant evidence that the effect of alcohol on gastric secretion is obtained independently of extrinsic nerves to the stomach, at least some data indicate that alcohol may stimulate gastric secretion via the central nervous system and the vagus nerve (Hirschowitz et al. 1956). This appears to be the predominant mode of action when alcohol is given intravenously. In summary, then, alcohol appears to act through two or more mechanisms when it stimulates gastric secretion.

Alcohol does not only stimulate gastric acid secretion, but it may also inhibit it. This mechanism becomes prominent when isolated mucosal glands are studied (Reichstein et al. 1984). Nonspecific hyperosmotic effects (Sernka and Jackson 1976), an increase in the apparent permeability of paracellular pathways (Dawson and Cooke 1978) and in the back diffusion of hydrogen ions (Davenport 1967), stimulation of the formation of prostaglandins (Puurunen 1978; Robert et al. 1983), and an interference with the cyclic AMP system (Reichstein et al. 1984) have been proposed as mechanisms of action. In persons who chronically abuse alcohol, gastric acid secretion appears to be diminished (Chey 1972) in spite of an increase in parietal cell mass observed in animals (Lillibridge et al. 1973); but the results may vary depending upon the conditions of study (Kuo and Shanbour 1983).

Gastric Motility and Emptying

Alcohol may lower the pressure in the pylorus (Phaosawasdi et al. 1979). In the stomach the motor effects of alcohol vary depending upon the ethanol concentration and accompanying meals (Barboriak and Meade 1970; Cooke 1972). In general, high alcohol concentrations appear to inhibit gastric emptying; this is probably due in part to hyperosmolarity (Iber 1971). The duodenal osmoreceptor, however, is relatively insensitive to ethanol (Kaufmann and Kaye 1979).

Therefore, the relationship between energy density and gastric emptying rate does not hold in the case of ethanol. On the whole, ethanol apparently is not very important in gastric emptying (Moore et al. 1981).

Mucosal Blood Flow

After an initial increase, gastric mucosal blood flow declines as ethanol application proceeds and back diffusion of H^+ ions appears (Augur 1970). These observations are consistent with protein precipitations and other morphological alterations of the capillaries in the gastric mucosa that have been observed after topical ethanol application (Dinoso et al. 1976a).

Mucosal Damage

Mucosal congestion after alcohol abuse has been described already by Beaumont in 1833 (Beaumont 1959). Erythema, mucosal hemorrhage, and eosinophilic infiltration occur within 3 h following alcohol ingestion (Gottfried et al. 1978). These alterations are reversible within 3 days if alcohol consumption is stopped. Ethanol also causes a sharp drop in gastric mucosal potential difference (Caspary 1975). The initial decrease in transmucosal potential difference correlates with the extent of accompanying mucosal lesions (Morris et al. 1984). A disruption of the gastric mucosal barrier is likely to be the cause for the observed morphologically changes in the gastric mucosa (Dawson and Cooke 1978). Mucosal disruption alters not only the morphology of the stomach but also its function (Dinoso et al. 1976b). This is indicated by an increased back diffusion of H^+ ions (Davenport 1967) and by a loss of plasma albumin through the gastric mucosa (Brassine 1979).

Early in the course of the mucosal damage bicarbonate appears in the stomach, which is explained by a diffusion through the leaky mucosa (Dayton et al. 1983) and may be an important mechanism in the repair process (Takeuchi and Okabe 1983). It is common clinical belief that those who take alcohol and then a salicylate are at special risk of gastric hemorrhage. Examination of the available data does not support this concept. On the contrary, ethanol may even protect the gastric mucosa from further damage (Miller and Henagan 1984). This has been explained by an increased synthesis and release of cytoprotective prostaglandins under the influence of alcohol (Robert et al. 1983).

Acute Gastritis

Excessive alcohol ingestion appears to be an important cause of severe hemorrhagic-erosive gastritis (Dagradi et al. 1973) (Fig. 5). Clinically, it presents as or is

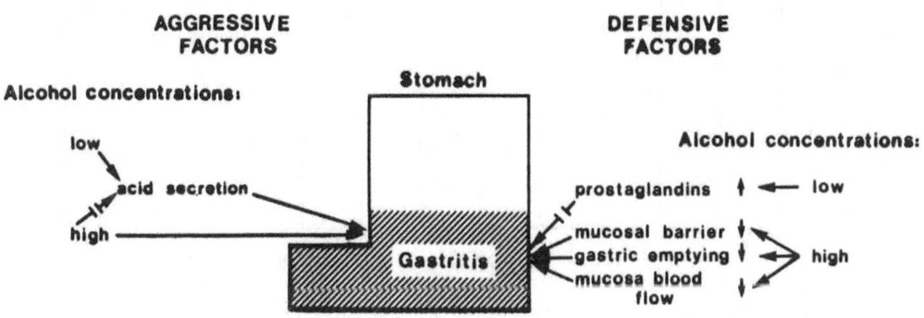

Fig. 5. Possible pathogenetic factors in the development of gastritis in the alcoholic

associated with anorexia, abdominal pain, and at times, hematemesis. Patients with preexisting chronic gastritis are particularly prone to bleeding (Dinoso et al. 1972b). Portal hypertension also predisposes to erosive injury after ethanol (Sarfeh et al. 1983). The alcohol-induced acute gastric lesions may be localized in the antrum and/or body of the stomach (Bode 1980). The type of lesions is indistinguishable in appearance and distribution from acute mucosal erosions induced by drugs or stress. After abstinence all acute lesions are rapidly reversible.

Chronic Lesions

Chronic Gastritis

Chronic feeding of ethanol to dogs leads to chronic gastritis and glandular atrophy (Chey 1972; Gordon et al. 1974). In man the question of chronic gastritis in long-term alcohol abuse was a matter of controversy for a long time. In a more recent study, Roberts (1972) clearly demonstrated an increased incidence of chronic gastritis in alcoholics, but in this study nutritional deficiency cannot be ruled out as an additional cause. When alcoholics were kept on a nutritionally balanced diet, superficial and atrophic gastritis were still more common than in controls (Dinoso et al. 1972a). Circumstantial evidence therefore indicates that excessive long-term alcohol ingestion may cause chronic gastritis.

Peptic Ulcer Disease

In small- and large-scale epidemiological studies no association between alcohol consumption and the prevalence of peptic ulceration was found (Friedman et al. 1974; Würsch et al. 1978). In males the ulcer rate has even been found to decrease with the amount of alcohol consumed (Fig. 6). Data on the frequency of ulcer disease

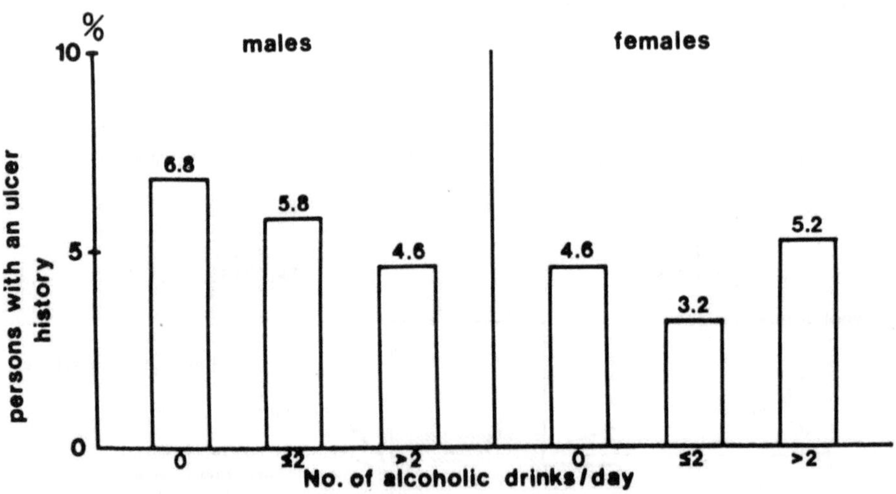

Fig. 6. Prevalence of peptic ulcer in relation to the average alcohol consumption in nonsmokers. (Modified from Friedman et al. 1974)

in chronic alcoholics in comparison to adequate control groups are not available (Bode 1980). It therefore appears that alcohol has no detrimental effect on peptic ulceration. Moderate alcohol intake may even lead to faster duodenal ulcer healing than abstinence (Sonnenberg et al. 1981).

Gastric Cancer

Chronic alcohol intake causes an increase in the parietal cell mass (Lillibridge et al. 1973). In the rat, chronic ethanol ingestion results in a significant increase in DNA synthesis of the gastric mucosa, which is thought to represent a cellular hyperregenerative response to mucosal injury (Seitz et al. 1983b).

For a long time gastric cancer did not appear to be related to alcohol consumption (Wienbeck and Stefenelli 1978). Two recent epidemiological studies, however, indicate that an association may exist. The incidence of carcinoma of the stomach was found to be related to beer consumption in Hawaii (Hinds et al. 1980) and to wine drinking in France (Hoey et al. 1981). But in general the association is rather weak, and alcoholism is much less of a risk factor for the development of gastric than of esophageal cancer.

Alcohol-Associated Problems

Alcohol Absorption After Gastric Resection

Evacuation of the stomach is a major regulator of alcohol absorption, and differences in the speed of gastric emptying might at least partially explain individual variations of the shapes of blood alcohol curves following a definite oral dose of alcohol (Stefenelli et al. 1977). Particular problems may arise after gastrectomy. Following gastric resection the blood alcohol level rises faster and to a higher maximal level than in normal persons (Elmslie et al. 1965; Kürzinger 1970). It is not known yet whether these higher blood levels of alcohol enhance the occurrence of mucosal damage in the gastrointestinal tract.

Constituents of Alcoholic Beverages

Beer and wine ingestion increase the serum gastrin levels and gastric acid secretion to a much greater extent than the corresponding concentrations of pure alcohol (Lenz et al. 1983; Singer et al. 1983). This effect has been ascribed to yet unknown constituents of certain alcoholic beverages. It would be of interest to study these postulated constituents in more detail, since it is conceivable that they may be responsible for some of the other alcohol effects in the stomach, too.

References

Audigier JC, Tuyns AJ, Lambert R (1975) Epidemiology of oesophageal cancer in France. Increasing mortality and persistent correlation with alcoholism. Digestion 13: 209–219

Augur NA (1970) Gastric mucosal blood flow following damage by ethanol, acetic acid, or aspirin. Gastroenterology 58: 311–320

Barboriak JJ, Meade RC (1970) Effect of alcohol on gastric emptying in man. Am J Clin Nutr 23: 1151–1153

Beaumont W (1959) Experiments and observations on the gastric juice and the physiology of digestion. Dover, New York

Berges W, Wienbeck M, Strohmeyer G (1980) Endoskopische Aspekte der intramuralen Pseudodivertikulose. In: Henning H (ed) Fortschritte in der gastroenterologischen Endoskopie XI. Witzstrock, Baden-Baden, pp 83–85

Bieg H, Berges W, Borchard F, Teschke R, Wienbeck M (1983) Alkohol und Epithelregeneration im Rattenmagen nach experimenteller Läsion. Z Gastroenterol 21: 111–115

Bode JCh (1980) Alcohol and the gastrointestinal tract. Ergeb Inn Med Kinderheilkd 45: 1–75

Brassinne A (1979) Effects of ethanol on plasma protein shedding in the human stomach. Am J Dig Dis 24: 44–47

Bucher P, Lepsien G, Sonnenberg A, Blum AL (1978) Verlauf und Prognose der Refluxkrankheit bei konservativer und chirurgischer Behandlung. Schweiz Med Wochenschr 108: 2072–2078

Caspary WF (1975) Einfluß von Aspirin, Antacida, Alkohol und Gallensäuren auf die transmurale elektrische Potentialdifferenz des menschlichen Magens. Dtsch Med Wochenschr 100: 1263–1268

Chey WY (1972) Alcohol and gastric mucosa. Digestion 7: 239–251

Chung RSK, Johnson GM, DenBesten L (1977) Effect of sodium taurocholate and ethanol on hydrogen ion absorption in rabbit esophagus. Am J Dig Dis 22: 582–588

Cooke AR (1972) Ethanol and gastric function. Gastroenterology 62: 501–502

Cooke AR, Birchall A (1969) Absorption of ethanol from the stomach. Gastroenterology 57: 269–272

Dagradi AE, Lee ER, Bosco Dl, Stempien SJ (1973) The clinical spectrum of hemorrhagic erosive gastritis. Am J Gastroenterol 60: 30–46

Davenport HW (1967) Ethanol damage to canine glandular mucosa. Proc Soc Exp Biol Med 126: 657–662

Dawson DC, Cooke AR (1978) Parallel pathways for ion transport across rat gastric mucosa: effect of ethanol. Am J Physiol 235: E 7–19

Dayton MT, Kauffman GL, Schlegel JF, Code CF, Steinbach JH (1983) Gastric bicarbonate appearance with ethanol ingestion. Mechanism and significance. Dig Dis Sci 28: 449–455

Dinoso VP Jr, Chey WY, Braverman SP, Rosen AP, Ottenberg D, Lorber SH (1972a) Gastric secretion and gastric mucosal morphology in chronic alcoholics. Arch Intern Med 130: 715–719

Dinoso VP Jr, Meshkinpour H, Lorber SH (1972b) Studies on the acute effects of ethanol on the gastric mucosa in man. Gastroenterology 62: 843

Dinoso VP Jr, Chuang J, Murthy SNS (1976a) Changes in mucosal and venous histamine concentrations during instillation of ethanol in the canine stomach. Am J Dig Dis 21: 93–97

Dinoso VP Jr, Ming SC, McNiff J (1976b) Ultrastructural changes of the canine gastric mucosa after topical application of graded concentrations of ethanol. Am J Dig Dis 21: 626–632

Doll R, Peto R (1981) The causes of cancer: quantitative estimates of avoidable risks of cancer in the United States today. JNCI 66: 1191–1308

Donner MW (1976) Spezielle Röntgendiagnostik von Funktionsstörungen der Speiseröhre. In: Siewert R, Blum AL, Waldeck F (eds) Funktionsstörungen der Speiseröhre. Springer, Berlin Heidelberg New York, pp 80–100

Eastwood GL, Castell DO, Higgs RH (1975) Experimental esophagitis in cats impairs lower esophageal sphincter pressure. Gastroenterology 69: 146–153

Elmslie RG, Davis RA, Magee DF, White TT (1965) Absorption of alcohol after gastrectomy. Surg Gynecol Obstet 119: 1256–1258

Enstrom JE (1975) Cancer mortality among Normans. Cancer 36: 825–841

Freund G (1979) Possible relationship of alcohol in membranes to cancer. Cancer Res 39: 2899–2901

Friedman GD, Siegelaub AS, Seltzer C (1974) Cigarettes, alcohol, coffee and petic ulcer. N Engl J Med 290: 469–473

Genta VM, Kaufmann DG, Harris CC (1974) Vitamin A deficiency enhances binding of benzo(a)pyrene to tracheal epithelial DNA. Nature 247: 48–49

Gibel W, Wildner GP, Lohs K (1968) Untersuchungen zur Frage einer kanzerogenen und hepatotoxischen Wirkung von Fuselöl. Arch Geschwulstforsch 32: 115–125

Gordon MJ, Skillman JJ, Edwards BG, Silen W (1974) Effect of ethanol, acetylsalicylic acid, acetaminophen, and ferrous sulfate on gastric mucosal permeability in man. Surgery 76: 405–412
Gottfried EB, Korsten MA, Lieber CS (1978) Alcohol-induced gastric and duodenal lesions in man. Am Gastroenterol 70: 587–592
Helm JF, Dodds WJ, Riedel DR, Teeter BC, Hogan WJ, Arndorfer RC (1983) Determinants of esophageal acid clearance in normal subjects. Gastroenterology 85: 607–612
Hinds MW, Kolonel LN, Lee J, Hirohata T (1980) Association between cancer incidence and alcohol/cigarette consumption among five ethnic groups in Hawaii. Br J Cancer 41: 929–940
Hirschowitz BI, Pollard HM, Hartwell SW Jr, London J (1956) The action of ethyl alcohol on gastric acid secretion. Gastroenterology 30: 244–256
Hoey J, Montvernay C, Lambert C (1981) Wine and tobacco: risk factors for gastric cancer in France. Am J Epidemiol 113: 668–674
Hogan WJ, Viegas de Andrade SR, Winship DH (1972) Ethanol-induced acute esophageal motor dysfunction J Appl Physiol 32: 755–760
Iber FL (1971) Alcohol and the gastrointestinal tract. Gastroenterology 61: 120–123
Kamionkowski MD, Fleshler B (1965) The role of alcoholic intake in esophageal carcinoma. Am J Med Sci 249: 696–700
Kaufmann SE, Kaye MD (1978) Induction of gastro-oesophageal reflux by alcohol. Gut 18: 336–338
Kaufmann SE, Kaye MD (1979) Effect of ethanol upon gastric emptying. Gut 20: 688–692
Kissin B, Kaley MM (1974) Alcohol and cancer. In: Kissin B, Begleiter H (eds) The biology of alcoholism, vol 3. Plenum, New York, pp 481–511
Kjellen G, Tibbling L (1978) Influence of body position, dry and water swallows, smoking, and alcohol on esophageal acid clearing. Scand J Gastroenterol 13: 283–288
Kouros M, Mönch W, Reiffer FJ, Dehnen W (1983) The influence of various factors on the methylation of DNA by the oesophageal carcinogen *N*-nitrosomethylbenzylamine. The importance of alcohol. Carcinogenesis 4: 1081–1084
Kürzinger R (1970) Bemerkungen über die Alkoholelimination magenresezierter Personen. Dtsch Gesundheitsw 25: 2022–2027
Kuo Y-J, Shanbour LL (1983) Route of ethanol administration and gastric acid output during chronic conditions. Dig Dis Sci 28: 820–826
Lenz HJ, Ferrari-Taylor J, Isenberg JI (1983) Wine and five percent ethanol are potent stimulants of gastric acid secretion in humans. Gastroenterology 85: 1082–1087
Lieber CS, Seitz HK, Garro AJ, Worner TM (1979) Alcohol-related diseases and carcinogenesis. Cancer Res 39: 2863–2886
Lijinsky W, Epstein SS (1970) Nitrosamines as environmental carcinogens. Nature 225: 21–23
Lillibridge CB, Yoshimori M, Chey WY (1973) Observations on the ultrastructure of oxyntic cells in alcohol-fed dogs. Am J Dig Dis 18: 443–454
Lowenfels AB (1979) Alcohol and cancer. A review and update. Br J Alcohol Alcoholism 14: 148–163
Mallory GK, Weiss S (1929) Hemorrhages from lacerations of the cardiac orifice of the stomach due to vomiting. JAMA 178: 506–515
Martinez I (1969) Factors associated with cancer of the esophagus, mouth, and pharynx in Puerto Rico. JNCI 42: 1069–1094
Martini GA, Wienbeck M (1974) Begünstigt Alkohol die Entstehung eines Barrett-Syndroms (Endobrachyösophagus)? Dtsch Med Wochenschr 99: 434–439
Mayer EM, Grabowski CJ, Fisher RS (1978) Effects of graded doses of alcohol upon esophageal motor function. Gastroenterology 75: 1133–1136
McLashan ND (1969) Oesophageal cancer and alcoholic spirits in Central Africa. Gut 10: 643–650
Mellow MH, Layne EA, Lipman TO, Kaushik M, Hostetler C, Smith JC (1983) Plasma zinc and vitamin A in human squamous carcinoma of the esophagus. Cancer 51: 1615–1620
Mendl K, McKay JM, Tanner CH (1960) Intramural diverticulosis of the oesophagus and Rokitansky-Aschoff sinuses in the gallbladder. Br J Radiol 33: 496–501
Messian RA, Hermos JA, Robbins AH, Friedlander DM, Schimmel EM (1978) Barrett's esophagus. Clinical review of 26 cases. Am J Gastroenterol 69: 458–466

Mettlin C, Graham S, Priore R, Marshall J, Swanson M (1981) Diet and cancer of the esophagus. Nutr Cancer 2: 143–147
Miller TA, Henagan JM (1984) Indomethacin decreases resistance of gastric barrier to disruption by alcohol. Dig Dis Sci 29: 141–149
Moore JG, Christian PE, Datz FL, Coleman RE (1981) Effect of wine on gastric emptying in humans. Gastroenterology 81: 1072–1075
Morris GP, Wallace JL, Hardwig PL, Krausse EJ, Lolle SJ (1984) Correlations between changes in indicators of gastric mucosal barrier integrity at time of exposure to "barrier breakers" and extent of hemorrhagic erosions one hour later. Dig Dis Sci 29: 6–11
Nier H, Wienbeck M, Berges W, Kremer K (1983) Syndrome nach Gastrektomie unter besonderer Berücksichtigung der Refluxösophagitis. Langenbecks Arch Chir 360: 71–80
Palmer DL (1978) Alcohol consumption and cellular immunocompetence. Laryngoscope 88: 13–17
Pestallozzi DM, Bühler R, von Wartburg JP, Hess M (1983) Immunohistochemical localization of alcohol dehydrogenase in the human gastrointestinal tract. Gastroenterology 85: 1011–1016
Phaosawasdi K, Tolin R, Mayer E, Fisher RS (1979) Effects of alcohol on pyloric sphincter. Dig Dis Sci 24: 934–939
Pottern LM, Morris LE, Blot WJ, Zeigler RG, Fraumeni JF Jr (1981) Esophageal cancer among black men in Washington DC. I. Alcohol, tobacco, and other risk factors. JNCI 67: 777–783
Puurunen J (1978) Studies on the mechanism of the inhibitory effect of ethanol on the gastric acid output in the rat. Naunyn Schmiedebergs Arch Pathol 303: 87–93
Reichstein BJ, Okamoto C, Forte JG (1984) The effect of ethanol on acid secretion by isolated gastric glands from rabbit. Gastroenterology 86: 1218
Robert A, Nezamis JE, Lancaster C, Davis JP, Field SO, Hanchar AJ (1983) Mild irritants prevent gastric necrosis through "adaptive cytoprotection" mediated by prostaglandins. Am J Physiol 245: G 113–121
Roberts DM (1972) Chronic gastritis, alcohol, and non-ulcer dyspepsia. Gut 13: 768–774
Salo JA (1983) Ethanol-induced mucosal injury in rabbit oesophagus. Scand J Gastroenterol 18: 713–721
Salvesen HA, Kolberg A (1958) Alcohol absorption in idiopathic steatorrhoea. Acta Med Scand 161: 135–142
Sarfeh IJ, Tanarwski A, Malki A, Mason GR, Mach T, Ivey KJ (1983) Portal hypertension and gastric mucosal injury in rats. Effects of alcohol. Gastroenterology 84: 987–993
Savary M, Miller G, Roethlisberger B (1981) Spezielle Probleme des Endobrachyösophagus. In: Blum AL, Siewert JR (eds) Refluxtherapie. Springer, Berlin Heidelberg New York, pp 437–461
Schottenfeld D (1979) Alcohol as a co-factor in the etiology of cancer. Cancer 43: 1962–1966
Seitz HK, Simon B, Czygan P, Veith S, Kommerell B (1983a) Effect of chronic ethanol ingestion on the cyclic AMP system of the upper gastrointestinal tract in the rat. Alcoholism 7: 369–371
Seitz HK, Czygan P, Kienapfel H, Veith S, Schmidt-Gayk H, Kommerell B (1983b) Changes in gastrointestinal DNA synthesis produced by acute and chronic ethanol consumption in the rat: a biochemical study. Z Gastroenterol 21: 79–84
Sernka ThJ, Jackson AF (1976) Responses to ethanol in rat stomach. Nutr Rep Intern 14: 415
Shay S, Johnson LF (1983) Mallory-Weiss laceration and esophageal hematoma. In: Castell DO, Johnson LF (eds) Esophageal function in health and disease. Elsevier, New York, pp 323–331
Shirazi SS, Platz CE (1978) Effect of alcohol on canine esophageal mucosa. J Surg Res 25: 373–379
Sinar DR, Fletcher JR, Cordova CM, Castell DO (1981) Acute acid-induced esophagitis impairs esophageal peristalsis in baboons. Gastroenterology 80: 1286
Singer MV, Eysselein V, Goebell H (1983) Beer and wine but not whiskey and pure ethanol do stimulate release of gastrin in humans. Digestion 26: 73–79
Sonnenberg A, Müller-Lissner SA, Vogel E, Schmid P, Gonvers JJ, Peter P, Strohmeyer G, Blum AL (1981) Predictors of duodenal ulcer healing and relapse. Gastroenterology 81: 1061–1067
Stefenelli N, Prießnitz E, Klotz H, Hahn S (1977) Die Beeinflußbarkeit der Alkoholresorption durch Änderung der Magenentleerung. Wien Klin Wochenschr 89: 161–164

Takeuchi K, Okabe S (1983) Role of luminal alkalinization in repair process of ethanol-induced mucosal damage in rat stomach. Dig Dis Sci 28: 993–1000
Tuyns AJ (1979) Epidemiology of alcohol and cancer. Cancer Res 39: 2840–2843
Tuyns AJ, Pequignot G, Jensen OM (1977) Le cancer de l'oesophage en Ille-et-Vilaine en fonction des niveaux de consommation d'alcool et de tabac. Bull Cancer 64: 45–60
Wienbeck M, Stenelli N (1978) Alkohol und Gastrointestinaltrakt. Leber Magen Darm 8: 299–303
Willett WC, McMahon B (1984) Diet and cancer. An overview I. N Engl J Med 310: 633–638
Winship DH, Caflish CR, Zboralske FF, Hogan WJ (1968) Deterioration of esophageal peristalsis in patients with alcoholic neuropathy. Gastroenterology 55: 173–178
Würsch TG, Hess H, Walser R, Koelz HR, Pelloni S, Vogel E, Schmid P, Blum AL (1978) Die Epidemiologie des Ulcus duodeni. Untersuchungen an 1105 Patienten in Zürich. Dtsch Med Wochenschr 103: 613–619
Wynder EL, Chan PC (1970) The possible role of riboflavin deficiency in epithelial neoplasia. Cancer 26: 1221–1224
Yang CS (1980) Research on esophageal cancer in China: a review. Cancer Res 40: 2633–2644

21 Acute and Chronic Actions of Alcohol on Pancreatic Exocrine Secretion in Humans and Animals

M. V. Singer and H. Goebell

Medizinische Klinik und Poliklinik, Abteilung für Gastroenterologie, Universitätsklinikum, Essen, Hufelandstrasse 55, 4300 Essen 1, Federal Republic of Germany

Introduction

At first glance the literature on the action of alcohol (ethanol, if not otherwise stated) on the pancreas seems to be full of contradictory results and rather confusing. A somewhat clearer picture emerges if one analyzes the results by taking into account the different experimental conditions under which they have been obtained. In order to make a proper assessment of the data concerning the secretory action of ethanol on the pancreas, several experimental variables have to be considered:

1. The duration of alcohol consumption. The action of ethanol is different and even opposite in humans and laboratory animals which are accustomed to a regular daily intake of alcohol from that in humans and animals receiving alcohol only episodically.
2. The route of administration of alcohol. There are different effects when alcohol is given orally or intravenously. When alcohol is given orally its action on the pancreas depends on whether or not the gastric juice is prevented from entering the duodenum. Its action is furthermore influenced by whether or not a meal is given simultaneously.
3. The secretory state of the gland. There are different effects of alcohol on basal and stimulated pancreatic exocrine secretion.
4. The state of consciousness of the animal, i.e., anesthetized or conscious. The studies cited in this review will mainly refer to experiments in conscious animals.
5. There may be differences in the pancreatic response to alcohol depending on the species of animal. Most of the alcohol studies have been done in man, dog, rat, rabbit, pigeon, cat, and pig.
6. The effect of alcohol on the pancreas is influenced by the dietary regimen. It is well known that pancreatic enzyme secretion is sensitive to dietary influences, and dietary adaptation is well established.
7. Another important question is whether or not control subjects or animals were employed. Especially in studies in which the long-term effects of alcohol on the pancreas are investigated, control animals which receive the same diet and live under the same conditions as the test animals should be included. Control animals and test animals should be tested at the same time intervals, since it is known that pancreatic secretion is quite variable not only between animals but also in the same animal studied on different days. Many of these criteria are difficult to fulfill

Alcohol Related Diseases in Gastroenterology
Edited by H. K. Seitz and B. Kommerell

in studies in human chronic alcoholics; but special care should be given to the diet regimen and smoking habits of the nonalcoholic subjects who are chosen as controls. In addition, in many published studies the alcohol consumption of the "normal" nonalcoholic controls is not mentioned.

8. There still exists a considerable confusion in the literature about the appropriate control solution for ethanol. In some studies there was either no control or equimolar solutions of urea, mannitol, or sodium chloride were used as an osmotic control for ethanol given orally, intragastrically, or intraduodenally, According to recent studies (for review see Singer 1983) distilled water seems to be the most suitable osmotic control for oral ethanol because it exerts nearly the same effective osmotic pressure on the gastrointestinal membrane as ethanol; that is, the two agents have a similar Stavermann reflection coefficient. In studies in which ethanol is administered intravenously, isotonic sodium chloride seems to be the appropriate control. In only a few acute studies has it been taken into acount that ethanol is also a source of energy. Therefore, in addition to distilled water isocaloric glucose solutions should be given when the action of oral ethanol on the gastrointestinal tract is investigated.

Having in mind these different experimental conditions and variables, we shall first review the secretory effects of alcohol on the pancreas in humans and animals not accustomed to alcohol, then we shall describe the action of prolonged intake of alcohol on the pancreas. In the last part of this review the possible mechanisms by which alcohol acts on the pancreas will be discussed. Since the earlier literature on the action of ethanol has been reviewed by different investigators (Bordalo et al. 1977; Cameron et al. 1975; Cueto et al. 1967; Davis and Pirola 1966; Dreiling and Bordalo 1977; Dreiling et al. 1952; Krager et al. 1972; Kalant 1969; Kimura et al. 1980; Pirola and Davis 1967; Sarles 1973; Sarles and Tiscornia 1974; Schapiro et al. 1966; Walton et al. 1960), we shall mainly refer to studies which have been published since 1965.

Effects of Acute Administration of Ethanol ("Acute Alcoholism") on the Pancreatic Exocrine Secretion of Nonalcoholic Humans or Animals

"Acute alcoholism" will be defined as the single or episodic administration of alcohol given by various routes to humans who are either teetotallers or drink less than 40 g pure ethanol daily and to animals which normally do not receive alcohol. Ingestion of alcohol stimulates gastric acid secretion and gastrin release in the dog (Elwin 1969; Eysselein et al. 1981, 1984; Schapiro et al. 1968; Treffot et al. 1975; Woodward et al. 1957) but has little or none of these effects in man (Becker et al. 1974; Cooke 1970, 1977; Hirschowitz et al. 1956; Korman et al. 1971; Singer et al. 1983a, c) (Table 1). The studies of Cooke (1970, 1972) suggest that in man ethanol only stimulates gastric acid secretion after it has left the stomach and has been absorbed from the small bowel. Cooke (1972) reported that an ethanol test meal in concentrations of 5.6%,

Table 1. The effects of acute administration of ethanol on gastric acid output in nonalcoholic human subjects and animals

Route of administration	Dose	Effect	Species	Author
Intragastric	350 ml 5.6%, 8%, 12%, and 16% (w/v)	No increase	Man	Cooke (1922)
Intragastric	500 ml 1.4% and 4% (v/v)	Small increase (about 20%–33% of MAO)	Man	Singer et al. (1983c)
	500 ml 5%, 6%, 7%, 8%, 10% (v/v) 250 or 125 ml 20% or 40% (v/v)	No increase	Man	Singer et al. (1983c)
	500 ml beer, red wine, white wine	As great as MAO	Man	Singer et al. (1983c)
	125 ml cognac, whisky	No effect	Man	Singer et al. (1983c)
Intravenous	10, 20, 40 ml 9% (v/v) ethanol in 250 ml 150 m*M* NaCl for 30 min	Increase	Man	Hirschowitz et al. (1956)
	300 mg/kg for 30 min plus 3 mg/kg/min for 3 h	Increase	Man	Demol et al. (1983a)
Oral/intragastric		Increase	Dog Dog	Elwin (1969) Woodward et al. (1957)
Intravenous		Increase	Dog	Woodward et al. (1957)

MAO, maximal acid output as determined by exogenous pentagastrin

8%, 12%, and 16% (w/v) was no more effective than an equal volume (350 ml) of water in stimulating acid output in man. Serum gastrin levels did not rise. Intravenous infusion of ethanol stimulates gastric acid secretion both in humans (Demol et al. 1983a; Hayakawa et al. 1982) and dogs (Kondo and Magee; Tiscornia et al. 1975b).

Recent studies by Singer et al. (1983c) showed that in healthy, nonalcoholic students, intragastric instillation of 500 ml ethanol in concentrations below 5% (v/v) caused a small (about 30% of the maximal response to pentagastrin) increase in gastric acid secretion as measured by intragastric titration. Ethanol in concentrations of 10%, 20%, and 40% (v/v) did not stimulate gastric acid output. Beer and red and white wine were potent stimulants of gastric secretion and release of gastrin; cognac and whisky had no effect (Singer et al. 1983c). Similar findings have been reported by Lenz and Isenberg (1984). Thus it may be speculated that in humans beer and wine and pure ethanol in concentrations below 5% (v/v) may influence pancreatic exocrine secretion indirectly by their stimulatory action on the stomach.

So far, there has only been one human study on the action of intragastric ethanol on basal (= interdigestive) pancreatic exocrine secretion when the gastric juice was allowed to enter the duodenum (see Table 2). Demol and co-workers (1982, 1983b) observed that an intragastric instillation of 100 ml 40% (v/v) ethanol caused a significantly higher increase in duodenal bicarbonate but not amylase output than water plasma levels of gastrin and secretin not being altered. Ethanol also changed the interdigestive motor complex by inducing a fed pattern and by lengthening the interval between two duodenal phases III of the interdigestive motor complex. The mechanisms by which oral ingestion of 40% ethanol stimulates pancreatic bicarbonate output are not clear, since gastric acid secretion was not stimulated by ethanol in these experiments nor was there a release of secretin and gastrin. Either release of other gastrointestinal hormones or activation of neural reflexes must be considered. Up to now the study by Demol and co-workers (1983b) is the only available study on the action of intragastric ethanol on interdigestive pancreatic exocrine secretion in humans in which the close interrelationship between fasting pancreatic exocrine secretion and the interdigestive motor activity of the upper gastrointestinal tract has been taken into account. Since interdigestive pancreatic enzyme secretion is lowest in the phase I of the interdigestive migrating complex, ethanol and several control solutions (water, glucose) were given during this time period.

The action of alcoholic beverages on human pancreatic exocrine secretion has not been studied so far.

When the gastric juice was prevented from entering the duodenum, intragastric infusion of 150 ml of 40% (v/v) ethanol substantially reduced pancreatic bicarbonate and enzyme secretion stimulated with intravenous secretin in nonalcoholism humans (Mott et al. 1972).

In the pig (Wheatley et al. 1975), intragastric instillation of ethanol produced a subtained increase in bicarbonate and protein output for about 6 h (the duration of the study) when gastric juice was not prevented from entering the duodenum. Peak pancreatic secretion occurred simultaneously with peak ethanol blood levels.

In the rat, intragastric ethanol also caused a stimulation of pancreatic flow rate and protein output when the gastric juice was allowed to enter the duodenum

Table 2. The acute effects of ethanol given orally, intragastrically, or intraduodenally on pancreatic exocrine secretion of nonalcoholic human subjects and animals

Route of administration	Dose of ethanol	Secretory state of the gland	Bicarbonate output	Enzyme output	Species	Author
No diversion of gastric acid						
Intragastric	100 ml 40% (v/v)	Basal	Small increase of duodenal bicarbonate output	No significantly higher output of duodenal amylase output than after water	Man	Demol et al. (1982, 1983c)
Intragastric	ca. 30 g/h (5% w/v)	Secretin plus CCK	Sustained increase	Sustained increase	Pig	Wheatly et al. (1975)
Intragastric	Different doses	Secretin or secretin plus CCK	Either stimulation or no effect	Either stimulation or no effect	Rat	Different investigators
Intragastric	1 g/kg (12% v/v)	Secretin alone	Increase	Increase	Dog	Tiscornia et al. (1974b)
Intragastric	Test meal + ethanol (2 g/kg bw)		Inhibition by 34% as compared to test meal alone	No effect	Dog	Tiscornia et al. (1977a)
Diversion of gastric acid						
Intragastric	150 ml 40% (v/v)	Secretin	Decreased by about 50%	Decreased by about 50%	Man	Mott et al. (1972)
Intragastric	2 g/kg (20% v/v)	Basal	Not tested	No effect	Rat	Demol et al. (1980)
Intragastric	12% (v/v)	Basal	No effect	Inhibition	Dog	Tiscornia et al. (1974b)
Intragastric	40%	Basal	Decrease	No effect	Dog	Tiscornia et al. (1974b)
Intragastric	80%	Basal	Decrease	Increase	Dog	Tiscornia et al. (1974b)
Intraduodenal	500 ml 12% (v/v)	Secretin plus CCK	Moderate increase	No alteration	Man	Capitaine et al. (1971)
Intraduodenal		Basal	No increase	No increase	Dog	Tiscornia et al. (1974a)
Intraduodenal		Basal	Strong stimulation	Strong stimulation	Rat	Demol et al. (1980)
Intrajejunal	1.4% (v/v)	Basal	No effect	No effect	Man	Marin et al. (1973)
Intrajejunal	1 g/kg bw	Basal	Strong reduction	Strong reduction	Man	Marin et al. (1973)

CCK, cholecystokinin

(Demol et al. 1980). When the pylorus was ligated, intragastric ethanol either caused a small, transient increase in pancreatic exocrine secretion followed by an inhibition of pancreatic flow rate (Cavarzan et al. 1975) or did not alter pancreatic exocrine secretion (Demol et al. 1980).

In the anesthetized dog, intragastric instillation of 12% ethanol inhibited basal pancreatic protein but not bicarbonate secretion when gastric secretion was diverted from the duodenum (Tiscornia et al. 1974b). Ethanol given in higher concentrations (40% or 80%) decreased water and bicarbonate output and did not modify (40%) or even increased (80%) pancreatic protein output (Tiscornia et al. 1974b). In antrectomized dogs, the intragastric instillation of 80% ethanol did not stimulate pancreatic enzyme secretion, thus suggesting that pancreatic protein response to orally ingested alcohol greatly depends on the release of antral gastrin (Tiscornia et al. 1974b). Gastrin is a weak stimulant of pancreatic enzyme secretion and a potent stimulant of gastric acid secretion in the dog (Stening and Grossman 1969). At least in the dog, oral administration of alcohol may stimulate pancreatic secretion via two different mechanisms: release of gastrin and increased gastric acid output. The latter mechanism is only effective when gastric juice passes to the duodenum, where it causes the release of secretin or other hormones or activates neural reflexes which in turn stimulate the pancreas to secrete bicarbonate and enzymes.

When in conscious dogs ethanol (1.5 or 2 g/kg) was given intragastrically together with a test meal, pancreatic bicarbonate output and flow rate were depressed by 34% during the first 60 min, protein output not being changed; 80 min after the test meal, however, no difference in flow rate and bicarbonate output between control studies and studies with alcohol was found (Tiscornia et al. 1975 b). The depression by alcohol of bicarbonate and colume output was probably due to delayed gastric emptying induced by alcohol (Tiscornia et al. 1975b). Recently, Diaz et al. (1983) confirmed these findings and those of Tiscornia et al. (1977b) and, in addition, showed that intragastric ethanol (1.5 g/kg) inhibited basal pancreatic bicarbonate and protein output of nonalcoholic dogs. I am not aware of similar studies in nonalcoholic humans.

Acute Alcoholism and Release of Gastrointestinal Hormones

With the exception of gastrin, the effect of alcohol on the release of gastrointestinal hormones has not been established (see Table 3). Since gastrin stimulates pancreatic enzyme secretion in humans and animals, the action of ethanol on release of gastrin is of interest. In humans, either no (Cooke 1972) or only little (Becker et al. 1974) release of gastrin by oral or intravenous ethanol has been found. Becker et al. (1974) found a higher 150-min integrated gastrin response to a drink of 80 ml of 18% or 50% (v/v) ethanol than to 80 ml of distilled water. Cooke (1972) found no increase in plasma gastrin levels in healthy subjects who drank 150 ml of Scotch whisky. In two studies (Korman et al. 1971; Straus et al. 1975) oral ingestion of vodka had no effect on gastrin release; in one uncontrolled study (Llanos et al. 1977) a small increase in plasma gastrin levels was observed after oral ingestion of 60 ml of 80% proof (40% ethanol) vodka. Recent studies by Singer et al. (1983a) in healthy students have shown that beer and red and white wine but not whisky and pure ethanol in different

Table 3. The effect of acute administration of ethanol on release of gastrointestinal hormones in nonalcoholic human subjects and animals

Hormone	Route of administration	Dose	Effect	Species	Author
Gastrin	Oral	18% or 50% (v/v)	Small increase (not compared with a meal)	Man	Becker et al. (1974)
	Oral	5.6%, 8%, 12%, 16% (w/v)	No effect	Man	Cooke (1972)
	Oral	4%, 10%, 20%, 40% (v/v)	No effect	Man	Singer et al. (1983a)
	Oral	Beer, white wine, red wine	Potent stimulation, 50% of the response to a steak meal	Man	Singer et al. (1983a)
		Whisky	No effect	Man	Singer et al. (1983a)
		Whisky	No effect	Man	Cooke (1972)
		Vodka	No effect	Man	Korman et al. (1971) Straus et al. (1975)
		Vodka	Small increase	Man	Llanos et al. (1977)
	Intravenous	300 mg/kg for 30 min plus 3 mg/kg/min for 3 h	No effect	Man	Singer et al. (1983a)
	Intraduodenal		No effect	Man Dog Cat	Llanos et al. (1977)
	Intragastric	40 ml 45% (v/v)	Small increase	Dog	Becker et al. (1974) Llanos et al. (1977)
		200 ml 1.4%, 10%, 40% (v/v)	Dose-dependent increase, 40% ethanol as potent as a mixed meat meal	Dog	Eysselein et al. (1981, 1983)

Table 3 (continued)

Hormone	Route of administration	Dose	Effect	Species	Author
Secretin	Intragastric	150 ml 40% (v/v) ethanol	No effect	Man	Demol et al. (1982, 1983b)
	Oral	60 ml vodka (86% proof)	Small increase	Man	Straus et al. (1975)
	Oral	60 ml vodka (80% proof) (40% v/v ethanol)	Small increase	Man	Llanos et al. (1977)
	Intragastric	Ethanol	No effect	Dog	Llanos et al. (1977)
		Ethanol	No effect	Cat	
	Intraduodenal	60 ml vodka (86% proof)	No effect	Man	Fahrenkrug and Schaffalitzky de Muckadell (1977)
CCK	Oral	60 ml 45% (v/v) ethanol	No effect	Man	Fried et al. (1983)
	Intraduodenal	40 ml 45% (v/v) ethanol	No effect	Dog	Fried et al. (1983)
PP	Oral	250 ml 4%, 10%, 20% (v/v); 125 ml 40%	No effect	Man	Singer et al. (1983b)
		250 ml beer, red wine, white wine 125 ml whisky	No effect	Man	Singer et al. (1983b)
	Intravenous	600 mg/kg (10% v/v)	Small increase	Man	Staub et al. (1981)
		300 mg/kg (10% v/v)	No effect	Man	Singer et al. (1983b)
		400 mg/kg (11% v/v)	No effect	Dog	Singer et al. (1983b)
	Intragastric	2.7 g/kg (40% v/v) but not lower doses	Small increase (about 12% of the response to a meal)	Dog	Singer et al. (1983b)
	Intraduodenal	5.5%	No effect	Dog	Singer et al. (1983b)
VIP	Intravenous	600 mg/kg	Small increase	Man	Staub et al. (1981)
Somatostatin-like activity	Intravenous	600 mg/kg	No effect	Man	Staub et al. (1981)

CCK, cholecystokinin; PP, pancreatic polypeptide; VIP, vasoactive intestinal polypeptide

concentration (4%, 10%, 20%, 40% v/v) are potent stimulants of gastrin. A drink of 250 ml of beer or red or white wine caused a 60-min integrated gastrin response which was 50% of the response to a steak meal, a potent releaser of gastrin.

Intraduodenal administration of alcohol failed to release gastrin in humans, dogs, and cats (Llanos et al. 1977).

The effect of oral alcohol on secretin release has been studied in man by Straus et al. (1975), who found that secretin was released within 10 min after the ingestion of vodka. Intragastric instillation of 150 ml 40% (v/v) ethanol did not cause an increase in plasma levels of secretin when the gastric juice was allowed to enter the duodenum (Demol et al. 1982). Llanos et al. (1977) reported that the man oral administration of alcohol (60 ml 80% proof vodka) released secretin only 60 min after the ingestion of alcohol. Intragastric administration of alcohol did not increase plasma concentrations of secretin in dogs and cats (Llanos et al. 1977). When alcohol was given intraduodenally no release of secretin was found in man, dog, and cat (Llanos et al. 1977). Fahrenkrug and Schaffalitzky de Muckadell (1977) observed no effect of intraduodenal vodka (60 ml, 86% proof) on release of secretin in humans.

Whether alcohol releases cholecystokinin (CCK) is difficult to assess, since reliable radioimmunoassays for CCK are still not available. In a recent report (Fried et al. 1983), published only in abstract form, no increase in plasma levels of CCK by oral ethanol (60 ml 45%) in man or by intraduodenal ethanol (40 ml 45%) in conscious dogs could be detected.

In humans, no release of pancreatic polypeptide (PP) by oral pure ethanol (4%, 10%, 20%, 40% v/v) or beer, red or white wine, or whisky could be detected (Singer et al. 1983b). An intravenous infusion of 300 mg kg^{-1} of pure ethanol (as a 10% v/v solution) also caused no release of PP, whereas a short-lasting increase in plasma PP levels could be induced by an intravenous infusion of 600 mg kg^{-1} of ethanol (Staub et al. 1981). In dogs, neither intragastric nor intraduodenal infusion of ethanol altered plasma PP levels. In dogs, only a high dose (2.7 g kg^{-1}) of intragastric ethanol caused an increase in plasma PP levels, which was only 12% of the PP response to a meal (Singer et al. 1983b). Thus, in humans and dogs, PP is unlikely to be the hormonal mediator of the inhibition of pancreatic exocrine secretion observed after intravenous ethanol.

Whether alcohol affects the pancreas by release of other gastrointestinal hormones or peptides such as neurotensin, bombesin, or enkephalins has not been studied. It is theoretically possible that the small increase in pancreatic volume and bicarbonate output in response to intraduodenal alcohol reported in man (Capitaine et al. 1971), dog (Brooks and Thomas 1953), and rat (Demol et al. 1980) is mediated by hormones other than secretin, e.g., vasoactive intestinal polypeptide (VIP). Intravenous ethanol has been found to increase plasma levels of VIP in humans (Staub et al. 1981), but unfortunately no control experiment was performed in that study. Plasma levels of somatostatin-like activity were not altered by ethanol in the same study (Staub et al. 1981).

Acute Alcoholism and the Sphincter of Oddi

In man (Capitaine and Sarles 1971), dog (Menguy et al. 1958), and rabbit (Sarles and Midejean 1973; Sarles et al. 1981) acute alcoholism increases the tone of the sphincter of Oddi. So far only a limited number of patients has been studied and further human data are needed. In humans, intragastric (Capitaine and Sarles 1971), intraduodenal (Nossel and Efran 1955), and intravenous (Pirola and Davis 1968) ethanol caused a moderate increase in biliary and pancreatic pressures. The rise in common bile duct pressure after intravenous ethanol was similar to that obtained with morphine (Pirola and Davis 1968), Bergh (1942), however, reported little change in resistance to flow through the sphincter of Oddi in humans after intraduodenal instillation of ethanol. Any action of intraduodenal ethanol on biliary and pancreatic pressure could as well be the effect of nonspecific irritation as of some more specific pharmacologic actions. In contrast to earlier studies, Viceconte (1983) recently observed that intragastric or intravenous ethanol infusion caused an inhibition of sphincter of Oddi activity in nonalcoholic humans. Alcohol affected neither the frequency of the sphincter concentrations nor the duodenal motor activity. The author speculated that alcohol favours reflux through an insufficiency of the sphincter of Oddi.

In dogs, Menguy et al. (1958) observed that intraduodenal instillation of ethanol caused pancreatic and biliary intraductal pressures to increase, the former more constantly than the latter. The resistance to flow through the respective sphincter was greatly increased. Walton et al. (1965) confirmed the action of intraduodenal ethanol, but did not find any effect of intravenous ethanol on pancreatic ductal pressure.

In the rabbit, intravenous ethanol caused a rise in biliary pressure (Sarles and Midejean 1973; Sarles et al. 1981).

With the exception of the study by Viceconte (1983), the human and animal reports seem to indicate that ethanol raises biliary and pancreatic pressures by increasing the sphincteric resistance. However, in many studies, the possibility that the rises in pressure were due to changes in flow could not be excluded. In addition, in some studies on the action of alcohol on human duodenal aspirates no attempt was made to exclude changes in the contribution of the bile or to introduce control periods which were strictly comparable to the experimental ones.

Satisfactory data supporting the concept that the increased tone of the sphincter of Oddi leads to acute or chronic pancreatitis in humans have not been produced (McCutcheon 1968). Sarles et al. (1965) noted that the moderate increase in the tone of the sphincter of Oddi after ethanol is clearly less than that seen during the course of odditis (benign Vaterian stenosis), a disease which indeed causes chronic pancreatitis. The morphological characteristics of chronic alcoholic pancreatitis and chronic pancreatitis due to odditis are clearly different (Sarles et al. 1945). However, an increased tone of the sphincter of Oddi due to acute ingestion of alcohol together with intraductal protein plugs could play a role in the acute attacks of chronic alcoholic pancreatitis in humans (Sarles and Laugier 1981).

Effect of Acute Intestinal Ethanol Administration of the Pancreas

Intraduodenal infusion of ethanol has been shown to induce a small (approximately three times the basal output) increase in hormonally stimulated pancreatic bicarbonate output and not to affect enzyme output in nonalcoholic humans (Capitaine et al. 1971). Intrajejunal perfusion of a high (1 g kg^{-1} body wt.) but not a low (0.05 g kg^{-1} body wt.) dose of ethanol caused marked reductions in basal outputs of pancreatic enzymes and bicarbonate in normal subjects and in two patients with "alcoholic" pancreatitis (Marin et al. 1973). In dogs, no increase in protein output was observed after intraduodenal ethanol (Brooks and Thomas 1953; Tiscornia et al. 1974b). Only in the rat did intraduodenal instillation of ethanol result in a strong stimulation of volume and protein secretion of the pancreas (Demol et al. 1980).

Acute Alcoholism and Biliary Secretion

Since intraduodenal bile has been shown to affect pancreatic exocrine secretion in humans (Forell et al. 1971) and animals (Thomas and Crider 1943), any ethanol-induced change in biliary secretion might influence pancreatic secretion. In nonalcoholic humans, jejunal perfusion of 1.40% alcohol had no effect on biliary secretion (Marin et al. 1973). Jejunal or intravenous alcohol administration in doses of 1 g kg^{-1} body weight, however, caused a marked reduction in output of bile salts into the duodenum (Marin et al. 1973).

In nonalcoholic dogs, the output of bile salts was inhibited by an intravenous infusion of ethanol (1.3 g kg^{-1}) more than the bile flow rate was stimulated by an intravenous infusion of secretin, CCK, and sodium taurocholate (Dzieniszewski et al. 1976). Similar results have been reported in rats (Maddrey and Boyer 1973).

In humans, a maximal dose of ethanol, 210 ml 100% proof vodka administered orally (blood alcohol 148 mg%), did not stimulate gallbladder emptying, despite rapid emptying of alcohol from the stomach (Brody et al. 1982).

Since in many investigations no attempt was made to divert the bile from the duodenum, more studies are needed to separate the direct actions of ethanol on the pancreas from indirect ones which might be mediated by the bile flow into the duodenum.

Direct Actions of Acute Alcoholism on In Vivo Exocrine Pancreatic Secretion

Intravenous administration appears to be the most adequate means of studying direct effects of alcohol on the pancreas in the intact animal. Gastric acid and bile should be diverted from the duodenum. In contrast to older studies (Kalant 1971; Lowenfels et al. 1968), recent studies have shown that intravenous alcohol *inhibits* the pancreatic secretion of volume, bicarbonate, and protein in man (Davis and Pirola 1966; Marin et al. 1973; Mott et al. 1972; Planche et al. 1982; Valenzuela et al. 1969), dog (Bayer et al. 1972; Tiscornia et al. 1973a, 1975b), cat (Hartmann et al. 1980; Hotz et al.

Table 4. The effects of acute intravenous administration of ethanol on pancreatic exocrine secretion in nonalcoholic human subjects and animals

Dose of ethanol	Secretory state of the gland	Bicarbonate output	Enzyme output	Species	Author
300 mg/kg for 30 min plus 3 mg/kg/min for 3 h	Basal	Not investigated	Decreased	Man	Demol et al. (1983a)
	Stimulation with secretin and/or cholecystokinin	Inhibition	Inhibition	Man	Davis and Pirola (1966) Marin et al. (1973) Mott et al. (1972) Planche et al. (1982) Valenzuela et al. (1969)
0.7 or 1.5 g/kg bw	Stimulation with secretin and/or cholecystokinin	Inhibition	Inhibition	Dog	Tiscornia et al. (1973a)
560 mg/kg bw (ED_{50} for inhibition of bicarbonate output)	Stimulation with secretin and/or cholecystokinin	Inhibition	No effect	Dog	Bayer et al. (1972)
	Stimulation with secretin and/or cholecystokinin	Inhibition	Inhibition	Cat	Hotz et al. (1978)
	Stimulation with secretin and/or cholecystokinin	Inhibition	No effect	Rabbit	Solomon et al. (1974)
	Stimulation with secretin and/or cholecystokinin	Inhibition	Inhibition	Rat	Demol et al. (1983a)
	Stimulation with secretin and/or cholecystokinin	Inhibition	Inhibition	Pig	Wheatley et al. (1975)

1978), rabbit (Solomon et al. 1974), rat (Demol et al. 1980), and pig (Wheatley et al. 1975) (see Table 4). Inhibition of protein secretion by alcohol has not been observed in all studies (Bayer et al. 1972; Solomon et al. 1974).

In nonalcoholic humans, *N*-methyl-hyoscine, an anticholinergic drug, smoothed but did not completely prevent the inhibition by intravenous ethanol of hormonally stimulated pancreatic bicarbonate and lipase output (Meullenet et al. 1979).

In these previous human studies, the action of intravenous ethanol on pancreatic secretion already stimulated by secretin or CCK or both was studied. In a very recent study, Demol et al. (1983a) examined the action of intravenous ethanol on basal pancreatic exocrine secretion in relation to the interdigestive motor activity of the upper gastrointestinal tract in humans. They found that intravenous ethanol (300 mg kg^{-1} for 30 min followed by 3 mg kg^{-1} min^{-1} for 3 h) depressed interdigestive pancreatic amylase output into the duodenum without altering the interdigestive motor activity of the upper gastrointestinal tract, gastric acid secretion being stimulated.

In conscious dogs with gastric and pancreatic Thomas fistulas, Tiscornia et al. (1973a) found that the intravenous infusion of 0.7 or 1.5 g ethanol per kilogram diminished the pancreatic bicarbonate and protein response to secretin alone or to secretin plus CCK. The same group (Tiscornia et al. 1975b) reported that intravenous ethanol inhibited the enzyme but not the bicarbonate output of the canine pancreas stimulated with an intravenous infusion of secretin plus gastrin. Bayer et al. (1972), also using conscious dogs, observed a dose-dependent inhibition by intravenous alcohol of colume and bicarbonate output during a continuous infusion of secretin alone, but failed to see a significant inhibition of protein concentration (protein output was decreased). The effective dose of ethanol producing 50% inhibition (ED_{50}) of volume and bicarbonate output was found to be 560 mg per kilogram body weight. Both groups prevented acid from entering the duodenum. In recent studies (Kubota et al. 1982, 1983; Noel-Jorand and Sarles 1983) the Marseille group reported that in nonalcoholic dogs a high dose of intravenous ethanol first inhibits and then stimulates hormonally stimulated pancreatic enzyme and bicarbonate output, low doses only causing stimulation. Further studies are needed to confirm this observation.

In the presence of an intravenous infusion of atropine (75 μg kg^{-1} h^{-1}), intravenous ethanol (1.5 g kg) did not further reduce pancreatic volume, bicarbonate, and protein output in response to a continuous intravenous infusion of secretin in conscious dogs (Tiscornia et al. 1973a, 1972).

The intravenous infusion of pentolinium (750 μg kg^{-1} h^{-1}), a nicotinic ganglion blocker, produced a significant reduction of (about 46%) protein output and no change in volume and bicarbonate output in response to a continuous intravenous infusion of secretin plus CCK. When intravenous ethanol (1.5 g/kg) was added, no additional reduction of protein output or changes in volume and bicarbonate were seen (Tiscornia et al. 1972, 1973c).

The only conclusions which can be drawn from these studies are that ethanol caused no further decrease in pancreatic secretion already reduced by atropine or pentolinium. The reverse studies (first ethanol and then atropine or pentolinium) need to be done to clarify further the role of the nerves in mediating the effect of ethanol on the pancreas.

The effect of intravenous ethanol on the pancreas has not been studied before and after truncal vagotomy in the same animals. In vagotomized dogs intravenous ethanol (1.5 g/kg) did not significantly depress bicarbonate and protein output in response to a continuous intravenous infusion of secretin plus CCK. Bicarbonate and protein concentrations were only slightly, not significantly, decreased (Tiscornia et al. 1973c).

Whether complete division of the splanchnic nerves alters the pancreatic secretory response to ethanol has not been studied. The inhibitory effect of intravenous ethanol on pancreatic secretion in the conscious dog was not modified by pretreatment with reserpine (0.1 mg/kg per day) started 2 days before the test (Voirol et al. 1976a, c).

Contradictory results of the effect of intravenous ethanol on pancreatic exocrine secretion have been reported in conscious rats (Dagorn et al. 1974; Demol et al. 1980; Joffe et al. 1979; Korsten et al. 1981; Roze et al. 1980). Either no (Demol et al. 1980) or a stimulatory (Korsten et al. 1981) action on basal pancreatic secretion has been reported. Korsten et al. (1981) found that an intravenous infusion of ethanol (1.5 g/kg) significantly increased the basal pancreatic flow rate, protein output, and bicarbonate output. This response persisted despite elimination of gastric acid production by infusion of the H_2-receptor antagonist cimetidine. In contrast, an intravenous infusion of ethanol (1.5 g/kg) inhibited the pancreatic flow rate and bicarbonate output in response to secretin. This study is clear evidence for the fact that the action of intravenous ethanol depends on the secretory state of the pancreas.

Action of Acetaldehyde on Exocrine Pancreatic Secretion

Whether the inhibitory action of ethanol on the pancreas is mediated by its first and main metabolic product, acetaldehyde, has not been studied systematically. In the isolated canine pancreas, intraarterial injection of acetaldehyde (47 μM) was found to inhibit bicarbonate and protein output (Fitzpatrick et al. 1975a, b). In consious dogs, Andersen et al. (1980) found that intravenous infusion of different doses of acetaldehyde (50, 100, and 200 mg kg^{-1} h^{-1}) did not modify the pancreatic secretory response to intravenous secretin and cerulein. The authors concluded that physiological plasma concentrations of acetaldehyde are unlikely to interfere with pancreatic exocrine secretion in the dog. In nonalcoholic rats, Demol et al. (1980) observed an inhibitory action of an intravenous infusion of 500 but not 100 mg kg^{-1} h^{-1} of acetaldehyde on pancreatic fluid and protein secretion.

In Vitro Studies on the Acute Action of Ethanol on the Pancreas

Results of in vitro studies in which the acute effect of ethanol on the pancreas was studied have been conflicting. Solomon et al. (1974) demonstrated in the rabbit that only high (3% w/v) concentrations of ethanol directly inhibited pancreatic secretion. Associated with ethanol-induced inhibition of secretion was a decrease in ATP content of pancreatic tissue. Glazer et al. (1976) however, also using the superfused rabbit pancreas, found that ethanol (3%) produced a rise in volume and protein content of pancreatic juice and that this effect continued after removal of the stimulus. In the isolated rat pancreas (Clemente et al. 1977; Estival et al. 1981), ethanol did not alter pancreatic secretion.

When pigeon pancreas slices (Valenzuela et al. 1974) were incubated with ethanol and/or CCK, spontaneous release of amylase into the incubation media or stimulated CCK were unaffected by the addition of ethanol

In dispersed acini prepared from guinea pig pancreas it was found that ethanol (0.8 *M*) inhibited the increase in amylase secretion caused by CCK, carbachol, secretin, or VIP (Uhlemann et al. 1979). These findings indicate that the ability of ethanol to inhibit pancreatic protein secretion in the intact animals may reflect, at least in part, a direct effect of the alcohol on pancreatic acinar cells. The same authors found that ethanol induced a 50% increase in the basal rate of amylase release from pancreatic acinar cells. Other investigators (Calderon-Attas et al. 1980) however, observed on effect of low doses (0.5–200 m*M*) of ethanol on basal amylase output in dispersed guinea pig acini.

Effects of "Chronic Alcoholism" on the Pancreas

Chronic alcohol abuse in humans is at least 80 g and usually more than 160 g distilled alcoholic beverage or its equivalent per day for years. In an attempt to elucidate the mechanisms of alcohol-induced chronic pancreatitis, mainly dogs (Sarles and Tiscornia 1974; Schmidt 1983) and rats (Sarles et al. 1971a) have been made "alcoholic." The most extensive studies have been done in Marseille (France) by Sarles and co-workers in dogs equipped with gastric and duodenal Thomas cannulae and receiving 2 g ethanol per kilogram body weight intragastrically together with a diet rich in fat and protein (not further specified).

Chronic Alcoholism and Gastric Acid Secretion

Chronic alcoholism stimulates gastric acid secretion in dogs. Chey et al. (1972) studied the effects of the daily administration of 4.4 g ethanol per kilogram (as a 40% solution) on the acid secretion from Heidenhain pouches in dogs for up to 14 months. They found that the mean daily secretion of acid increased during the first months and remained elevated throughout the entire period of observation. The mean maximal acid output in response to histamine increased by 35% within the first months after starting treatment with ethanol, tended to decrease with time, and at the end of 6 months was no longer different from that of the preethanol period. Light microscopic studies of the gastric mucosa of these dogs showed hypertrophy of the parietal cells and a small increase (hyperplasia) of about 14% of the total parietal cell mass after 3 months of daily administration of ethanol (Lillibridge et al. 1973). Electron microscopic studies revealed mitochondrial hypertrophy and vesicotubular hyperplasia of the oxyntic cells in ethanol-fed dogs. The authors felt that these alterations of the oxyntic cells were the cause of the increased mean maximal acid output of chronically alcoholic dogs (Yoshimori et al. 1972).

In humans, chronic ingestion of alcohol does not seem to stimulate gastric acid secretion (Bank et al. 1966; Chey et al. 1968; Dinoso et al. 1972; Gupta and Rao 1975; Sarles et al. 1975). In chronic alcoholics without clinical and laboratory evidence of pancreatic disease and in patients with chronic pancreatitis associated with chronic alcoholism, Chey et al. (1968) found hyposecretion of gastric acid or achlorhydria. The mean maximal acid outputs of chronic alcoholics with and without chronic pancreatitis were similar and were less than those of healthy controls (Chey et al. 1968). The authors suggested that the reduced gastric acid secretion in patients with chronic alcoholic pancreatitis was rather due to the gastric mucosal damage produced by alcoholic beverages than to the pancreatitis per se.

Gullo et al. (1983) found that half of their patients with alcoholic pancreatitis had gastric acid secretion within the normal range, more than one-third had acid hypersecretion, and only 10% had achlorhydria.

In another recent study (Piubello et al. 1982) the mean basal acid output of patients with chronic alcoholic pancreatitis was significantly higher than that of chronic alcoholics without evidence of chronic pancreatitis and healthy controls. The mean peak acid outputs of chronic alcoholic patients with and without clinical pancreatitis were similar and significantly higher than in controls. There was a large overlap of individual values between the different groups studied.

In the same study, chronic gastritis was not more often observed in patients with chronic alcoholic pancreatitis and chronic alcoholics than in healthy subjects and patients with duodenal ulcers (Piubello et al. 1982). Cheli et al. (1981) did not find any significant differences in the frequency and localization of chronic gastritis between chronic alcoholics and teetotallers when the data were related to the age of the subjects tested. This study suggests that chronic histological alterations of the gastric mucosa by alcohol do not play an important role in the pathogenesis of human chronic alcoholic pancreatitis. However, this does not exclude an effect of repeated acute gastritis on the development of chronic alcoholic pancreatitis, since it is well established that large quantities of ethanol administered acutely cause erythema followed by mucosal erosion and inflammation of the stomach. A merely hypothetical possibility is that the acutely inflamed gastric mucosa releases a substance or substances (e.g., hormones) which in turn damage the pancreas. This putative hormone is probably not gastrin, since a single oral administration of pure ethanol and some alcoholic beverages, such as whisky, vodka, and cognac, do not release gastrin in humans (Singer et al. 1983a).

Chronic Alcoholism and Release of Gastrointestinal Hormones

Besides a direct action at the parietal cell, chronic alcoholism probably stimulates gastric acid secretion in the dog by the release of gastrin. Gastrin levels in response to a test meal were significantly higher in chronic alcoholic than in nonalcoholic dogs (Treffot et al. 1975). The mean basal gastrin blood levels, however, did not differ significantly between the two groups (Treffot et al. 1975).

In humans most investigators (Bestermann et al. 1982; Piubello et al. 1982; Gullo et al. 1980) did not observe an alteration in fasting and meal-stimulated serum gastrin levels by chronic ingestion of alcohol. Treffot et al. (1980) however, observed that the plasma gastrin response to a meal with 100 ml 40% (v/v) ethanol was significantly higher in patients with alcoholic chronic calcifying pancreatitis and in alcoholics without clinical signs of pancreatitis than in cirrhotics or nonalcoholic controls. Basal (prestimulatory) plasma gastrin levels were significantly higher in patients with alcoholic cirrhosis than in patients with alcoholic chronic pancreatitis, alcoholics without pancreatic disease, and normal subjects. Piubello et al. (1982) did not detect any significant difference in basal and meal-stimulated plasma gastrin levels between patients with chronic alcoholic pancreatitis, chronic alcoholics without pancreatic disease, patients with alcoholic liver cirrhosis, and nonalcoholic controls. Similar findings were made by Bestermann et al. (1982). Gullo et al. (1980) found no relationship between serum fasting and meal-stimulated gastrin levels and steatorrhea in patients with chronic alcoholic pancreatitis.

The lack of significant action of chronic alcoholism on gastric acid secretion and release of gastrin in humans does not support the concept that the stomach plays an important role in the pathogenesis of chronic alcoholic pancreatitis in humans. In dogs, however, the enhanced release of gastrin and the increased gastric output in chronic alcoholism may be involved in the pathogenesis of chronic pancreatitis. Gastrin may exert direct trophic actions on the stomach and pancreas. The increased gastric acid output may lead to enhanced release of gastrointestinal hormones, e.g., secretin and CCK, which in turn could have trophic and stimulatory secretory actions on the pancreas. Chronic hormonal overstimulation of the pancreas might cause pancreatitis, as has been shown by Adler et al. (1982) in rats.

Blood levels of immunoreactive secretin in response to intraduodenal infusion of HCl (0.1 N, 50 ml) were significantly lower in dogs receiving 2 g 50% ethanol per kilogram daily for 3 years than in nonalcoholic control dogs (Bretholz et al. 1978). Basal plasma secretin levels did not differ between the two groups. Whether chronic alcoholism alters the release of secretin in humans is unknown.

Blood levels of CCK (measured by a bioassay on in vitro rabbit gallbladder) in response to intraduodenal injection of 20 ml oleate did not significantly differ between chronically alcoholic and nonalcoholic dogs (Planche et al. 1977). Whether chronic alcoholism alters release of CCK in humans is unknown.

In chronic alcoholics (Fink et al. 1983) and in patients with chronic pancreatitis (Besterman et al. 1982), normal plasma levels of insulin, pancreatic glucagon, and gastric inhibitory polypeptide was found. Patients with chronic pancreatitis, with or without exocrine insufficiency, has two- to threefold higher plasma levels of motilin and enteroglucagon than controls (Besterman et al. 1982).

Patients with advanced chronic pancreatitis very often have diminished fasting levels and postprandial rises of PP (Stasiewicz et al. 1980; Valenzuela et al. 1974). In a recent study (Fink et al. 1983) significantly higher basal and postprandial plasma PP levels were found in human chronic alcoholics than in controls. Whether these alcoholics suffered from chronic pancreatitis was not mentioned in the paper.

Whether the release of other gastrointestinal hormones, such as VIP, bombesin, and somatostatin, is altered by chronic alcoholism has not been reported so far.

Chronic Alcoholism and Pancreatic Response to a Test Meal

Three months of intragastric alcohol feeding (2 g/kg per day) induced in dogs a significant reduction of about 45% of the pancreatic water and bicarbonate response to a test meal, when compared to the response to the same test meal in the prealcoholic period (Tiscornia et al. 1977a). Protein output was not altered. This decreased volume and bicarbonate response could be explained either by ethanol-induced delay of gastric emptying or by impaired release of secretin or other hormones from the small intestine. After 24 months of ethanol feeding (2 g/kg per day) no inhibitory action of the acute addition of ethanol (1.5 g kg^{-1}) to a test meal could be detected in chronically alcoholic dogs as compared to a group of nonalcoholic dogs (Diaz et al. 1983).

In alcoholic humans without evidence of pancreatic disease basal duodenal chymotrypsin concentration but not output was significantly higher than in nonalcoholic controls. There were no significant differences in the chymotrypsin response to a modified Lundh meal between the two groups studied (Planche et al. 1982). Whether the pancreatic response to a mixed solid-liquid test meal in chronically alcoholic humans without pancreatic disease is different from that in teetotallers has not been studied so far.

Chronic Alcoholism and Pancreatic Response to Acute Intragastric or Intestinal Administration of Ethanol

In conscious dogs receiving ethanol (2 g kg^{-1} day^{-1}) for 24 months, the acute intragastric administration of ethanol (1.5 g/kg) did not alter basal pancreatic bicarbonate and protein output as compared to nonalcoholic controls (Diaz et al. 1983). In nonalcoholic dogs, however, bicarbonate and protein outputs were significantly decreased by the acute intragastric administration of ethanol (Diaz et al. 1983).

In conscious rats, receiving ethanol ad libitum for 13 months, acute intragastric instillation of 2 g ethanol per kilogram stimulated pancreatic protein output, whereas it inhibited pancreatic secretion in nonalcoholic rats (Cavarzan et al. 1975). The pylorus was ligated in both groups. A more recent study by Demol et al. (1980) in chronically alcoholic and nonalcoholic conscious rats did not confirm these results but showed that intragastric ethanol (2 g/kg) stimulated pancreatic volume and protein output to about the same degree when gastric juice was allowed to pass into the duodenum. No effect of intragastric ethanol on the pancreas was observed when the pylorus was ligated in either group.

Intraduodenal ethanol stimulated pancreatic volume and protein output about equally in nonalcoholic and chronically alcoholic rats (Demol et al. 1980). In four chronically alcoholic subjects with pancreatitis, intrajejunal administration of ethanol, 1 g/kg body weight, resulted in a less pronounced decrease in duodenal outputs of bicarbonate and amylase than in normal controls (Marin et al. 1973). Whether acute intragastric or intraduodenal administration of ethanol alters pancreatic exocrine secretion in alcoholic human subjects without pancreatic disease is unknown.

Chronic Alkoholism and Pancreatic Response to Intraduodenal Stimulants Other than Ethanol

Intestinal absorption of several nutrients may be altered by chronic alcoholism (Mezey and Potter 1976; Mezey et al. 1970). Dose-response studies of the pancreatic secretory response to intraduodenal instillation of HCl, amino acids, and fatty acids might be helpful in assessing the impaired or unimpaired release of gastrointestinal hormones or/and alterations in nervous connections between the intestine and the pancreas, but this kind of study has not been performed in humans. These studies might help to decide whether a loss of pancreatic function is due to malabsorption of nutrients or altered stimulatory mechanisms.

An intraduodenal bolus instillation of 20 ml oleic acid caused a higher protein output in chronic alcoholic dogs than in control dogs (Palasciano et al. 1974). Voirol et al. (1976) confirmed this finding and, in addition, showed that a large dose of atropine (750 $\mu g\ kg^{-1}\ h^{-1}$ IV) greatly inhibited the pancreatic protein response to intraduodenal oleic acid in nonalcoholic and chronically alcoholic dogs. Plasma levels of CCK and other hormones were not measured in either study.

Chronic Alcoholism and the Sphincter of ODDI

We did not find any report on the action of chronic alcoholism on the intraductal pressures and sphincteric resistance in pancreatic and biliary ducts in humans and experimental animals in the literature. In addition, we have not seen data on the action of acute administration of ethanol on the tone of the sphincter of Oddi in chronically alcoholic humans or animals.

Chronic Alcoholism and Biliary Secretion

In chronically alcoholic humans without signs of hepatic, pancreatic, or any other disease, basal concentrations of bile salts, cholesterol, and phospholipids in duodenal juice were not significantly different from those of nonalcoholic healthy controls (Planche et al. 1982). An intravenous infusion of 10% (600 mg/kg) ethanol caused a decrease in the three parameters studied in both groups, the decrease in duodenal cholesterol and phospholipid concentrations being more marked in nonalcoholic than in chronically alcoholic humans (Planche et al. 1982). Marin et al. (1973) observed no alteration in biliary secretion in normal subjects and in four patients with alcoholic pancreatitis after jejunal perfusion of 1.39% (v/v) ethanol. Intrajejunal or intravenous alcohol administration in doses of 1 g/kg body weight, however, caused a marked reduction in output of bile salts in both groups. The basal collections of bile salts were significantly higher in the four alcoholic patients than in the normal controls (Marin et al. 1973). Further studies are needed, since only a limited number of alcoholic patients has been studied so far.

Contrary to the case in nonalcoholic dogs, in chronically alcoholic dogs an intravenous infusion of ethanol (1.3 g/kg) inhibited only bile flow rate but not bile salts output during an intravenous infusion of secretin, CCK, and sodium taurocholate (Dzieniszewski et al. 1976). In ethanol-fed rats a depressant action of acute administration of alcohol upon bile flow and bile salts concentration has also been reported (Maddrey and Boyer 1973).

Chronic Alcoholism and Basal Pancreatic Secretion

Three different groups of investigators (Hayakawa et al. 1982; Renner et al. 1980; Sahel and Sarles 1979) have reported an increased "basal" pancreatic enzyme secretion in chronically alcoholic men as compared to nonalcoholics. Studies in humans in whom pure pancreatic juice was obtained by endoscopic cannulation of the pancreatic duct have revealed a significantly higher protein concentration in the few first minutes after intravenous injection of either secretin or CCK, or both, in chronic alcoholics as compared to normal subjects (Renner et al. 1978, 1980; Rinderknecht et al. 1979a, b, 1978; Sahel and Sarles 1979). Since the volume of pancreatic juice was similar in controls and alcoholics, these findings probably indicate true hypersecretion of proteins in chronic alcoholics. The first 5-min samples after secretin or CCK probably correspond to the secretion preloaded in pancreatic ducts and therefore give an idea of the resting secretion of the pancreas at a given time.

The difference in protein content of pancreatic juice between alcoholics and normal subjects was greater when the juice was collected immediately after alcohol abuse (Renner et al. 1978; Rinderknecht et al. 1979b) than after 1–3 weeks of abstinence in hospital (Sahel and Sarles 1979). Sahel and Sarles (1979) as well as Renner et al. (1978) observed protein precipitates more frequently in the pure pancreatic juice of alcoholics than in normal controls.

Bicarbonate concentration was significantly lower in alcoholic patients than in nonalcoholic controls in two samples following secretin injection (Sahel and Sarles 1979). Sarles and co-workers feel that the increased concentration of proteins associated with a decreased concentration of bicarbonate indicates a change of ionic environment of proteins and this could lead to precipitation of proteins.

In studies in which pancreatic juice was collected by duodenal suction, an increased duodenal chymotrypsin concentration but not output was observed in chronic alcoholics as compared to controls (Planche et al. 1982).

In a very recent study, Brugge et al. (Brugge and Burke 1982; Brugge et al., to be published), simultaneously monitoring duodenal motility and basal pancreatic secretion into the duodenum, observed a significantly higher basal output of trypsin during duodenal phase II of the interdigestive motor complex in alcoholics without pancreatitis and in alcoholics with early chronic relapsing pancreatitis than in normal volunteers. The basal total protein output was significantly higher only for alcoholics with early chronic relapsing pancreatitis. Alcoholics without and with early pancreatitis had an increased basal duodenal contraction rate compared to normal subjects, which did not change during the basal period. The authors had the chance to restudy two subjects with early alcoholic pancreatitis 5 months later, after they had reportedly abstained from alcohol. In both subjects high levels of duodenal basal trypsin output could no longer be detected. To my knowledge the study of Brugge et al. represents the first investigation in human alcoholics in which the relationship between upper intestinal motility and basal pancreatic secretion has been considered and in which the action of chronic alcoholism on interdigestive duodenal motility has been reported.

Further investigations on the action of chronic alcoholism on interdigestive pancreatic secretion under optimal conditions, that is simultaneously measuring gastrointestinal motility and pancreatic exocrine secretion, are needed to confirm the "basal hypersecretion of proteins" in human alcoholics.

The results of the action of chronic alcoholism on basal pancreatic secretion in animals are inconclusive. In dogs receiving 2 g ethanol per kilogram (50% v/v) per day intragastrically for 12–15 months, the basal pancreatic flow rate, bicarbonate concentration, bicarbonate output and conductivity were significantly lower than in nonalcoholic dogs (Noel-Jorand et al. 1981). Basal protein concentration but not protein output was significantly increased in alcoholic dogs as compared to controls. Since total calcium, magnesium, and zinc in pancreatic juice were identical in the two groups, the ratios of calcium, magnesium, and zinc to protein were decreased in alcoholic versus control dogs. On some experimental days protein plugs were observed in basal pancreatic juice of chronic alcoholic dogs but not of controls. The analysis of these plugs mainly identified three components: (a) desquamated acinar and duct cells, (b) crystallized calcium, and (3) amorphous protein material. The authors felt that the change in ionic distribution in pancreatic juice of alcoholic dogs which caused a new Donan equilibrium could lead to modifications of protein and calcium solubilities with formation of precipitates (Noel-Jorand et al. 1981).

This careful and long overdue study by Noel-Jorand et al. (1981) has two limitations. Only four chronically alcoholic dogs have been studied and the association between pancreatic exocrine secretion and the interdigestive motor activity of the upper intestine has not been taken into account. It is therefore questionable that a 2-h collection period of pancreatic juice which they performed represents true basal secretion. In addition, basal pancreatic flow rate in dogs is rather low (below 2 ml h^{-1} in the study of Noel-Jorand et al.), which creates collection problems when a Thomas or Scott cannula is used. Despite these limitations, the study of Noel-Jorand et al. is the first one in which "basal" pancreatic secretion of alcoholic dogs has been studied systematically.

In conscious rats receiving ethanol ad libitum for 3 months, a decreased basal protein secretion was found (Cavarzan et al. 1975). In another study of the same group (Sarles et al. 1971b) basal protein secretion was significantly higher after 30 months of oral ethanol than in a control group. Huttunen et al. (1976) did not observe any significant differences in basal pancreatic secretion between chronically alcoholic (for more than 10 weeks) and nonalcoholic rats. Laugier and Sarles (1977) observed an increase in basal flow rate in rats after 8 months of alcoholism but no more after 12 months. In another study (Demol et al. 1980) in conscious rats an increased basal flow rate was found after 18 months of oral ethanol when compared to nonalcoholic rats. Basal pancreatic protein output did not differ between the two groups. Chariot et al. (1980) found a depressed basal secretion of bicarbonate and protein in alcoholic rats as compared to controls.

The differences in the results of the effect of chronic oral administration of ethanol on the basal pancreatic secretion may be explained partly by the different duration of oral ethanol administration and partly by the diet which the animals received. In rats, chronic intake of ethanol augments the concentration of enzymes in basal pancreatic secretion if the diet is rich in protein and fat, but diminishes it if the diet is low in these elements (Sarles et al. 1971a).

Another explanation (which is true for all studies of basal pancreatic secretion) might be the cyclic pattern of interdigestive pancreatic secretion. Recent studies by the Mayo Clinic group (Keane et al. 1980, 1981) demonstrated a close association of pancreatic secretion and interdigestive motor activities in humans and in dogs. Depending on the phase of duodenal motor activity at which the study of basal pancreatic secretion is performed, either a high or a low protein secretion may be observed. Except for the study by Brugge et al. (Brugge and Burke 1982; Brugge et al., to be published) in human alcoholics, in no human or animal study on the action of chronic alcoholism has the close interrelationship between exocrine pancreatic interdigestive secretion and motility of the upper gastrointestinal tract been taken into account. Therefore, to be meaningful, further studies should consider this association.

Chronic Alcoholism and Hormone-Stimulated Pancreatic Secretion

Whether, in humans, the sensitivity of the pancreatic duct and acinar cells to exogenous hormones is changed by chronic alcoholism is not known, since no pancreatic dose-response curves to secretin or CCK have been reported in chronically alcoholic men without pancreatic disease. An increase in secretin-stimulated volume and bicarbonate output was demonstrated during the early stages of alcoholism, which, however, declined with continued alcohol intake (Dreiling et al. 1973). This was ascribed to an increase in pancreatic mass due to regeneration in association with early inflammation of the pancreas. In that study duodenal juice was aspirated, and many of the patients had simultaneous liver cirrhosis, which is known to cause hypersecretion of water and electrolytes into the duodenum.

When pure pancreatic juice was collected by endoscopic retrograde catheterization of the papilla, the flow rate in response to an intravenous bolus injection of secretin did not differ between chronic alcoholics and normal controls but was significantly lower in patients with alcoholic pancreatitis than in controls. Bicarbonate concentration but not output was significantly lower in the first and third 1-min samples after secretin in alcoholics than in healthy controls (Sahel and Sarles 1979).

Alcoholics without pancreatic disease and alcoholics with early pancreatitis had similar duodenal trypsin and protein outputs in response to an intravenous injection of CCK octapeptide (40 ng kg^{-1}) to healthy controls (Brugge and Burke 1982; Brugge et al., to be published).

In animals, pancreatic exocrine secretion in response to exogenous hormones is influenced by the duration of chronic alcoholism. Long-term follow-up studies have been done by Sarles and co-workers (Sarles and Tiscornia 1974; Sarles et al. 1977) in dogs receiving 2 g ethanol per kilogram (50% v/v) per day intragastrically together with a diet rich in protein and fat for more than 2 years. During the period of observation of about 2–3 years the following sequential changes in pancreatic secretion in response to intravenous secretin or CCK were observed:

1. As early as after 6 months of chronic ethanol administration, maximal flow rate and bicarbonate output in response to different doses of secretin were significantly higher in chronically alcoholic than in nonalcoholic dogs (Schmidt et al. 1982a). Maximal outputs of volume and bicarbonate were obtained at 1.2 and 0.6 μg kg^{-1} h^{-1} of secretin in the ethanol and control animals respectively. Administration of ethanol for 12 months did not further increase the pancreatic output from values obtained after 6 months of ethanol administration (Schmidt et al. 1982a). The D_{50} of secretin for water and bicarbonate stimulation was similar in both nonalcoholic and alcoholic animals. In another study it was found that after 2 years of ethanol administration maximal volume and bicarbonate responses to secretin were still significantly higher in alcoholic than in nonalcoholic dogs (Sarles et al. 1977). The increased maximal response of volume and bicarbonate to secretin is probably not due to an increased sensitivity of ductal cells to secretin but to an increased number of ductal cells, since ductal reduplication has been found in pancreatic biopsies after 2 years of ethanol administration in these dogs (Sarles et al. 1977).
2. Atropine (5 μg kg^{-1} h^{-1} IV) did not significantly alter the volume and bicarbonate response in normal dogs and in alcoholic dogs after 6 and 12 months of alcohol feeding.
3. A most interesting finding of the study by Schmidt et al. (1982a) is that the pancreatic protein output increased after 0.15 and 0.30 μg kg^{-1} h^{-1} of secretin in the ethanol animals after 6 months of alcohol feeding but did not change in the untreated control dogs. The administration of atropine (5μg kg^{-1} h^{-1}) decreased protein output in response to all doses of secretin in control dogs, while it only inhibited protein output at the lowest doses (0.15–0.60 μg kg^{-1} h^{-1}) of secretin in the ethanol animals. The finding that secretin did not stimulate protein output and that atropine inhibited the protein output during secretin in normal dogs confirms the observations of other investigators (Singer et al. 1981). The mechanism of the stimulation of pancreatic protein output by secretin in alcoholic dogs is unknown and remains to be determined.
4. Dose-response curves for protein output to cerulein or CCK were not altered by 9, 12, or 24 months of daily ethanol administration. There was no statistically significant difference in output whether the responses were compared to the preethanol values in the same dog or to responses of a control group (Sarles et al. 1977; Schmidt et al. 1981, 1983b).
5. Intravenous infusion of atropine (5 or 40 μg kg^{-1} h^{-1}) enhanced the incremental protein response to 100 ng kg^{-1} h^{-1} (but not higher or lower doses) of cerulein in alcohol-fed (for 9 months) dogs (Schmidt et al. 1981). The enhancement by atropine (5 μg kg^{-1} h^{-1}) of the protein response to cerulein was smaller in alcoholic than in nonalcoholic dogs. After 36 months of chronic alcoholism no enhancement by atropine of the protein response to cerulein could be detected (Schmidt et al. 1981). The authors postulated the existence of an atropine-sensitive factor inhibiting the protein response, which is diminished by chronic alcohol feeding. Contrary to the results of Schmidt et al. (1981), an enhancing effect of atropine on pancreatic protein response to graded doses of cerulein could not be found by other investigators in normal dogs (Singer et al. 1980).

6. The pancreatic response to a constant intravenous infusion of secretin plus CCK varied most during 24 months of chronic ethanol feeding. Protein and bicarbonate responses were significantly higher from the 6th week up to the end of the 14th week when compared with the respective values during the prealcoholic phase (Sarles et al. 1973). From the 15th week up to 24 months no significant changes in bicarbonate and protein in comparison with the prealcoholic response were found.
7. In dogs receiving alcohol for 3 or 12 months, an intravenous infusion of 400 $\mu g\,kg^{-1}\,h^{-1}$ (but not lower doses) of bethanechol increased pancreatic protein output above basal. In control dogs, also lower (50–200 $\mu g\,kg^{-1}\,h^{-1}$) doses of bethanechol stimulated pancreatic protein secretion. Atropine (5 $\mu g\,kg^{-1}\,h^{-1}$) inhibited the protein response to 400 $\mu g\,kg^{-1}\,h^{-1}$ of bethanechol in control dogs and in dogs after 3 and 12 months of chronic alcoholism (Schmidt 1983; Schmidt et al. 1982b). The authors concluded that after as little as 3 months chronic alcohol feeding leads to a diminished pancreatic responsiveness to cholinergic stimulation and inhibition of protein output, as evidenced by at least a fourfold increase in the minimum effective dose of bethanechol and diminished inhibitory action of atropine (Schmidt et al. 1982b).
8. After 1–2 years of chronic alcoholism the pancreas of these animals may be completely normal or may have lesions (protein precipitates in the ducts, periductal fibrosis, hyperplasia and duplication of the ducts) but extremely rarely shows atrophy of the acini (Sarles et al. 1977).

Effect of Intravenous Ethanol on Pancreatic Secretion in Chronically Alcoholic Humans and Laboratory Animals

Intravenous ethanol has a different action on the exocrine pancreatic secretion of alcoholics and nonalcoholics. When the pancreas was stimulated with a submaximal dose of secretin and CCK, an intravenous infusion of 600 $mg\,kg^{-1}$ of ethanol (10% v/v) decreased pancreatic enzyme and bicarbonate secretion by at least 50% in nonalcoholic men, but did not modify flow rate and output of chymotrypsin into the duodenum in alcoholic humans (Planche et al. 1982). So far, the action of intravenous ethanol on basal pancreatic secretion in relationship to the interdigestive motor activity of the upper intestine has not been studied in alcoholic humans.

Pancreatic response to intravenous ethanol is different in nonalcoholic dogs and depends on the duration of chronic alcoholism. In dogs receiving 2 g ethanol per kilogram daily intragastrically, an intravenous infusion of ethanol (0.7 or 2.5 $g\,kg^{-1}$) no longer inhibited pancreatic bicarbonate and protein response to secretin and CCK, but stimulated bicarbonate and protein output after 1 year of chronic alcoholism (Tiscornia et al. 1973a). These results have been confirmed by Schmidt et al. (1982a), who in addition demonstrated that as few as 6 months of alcohol feeding are sufficient to prevent the inhibitory action of intravenous ethanol on pancreatic exocrine secretion of normal dogs.

Just as it prevented the ethanol-induced inhibition of pancreatic secretion in the nonalcoholic period, atropine (1 mg/h IV) (Tiscornia et al. 1975a) prevented the stimulatory action of intravenous ethanol on the pancreas after 1 year of chronic alcoholism.

Pentolinium (750 μg kg^{-1} IV) and truncal vagotomy (Tiscornia et al. 1974a) are claimed not to abolish the stimulatory action of intravenous ethanol on pancreatic secretion in chronically alcoholic dogs, but data have not been published so far.

In contrast to previous studies, Demol et al. (1980) did not find different actions of intravenous ethanol on pancreatic secretion of alcohol-fed (for 18 months) and nonalcoholic Sprague-Dawley rats. Only a high dose of intravenous ethanol (4 g kg^{-1} but not 2 g kg^{-1}) caused a significant inhibition of pancreatic output in both groups.

In Vitro Studies on the Action of Chronic Ethanol Feeding on the Pancreas

Most in vitro studies on the action of chronic alcoholism have been performed in the rat with contradictory results (for review see Sarles and Laugier 1981; Singh 1983; Singh et al. 1982). In an earlier study, Sarles et al. (1971b) observed in rats receiving 20% (v/v) ethanol per day ad libitum for 2 years lesions resembling human chronic pancreatitis: intraductal protein precipitates, dilatation and neogenesis of ducts. atrophy of acini. These lesions were sometimes observed in normal animals but were significantly more frequent in alcoholic rats (Sarles et al. 1971b). In the alcoholic animals, the protein concentration of the juice was significantly increased and the transit time of labeled proteins from zymogen granules to the lumen of the acinus was lower than in nonalcoholic controls, demonstrating a hypersecretion (Lechene de la Porte and Sarles 1974). Other investigators, however, could not confirm these findings (Singh 1983; Singh et al. 1982).

In a well-controlled recent study, Singh et al. (1982) examined the action of prolonged ethanol administration on the morphology and protein metabolism in the rat pancreatic acinar cells. Weight-matched triplets of Sprague-Dawley rats were fed Lieber-DeCarli diet containing 5% (w/v) concentration of ethanol, isocaloric amounts of Lieber-DeCarli diet, or rat chow ad libitum for 6, 12, and 18 months. In the ethanol-fed group, histologic studies by light microscopy showed absence of protein plugs in the pancreatic ducts and/or pancreatitis, but electron microscopic evaluation revealed progressive accumulation of lipid droplets in acinar and ductal cells. No definite changes in the mitochondria and endoplasmic reticulum were noticed.

The biochemical data also suggested that ethanol altered neither structural proteins nor exportable proteins because incorporation into microsomal proteins and deoxycholate soluble subfractions of microsomal vesicles showed an aging effect but no ethanol effect. The findings by Singh et al. (1982) are – in some aspects – in contrast to those of Sarles et al. (1971), which indicated an increased concentration of exportable proteins in the pancreas of ethanol-fed animals, and those of Sardesai and Orten (1968) of an inhibitory effect not only on total protein synthesis, but also on that of exportable proteins, including trypsin and ribonuclease.

Singh et al. (1982) also observed an increased specific activity of trypsinogen, chymotrypsinogen, and lipase and decreased specific activity of amylase. Trypsin-inhibiting capacity was decreased in the tissue and in the medium in a progressive fashion.

In addition, they observed that prolonged ethanol feeding produced a slightly diminished protein secretion as well as an absence of protein plug formation as described by Sarles et al. (1971b) and Kagaya et al. (1979). Sarles et al. had suggested that these plugs were due to hyperconcentration of proteins, whereas Kagaya et al., by using the same strain of animals given ethanol in the same way, found decreased secretin and CCK-like activity in the duodenal mucosa and significantly lowered protein concentrations in the pancreatic juice. Reber et al. (1976) were also unable to observe any protein plug formation in female Wistar rats maintained on ethanol for 2 years. Chariot et al. (1977) reported decreased protein secretion in Wistar rats fed ethanol for 3 months and suggested that apparent discrepancies of decreased or increased protein secretion might be due to biphasic changes associated with various degrees of ethanol injury over different periods of time. This is not supported by the study of Singh et al. (1982), as both basal and stimulated protein secretion in vitro studied sequentially did not vary due to ethanol administration lasting 6–18 months.

Singh et al. (1982) concluded from their study that prolonged ethanol feeding resulted in significant alterations of digestive enzymes, including an increase in tissue and medium content of trypsinogen and chymotrypsinogen and a decreased trypsin-inhibiting capacity of tissue and medium. In another study Singh (1983) found an increased basal secretion of pancreatic enzymes of ethanol-fed rats at 6, 12, and 18 months. Secretion of pancreatic secretory trypsin inhibitor was stimulated by bethanechol, whereas the secretion of pancreatic enzymes was not stimulated in the ethanol-fed rats. In the ethanol-fed rats the dose-response curve of amylase and lipase secretion was shifted upward with an increase in ED_{50}. These data indicated a subsensitivity of the cholinergic receptors in the ethanol-fed rats.

In a recent acute in vitro study of rabbit pancreas, trypsin-inhibiting capacity in the bathing medium was significantly decreased in the 2 h after adding 1% ethanol and in both hours after adding 3% ethanol. It was concluded that this could lead to premature activation of proteolytic enzymes and the development of acute pancreatitis (Steer et al. 1979). The relevance of increased proenzymes and decreased trypsin-inhibiting capacity in the in vitro model exposed to high concentrations of ethanol to the pathogenesis of pancreatitis remains plausible but speculative.

In other animal models of chronic alcoholism, both increases and decreases in the content of specific digestive enzymes were reported (Chariot et al. 1977; Jalovaara and Hultunen 1977; Sardesai and Orten 1968; Sarles et al. 1971a). The reasons for these conflicting data are not well understood. In many studies, 15%–20% ethanol in water was used as the source of drinking fluid, with the result that total intake of ethanol was variable and hard to quantitate. Moreover, prolonged ethanol administration by this method leads necessarily to variations in caloric intake that could modify the function of the organ under study. To develop a reliable model for study of ethanol effect on pancreas, Singh et al. used a liquid diet developed by Lieber and DeCarli with which to administer ethanol to the extent of 36% of total calories in the absence of dietary deficiency.

The action of ethanol on the rat pancreas is dependent on many variables: 2 months' alcoholism increases the enzyme content of the fasting rat pancreas when associated with a high-fat, high-protein diet but decreases it when associated with either low-fat or low-protein diets (Sarles et al. 1971a). With a high-protein, low-fat diet, ethanol increases protein secretion at the end of 6 months but decreases it at the end of 12 months, with the exception of some animals which maintained a high secretory rate (Laugier and Sarles 1977).

Dürr et al. (1979), studying enzyme secretion of isolated pancreatic acini of rats after different periods of time of ethanol feeding (1–3 months), observed no difference in basal outputs of different enzymes between control rats and those fed alcohol for 1, 2, or 3 months. Cerulein induced a higher chymotrypsin output after 2 and 3 months and a higher trypsin output after 1, 2, and 3 months of ethanol feeding, amylase output not being changed. The authors concluded that, with the exception of emylase, enzyme outputs of isolated pancreatic acini of rats are stimulated by oral intake of alcohol for various periods of time. This effect is more pronounced for lipase than for trypsin and chymotrypsin.

Possible Mechanisms of the Action of Ethanol on the Pancreas

On the basis of the evidence of the above-mentioned experimental studies in humans and animals the following hypotheses of the mechanisms of action of ethanol on pancreatic exocrine secretion can be discussed.

1. Ethanol might influence the pancreatic secretion by a direct effect on the secretory cells. This direct effect might be localized to events in membrane transport, since it has been found that ethanol affects ion movements in a variety of organs (Kalant 1971). Ethanol could impair the energy supply of the secretory cells (Solomon et al. 1974). Furthermore, ethanol influences the mechanical features of the cells such as motility, pinocytosis, and exocytosis (Kalant 1971). Since only low concentrations of alcohol dehydrogenase activity are found in the pancreas (Schmidt and Schmidt 1960), a toxic effect due to metabolic changes as in the liver is not probable. Nevertheless, ethanol is metabolized to some extend by the pancreas. Whether the metabolites of alcohol mediate the pancreatic response to ethanol has not been studied thoroughly. Acetaldehyde, the main metabolite of alcohol, does not appear to cause pancreatic lesions (Andersen et al. 1980; Fitzpatrick et al. 1975b).
2. Atropine, pentolinium, and truncal vagotomy alter the pancreatic response to ethanol in the intact animal. It has been postulated that in the nonalcoholic animal alcohol inhibits the pancreatic secretion by its effect on the intracranial ganglia of the vagal nerves and that this action would be transmitted to the pancreas via inhibitory vagal fibers (Sarles and Tiscornia 1974). In the chronically alcoholic dog, however, it has been proposed that ethanol augments pancreatic protein secretion by its action at the intrapancreatic ganglia of the vagal nerves, augmenting the "cholinergic tone" of the gland (Sarles and Tiscornia 1974). The finding that acetylcholinesterase of the intrapancreatic nerve endings is diminished in the chronically alcoholic dog while choline acetyltransferase is increased has been taken as evidence for this hypothesis (Celener et al. 1977; Perec et al. 1979; Yahata et al. 1982).

However, on the basis of the available experimental evidence, it cannot be decided at present whether these hypotheses are correct. For example, the existence of inhibitory fibers in the vagal nerves has never been proven.
Another means by which the vagal nerves could mediate the stimulatory action of intraduodenal ethanol on pancreatic bicarbonate secretion has not been studied: release of VIP. It is known that electric stimulation of the vagal nerves releases VIP (Schaffalitzky de Muckadell et al. 1975). VIP has been shown to stimulate pancreatic bicarbonate secretion in different species.

3. The possible mediation of the ethanol action on the pancreas by the peptidergic and adrenergic innervation of the pancreatic ganglia has not been tested so far. Immunocytochemical studies have revealed the presence of nerves showing not only VIP immunoreactivity but also CCK, substance P, and enkephalin immunoreactivity in the pancreas of different species, including cat, dog, pig, and man (Larsson and Rehfeld 1979). The peptidergic and adrenergic nerve fibers are detected in ganglia of the pancreas, and only occasionally in the vicinity of pancreatic exocrine cells. Since the intrapancreatic ganglia of most species seem to represent important integration centers for incoming peptidergic and adrenergic innervation, it seems essential to develop specific means for studying the action of the ethanol on the function of the ganglia.
4. Another unexamined possibility is that ethanol might influence the function of the duct and acinar cells by release of a pancreatic peptide acting as a paracrine transmitter. Somatostatin and PP are candidates for paracrine transmitters, since they have been found in the islets of Langerhans and have been shown to inhibit pancreatic exocrine secretion (Taylor et al. 1978). The observation (Singer et al. 1983) that ethanol and alcoholic beverages did not increase plasma levels of PP in humans and dogs does not rule out a small release of PP sufficient to influence the adjacent cells via paracrine (that is without entering the circulation) transmission. Up to now we do not have methods of measuring paracrine secretion. Thus paracrine function is still a hypothetical concept, not an established mechanism.
5. Besides its action on the extrinsic and intrinsic innervation of the pancreas, administration of ethanol might influence the pancreas by release of gastrointestinal hormones. This action of ethanol might include release of hormones which stimulate the pancreas such as gastrin, and of inhibiting hormones such as somatostatin or neurotensin. With the availability of radioimmunoassays these possibilities should be clarified in the near future.
6. Another possibility proposed by Korsten et al. (1981) is that ethanol inhibits hormonally induced pancreatic secretion by blocking the pancreatic response to a given secretagogue. Studies utilizing isolated guinea pig pancreatic cells are consistent with this interpretation (Uhlemann et al. 1979): ethanol increased the release of amylase from cells in the basal state but had an inhibiting effect on cells that were stimulated with secretin. Uhlemann et al. (1979) have proposed that ethanol inhibits the pancreatic response to secretagogues by distorting the receptors, but it is unknown whether ethanol could disturb membrane structure at the lower concentrations used in the study of Korsten et al. (1981). It is also conceivable that intravenous ethanol releases a polypeptide or another neurohormonal substance that acts as a competitive agonist of pancreatic

secretion. Such an agonist would increase basal state secretion but inhibit secretin-stimulated secretion if its efficacy (effectiveness of the agonist/receptor complex) were less than that of secretin.

Summary and Conclusion

There is considerable experimental evidence that acute and chronic administration of ethanol induces secretory modifications of the pancreas in humans and laboratory animals. Acute intravenous administration of ethanol inhibits secretin and cholecystokinin-stimulated pancreatic secretion of water, bicarbonate, and protein and nonalcoholic humans and most species of animals tested. In vitro studies suggest a direct toxic action of ethanol on the pancreatic acinar and ductal cells. The precise mechanism operative at the molecular level has not been identified. Since atropine, pentolinium, and truncal vagotomy diminished the ethanol-induced inhibition of pancreatic exocrine secretion in the intact animal, it was postulated that the action of ethanol on the pancreas is at least partly mediated by inhibitory cholinergic mechanisms (e.g., inhibitory fibers of the vagal nerves). Whether the extrinsic and intrinsic peptidergic nerves and paracrine transmitters are involved is not known. It is also conceivable that intravenous ethanol, inhibits hormonally induced pancreatic secretion by blocking the pancreatic response to these secretagogues (e.g., by distorting their receptors).

Intragastric (with and without diversion of gastric juice) and intraduodenal ethanol have been shown to produce either no change or a slight and highly variable increase in basal (interdigestive) and hormonally induced pancreatic exocrine secretion in humans and different species of animal. In dogs, the stimulatory action of intragastric instillation of ethanol is probably due to release of gastrin by ethanol and subsequent direct (gastrin itself) and indirect (acid secreted in response to gastrin causing secretin release) effects on the pancreas. In humans, however, it is uncertain whether the mechanism is operative. The stimulatory action of intraduodenal ethanol on the pancreas might be mediated either by neural reflexes or release of gastrointestinal hormones.

Little is known about the releasing action of ethanol on gastrointestinal hormones. In humans, pure ethanol in different concentrations and whisky do not stimulate release of gastrin. Beer and red and white wine, however, are potent stimulants of gastrin. Neither pure alcohol nor the above-mentioned alcoholic beverages release pancreatic polypeptide in humans. In dogs, only a high dose (2.7 g kg^{-1} body wt.) of intragastric ethanol caused a small increase in plasma levels of pancreatic polypeptide, which was only 12% of the response seen after a meal. Therefore, it is unlikely that pancreatic polypeptide is the hormonal mediator of the ethanol-induced inhibition of exocrine pancreatic secretion in humans and dogs. Whether the inhibitory action of intravenous ethanol is mediated by release of other gastrointestinal hormones, e.g., neurotensin, has not been studied.

The results of the action of chronic alcoholism on interdigestive (basal) pancreatic secretion in humans and animals are inconclusive. There is some evidence that in chronically alcoholic humans and dogs basal secretion of pancreatic enzymes is increased, whereas bicarbonate output and water output are decreased.

Pancreatic response to intravenous ethanol is different in nonalcoholic and alcoholic dogs and depends on the duration of chronic alcoholism. In dogs receiving 2 g ethanol per kilogram daily intragastrically for at least 12 months, an intravenous infusion of ethanol (1.5 g kg^{-1}) no longer inhibited but stimulated hormonally stimulated pancreatic bicarbonate and protein output. In alcoholic men, the reversal of the action of intravenous ethanol was not observed, but the inhibitory action of alcohol disappeared. As it prevented the ethanol-induced inhibition of pancreatic secretion in the nonalcoholic period, atropine prevented the stimulatory action of intravenous ethanol on the dog pancreas after 1 year of chronic alcoholism.

After 12 and 24 months of alcoholism, protein is not modified in response to caerulein or cholecystokinin. As early as 6 months after the beginning of chronic ethanol feeding, maximal volume and bicarbonate responses to secretin are higher in chronically alcoholic than in nonalcoholic dogs. This is still the case after 12 and 24 months of ethanol feeding. Atropine did not alter the bicarbonate response in alcoholic and nonalcoholic dogs. Paradoxically, secretin stimulated protein output in alcoholic but not in nonalcoholic dogs.

After as little as 6 weeks of chronic alcoholism, dogs secreted precipitates through the pancreatic cannula. According to Sarles, these protein precipitates consist of proteins normally found in pancreatic juice and are believed to be due to ethanol-induced hyperconcentration of pancreatic proteins. The intraductal proteins form the matrix of calcified pancreatic calculi. The three-dimensional anatomical studies of the pancreas suggest that protein precipitates and calculi form the first lesion of chronic pancreatitis in man (Nakamura et al. 1972; Tasso et al. 1973). In alcoholics without pancreatic disease, the concentration of protein in pancreatic juice was found to be higher than in nonalcoholic controls. The protein precipitates as well as the pancreatic stones have been shown to be dissolved in vivo and in vitro by citrate and other calcium chelators (Lohse et al. 1979, 1971 a, b, 1983).

Recently, the group of Sarles has found a protein which they call "pancreatic stone protein" and which constitutes a stable fraction of juice protein in controls, in alcoholics without pancreatic lesions, and in patients with acute pancreatitis amounting to 18% of total pancreatic juice protein (de Caro et al. 1979; Guy et al. 1983; Multigner et al. 1983). From preliminary observations it would appear that the concentration of this protein is low in congenital and in alcoholic calcified pancreatitis. The authors speculated that this reduced concentration of pancreatic stone protein could explain the spontaneous development of hereditary pancreatitis and the individual risk of developing alcoholic pancreatitis.

Littls is known about the basal and stimulated plasma concentrations of gastrointestinal hormones in chronic alcoholic humans and animals. Normal, decreased, and increased basal levels of gastrin have been found in chronic alcoholic humans. The same is true for basal and stimulated gastric acid secretion in patients with chronic alcoholic pancreatitis. In one study, plasma gastrin levels in response to a meal were higher in patients with chronic alcoholic pancreatitis or in "normal" alcoholics than in nonalcoholic controls. In chronically alcoholic dogs lower plasma levels of secretin in response to intraduodenal HCl have been found than in nonalcoholic control dogs. Whether the release of other gastrointestinal hormones such as cholecystokinin, vasoactive intestinal polypeptide, and somatostatin and pancreatic polypeptide is altered by chronic alcoholism has not been studied systematically.

In chronic alcoholism the dietary regimen associated with alcohol feeding seems to be important for the action of alcohol on the pancreas (Pitchumoni et al. 1980; Sarles 1973). In rats, ethanol feeding together with a diet rich in protein and fat resulted in an increased pancreatic enzyme secretion, whereas the opposite action was seen with a diet low in fat and protein. Contrary to the results in dog, studies on the long-term effect of ethanol feeding on the pancreas of rats have given contradictory results and are therefore inconclusive. In humans, it has been shown that the risk of developing chronic pancreatitis increases logarithmically with rising alcohol consumption (Durbec and Sarles 1978). High protein intake and both high and low fat intake clearly increase the risk. There is a logarithmic increase in the risk with high protein intake, but the effect of protein is much less than that of alcohol. In the case of fat intake, the lowest risk was observed with an average diet (80–110 g/day), the highest with a high-fat diet, and an intermediate risk with a low-fat diet. The action of alcohol, protein, and fat was additive. While in Europe chronic alcoholic pancreatitis is often associated with a high-alcohol, high-fat, high-protein diet, a recent North American study revealed low fat and protein intakes but an alcohol consumption which was much higher than in European series (Durbeck and Sarles 1978; Sarles 1973; Sarles et al. 1979).

In conclusion, despite the many animal and the few human studies, the mechanisms by which chronic oral ingestion of alcohol influences the pancreas are still not known. Whether ethanol-induced changes in release of gastrointestinal hormones and/or alterations of the neural tone of the pancreas are responsible for the effects of chronic alcoholism on the pancreas is still speculative. In addition, there is still a lot of descriptive work to do. Thus under well-controlled conditions dose-response curves of basal and hormonally stimulated pancreatic response to increasing doses of ethanol given by different routes need to be drawn. There are also no studies on the action of different doses of ethanol on pancreatic dose-response curves to intestinal stimulants such as HCl, amino acids, and fatty acids. The action of acute and chronic ethanol on the different phase of basal pancreatic secretion needs to be clarified. The close interrelationship between the interdigestive motor complex and pancreatic exocrine secretion needs to be considered in studies of that kind. Since in the animal species studied so far pancreatic lesions similar to chronic alcoholic human pancreatitits could not be produced, we should search for a more suitable animal in order to elucidate the action of chronic alcohol feeding.

The acute action of alcoholic beverages on exocrine pancreatic secretion in humans is completely unknown. Since humans rarely ingest pure ethanol and since the congeners of alcoholic beverages might be important for the action of ethanol on the pancreas, commonly ingested alcoholic beverages such as beer, wine, cognac, vodka, and whisky should be tested.

References

Adler G, Rohr G, Kern HF (1982) Alteration of membrane fusion as a cause of acute pancreatitis in the rat. Dig Dis Sci 27: 993–1002

Andersen BN, Demol P, Treffot MJ, Sarles H (1980) The effect of acetaldehyde on pancreatic and gastric secretion. Scand J Gastroenterol 15: 805–809

Bank S, Marks IN, Groll A (1966) Gastric acid secretion in pancreatic disease. Gastroenterology 51: 649–655

Bayer M, Rudick J, Lieber CS, Janowitz HD (1972) Inhibitory effects of ethanol on canine exocrine pancreatic secretion. Gastroenterology 63: 619–626

Becker HD, Reeder DD, Thompson JC (1974) Gastrin release by ethanol in man and dogs. Ann Surg 906: 1979

Bergh GS (1942) The sphincter mechanism of common bile duct in human subjects: its relation to certain types of stimulation. Surgery 11: 299–330

Besterman HS, Adrian TE, Bloom SR et al. (1982) Pancreatic and gastrointestinal hormones in chronic pancreatitis. Digestion 24: 195–208

Bordalo O, Goncalves D, Noronha M, Christina ML, Saldadinho A, Dreiling DA (1977) Newer concept for the pathogenesis of chronic alcoholic pancreatitis. Am J Gastroenterol 68: 278–285

Bretholz A, Levesque D, Voirol M, Tiscornia OM, Bloom SR, Sarles H (1978) Impaired secretin release in chronic alcoholic dogs. Digestion 17: 437–440

Brody S, Malmud LS, DeVegvar ML et al. (1982) Effects of alcohol on gallbladder emptying. Gastroenterology 82: 1026 (abstract)

Brooks FP, Thomas JP (1953) The effect of alcohol on canine external pancreatic secretion. Gastroenterology 23: 36

Brugge WR, Burke C (1982) Basal trypsin outputs in alcoholic pancreatic disease. Abstract, Meeting of the American Pancreatic Association

Calderon-Attas P, Furnelle J, Christophe J (1980) In vitro effects of ethanol and ethanol metabolism in the rat pancreas. Biochim Biophys Acta 620: 387–399

Cameron JL, Zuidema GD, Margolis S (1975) A pathogenesis for alcoholic pancreatitis. Surgery 77: 755–763

Capitaine Y, Sarles H (1971) Action de l'éthanol sur le tonus du sphincter d'Oddi chez l'homme. Biol Gastroenterol 3: 231–236

Capitaine Y, De Barros Mott C, Gullo L, Sarles H (1971) Action de l'ethanol sur la sécrétion pancréatique chez l'homme. Biol Gastroenterol 3: 193–198

Cavarzan A, Teixeira AS, Sarles H, Palasciano G, Tiscornia O (1975) Action of intragastric ethanol on the pancreatic secretion of conscious rats. Digestion 13: 145–152

Celener D, Lechene de la Porte P, Tiscornia O, Sarles H (1977) Histochemical study of cholinergic activities in exocrine pancreas of dogs: modifications related to chronic alcoholism. Biomedicine 27: 161–165

Chariot J, Roze C, de la Tour J, Souchard M, Hollande E, Vaille C, Debray C (1977) Effects of chronic ethanol consumption on pancreatic response to central vagal stimulation by 2-deoxy-D-glucose in the rat. Digestion 15: 425–437

Chariot J, Le Bourhis B, Roze C (1980) Response of rat exocrine pancreas to chronic pulmonary ethanol intoxication. Dig Dis Sci 25: 656–659

Cheli R, Giacosa A, Marenco G, Canera M, Dante GL, Ghezzo L (1981) Chronic gastritis and alcohol. Z Gastroenterol 19: 459–463

Chey WY, Kusakioglu O, Dinoso V, Lorber SH (1968) Gastric secretion in patients with chronic pancreatitis and in chronic alcoholics. Arch Intern Med 122: 399–403

Chey WY, Kosay S, Lorber SH (1972) Effects of chronic administration of ethanol on gastric secretion of acid in dogs. Am J Dig Dis 17: 153

Christophe J, Furnelle J, Calderon-Attas P (1980) Acute in vitro effects of ethanol on the rat pancreas. In: Stock C, Sarles H (eds) Alcohol and the gastrointestinal tract. INSERM, Paris, pp 243–248

Clemente F, Durand S, Laval J, Thouvenot JP, Ribet A (1977) Métabolisme de l'éthanol par le pancréas de rat. I. Recherche des éventuelles enzymes impliquées et de leur variation de niveau. Gastroenterol Clin Biol 1: 39–48

Cooke AR (1970) The simultaneous emptying and absorption of ethanol from the human stomach. Am J Dig Dis 15: 449
Cooke AR (1972) Ethanol and gastric acid secretion. Gastroenterology 62: 501–502
Cueto J, Tajen N, Zimmermann B (1967) Studies of experimental alcoholic pancreatitis in the dog. Surgery 62: 159–166
Dagorn J, Figarella C, Michel R, Sarles H (1974) Effet de l' alcoolisme chronique sur la synthèse des enzymes pancréatiques chez le rat avant et après stimulation éndogène de la sécrétion. Biol Gastroenterol (Paris) 7: 201
Darle N, Ekholm R, Edlund Y (1970) Ultrastructure of the rat exocrine pancreas after long term intake of ethanol. Gastroenterology 58: 62–72
Davis AE, Pirola RC (1966) The effects of ethyl alcohol on pancreatic exocrine function. Med J Aust 2: 757–760
de Caro A, Lohse J, Sarles H (1979) Characterisation of a protein isolated from pancreatic calculi of men suffering from chronic calcifying pancreatitis. Biochem Biophys Res Commun 87: 1176–1182
Demol P, Andersen BN, Sarles H (1980) Chronic ethanol consumption and exocrine pancreatic response to ethanol and acetaldehyde in the rat. Digestion 20: 85–94
Demol P, Singer MV, Hotz J, Hoffmann U, Goebell H (1982) Wirkung von intragastralem Alkohol auf die exokrine Pankreassekretion und die interdigestive Motilität des oberen Gastrointestinaltraktes beim Menschen. Z Gastroenterol 20: 531
Demol P, Singer MV, Goebell H (1983a) Effect of intravenous ethanol on basal (= interdigestive) secretion of pancreatic enzymes and gastric acid in humans. Digestion 28: 21–22 (abstract)
Demol P, Singer MV, Hotz J, Hoffmann U, Goebell H (1983b) Effect of intragastric (ig.) ethanol on upper interdigestive motility and pancreatic exocrine secretion in humans. Gastroenterology 84: 1135 (abstract)
Diaz GR, Devaux MA, Sarles H (1984) Acute and chronic oral administration of ethanol inhibits water and bicarbonate pancreatic secretion.
Dinoso VP, Chey WY, Braverman SP, Rosen AP, Ottenberg D, Lorber SH (1972) Gastric secretion and gastric mucosa morphology in chronic alcoholics. Arch Intern Med 130: 715–719
Dreiling DA, Bordalo O (1977) A toxic-metabolic hypothesis of pathogenesis of alcoholic pancreatitis. Clin Exp Res 1: 293–299
Dreiling DA, Richman A, Fradkin NF (1952) The role of alcohol in the etiology of pancreatitis: a study of the effect of intravenous ethyl alcohol on the external secretion of the pancreas. Gastroenterology 20: 636–646
Dreiling DA, Grenstein A, Bordalo O (1973) The hypersecretory state of the pancreas. Am J Gastroenterology 59: 505–511
Dürr HK, Bode C, Schäfer HP, Bode JCh (1979) Enzyme secretion of isolated pancreatic acini of rats. Alternations by acute and long term intake of alcohol. Dan Med Bull 26 [Suppl 1]: 3
Durbec JP, Sarles H (1978) Multicenter survey on the etiology of pancreatic diseases. Relationship between the relative risk of developing chronic pancreatitis and alcohol, protein and lipid consumption. Digestion 18: 337–350
Dzieniszewski J, Tiscornia OM, Palasciano G, Domingo N, Carvazan A, Teixeira AS, Sarles H (1976) The effects of acute and chronic ethanol administration on canine bile secretion. Am J Dig Dis 21: 1037–1043
Elwin CE (1969) Stimulation of gastric acid secretion by irrigation of the antrum with some alipathic alcohols. Acta Physiol Scand 75: 1
Estival A, Clemente F, Ribet A (1981) Ethanol metabolism by the rat pancreas. Toxicol Appl Pharmacol 61: 155–165
Eysselein V, Singer MV, Wentz H, Goebell H (1981) Effect of intragastric (ig.) ethanol on release of pancreatic polypeptide (PP) and gastrin in dogs. Gastroenterology 80: 1145
Eysselein V, Singer MV, Wentz H, Goebell H (1984) Action of ethanol on gastrin release in the dog. Dig Dis Sci 29: 12–18
Fahrenkrug J, Schaffalitzky de Muckadell OB (1977) Plasma secretin concentration in man: effect of intraduodenal glucose, fat, amino acids, ethanol, HCl, or ingestion of a meal. Eur J Clin Invest 7: 201–203
Fink RS, Adrian TE, Margot DH, Bloom SR (1983) Increased plasma pancreatic polypeptide in chronic alcohol abuse. Clin Endocrinol 18: 417–421

Fitzpatrick J, Fitzgerald D, Fitzgerald P, McGeeney KF (1975a) The effect of acetaldehyde on the totally isolated canine pancreas. Biol Gastroenterol 64: 363

Fitzpatrick JM, Fitzgerald O, Fitzgerald P, McGeeney KF (1975b) A study of ethanol action on the totally isolated canine pancreas. Bull Soc Int Chir 6: 593–599

Forell MM, Otte M, Kohl HJ, Lehnert P, Stahlheber HP (1971) The influence of bile and pure bile salts on pancreatic secretion in man. Scand J Gastroenterol 6: 261–266

Fried GM, Ogden WD, Zhu XG, Greeley GH, Thompson JC (1983) Effect of alcohol on the release of cholecystokinin and pancreatic enzyme secretion. AGA Meeting, May 23–25, Washington DC

Glazer C, Baldwin C, Steer M (1976) The effect of ethanol on enzyme secretion from the rabbit pancreas in vitro. Br J Surg 63: 663

Goslin J, Hong SS, Magee DF, White TT (1965) Relationship between diet, ethyl alcohol consumption and some activities of the exocrine pancreas in the rat. Arch Intern Pharmacodyn 157: 462

Gullo L, Vezzadini P, Ventrucci M et al. (1980) Serum gastrin in chronic pancreatitis. Am J Gastroenterol 73: 33–36

Gullo L, Corinaldesi R, Casadio R, Vezzadini P, Tomassetti P, Ventrucci M, Priori P, Labo G (1983) Gastric acid secretion in chronic pancreatitis. Hepatogastroenterology 30: 60–62

Gupta S, Rao TRG (1975) Gastric acid secretion in pancreatic disease. Digestion 12: 189–191

Guy O, Robles-Diaz G, Adrich Z, Sahel J, Sarles H (1983) Protein content of precipitates present in pancreatic juice of alcoholic subjects and patients with chronic calcifying pancreatitis. Gastroenterology 84: 102–107

Hartmann W, Hotz J, Ormai S, Aufgebauer J, Schneider F, Goebell H (1980) Stimulation of bile and pancreatic secretion by duodenal perfusion with Na-taurocholate in the cat compared with jejunal and ileal perfusion. Scand J Gastroenterol 15: 433–442

Hayakawa T, Noda A, Kondo T (1982) Changes in serum pancreatic enzymes during 2 months' abstinence in asymptomatic chronic alcoholics. Am J Gastroenterol 77: 625–629

Hirschowitz BI, Pollard HM, Hartwell SW et al. (1956) The action of ethyl alcohol on gastric acid secretion. Gastroenterology 30: 244

Hotz J, Renner R, Hartmann W, Goebell H (1978) Hemmung der maximal stimulierten Pankreasenzymsekretion durch i. v. Gabe von Alkohol bei der Katze. Z Gastroenterol 14: 156

Huttunen R, Huttunen P, Jalovaara P (1976) The effect of chronic intragastric alcohol on the pancreatic secretion of the rat. Scand J Gastroenterol 11: 103–106

Jalovaara P (1979) Pancreatic exocrine secretion in the rat after chronic alcohol ingestion: nonparallel secretion of proteins and pancreatic secretory trypsin inhibitor. Scand J Gastroenterol 14: 57–63

Jalovaara P, Huttunen R (1977) The effect of chronic ethanol ingestion on the pancreatic proteolytic enzymes and their inhibitors in the rat. Scand J Gastroenterol 12: 785–792

Joffe SN, Ferri MM, O'Hare R (1979) The effect of ethanol ingestion on plasma biochemistry and the pancreas in the rat. Acta Hepatogastroenterol 26: 323

Kagaya T, Takebe T, Koizumi M, Kataoka S, Kamei T, Oyama K (1979) Effect of long-term alcohol feeding on the pancreas in rat. Gastroenterol Jpn 14: 428–435

Kager L, Lindberg S, Agren G (1972) Alcohol consumption and acute pancreatitis in men. Scand J Gastroenterol 7 [Suppl 15]: 1–38

Kalant H (1969) Alcohol, pancreatic secretion, and pancreatitis. Gastroenterology 59: 380–384

Kalant H (1971) Absorption, diffusion, distribution, and elimination of ethanol: effects of biological membranes. In: Kissin B, Begleiter H (eds) Biochemistry. Plenum, New York (The biology of alcoholism, vol 1)

Keane FB, DiMagno EP, Dozois RR, Go VL (1980) Relationships among canine interdigestive exocrine pancreatic and biliary flow, duodenal motor activity, plasma pancreatic polypeptide and motilin. Gastroenterology 78: 310

Keane FD, Dozois RR, Goul W, DiMagno EP (1981) Interdigestive canine pancreatic juice composition and pancreatic reflux and pancreatic sphincter anatomy. Dig Dis Sci 26: 577

Kimura T, Zuidema GD, Cameron JL (1980) Experimental pancreatitis: influence of secretory state of the pancreas. Surgery 88: 661

Kondo T, Magee DF (1977) The action of intravenous ethanol on gastric secretion. Proc Soc Exp Biol Med 156: 299
Korman MG, Soveny C, Hansky J (1971) Effect of food on serum gastrin evaluated by radioimmunoassay. Gut 12: 619
Korsten MA, Seitz H, Hodes SF, Klingenstein J, Lieber CS (1981) Effect of intravenous ethanol on pancreatic secretion in the conscious rat. Dig Dis Sci 26: 790–795
Kubota K, Magee D, Sarles H (1982) Action biphasique de l'éthanol intraveineux sur le pancréas exocrine du chien. Gastroenterol Clin Biol 6: 72A
Kubota K, Magee DF, Sarles H (1983) Biphasic action of intravenous ethanol on dog exocrine pancreatic secretion. Dig Dis Sci 28: 1116–1120
Larsson LT, Rehfeld JF (1979) Innervation of pancreatic ganglia. Scand J Gastroenterol 14: 433–437
Laugier R, Sarles H (1977) Effets de la consommation chronique d'alcool sur la sécrétion pancréatique exocrine du rat. Variations en fonction de la durée de consommation. Gastroenterol Clin Biol 1: 767–774
Lechene de la Porte P, Sarles H (1974) Etude par autoradiographie du transport et de l'excrétion des enzymes pancréatiques dans la cellule acineuse de rats alcooliques. Biol Gastroenterol 21: 263–267
Lechene de la Porte P, Tasso F, Cros R, Clop J, Sarles H (1979) Comparative studies of the action of long term duration alcohol drinking on rat pancreatic exocrine cells. Biomedicine 30: 270–278
Lenz HJ, Ferrari-Taylor J, Isenberg JI (1983) White wine and five percent ethanol are potent stimulants of gastric acid secretion in man. Gastroenterology 85: 1082–1087
Lillibridge CB, Yoshimori M, Chey WY (1973) Observations on the ultrastructure of oxyntic cells in alcohol-fed dogs. Am J Dig Dis 18: 443
Llanos OL, Swierczek JS, Teichmann RK, Rayford PL, Thompson JC (1977) Effect of alcohol on the release of secretin and pancreatic secretion. Surgery 81: 661
Lohse J, Charbit J, Sahel J, Schmidt D, Verine H, Sarles H (1979) Citrate secretion in human and canine pancreatic juice. Gastroenterol Clin Biol 3: 285
Lohse J, Verine H, Sarles H (1981 a) Studies on pancreatic stones. I. In vitro dissolution. Digestion 21: 125
Lohse J, Verine H, Charbit JJ, Sarles H (1981b) Studies on pancreatic stones. II. Citrate secretion in the canine pancreatic juice. Digestion 21: 198
Lohse J, Schmidt D, Sarles H (1983) Pancreatic citrate and protein secretion of alcoholic dogs in response to graded doses of caerulein. Pfluegers Arch 397: 141–143
Lowenfels AB, Masch B, Lee TCY, Rohmann M (1968) Effect of intravenous alcohol on the pancreas. Arch Surg 98: 440–441
Maddrey WC, Boyer JL (1973) The acute and chronic effects of ethanol administration on bile secretion in the rat. J Lab Clin Med 82: 215–225
Marin GA, Ward NL, Fischer R (1973) Effect of ethanol on pancreatic and biliary secretions in humans. Am J Dig Dis 18: 825–833
McCutcheon AD (1968) A fresh approach to the pathogenesis of pancreatitis. Gut 9: 296–310
Menguy RB, Hallenbeck GA, Bollmann JL, Grindlay JH (1958) Intraductal pressures and sphincteric resistance in canine pancreatic biliary ducts after various stimuli. Surg Gynecol Obstet 106: 306–310
Meullenet J, Baratta H, Sarles H (1979) Alcool et anticholinergiques: effets sur la sécrétion pancréatique exocrine de l'homme. Gastroenterol Clin Biol 3: 885–892
Mezey E, Potter JJ (1976) Changes in exocrine pancreatic function produced by altered dietary protein intake in drinking alcoholics. Johns Hopkins Med J 138: 7–12
Mezey E, Jow E, Slavin RE, Fobon F (1970) Pancreatic function and intestinal absorption in chronic alcoholism. Gastroenterology 59: 657
Mott C, Sarles H, Tiscornia O, Gullo L (1972) Inhibitory action of alcohol on human exocrine pancreatic secretion. Am J Dig Dis 17: 902–910
Multigner L, de Caro A, Lombardo D, Campese D, Sarles H (1983) Pancreatic stone protein, a phosphoprotein which inhibits calcium carbonate precipitation from human pancreatic juice. Biochem Biophys Res Commun 110: 69–74

Nakamura K, Sarles H, Payan H (1972) Three dimensional reconstruction of the pancreatic ducts in chronic pancreatitis. Gastroenterology 62: 942–949
Noel-Jorand MC, Sarles H (1983) Simultaneous mechanisms on exocrine pancreatic secretion initiated by alcohol in the conscious dog. Dig Dis Sci 28: 879–888
Noel-Jorand MC, Colomb E, Astier JP, Sarles H (1981) Pancreatic basal secretion in alcohol-fed and normal dogs. Dig Dis Sci 26: 783–789
Nossel HL, Efron G (1955) The effect of morphine on the serum and urinary amylase and the sphincter of Oddi: with some preliminary observations on the effect of alcohol on the serum amylase and the sphincter of Oddi. Gastroenterology 29: 409–416
Palasciano G, Tiscornia OM, Hage G, Sarles H (1974) Chronic alcoholism and endogenous cholecystokinine-pancreozymine (CCK-PZ). Biomedicine 20: 94–97
Perec CJ, Celener D, Tiscornia OM, Baratti C (1979) Effects of chronic ethanol administration on the automatic innervation of salivary glands, pancreas and heart. Am J Gastroenterol 72: 46–59
Pirola RC, Davis AE (1967) The role of ethyl alcohol in the etiology of pancreatitis. Gut 8: 526–536
Pirola RC, Davis AE (1968) Effects of ethyl alcohol on sphincter resistance at the choledochal-duodenal junction in man. Gut 9: 557–560
Pitchumoni CS, Sonnenshein M, Candido FM, Panchacharam P, Cooperman JM (1980) Nutrition in the pathogenesis of alcoholic pancreatitis. Am J Clin Nutr 33: 631–636
Piubello W, Vantini I, Scuro LA, Novelli P, Benini L, Brocco G, Cavallini G (1982) Gastric secretion, gastroduodenal histological changes, and serum gastrin in chronic alcoholic pancreatitis. Am J Gastroenterol 77: 105–110
Planche NE, Chau-Huu T, Lai PO, Sarles H (1977) Chronic alcohol consumption does not modify cholecystokinin blood levels by bioassay in the dog. Digestion 16: 194–198
Planche NE, Palasciano G, Meullenet J, Laugier R, Sarles H (1982) Effects of intravenous alcohol on pancreatic and biliary secretion in man. Dig Dis Sci 27: 449–453
Reber HA, Johnson FE, Montgomery CD, Carl W, Wong NF (1976) The effect of chronic alcohol ingestion on rat pancreas. Gastroenterology 72: 94–117
Renner G, Rinderknecht H, Douglas AP (1978) Profiles of pure pancreatic secretions in patients with acute pancreatitis: the possible role of proteolytic enzymes in pathogenesis. Gastroenterology 75: 1090–1098
Renner IG, Rinderknecht H, Valenzuela JE, Douglas AP (1980) Studies of pure pancreatic secretions in chronic alcoholic subjects without pancreatic insufficiency. Scand J Gastroenterol 15: 421–424
Rinderknecht H, Renner IG, Douglas AP, Adham NF (1978) Profiles of pure pancreatic secretion by direct pancreatic duct cannulation in normal human subjects. Gastroenterology 75: 1083–1089
Rinderknecht H, Renner IG, Carmack C (1979a) Trypsinogen variants in pancreatic juice of healthy volunteers, chronic alcoholics and patients with pancreatitis and cancer of pancreas. Gut 20: 886–891
Rinderknecht H, Renner IG, Koyama HH (1979b) Lysosomal enzymes in pure pancreatic juice from normal healthy volunteers and chronic alcoholics. Dig Dis Sci 24: 180–186
Roze C, Chariot J, Souchard M, Vaille C (1980) Extra pancreatic factors in pancreatic secretory changes occurring after three months of chronic alcoholization in rats. In: Stock Ch, Sarles H (eds) Alcohol and the gastrointestinal tract. INSERM, Paris, pp 283–288
Sahel H, Sarles H (1979) Modifications of pure pancreatic juice induced by chronic alcohol consumption. Dig Dis Sci 24: 897–905
Sardesai VM, Orten JM (1968) Effect of prolonged alcohol consumption in rats on pancreatic protein synthesis. J Nutr 96: 241–246
Sarles H (1973) An international survey on nutrition and pancreatitis. Digestion 9: 389–403
Sarles H (1974) Chronic calcifying pancreatitis – chronic alcoholic pancreatitis. Gastroenterology 66: 604–616
Sarles H, Laugier R (1981) Alcoholic pancreatitis. Clin Gastroenterol 10: 401–415
Sarles JC, Midejean A (1973) Electromyographic study of the action of alcohol upon the Oddi's sphincter. Digestion 9: 83
Sarles H, Tiscornia O (1974) Ethanol and chronic calcifying pancreatitis. Med Clin North Am 58: 1333–1346

Sarles H, Sarles JC, Camatte R, Muratore R, Gaini M, Guien C, Pastor J, Le Roy F (1965) Observations on 205 confirmed cases of acute pancreatitis, recurring pancreatitis and chronic pancreatitis. Gut 6: 545–559
Sarles H, Figarella C, Clemente F (1971 a) The interaction of ethanol, dietary lipids, and proteins on the rat pancreas. Digestion 4: 12–22
Sarles H, Lebreuil G, Tasso F et al. (1971 b) A comparison of alcoholic pancreatitis in rat and man. Gut 12: 377–388
Sarles H, Tiscornia O, Palasciano G, Brasca A, Hage G, Devaux MA, Gullo L (1973) Effects of chronic intragastric ethanol administration on canine exocrine pancreatic secretion. Scand J Gastroenterol 8: 85–96
Sarles H, Sahel J, Lebreuil G, Tiscornia O (1975) Pancréatite alcoolique chronique expérimentale du chien. Étude anatomo-pathologique. Biol Gastroenterol 8: 152
Sarles H, Tiscornia O, Palasciano G (1977) Chronic alcoholism and canine exocrine pancreas secretion: a long term follow-up study. Gastroenterology 72: 238–248
Sarles H, Cros RC, Bidart JM, the International Group for the Study of Pancreatic Diseases (1979) A multicenter inquiry into the etiology of pancreatic diseases. Digestion 19: 110–125
Sarles JC, Delecourt P, Castello H, Gaeta L, Macchiero M, Amoros JP, Devaux MA, Awad R (1981) Action of gastrointestinal hormones on the myoelectric activity of the sphincter of Oddi in living rabbit. Regul Pept 2: 113–124
Schaffalitzky de Muckadell OB, Fahrenkrug J, Holst JJ (1975) Release of vasoactive intestinal polypeptide (VIP) by electric stimulation of the vagal nerves. Gastroenterology 72: 373
Schapiro H, Wruble LD, Britt LG (1966) The possible mechanism of alcohol in the production of acute pancreatitis. Surgery 60: 1108–1111
Schapiro H, Wruble LD, Estes JW, Britt LG (1968) Pancreatic secretion stimulated by the action of alcohol on the gastric antrum. Am J Dig Dis 13: 536–539
Schmidt D (1983) Studies on neurohormonal control of exocrine pancreas in dogs. Effects of chronic ethanol feeding. Thesis, Stockholm
Schmidt E, Schmidt FW (1960) Enzym-Muster menschlicher Gewebe. Klin Wochenschr 38: 957–961
Schmidt DN, Sarles H, Biedzinski TM, Devaux MA (1981) Cholinergic secretory and inhibitory mechanisms in canine pancreatic secretion. Scand J Gastroenterol 16: 341–352
Schmidt DN, Sarles H, Devaux MA (1982a) Early increased pancreatic secretory capacity during alcohol adaptation in the dog. Scand J Gastroenterol 17: 49–55
Schmidt DN, Devaux MA, Biedzinski TM, Sarles H (1982 b) Disappearance of an inhibitory factor of exocrine pancreas secretion in chronic alcoholic dogs. Scand J Gastroenterol 17: 761–768
Schmidt DN, Devaux MA, Biedzinski TM, Sarles H (1983 b) Cholinergic stimulation and inhibition of pancreatic secretion in alcohol-adapted dogs. Scand J Gastroenterol 18: 425–431
Singer MV (1983) Welche Kontrollösungen sollen bei Studien über die Wirkung von Alkohol auf den Gastrointestinaltrakt gegeben werden? Z Gastroenterol 21: 567–573
Singer MV, Solomon TE, Grossman MI (1980) Effect of atropine on secretion from intact and transplanted pancreas in dog. Am J Physiol 283: G18–G22
Singer MV, Solomon TE, Rammert H, Caspary F, Niebel W, Goebell H, Grossman MI (1981) Effect of atropine on pancreatic response to HCl and secretin. Am J Physiol 240: G376–380
Singer MV, Eysselein V, Goebell H (1983a) Beer and wine but not whisky and pure ethanol do stimulate release of gastrin in humans. Digestion 26: 73–79
Singer MV, Eysselein V, Goebell H (1983 b) Pancreatic polypeptide response to intragastric ethanol in humans and dogs. Regul Pept 6: 13–23
Singer MV, Calden H, Eysselein V, Goebell H (1983 c) Bier stimuliert stark, seine Inhaltsstoffe Äthanol und Aminosäuren hingegen nur schwach die Magensäuresekretion des Menschen. Z Gastroenterol 21: 439–440
Singh M (1983) Effect of chronic ethanol feeding on pancreatic enzyme secretion in rats in vitro. Dig Dis Sci 28: 117–123
Singh M, LaSure MM, Bockman DE (1982) Pancreatic acinar cell function and morphology in rats chronically fed in ethanol diet. Gastroenterology 82: 424–435
Sive AA, Vinik AI, Lund A, Van Tonder S (1978) Impaired pancreatic polypeptide secretion in chronic pancreatits. J Clin Endocrinol Metab 47: 556–559

Solomon N, Solomon TE, Jacobson ED, Shanbour LL (1974) Direct effects of alcohol on in vivo and in vitro exocrine pancreatic secretion and metabolism. Am J Dig Dis 18: 253–260

Stasiewicz J, Adler M, Delcourt A (1980) Pancreatic and gastrointestinal hormones in chronic pancreatitis. Hepatogastroenterology 27: 152–160

Staub JL, Sarles H, Chayvialle JA, Descos F, Lassmann V, Vague P (1981) Relationship between intravenous ethanol, alcohol-induced inhibition of pancreatic secretion and plasma concentration of immunoreactive pancreatic polypeptide, vasoactive intestinal peptide, and somatostatin in man. Regul Pept 2: 61–68

Steer ML, Glazer G, Manabe T (1979) Direct effects of ethanol on exocrine secretion from the in vitro pancreas. Dig Dis Sci 24: 769–774

Stening GF, Grossman MI (1969) Gastrin-related peptides as stimulants of pancreatic and gastric secretion. Am J Physiol 217: 262

Straus E, Urbach HJ, Yalow RS (1975) Alcohol-stimulated secretion of immunoreactive secretion. N Engl J Med 293: 1031–1032

Tasso F, Stemmelin N, Sarles H, Clop J, Durbec JP, Cros RC, Cornee J (1973) Comparative morphometric study of the human pancreas in its normal state and in primary chronic calcifying pancreatitis. Biomedicine 18: 134–144

Taylor IL, Solomon TE, Walsh JH, Grossman MI (1978) Physiological blood levels of pancreatic polypeptide inhibit pancreatic secretion. Gastroenterology 1: 26

Thomas JE, Crider JO (1943) The effect of bile in the intestine on the secretion of pancreatic juice. Am J Physiol 138: 548–552

Tiscornia O, Brasca AF, Hage G, Palasciano G, Devaux MA, Sarles H (1972) Les effets de l' atropine, du penthonium, de la vagotomie et du fluothane sur la sécrétion pancrétique du chien. Biol Gastroenterol 5: 249–256

Tiscornia O, Gullo L, Sarles H (1973a) The inhibition of canine exocrine pancreatic secretion by intravenous ethanol. Digestion 9: 231–240

Tiscornia O, Hage G, Palasciano G, Brasca AP, Devaux MA, Sarles H (1973c) The effect of pentolinium and vagotomy on the inhibition of canine exocrine pancreatic secretion by intravenous ethanol. Biomedicine 18: 159–163

Tiscornia OM, Palasciano G, Sarles H (1974a) Pancreatic changes induced by chronic (two years) ethanol treatment in the dog. Gut 15: 839 (abstract)

Tiscornia O, Gullo L, Sarles H, Mott CB, Brasca A, Devaux MA, Palasciano G, Hage G (1974b) Effects of intragastric and intraduodenal ethanol on canine exocrine pancreatic secretion. Digestion 10: 52–60

Tiscornia OM, Palasciano G, Sarles H (1975a) Atropine and exocrine pancreatic secretion in alcoholic fed dogs. Am J Gastroenterol 63: 33–36

Tiscornia O, Palasciano G, Dzieniszewski J, Sarles H (1975b) Simultaneous changes in pancreatic and gastric secretion induced by acute intravenous ethanol infusion. Effects of atropine and reserpine. Am J Gastroenterol 63: 389–395

Tiscornia O, Singer M, Mendes de Oliveira JP, Demo P, Sarles H (1977a) Exocrine pancreas response to a test meal in the dog: changes induced by 3 months' ethanol feeding. Am J Dig Dis 22: 769–774

Tiscornia O, Levesque D, Sarles H, Bretholz A, Voirol M, Mendes de Oliveira JP, Singer M, Demol P (1977b) Canine exocrine pancreatic secretory changes induced by intragastric ethanol associated to a test meal. Am J Gastroenterol 67: 121–130

Treffot MJ, Tiscornia O, Palasciano G, Hage G, Sarles H (1975) Chronic alcoholism and endogenous gastrin. Am J Gastroenterol 63: 29–32

Treffot MJ, Laugier R, Bretholz A, Voirol M, Sarles H (1980) Increased gastrin release in chronic calcifying pancreatitis and in chronic alcoholism. Horm Metab Res 12: 240–242

Uhlemann ER, Robberecht P, Gardner JD (1979) Effects of alcohol on the actions of VIP and secretin on acinar cells from guinea pig pancreas. Gastroenterology 76: 917–925

Valenzuela JE, Ugarte G, Peterman M (1969) Effect of intravenous ethanol on pancreatic secretion. Gastroenterology 56: 1203

Valenzuela JE, Salinas J, Petermann M (1974) Effect of ethanol on pancreatic enzyme secretion. Gastroenterology 67: 991–993

Valenzuela JE, Taylor IL, Walsh JH (1979) Pancreatic polypeptide response in patients with chronic pancreatitis. Dig Dis Sci 24: 862–864

Viceconte G (1983) Effects of ethanol on the sphincter of Oddi: an endoscopic manometric study. Gut 24: 20–27

Voirol M, Tiscornia OM, Levesque D, Bretholz A (1976a) Reserpine-induced dissociation of canine pancreatic secretion. Am J Gastroenterol 66: 446–448

Voirol M, Bretholz A, Levesque D, Tiscornia O, Sarles H (1976b) Atropine-induced inhibition of the enhanced CCK release observed in alcoholic dogs. Digestion 14: 174–178

Voirol M, Tiscornia O, Levesque D, Bretholz A, Dzieniszewski J, Palasciano G, Laugier R, Sarles H (1976c) Evidence of a dissociation in pancreatic secretion when catecholamine storage is depleted. Biol Gastroenterol 9: 73

Walton BE, Shapiro H, Woodward ER (1962) The effect of alcohol and histamine on pancreatic secretion. Am Surg 28: 443

Walton BE, Shapiro H, Yeung T, Woodward ER (1965) Effect of alcohol on pancreatic duct pressure. Am Surg 31: 142–144

Wheatley IC, Barbezat GO, Hickman R, Terblanche J (1975) The effect of acute ethanol administration on the exocrine pancreatic secretion of the pig. Br J Surg 62: 707

Woodward ER, Robertson C, Ruttenberg HD, Shapiro H (1957) Alcohol as a gastric secretory stimulant. Gastroenterology 32: 727

Yahata K, Ishikawa I, Kubota K, Kuroda A (1982) Role of intrapancreatic cholinergic innervation in pathogenesis of alcohol-induced pancreatitis in dog. Secretory pattern and histochemical study. Scand J Gastroenterol 17 [Suppl 78] (abstract)

Yoshimori M, Chey W, Escoffery R et al. (1972) Effects on gastric secretion of acid and parietal cells in dogs. Gastroenterology 62: 732

22 Subject Index

MIX
Papier aus verantwortungsvollen Quellen
Paper from responsible sources
FSC® C105338

If you have any concerns about our products,
you can contact us on
ProductSafety@springernature.com

In case Publisher is established outside the EU,
the EU authorized representative is:
Springer Nature Customer Service Center GmbH
Europaplatz 3, 69115 Heidelberg, Germany

Printed by Libri Plureos GmbH
in Hamburg, Germany